Annals of the International Society of Dynamic Games
Volume 1

Series Editor
Tamer Başar

Annals of the International Society of Dynamic Games

Advances in Dynamic Games and Applications

Tamer Başar
Alain Haurie
Editors

Birkhäuser
Boston • Basel • Berlin

Tamer Başar
Coordinated Science Laboratory
University of Illinois
Urbana, IL 61801
USA

Alain Haurie
Department of Management Studies
University of Geneva
CH-1211 Geneva
Switzerland

Library of Congress Cataloging-In-Publication Data

Advances in dynamic games and applications / Tamer Başar, Alain Haurie, editors.
p. cm. -- (Annals of the International Society of Dynamic Games ; v. 1)
" Contains refereed and edited papers from the biannual symposium of ISDG, held in July 1992, Geneva, Switzerland"--Galley.
Includes bibliographical references.
ISBN 0-8176-3691-9
1. Game theory--Congresses. I. Başar, Tamer. II. Haurie, A. (Alain), 1940- . III. International Society of Dynamic Games. IV. Series.
QA269.A36 94-937
519.3--dc20 CIP

Printed on acid-free paper

ISBN 0-8176-3691-9
ISBN 3-7643-3691-9
Camera-ready copy by the Authors.
Printed and bound by Quinn-Woodbine, Woodbine, NJ.
Printed in the U.S.A.

9 8 7 6 5 4 3 2 1

Table of Contents

Part IV. Stochastic games: Differential, sequential and Markov Games

Part V. Applications

Preface

Recent years have witnessed a surge of activity in the field of dynamic games, in both theory and applications. Theoretical as well as practical problems in zero-sum and nonzero-sum games, continuous time differential games and discrete time multistage games, and deterministic and stochastic games are currently being investigated by researchers in diverse disciplines, such as engineering, mathematics, biology, economics, management science, and political science. This surge of interest has led to the formation of the *International Society of Dynamic Games* (ISDG) in 1990, whose primary goal is to foster the development of advanced research and applications in the field of game theory. One important activity of the Society is to organize biannually an international symposium which aims at bringing together all those who contribute to the development of this active field of applied science. In 1992 the symposium was organized in Grimentz, Switzerland, under the supervision of an international scientific committee and with the help of a local organizing committee based at University of Geneva. This book, which is the first volume in the new Series, *Annals of the International Society of Dynamic Games* (see the *Preface* to the Series), is based on presentations made at this symposium. It is however more than a book of proceedings for a conference. Every paper published in this volume has passed through a very selective refereeing process, as in an archival technical journal. Hence this first volume of the *Annals of Dynamic Games* is a quality publication that presents a timely account of the state of the art in this field.

The papers which have been included in this volume attest to the vitality and diversity of ongoing research in dynamic games and applications. The reader will find here important contributions in the following five areas which also constitute the groups according to which the twenty-three papers have been organized :

• *Robust control design and H^∞-theory.* This is an important recent achievement of zero-sum game theory, which has provided a unifying framework as well as a computational tools for the design of robust controllers for systems with uncertain inputs. Papers addressing recent developments on this major application of zero-sum dynamic game theory can be found in Part I of the volume, as chapters 2-4, following a general survey paper on the theory of zero-sum differential games.

• *Pursuit-evasion games and numerical schemes.* Following the seminal work of Kushner, the numerical analysis techniques which have been successfully implemented in the stochastic control framework are currently being extended to the case of zero-sum differential games, particularly for such discretization schemes in the context of pursuit-evasion games. The next two chapters discuss the characterization of singular surfaces in some special classes of differential games.

• *Numerical solutions of discrete time games, based on mathematical programming techniques.* Renewing with the early tradition in game theory where mathematical programming techniques were closely related to the development of the game concept (e.g. linear programming and saddle points of matrix games, or complementarity algorithms and equilibria in matrix games), the computation of Nash or Stackelberg equilibria is now performed for large-scale models through the use of a new family of mathematical programming algorithms. Some recent developments in this area are discussed in chapters 9, 10 and 11.

• *Sequential and Markov games.* This formalism offers a fundamental paradigm as well as a numerical framework. Both in the field of engineering and economics new models, based on a sequential game formalism, are proposed as a representation of total or partial conflict situations. Four such papers are included in part IV of this volume, as chapters 12-15.

• *Applications in ecology, environmental management, and biology.* These sciences are now benefiting from a variety of models based on game theoretic concepts, as can be seen from the papers comprising the fifth part of this volume, as chapters 16-23.

We would like to take this opportunity to thank the associate editors of this volume, for their efficient and expeditious handling of the review process; the reviewers for responding with diligence to our request for assistance in this evaluation process; and the authors for their fine contributions to this first volume of the *Annals.*

Tamer Başar
Urbana, Illinois, USA
April 12, 1993

Alain Haurie
Geneva, Switzerland
April 12, 1993

Series Preface

With this volume, we are launching a new series in dynamic game theory, to be called "*Annals of the International Society of Dynamic Games*," or in short "*ANNALS OF DYNAMIC GAMES*," which will be a yearly publication consisting of refereed papers representing the latest developments concerning the theory and applications of dynamic games. This is an outgrowth of the activities of "The International Society of Dynamic Games" (in short, *ISDG*), which was founded in 1990. The primary goals of *ISDG* are to promote interactions among researchers interested in the theory and applications of dynamic games; to facilitate dissemination of information on current activities and results in this area; and to enhance the visibility of dynamic games research and its vast potential applications.

This first volume of the *Annals of Dynamic Games* contains refereed and edited papers from the biannual symposium of *ISDG*, held in July 1992, in Geneva, Switzerland. Every two years such a symposium volume will appear in this *Series*, based on papers presented at the biannual meeting of *ISDG* the year before. In alternate years, the *Series* will publish volumes of invited papers dedicated to specific, fast developing, topics. Each volume will be put together by a guest editor or guest co-editors; plans for two such volumes (with tentative 1994 publication dates) are already underway.

Tamer Başar, *Editor*
Urbana, Illinois, USA
April 19, 1993

PART I

Zero-sum differential games: Theory and applications in worst-case controller design

A Theory of Differential Games

Leonard D. Berkovitz

1. Introduction

The differential games that we shall consider can be formulated intuitively as follows. The state of the game at time t is given by a vector $x(t)$ in R^n and is determined by a system of differential equations

$$\frac{dx}{dt} = f(t, x, u(t), v(t)) \quad x(t_0) = x_0, \tag{1.1}$$

where $u(t)$ is chosen by Player I at each time t and $v(t)$ is chosen by Player II at each time t. The choices are constrained by the conditions $u(t) \in Y$ and $v(t) \in Z$, where Y and Z are preassigned sets in euclidean spaces. The choice of $u(t)$ is governed by a set of rules or "strategy" U selected by Player I prior to the start of play and the choice of $v(t)$ is governed by a "strategy" V selected by Player II prior to the start of play. Play proceeds from the initial point (t_0, x_0) until the point $(t, \phi(t))$, where ϕ is the solution of (1.1), reaches some preassigned terminal set $\mathcal{T}$. The point at which $(t, \phi(t))$ reaches $\mathcal{T}$ is called the terminal point and is denoted by $(t_f, \phi(t_f))$, or (t_f, x_f). The payoff is

$$P(t_0, x_0; U, V) = g(t_f, x_f) + \int_{t_0}^{t_f} f^0(s, \phi(s), u(s), v(s))\, ds, \tag{1.2}$$

where g and f^0 are preassigned functions. Player I wishes to choose U so as to maximize P while Player II wishes to choose V so as to minimize P.

If

$$\sup_U \inf_V P(t_0, x_0; U, V) = \inf_V \sup_U P(t_0, x_0; U, V)$$

then we denote this number by $W(t_0, x_0)$ and say that the game has value equal to $W(t_0, x_0)$. A pair of strategies (U^*, V^*) is said to be a saddle point if for all U over which the sup is taken and all V over which the inf is taken

$$P(t_0, x_0; U, V^*) \leq P(t_0, x_0; U^*, V^*) \leq P(t_0, x_0; U^*, V).$$

The strategies U^* and V^* are called optimal strategies. It follows from the definitions that if a saddle point exists, then the game has a value $W(t_0, x_0)$ equal to $P(t_0, x_0; U^*, V^*)$.

The first problem associated with differential games is to formulate the intuitive notion of the game as a game in the mathematical sense of a zero-sum two person game. This involves defining the notions of strategy and strategy space precisely and defining the corresponding notion of payoff as a function defined on the strategy spaces of the players. The second problem is to determine whether for the given definitions of strategy spaces and payoff the game has a value. If the game has a value, then the third problem is to determine whether it has a saddle point. Lastly, if a saddle point exists, then the fourth problem is to find the optimal strategies.

The study of differential games was initiated by Rufus Isaacs in a series of RAND Corporation Memoranda in 1954. He incorporated this work and subsequent work in his book [16]. Isaacs did not give an adequate definition of strategy nor did he give a priori conditions that ensured the existence of value and of saddle points. He assumed that the game did have a continuously differentiable value, as a function of initial position, except on some singular surfaces and that the game did have a saddle point in a sense never precisely defined. He then proceeded to develop a technique that would give the optimal choices $U^*(t,x)$ and $V^*(t,x)$ if the game were at state x at time t.

The basic idea underlying Isaacs' approach is quite simple. To simplify the exposition we suppose that $f^0 \equiv 0$. We shall see in Section 3 below that there is no loss of generality in assuming that $f^0 \equiv 0$. If at time τ the position of the game is ξ, Player I chooses $u(\tau) = y$ and Player II chooses $v(\tau) = z$, then to within first order the new position of the game will be $\xi + dx = \xi + f(\tau, \xi, y, z)dt$ and the value at the new position will be $W + dW$, where

$$dW = \left[W_t + \langle W_x, f(\tau, \xi, y, z)\rangle\right] dt. \tag{1.3}$$

Here the argument of W, dW, W_t and W_x is (τ, ξ) and $\langle\ ,\ \rangle$ denotes inner product. Isaacs assumes that for each initial point (t,x) a saddle point exists for the game. For $\tau \le t \le t_f$ let $u^*(t)$ and $v^*(t)$ be the optimal choices for the game with initial point (τ, ξ) and let $\phi^*(\ ) = \phi^*(\ ,\tau,\xi)$ be the resulting trajectory. Then since $W(t, \phi^*(t,\tau,\xi)) = g(t_f, \phi^*(t_f,\tau,\xi))$ is constant along the trajectory $\phi^*(\ )$, we have from (1.3) that

$$W_t(\tau,\xi) + \langle W_x(\tau,\xi), f(\tau,\xi,u^*(\tau),v^*(\tau))\rangle = 0. \tag{1.4}$$

At (τ, ξ), Player I wishes to choose y so as to maximize dW and will then play optimally from the new position. Player II wishes to choose z so as to minimize dW and will then play optimally from the new position. It now follows that for each initial condition (τ, ξ), the game over $Y \times Z$ with payoff

$$W_t(\tau,\xi) + \langle W_x(\tau,\xi), f(\tau,\xi,y,z)\rangle$$

has a saddle point $(y^*, z^*) = (u^*(\tau), v^*(\tau))$ and has value zero.

These considerations led Isaacs to impose the following condition, now known as the *Isaacs condition.* For each initial position (t, x) and each s in R^n

$$\min_{y \in Y} \max_{z \in Z} \langle s, f(t, x, y, z) \rangle = \max_{z \in Z} \min_{y \in Y} \langle s, f(t, x, y, z) \rangle. \tag{1.5}$$

The value function is then a solution of

$$\begin{gathered} \max_{y \in Y} \min_{z \in Z} \left[W_t(t, x) + \langle W_x(t, x), f(t, x, y, z) \rangle \right] = 0 \\ W(t_f, x_f) = g(t_f, x_f) \end{gathered} \tag{1.6}$$

for all initial points (t, x) and points (t_f, x_f) on the boundary of $\mathcal{T}$. In some specific examples Isaacs solved (1.6) by the method of characteristics, and thus also obtained the optimal choices $u^*(\tau)$, $v^*(\tau)$ of the two players at position (τ, ξ).

In [1] a direct approach to the differential game problem, which put some of Isaacs' results on a rigorous mathematical basis, was given. In this approach one takes the strategies U and V to be functions of (t, x) with appropriate properties and substitutes these functions into (1.1) for $u(t)$ and $v(t)$ respectively. This approach was not entirely satisfactory since it applied to a special class of games, some of whose properties could not be verified a priori. Given the difficulties of a direct approach, Fleming, Friedman and Krasovskii and Subbotin [10], [11], [12], [13], [17] and [18] developed indirect approaches which first involve choices of control functions u and v on small intervals. The choices are made at discrete times and are based on the history of the game. A passage to the limit as the lengths of the small intervals tend to zero is then carried out. Somewhat different approaches were given by Elliott and Kalton and Varaiya and Lin in [9] and [21]. With the exception of Krasovskii and Subbotin [17], [18], these authors were primarily concerned with establishing the existence of value in some sense and in showing that the value satisfied the Isaacs equation in a generalized sense. The developments of the last decade in the theory of viscosity solutions of partial differential equations have done away with the need for large portions of this work.

In this paper we shall present a theory developed by this author and some of his students. We shall present the principal results and some of the ideas behind them and behind the arguments used to establish these results. We hope that this will serve as a readers' guide for those who wish to read the detailed proofs in the literature.

2. Assumptions and Notation

Let t denote time, let $x = (x^1, \ldots, x^n)$ denote a vector in n-dimensional real euclidean space R^n, let $y = (y^1, \ldots, y^r)$ denote a vector in R^r and let $z = (z^1, \ldots, z^s)$ denote a vector in R^s. We use superscripts to denote components of a vector and use subscripts to distinguish vectors. We shall denote the inner product of two vectors x and y by $\langle x, y\rangle$, and the euclidean norm of a vector x by $|x|$.

Let Y be a compact set in R^r, let Z be a compact set in R^s and let T_0 and T_1 be two real numbers satisfying $T_0 < T_1$. Let $f^0, f_1, \ldots, f^n$ be real valued functions defined on the set

$$\mathcal{D} = [T_0, T_1] \times R^n \times Y \times Z \tag{2.1}$$

and let $f = (f^1, \ldots, f^n)$. Let $\hat{f} = (f^0, f)$. Let g be a real valued function defined on $[T_0, T_1] \times R^n$ and bounded on bounded subsets of its domain. Let $\mathcal{T}$ be a closed set in (t, x)-space that is the closure of some region and such that

$$[T_0, \infty) \times R^n \supset \mathcal{T} \supseteq [T_1, \infty) \times R^n. \tag{2.2}$$

The function f will govern the dynamics as suggested by (1.1). The functions f^0 and g will enter the payoff as in (1.2) and the set $\mathcal{T}$ will be the terminal set. The precise interpretation of (1.2) will be given in Section 3.

Assumption I

(i) The function $\hat{f}$ is continuous on $\mathcal{D}$.

(ii) For every $R > 0$ there exists a constant $K_R > 0$ such that for all $t \in [T_0, T_1]$, x, $\bar{x}$ in R^n with $|x|$, $|\bar{x}| \leq R$, $y \in Y$ and $z \in Z$

$$|f(t, x, y, z) - f(t, \bar{x}, y, z)| \leq K_R|x - \bar{x}|.$$

(iii) There exist a constant $A > 0$ such that for all (t, x, y, z) in $\mathcal{D}$

$$|f(t, x, y, z)| \leq A(|x| + 1).$$

(iv) For every $R > 0$ there exists a constant $K'_R > 0$ such that for all t, $\bar{t}$ in $[T_0, T_1]$, x, $\bar{x}$ in R^n, $|x|$, $|\bar{x}| \leq R$,

$$|g(\bar{t}, \bar{x}) - g(t, x)| \leq K'_R(|t - \bar{t}| + |x - \bar{x}|).$$

3. Definition of Game

Our definition of strategy is related to the "K strategies" of Isaacs [16] and the lower strategies of Friedman [12], [13]. Our definition of payoff is related to that of Krasovskii–Subbotin [17], [18].

For any interval $[a, b] \subseteq [T_0, T_1]$, let $\mathcal{Y}[a, b]$ denote the set of measurable functions u on $[a, b]$ such that $u(t) \in Y$ a.e. Let $\mathcal{Z}[a, b]$ denote the set of measurable functions v on $[a, b]$ such that $v(t) \in Z$ a.e. Functions u in $\mathcal{Y}[a, b]$ will be called *controls* for Player I on $[a, b]$ and functions v in $\mathcal{Z}[a, b]$ will be called *controls* for Player II on $[a, b]$.

A *strategy* for Player I for the game starting at (t_0, x_0) is a choice of a sequence of partitions $\pi = \{\pi_n\}$ of $[t_0, T_1]$ and a choice of a sequence of maps $\Gamma_\pi = \{\Gamma_{\pi,n}\}$, where the $\Gamma_{\pi,n}$ are to be defined below. For typographic simplicity we shall suppress the dependence on π in the notation and shall write Γ for Γ_π and $\{\Gamma_n\}$ for $\{\Gamma_{\pi,n}\}$. We restrict the choice of sequences of partitions to those such that the norms of π_n, denoted by $\|\pi_n\|$, satisfy $\|\pi_n\| \leq A/n$ for $n = 1, 2, 3, \ldots$, where $A > T_1 - T_0$. Let the partition points of π_n be $t_0 < t_1 < \ldots < t_p = T_1$, where $p = p(n)$ depends on n, but the dependence on n is suppressed in the notation. Each Γ_n, $n = 1, 2, \ldots$ is a collection of maps $\Gamma_{n,1}, \ldots, \Gamma_{n,p}$ as follows. The map $\Gamma_{n,1}$ selects an element in $\mathcal{Y}[t_0, t_1)$. For $2 \leq j \leq p$ the map $\Gamma_{n,j}$ is a map from $\mathcal{Y}[t_0, t_{j-1}) \times \mathcal{Z}[t_0, t_{j-1})$ to $\mathcal{Y}[t_{j-1}, t_j)$.

A *strategy* for Player II is a choice of a sequence of partitions $\bar{\pi} = \{\bar{\pi}_n\}$ of $[t_0, T_1]$ such that $\|\bar{\pi}_n\| \leq \bar{A}/n$, for $n = 1, 2, 3, \ldots$ and a choice of a sequence of maps $\Delta = \{\Delta_n\}$. Here $\bar{A} > T_1 - T_0$. Let $\bar{\pi}_n$ have partition points $t_0 = s_0 < s_1 < \ldots < s_q = T_1$, where $q = q(n)$ depends on n. Each Δ_n is a collection of maps $\Delta_{n,1}, \ldots, \Delta_{n,q}$ as follows. The map $\Delta_{n,1}$ selects a function v in $\mathcal{Z}[s_0, s_1)$. For $2 \leq j \leq q$, $\Delta_{n,j}$ is a map from $\mathcal{Y}[t_0, s_{j-1}) \times \mathcal{Z}[t_0, s_{j-1})$ to $\mathcal{Z}[s_{j-1}, s_j)$. The collection of maps $\Gamma_n = (\Gamma_{n,1}, \ldots, \Gamma_{n,p})$ will be called an n-th *stage strategy* for Player I. The collection of maps $\Delta_n = (\Delta_{n,1}, \ldots, \Delta_{n,q})$ will be called an n-th *stage strategy* for Player II.

We next define the concept of a motion in the game. If w is a measurable function defined on $[t_0, T_1]$ and if $a \in (t_0, T_1)$ then by $w \mid a$ we shall mean the restriction of w to $[t_0, a)$. A pair of n-th stage strategies (Γ_n, Δ_n) determines a control u_n on $[t_0, T_1]$ for Player I and a control v_n for Player II on $[t_0, T_1]$ as follows. By considering the various possibilities that may occur in the natural ordering of the union of the partition points $\{t_i\}_{i=0}^p$ and $\{s_i\}_{i=0}^q$, the reader will convince himself that the functions u_n and v_n are indeed well defined on $[t_0, T_1]$ by

$$\begin{aligned} u_n(t) &= (\Gamma_{n,1})(t) \quad \text{for } t \in [t_0, t_1) \\ v_n(t) &= (\Delta_{n,1})(t) \quad \text{for } t \in [t_0, s_1). \end{aligned} \tag{3.1}$$

and by

$$\begin{aligned} u_n(t) &= \Gamma_{n,j}(u_n|t_{j-1}, v_n|t_{j-1})(t) \quad t \in [t_{j-1}, t_j) \;\; 2 \leq j \leq p \\ v_n(t) &= \Delta_{n,j}(u_n|s_{j-1}, v_n|s_{j-1})(t) \quad t \in [s_{j-1}, s_j) \;\; 2 \leq j \leq q. \end{aligned} \tag{3.2}$$

The controls (u_n, v_n) determined this way are called the n-th *stage outcomes* of (Γ_n, Δ_n).

In equation (1.1), if we replace $u(t)$ by $u_n(t)$, $v(t)$ by $v_n(t)$ and x_0 by x_{0n} we obtain the system of differential equations

$$\frac{dx}{dt} = f(t, x, u_n(t), v_n(t)) \quad x(t_0) = x_{0n}. \tag{3.3}$$

It follows from Assumption I that equation (3.3) has a unique solution $\phi_n(\ , t_0, x_{0n}, u_n, v_n)$ defined on $[t_0, T_1]$. We call this solution an n-th *stage trajectory.*

Any uniform limit of a subsequence of the n-th stage trajectories $\phi_n(\,, t_0, x_{0n}, u_n, v_n)$, $n = 1, 2, 3, \ldots$ where $x_{0n} \to x_0$ and (u_n, v_n) is the outcome of (Γ_n, Δ_n) will be called a *motion* or *motion of the game* corresponding to strategies $\Gamma = \{\Gamma_n\}$ and $\Delta = \{\Gamma_n\}$. Following Krasovskii and Subbotin [17] we denote a motion corresponding to (Γ, Δ) by $\phi[\,, t_0, x_0, \Gamma, \Delta]$.

It follows from Assumption I that corresponding to a pair of strategies (Γ, Δ) and a sequence of initial conditions $\{x_{0n}\}$ with $x_{0n} \to x_0$, the corresponding sequence of n-th stage trajectories $\{\phi_n(\,, t_0, x_{0n}, u_n, v_n)\}$ is uniformly bounded and equicontinuous. Hence motions $\phi[\,, t_0, x_0, \Gamma, \Delta]$ do exist. We shall denote the set of all motions corresponding to (Γ, Δ) by $\Phi[\,, t_0, x_0, \Gamma, \Delta]$.

By $\phi[t, t_0, x_0, \Gamma, \Delta]$ we shall mean the motion evaluated at the point t. By $\Phi[t, t_0, x_0, \Gamma, \Delta]$ we shall mean the set of all values $\phi[t, t_0, x_0, \Gamma, \Delta]$, where $\phi[\,, t_0, x_0, \Gamma, \Delta]$ ranges over $\Phi[\,, t_0, x_0, \Gamma, \Delta]$.

It is important to keep in mind that, in general, a motion $\phi[\,, t_0, x_0, \Gamma, \Delta]$ will not be obtained as a solution of

$$\frac{dx}{dt} = f(t, x, u(t), v(t)) \quad x(t_0) = x_0$$

for an appropriate choice of controls u, v. Nevertheless, motions do have properties similar to or analogous to properties of solutions of differential equations. For example, the continuous dependence on initial data has the following analogue, imprecisely stated. If (τ', ξ') is near (τ, ξ) then for each motion $\phi[\,, \tau, \xi, \Gamma, \Delta]$ there exists a motion $\phi[\,, \tau', \xi', \Gamma', \Delta']$ that is close to $\phi[\,, \tau, \xi, \Gamma, \Delta]$ in the sup norm. For the precise statement and proof of this property and others, see [1].

To complete the definition of the game we must define the payoff. We first consider games of fixed duration. In such games the terminal set $\mathcal{T}$ is given by

$$\mathcal{T} = [T, \infty) \times R^n,$$

where $T_0 < T \leq T_1$. The function g is a mapping $x \to g(x)$ and is defined on R^n. If $f^0 \equiv 0$ the game is said to have terminal payoff. There is no loss of generality in assuming $f^0 \equiv 0$. To see this let $\hat{x} = (x^0, x)$ and adjoin the equation

$$\frac{dx^0}{dt} = f^0(t, x, u(t), v(t)) \quad x^0(t_0) = 0 \tag{3.4}$$

to the system (1.1). The game with payoff (1.2) becomes a game of fixed duration in $[T_0, T_1] \times R^{n+1}$ with terminal payoff $\hat{g}(\hat{x}) = g(x) + x^0$.

For games of fixed duration with terminal payoff g the payoff corresponding to a pair of strategies (Γ, Δ) is set valued and is defined by

$$P(t_0, x_0, \Gamma, \Delta) = g(\Phi[T, t_0, x_0, \Gamma, \Delta]).$$

We next consider games with $\mathcal{T}$ as in (2.2) and $g \equiv 0$ in (1.2). Such games are often called games of generalized pursuit and evasion. We treat such games by adjoining (3.4) to (1.1) as in the preceding paragraph and study the resulting system in R^{n+1}. The sequence of n-th stage trajectories corresponding to a pair of strategies (Γ, Δ) is the sequence $\{\hat{\varphi}_n(\ , t_0, \hat{x}_{0n}, u_n, v_n)\}$, the solutions of

$$\frac{d\hat{x}}{dt} = \hat{f}(t, x, u_n(t), v_n(t)) \quad \hat{x}(t_0) = \hat{x}_{0n}, \tag{3.5}$$

where $\hat{x}_{0n} \to (0, x_0)$. By the capture time t_f of a motion $\hat{\varphi}[\ , t_0, x_0, \Gamma, \Delta]$ we mean the first time t_f such that $(t, \varphi[t, t_0, x_0, \Gamma, \Delta]) \in \mathcal{T}$. The payoff is now defined by

$$P(t_0, x_0, \Gamma, \Delta) = \cup\{\varphi^0[t_f, t_0, x_0, \Gamma, \Delta]\} \tag{3.6}$$

where the union is taken over all motions $\hat{\varphi}[\ , t_0, x_0, \Gamma, \Delta]$ resulting from (Γ, Δ). Note that in general t_f will be different for different motions.

Lastly we consider the game with the terminal set as in (2.2) and the payoff as in (1.2). Such games are often called games of survival. To define the payoff we again take the sequence of n-th stage trajectories corresponding to a pair of strategies (Γ, Δ) to be the sequence $\{\hat{\varphi}_n, (\ , t_0, \hat{x}_{0n}, u_n, v_n)\}$, where $\hat{\varphi}_n$ is the solution of (3.5). The capture time t_f of a motion is as in the preceding paragraph. The payoff $P(t_0, x_0, \Gamma, \Delta)$ resulting from a pair of strategies (Γ, Δ) is defined as

$$P(t_0, x_0, \Gamma, \Delta) \equiv \cup\big\{g(\varphi[t_f, t_0, x_0, \Gamma, \Delta]) + \varphi^0[t_f, t_0, x_0, \Gamma, \Delta]\big\},$$

where the union is taken over all motions $\hat{\varphi}[\ , t_0, x_0, \Gamma, \Delta]$ resulting from (Γ, Δ).

We conclude this section by presenting two other definitions of strategies. Let $\mathcal{Y}_c[a, b]$ denote the set of piecewise constant functions u on $[a, b]$ with values $u(t)$ in Y and let $\mathcal{Z}_c[a, b]$ denote the set of piecewise constant functions v on $[a, b]$ with values $v(t)$ in Z. We define strategies as before except that now $\Gamma_{n,1}$ selects a vector $y_{n,1}$ in Y and $u_n(t) = y_{n,1}$ for all $t \in [t_0, t_1)$. Similar $\Delta_{n,1}$ selects a vector $z_{n,1}$ in Z and $v_n(t) = z_{n,1}$ for all $t \in [t_0, s_1)$. For $2 \le j \le p$, $\Gamma_{n,j}$ is a map from $\mathcal{Y}_c[t_0, t_{j-1}) \times \mathcal{Z}_c[t_0, t_{j-1})$ to Y and for $t \in [t_{j-1}, t_j)$, $u_n(t) = (\Gamma(u_n|t_{j-1}, v_n|t_{j-1}))$. For $2 \le j \le q$, $\Delta_{n,j}$ is a map from $\mathcal{Y}_c[t_0, s_{j-1}) \times \mathcal{Z}_c[t_0, s_{j-1})$ to Z and for $t \in [s_{j-1}, s_j)$, $v_n(t) = (\Delta(u_n|s_{j-1}, v_n|s_{j-1}))$. Thus, at a decision point t_j or s_j an n-th stage strategy selects a constant valued control over the succeeding interval. Motions are now defined as before, except that we use trajectories obtained from these (u_n, v_n). Also defined as before are the various payoffs. We shall denote the game so defined as G_c.

To define the second version, we let $\mathcal{Y}_r[a, b]$ denote the set of relaxed controls on $[a, b]$ relative to the compact set Y. We let $\mathcal{Z}_r[a, b]$ denote the set of relaxed controls on $[a, b]$ relative to the compact set Z. For a summary of properties of relaxed controls see [1], [3]. For a full discussion see [20].

We define a strategy for Player I as before except that now $\Gamma_{n,1}$ selects a relaxed control on $[t_0, t_1)$ and $\Delta_{n,1}$ selects a relaxed control on$[t_0, s_1)$. For $2 \leq j \leq p$, $\Gamma_{n,j}$ is a map from $\mathcal{Y}_r[t_0, t_{j-1}) \times \mathcal{Z}_r[t_0, t_{j-1})$ to $\mathcal{Y}_r[t_{j-1}, t_j)$. Similarly, for $2 \leq j \leq q$, $\Delta_{n,j}$ is a map from $\mathcal{Y}_r[t_0, s_{j-1}) \times \mathcal{Z}_r[t_0, s_{j-1})$ to $\mathcal{Z}_r[s_{j-1}, s_j)$. A pair of n-th stage strategies (Γ_n, Δ_n) determined a pair of n-th stage relaxed controls (η_n, ζ_n). These relaxed controls determine a sequence of n-th stage relaxed trajectories via the differential equation

$$\frac{dx}{dt} = f(t, x, \eta_n(t), \zeta_n(t)) \equiv \int_Z \int_Y f(t, x, y, z) d\eta_n(t) d\zeta_n(t).$$

Motions and payoffs are defined as before using relaxed trajectories and strategies as defined here. For details see [3]. We shall call the game just described the game G_r.

4. The Upper and Lower Values, Saddle Points

Let

$$\begin{aligned} W^-(t_0, x_0) &= \sup_{\Gamma} \inf_{\Delta} P(t_0, x_0, \Gamma, \Delta) \\ W^+(t_0, x_0) &= \inf_{\Delta} \sup_{\Gamma} P(t_0, x_0, \Gamma, \Delta). \end{aligned} \tag{4.1}$$

The number $W^+(t_0, x_0)$ is called the upper value of the game with initial point (t_0, x_0) and $W^-(t_0, x_0)$ is called the lower value. It is always true that

$$W^-(t_0, x_0) \leq W^+(t_0, x_0). \tag{4.2}$$

If $W^-(t_0, x_0) = W^+(t_0, x_0)$ we denote this common value by $W(t_0, x_0)$ and say that the game has a value equal to $W(t_0, x_0)$.

Let A and B be two sets of real numbers. We say that $A \geq B$ if for every a in A and every b in B the inequality $a \geq b$ holds.

A pair of strategies (Γ^*, Δ^*) will be called *optimal strategies*, or will be said to constitute a *saddle point* of the game, if

$$P(t_0, x_0, \Gamma, \Delta^*) \leq P(t_0, x_0, \Gamma^*, \Delta^*) \leq P(t_0, x_0, \Gamma^*, \Delta) \tag{4.3}$$

for all strategies Γ for Player I and all strategies Δ for Player II. It follows from (4.1) and (4.2) that if a saddle point exists, then the game has a value and

$$W(t_0, x_0) = P(t_0, x_0, \Gamma^*, \Delta^*).$$

Analogous definitions of upper value, lower value, value and saddle point hold for the games G_c and G_r.

5. Games of Fixed Duration

In this section we summarize the principal results presented in [1], [3], [4], [5] and [22] as they pertain to games of fixed duration. Although we refer the reader to the above cited papers for proofs and details we will give an account of the order in which various results are obtained. The presentation here will incorporate some improvements over the results in the above cited papers.

We consider W^+ and W^- as functions of initial point (t, x) in $[T_0, T) \times R^n$. If we set $W^+(T, x) = g(x)$ and set $W^-(T, x) = g(x)$ and if Assumption I holds, then it can be shown that W^+ and W^- are uniformly continuous on bounded subsets of $[T_0, T] \times R^n$ and uniformly Lipschitz continuous in x on bounded subsets of $[T_0, T] \times R^n$.

We now return to the game with fixed initial point (t_0, x_0). Let

$$v^0 = W^+(t_0, x_0) \quad v_0 = W^-(t_0, x_0).$$

Let $W^+(T, x) = g(x)$ and let $W^-(T, x) = g(x)$. Let

$$C(v^0) = \{(\tau, \xi): t_0 \le \tau \le T,\ \xi \in R^n,\ W^+(\tau, \xi) \ge v^0\}$$
$$C(v_0) = \{(\tau, \xi): t_0 \le \tau \le T,\ \xi \in R^n,\ W^-(\tau, \xi) \le v_0\}.$$

Note that since W^+ and W^- are continuous functions, the sets $C(v^0)$ and $C(v_0)$ are closed.

It is an immediate consequence of the definitions of $C(v_0)$ and $W^-(\tau, \xi)$ that a point (τ, ξ) belongs to $C(v_0)$ if and only if for every $\varepsilon > 0$ and every strategy Γ for Player I there exists a strategy $\Delta(\Gamma)$ for Player II and a motion $\varphi[\ , \tau, \xi, \Gamma, \Delta(\Gamma)]$ such that $g(\phi[T, \tau, \xi, \Gamma, \Delta(\Gamma)] < v_0 + \varepsilon$. From this observation, the definition of $C(v_0)$ and the properties of motions, it is not hard to show the following. Let there exist a strategy Δ_0 such that for any strategy Γ the graphs of all motions $\phi[\ , t_0, x_0, \Gamma, \Delta_0]$ lie in $C(v_0)$. Then for any strategy Γ and all motions $\phi[\ , t_0, x_0, \Gamma, \Delta_0]$ the inequality

$$g(\phi[T, t_0, x_0, \Gamma, \Delta_0]) \le v_0 \equiv W^-(t_0, x_0) \tag{5.1}$$

holds.

Similarly, if there exists a strategy Γ_0 such that for all strategies Δ the graphs of all motions $\phi[\ , t_0, x_0, \Gamma_0, \Delta]$ lie in $C(v^0)$, then for any strategy Δ and all motions $\phi[\ , t_0, x_0, \Gamma_0, \Delta]$ the inequality

$$g(\phi[T, t_0, x_0, \Gamma_0, \Delta]) \ge v^0 \equiv W^+(t_0, x_0) \tag{5.2}$$

holds.

If we can find a strategy Δ_0 as in (5.1) (or a Γ_0 as in (5.2)) then the game will have a value. For then,

$$\sup_{\Gamma} g[T, t_0, x_0, \Gamma, \Delta_0] \le W^-(t_0, x_0)$$

and so

$$W^+(t_0, x_0) = \inf_{\Delta} \sup_{\Gamma} g[T, t_0, x_0, \Gamma, \Delta] \leq W^-(t_0, x_0).$$

If strategies Γ_0 and Δ_0 both exist then (5.1), (5.2) and the relation,

$$W^+(t_0, x_0) = W^-(t_0, x_0) = W(t_0, x_0)$$

imply that (Γ_0, Δ_0) is a saddle point.

We now outline a construction of a strategy Δ_0. Although the construction is such that it makes the property that the graphs of all motions $\phi[\ , t_0, x_0, \Gamma, \Delta_0]$ lie in $C(v_0)$ plausible, the actual proof is not simple. We start by defining a function V_e on $[t_0, T_0] \times \mathbb{R}^n$, which we shall call an extremal feedback function.

If (t^*, x^*) belongs to $C(v_0)$ we define $V_e(t^*, x^*)$ to be any element z^* in Z.

If (t^*, x^*) does not belong to $C(v_0)$ we define $V_e(t^*, x^*)$ as follows. Let $H(t^*)$ denote the hyperplane $t = t^*$ in R^{n+1}. Let $S(t^*) = H(t^*) \cap C(v_0)$. It can be shown that $S(t^*)$ is not empty. Since $C(v_0)$ is closed, so is $S(t^*)$. Let w^* be a point in $S(t^*)$ that is closest to x^*. Let $s^* = x^* - w^*$ and consider the game over $Y \times Z$ with payoff $\langle s^*, f(t^*, x^*, y, z)\rangle$. If we assume that the Isaacs condition (1.5) holds, then this game has a saddle point (y^*, z^*). (Recall that Y and Z are compact and $f(t^*, x^* \cdot, \cdot)$ is continuous on $Y \times Z$.) Now define $V_e(t^*, x^*) = z^*$.

We now define an *external strategy* Δ_e. The desired strategy Δ_0 will be Δ_e. Suppose τ_i is a partition point of π_n and that the n-th stage trajectory ϕ_n takes on the value ξ_i at τ_i. Then Δ_e assigns the constant control $u(t) = V_e(\tau_i, \xi_i)$ to be used by Player II on $[\tau_i, \tau_{i+1})$. Intuitively, Player II tries, as best he can, to point the tangent vector of the n-th stage trajectory into the set $C(v_0)$, assuming that Player I wishes to point the tangent vector away from the set as best he can. The reader will see that Δ_e is a strategy, since the position ξ_i at time τ_i is a function of $u_n \mid [t_0, \tau_i)$ and $v_n \mid [t_0, \tau_i)$.

The strategy $\Gamma_0 = \Gamma_e$ is constructed in a similar fashion.

The proof that the strategies $\Gamma_0 = \Gamma_e$ and $\Delta_0 = \Delta_e$ have the desired properties depends in a crucial way on the following lemma, which can be interpreted as playing the role of the principle of optimality in our considerations. This lemma also implies that the sets $S(t)$ are not empty for $t_0 \leq t \leq T$. Moreover Lemma 5.1 plays a crucial role in the proof of Lemma 5.2 below.

Lemma 5.1. *Let (τ, ξ) be a point of $C(v_0)$ such that $\tau < T$. Let t_1 satisfy $\tau < t_1 < T$ and let u be any control for Player I on $[\tau, t_1]$. Then there exists a relaxed control $\zeta = \zeta(u)$ such that the relaxed trajectory $\psi(\ , \tau, \xi, u, \zeta)$ has the property that $(t_1, \psi(t_1)) \in C(v_0)$.*

To make Lemma 5.1 plausible first note that if over the interval $[\tau, t_1]$ Player I uses a strategy Γ_u that always chooses a fixed control $u(\ )$ and Player II uses a strategy Δ, then over the interval $[\tau, t_1]$ the resulting motion with initial point (τ, ξ) will be a relaxed trajectory $\psi(\ , \tau, \xi, u, \zeta)$ for some relaxed control ζ. Thus, if the assertion of Lemma 5.1 were false, in the game starting from (τ, ξ) there would exist a strategy Γ_u such that for *any* strategy Δ of Player II the resulting motion $\phi[\ , \tau, \xi, \Gamma_u, \Delta]$ would be such that the point $(t_1, \varphi[t_1, \tau, \xi, \Gamma_u, \Delta])$ does not lie in $C(v_0)$. For each strategy Δ of Player II we can then redefine Γ_u on $[t_1, T]$ to get a strategy $\Gamma_u(\Delta)$ on $[\tau, T]$ such that the resulting motion $\varphi[\ , \tau, \xi, \Gamma_u(\Delta), \Delta]$ does not lie in $C(v_0)$. If we could choose the $\Gamma_u(\Delta)$ to actually be independent of Δ, say $\Gamma_u(\Delta) = \bar{\Gamma}$ for all Δ, then for all strategies Δ the motions $\varphi[\ , \tau, \xi, \bar{\Gamma}, \Delta]$ would not lie in $C(v_0)$. This would contradict the hypothesis that (τ, ξ) is in $C(v_0)$. We cannot choose a $\bar{\Gamma}$ as described, but the idea can be modified to give the desired result. See [6].

Our next result involves the lower Dini derivates of the function W^+ and W^-. Let L be a real valued Lipschitz continuous function defined on a set $[T_0, T_1) \times \mathcal{G}$, where $\mathcal{G}$ is an open set in R^n and let h be a vector in R^n. The upper Dini derivate of L at (t_0, x_0) in the direction h is denoted by $D^+L(t, x; 1, h)$ and is defined by

$$D^+L(t, x; 1, h) = \limsup_{\delta \downarrow 0}[L(t + \delta, x + \delta h) - L(t, x)]\delta^{-1}.$$

The lower Dini derivate of L in the direction $(1, h)$ at the point (t, x) is denoted by $D^-L(t, x; 1, h)$ and is defined by

$$D^-L(t, x; 1, h) = \liminf_{\delta \downarrow 0}[L(t + \delta, x + \delta h) - L(t, x)]\delta^{-1}.$$

If L is differentiable at (t, x) then for every h in R^n, $D^+L(t, x; 1, h) = D^-L(t, x; 1, h)$. If we denote this common value by $DL(t, x; 1, h)$ we have

$$DL(t, x; 1, h) = L_t(t, x) + \langle L_x(t, x), h \rangle.$$

Define

$$F_z(t, x, y) = co\{f(t, x, y, z) : z \in Z\}$$
$$F_y(t, x, z) = co\{f(t, x, y, z) : y \in Y\},$$

where $co(A)$ denotes the convex hull of A.

Lemma 5.2. *Let Assumption I hold. Then for each (t, x) in $[T_0, T) \times R^n$*

$$\begin{aligned} &\max_{y \in Y}\left[\min_{h \in F_z(t,x,y)} D^-W^-(t, x; 1, h)\right] \leq 0 \\ &\min_{z \in Z}\left[\max_{h \in F_y(t,x,z)} D^+W^+(t, x; 1, h)\right] \geq 0. \end{aligned} \tag{5.3}$$

If we set

$$\begin{aligned} H^+(t,x,p) &= \min_{z\in Z}\max_{y\in Y}\langle p, f(t,x,y,z)\rangle \\ H^-(t,x,p) &= \max_{y\in Y}\min_{z\in Z}\langle p, f(t,x,y,z)\rangle, \end{aligned} \tag{5.4}$$

then Lemma 5.2 implies the following:

Lemma 5.3. *The upper value W^+ is a viscosity subsolution of the Cauchy problem*

$$u_t + H^+(t,x,\nabla_x u) = 0$$

for (t,x) in $[T_0,T)\times R^n$ and

$$u(T,x) = g(x).$$

The lower value W^- is a viscosity supersolution of the Cauchy problem

$$\begin{aligned} &u_t + H^-(t,x,\nabla_x u) = 0 \\ &u(T,x) = g(x). \end{aligned}$$

For the definitions of viscosity solution, subsolution and supersolution see [7].

The Isaacs condition can be stated in terms of the functions H^+ and H^- defined in (5.4) as follows:

$$H^+(t,x,p) = H^-(t,x,p) \equiv H(t,x,p). \tag{5.5}$$

Except for the use of Lemma 5.1, the arguments and the results outlined in this section are not needed to establish Lemma 5.3. Thus, we could, if we so desired, appeal to a comparison theorem of Crandall and Lions [8] and show that if the Isaacs condition holds, then $W^+ = W^-$ and the game has value. This argument, however, gives no information concerning the existence of saddle points. Since we wish to establish the existence of saddle points in games of fixed duration when the Isaacs condition holds, and since that argument also gives the existence of value, we do not appeal to the theory of viscosity solutions to establish the existence of value.

We now state the principal result concerning games of fixed duration. For the proofs see [1], [4], [6], [22].

Theorem 5.1. *Let Assumption I hold and let the Isaacs condition hold. Then the game of fixed duration has value $W(t,x)$ for each initial position (t,x) in $[T_0,T)\times R^n$. The value function W is Lipschitz continuous on bounded subsets of $[T_0,T)\times R^n$. If $W(T,x)\equiv g(x)$, then W is Lipschitz continuous on bounded subsets of $[T_0,T]\times R^n$ and W is the unique viscosity solution of the Isaacs equation*

$$\begin{aligned} &u_t + H(t,x,\nabla_x u) = 0 \quad (t,x)\in[T_0,T)\times R^n \\ &u(T,x) = g(x). \end{aligned} \tag{5.6}$$

Moreover, for each initial point (t_0, x_0) *the game has a saddle point* (Γ^*, Δ^*)*, which may depend on* (t_0, x_0).

Corollary 5.1. *At points of differentiability, the function* W *satisfies*

$$W_t(t,x) + H(t,x,\nabla_x W(t,x)) = 0.$$

An examination of the arguments used to establish Theorem 5.1 shows that if we consider the game G_c, in which constant valued controls are chosen over each interval, then Theorem 5.1 holds for the game G_c. If we denote the value of this game by W_c, then it follows from a uniqueness theorem of [8] for viscosity solutions of (5.6) that $W_c = W$. This observation is useful when we approximate the game by its n-th stage discrete approximation.

If the Isaacs condition does not hold, then we use relaxed strategies and consider the game G_R. Let $\mathcal{Y}$ denote the set of probability measures on Y and let $\mathcal{Z}$ denote the set of probability measures on Z. Let

$$H_R^+(t,x,p) = \min_{dz\in\mathcal{Z}} \max_{dy\in\mathcal{Y}} \int_Z \int_Y \langle p, f(t,x,y,z)\rangle dy\, dz$$
$$H_R^-(t,x,p) = \max_{dy\in\mathcal{Y}} \min_{dz\in\mathcal{Z}} \int_Y \int_Z \langle p, f(t,x,y,z)\rangle dz\, dy.$$

Then the Isaacs condition holds in the sense that

$$H_R^+(t,x,p) = H_R^-(t,x,p).$$

We denote this common value by $H_R(t,x,p)$. It follows that the analogue of Theorem 5.1 holds for the game G_R, with value function W_R.

Remark 5.1. If the game as originally defined satisfies the Isaacs condition (5.5) then

$$H(t,x,p) = H_R(t,x,p).$$

Hence W and W_R are viscosity solutions of (5.1). It then follows from a uniqueness theorem of [8], that $W = W_R$.

For a detailed discussion of the game G_R, see [3].

Other definitions of a differential game have been proposed by Friedman [12], [13], Fleming [10], [11] and Elliott and Kalton [9]. It has been shown by various authors that for games of fixed duration all of these values are viscosity solutions of (5.6). It then again follows from a uniqueness theorem of [8] that all of these values are equal and are equal to our value.

6. Games of Generalized Pursuit and Evasion

In this section we summarize the results for games of generalized pursuit and evasion. For details and proofs see [2], [3], [4] and [19].

We make the following assumption concerning the terminal set $\mathcal{T}$.

Assumption II. Let $\mathcal{T}$ be as in (2.2) and let $\mathcal{T}_e = \mathcal{T} \cap \{[T_0, T_1) \times R^n\}$. Let $\partial\mathcal{T}_e$, the boundary of $\mathcal{T}_e$, be a $C^{(2)}$ manifold. At each point (t, x) of $\partial\mathcal{T}_e$ let (ν_0, ν) denote the normal to $\partial\mathcal{T}_e$ that points to the exterior of $\mathcal{T}_e$. Then

$$\min_z \max_y [\nu_0 + \langle \nu, f(t, x, y, z)\rangle] < 0, \tag{6.1}$$

where the min is taken over Z and the max is taken over Y.

For games of generalized pursuit and evasion the Isaacs condition is stated in terms of

$$\mathcal{H}(t, x, p^0, p, y, z) = p^0 f^0(t, x, y, z) + \langle p, f(t, x, y, z)\rangle.$$

Definition 6.1. The Isaacs condition is said to hold for a game of generalized pursuit and evasion if for all (t, x) in $[T_0, T_1]$ that are not in $\mathcal{T}$,

$$\max_y \min_z \mathcal{H}(t, x, 1, p, y, z) = \min_z \max_y \mathcal{H}(t, x, 1, p, y, z),$$

where the max is taken over Y and the min is taken over Z. In this event the common value is denoted by $\bar{H}(t, x, p)$.

We now state the principal result concerning games of generalized pursuit and evasion. Let

$$\mathcal{R} = \{[T_0, T_1] \times R^n\} - \mathcal{T}.$$

Theorem 6.1. *Let Assumptions I and II hold and let the Isaacs condition of Definition* 6.1 *hold. For all* (t, x, y, z) *in* $[T_0, T_1] \times R^n \times Y \times Z$ *let* $f^0(t, x, y, z) \geq \psi(t)$, *where* ψ *is a function that is integrable on* $[T_0, T_1]$. *Then for each* (t, x) *in* $\mathcal{R}$ *the game with initial point* (t, x) *has value* $W(t, x)$. *The function* W *is Lipschitz continuous on compact subsets of* $\mathcal{R}$ *and is a viscosity solution of the Cauchy problem*

$$\begin{aligned} &u_t + \bar{H}(t, x, \nabla_x u) = 0 \quad \text{on } \mathcal{R} \\ &u(t, x) = 0 \quad \text{on } \partial\mathcal{T}. \end{aligned} \tag{6.2}$$

If (6.1) *in Assumption II is replaced by the condition*

$$\nu_0 + \langle \nu, f(t, x, y, z)\rangle < 0,$$

then a saddle point exists.

The existence of a value W and the Lipschitz continuity of W are obtained by comparing the game of generalized pursuit and evasion, which for initial condition (t,x) we denote by $G(t,x)$, with a family of games $G_\mu(t,x)$, $0<\mu\le\mu_0$, of fixed duration. Since the payoff for a game $G(t_0,x_0)$ with initial point (t_0,x_0) is given by (3.6) and the zero-th components of the n-th stage trajectories φ_n^0 satisfy (3.4) with $u(t)$, $v(t)$ replaced by the appropriate $u_n(t)$, $v_n(t)$, the assumption $f^0\ge\psi$ can be replaced by the assumption $f^0\ge 0$ without loss of generality. Let $\mathcal{C}$ be a compact subset of $\mathcal{R}$. Then there exists an $R>0$ such that all n-th stage trajectories $\varphi_n(\,)$ initiating at points (t_0,x_0) in $\mathcal{C}$ satisfy $|\varphi_n(t)|\le R$ for $t_0\le t\le t_{n,f}$ and all motions $\varphi[\,]$ initiating at points (t_0,x_0) in $\mathcal{C}$ satisfy $|\varphi[t]|\le R$ for $t_0\le t\le t_f$. For (t,x) in $\mathcal{T}$ let $\rho(t,x)$ denote the distance from (t,x) to $\partial\mathcal{T}$. Let $\mathcal{T}_\mu=\{(t,x)\in\mathcal{T}$: $t<T_1$, $|x|<R$, $\rho(t,x)<\mu\}$. Then there exists a μ_0 such that for $\mu<\mu_0$, $\mathcal{T}_c-\mathcal{T}_\mu\ne\phi$. Let $\gamma(r)=1-r$ if $0\le r\le 1$ and let $\gamma(r)=0$ if $r>1$.

We now define the family of games of fixed duration. For each $0<\mu\le\mu_0$ let

$$f_\mu^0(t,x,y,z)=\begin{cases} f^0(t,x,y,z)\gamma(\rho(t,x)/\mu) & \text{if } (t,x)\in\mathcal{T}_e\\ f^0(t,x,y,z) & \text{if } (t,x)\notin\mathcal{T}_e.\end{cases}$$

The function f_μ^0 is continuous on $[T_0,T_1]\times R^n\times Y\times Z\times(0,\mu_0)$ and $f_\mu^0(t,x,y,z)=0$ if $(t,x)\in\mathcal{T}_e-\mathcal{T}_\mu$. For each μ in $(0,\mu_0)$ we consider the game of fixed duration with payoff

$$P(t_0,x_0,\Gamma,\Delta)=\cup\{\varphi^0[T_1,t_0,x_0,\Gamma,\Delta]\}$$

and dynamics

$$\frac{dx^0}{dt}=f_\mu^0(t,x,u(t),v(t))$$
$$\frac{dx}{dt}=f(t,x,u(t),v(t)).$$

Figure 1 gives a schematic representation of the preceding discussion.

The "payoff" to an n-th stage trajectory $\varphi(\,)$ (we omit the subscript n) in the game G_μ differs from the payoff of this trajectory in the game G by an amount

$$\varphi^0(T)-\varphi^0(t_f)=\int_{t_f}^{T} f_\mu^0(t,\varphi(t),u(t),v(t))dt=\int_{t_f}^{t^*} f_\mu(t,\varphi(t),u(t),v(t))dt,$$

where t_f is the time at which $\varphi^0(\,)$ hits the boundary of $\mathcal{T}_e$ and t^* is the time at which $\varphi^0(\,)$ exits from $\mathcal{T}_\mu$. Since $f_\mu\ge 0$ in $\mathcal{T}_e$ it follows that if $W^+(t_0,x_0,\mu)$ denotes the upper value of $G_\mu(t_0,x_0)$, then

$$W^+(t_0,x_0)\le W^+(t_0,x_0,\mu). \tag{6.3}$$

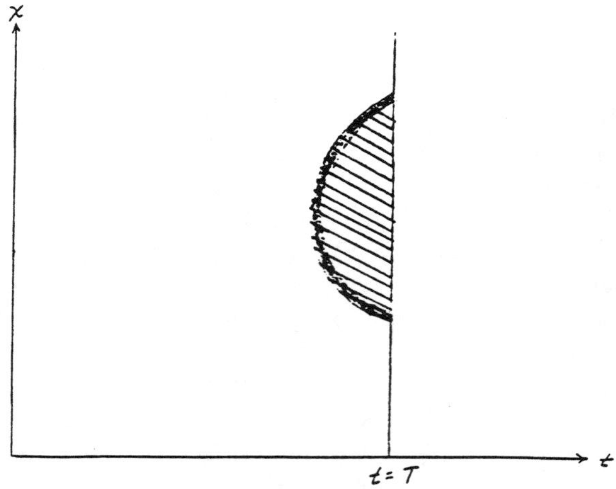

Figure 1. $\mathcal{T}_e \equiv$ cross hatched area; $\mathcal{T}_\mu \equiv$ shaded area; $f_\mu = 0$ in $\mathcal{T}_e - \mathcal{T}_\mu$; $f_\mu = f$ outside of $\mathcal{T}_e$.

Condition (6.1) guarantees that Player II can choose a control $u(t) = y^*$ for points in $\mathcal{T}_\mu$ that will cause an n-th stage trajectory to traverse $\mathcal{T}_\mu$ in time $0(\mu)$, where 0 is uniform for (t_0, x_0) in $\mathcal{C}$. Also, $f_\mu(t, x, y, z) = 0(\mu)$ on $\mathcal{T}_\mu$. These considerations can be made precise and used to show that there exists a constant C such that for all (t_0, x_0) in $\mathcal{C}$

$$W^-(t_0, x_0) \geq W^-(t_0, x_0, \mu) - C\mu. \tag{6.4}$$

The Isaacs condition of Definition (6.1) implies that the games G_μ of fixed duration satisfy the appropriate Isaacs condition, and thus have values $W(t_0, x_0, \mu)$. From this and from (6.3) and (6.4) we get

$$W^+(t_0, x_0) \leq W^+(t_0, x_0, \mu) = W^-(t_0, x_0, \mu) \leq W^-(t_0, x_0) + C\mu.$$

Letting $\mu \to 0$ gives the existence of value in $\mathcal{C}$ and hence on $\mathcal{R}$.

The relations (5.3) of Lemma 5.2 hold for each (t, x) in $\mathcal{R}$, and from this it again follows that W is a viscosity solution of (6.2).

The existence of a saddle point (Γ^*, Δ^*) is established as follows. Let k_0 be a positive integer such that $1/k_0 < \mu_0$. Then for $k > k_0$, the games $G(t_0, x_0; k) \equiv G_{1/k}(t_0, x_0)$ are well defined and have saddle points (Γ_k^*, Δ_k^*). Let $\{\pi_{k,n}^*\}$ be the sequence of partitions associated with Γ_k^* and let $\{\Gamma_{kn}^*\}$ denote the sequence of maps. Let $\{\bar{\pi}_{k,n}^*\}$ and $\{\Delta_{k,n}^*\}$ have similar meaning for Δ^*. It is shown in [19] that the diagonals $\{\pi_{k,k}^*\}_{k=k_0}^\infty$ and $\{\Gamma_{k,k}^*\}_{k=k_0}^\infty$ determine an optimal strategy Γ^* for Player I and that the diagonals $\{\bar{\pi}_{k,k}\}_{k=k_0}^\infty$ and $\{\Gamma_{k,k}^*\}_{k=k_0}^\infty$ determine an optimal strategy Δ^* for Player II.

7. Games of Survival

The principal result for games of survival is given in the following theorem.

Theorem 7.1. *Let Assumptions I and II hold and let*

$$\min_{y\in Y}\max_{z\in Z}[\nu^0 + \langle \nu, f(t,x,y,z)\rangle] < 0$$

at all points of $\partial\mathcal{T}_e$. Let the Isaacs condition as stated in Definition (6.1) *hold. Then for each (t,x) in $\mathcal{R}$ the game with initial point (t,x) has value $W(t,x)$ and a saddle point $(\Gamma^*(t,x), \Delta^*(t,x))$. The function W is Lipschitz continuous on compact subsets of $\mathcal{R}$. Moreover, W satisfies* (5.3) *on $\mathcal{R}$ and is a viscosity solution of the Cauchy problem*

$$\begin{aligned} u_t + \bar{H}(t,x,\nabla_x u) &= 0 \quad \text{on } \mathcal{R} \\ u(t,x) &= g(t,x) \quad \text{on } \partial\mathcal{T}. \end{aligned}$$

The properties of the value function are first established under the assumption that $g \equiv 0$. The arguments used are minor modifications of those used in [2]. Next, we consider the case in which g is $C^{(2)}$ on all of $(T_0, T_1]$. We suppose, without loss of generality, that $g(t_0, x_0) = 0$ and we consider the game with payoff

$$\int_{t_0}^{t_f} [g_t(s,\phi(s)) + \langle Dg(s,\phi(s)), f(s,\varphi(s),u(s),v(s))\rangle \\ + f^0(s,\varphi(s),u(s),v(s))]\, ds,$$

where φ is a solution of $dx/dt = f(t,x,u(t),v(t))$ and $Dg = (\partial g/\partial x_1, \ldots, \partial g/\partial x_n)$. This now becomes a game with $g \equiv 0$, to which the previous results apply.

If g is continuous and $\mathcal{B}$ is a compact set containing (t_0, x_0), then there exists a sequence of $C^{(2)}$ functions $\{g_n\}$ on $[T_0, T_1] \times R^n$ such that $|g_n(t,x) - g(t,x)| \le 1/n$ for all (t,x) in $\mathcal{B}$. Let P_n denote the payoff in the game $G_n(t_0, x_0)$ with initial point (t_0, x_0) and payoff g_n. Then for any Γ, Δ

$$P_n(t_0,x_0,\Gamma,\Delta) - \frac{1}{n} \le P(t_0,x_0,\Gamma,\Delta) \le P_n(t_0,x_0,\Gamma,\Delta) + \frac{1}{n}.$$

Since the games G_n have values W_n,

$$W_n(t_0,x_0) - \frac{1}{n} \le W^-(t_0,x_0) \le W^+(t_0,x_0) \le W_n(t_0,x_0) + \frac{1}{n}.$$

If we take lim sup as $n \to \infty$ on the left and lim inf on the right we get that $W(t_0, x_0)$ exists. Moreover, $W_n(t_0, x_0) \to W(t_0, x_0)$ uniformly on B. Thus W is continuous on $\mathcal{B}$ and hence on $\mathcal{R}$. For all of the details and for the arguments used to establish the other assertions of the theorem, see [5], [19].

8. Games with Lags

All of the games studied thus far were so-called games of perfect information. That is, it was assumed that at each time $t > t_0$ each player knew his choice of control on $[t_0, t]$ and his opponent's choice of control on $[t_0, t)$. Since each player is assumed to know the dynamics he knows the current state of the game. A more realistic model of conflict situations would be one in which information lags exist. That is, it is assumed that it takes some time to process information about the opponents' choice of control. Thus we assume that Player I makes his choice at time t based on his previous choices on the interval $[t_0, t]$ and on the choices of Player II over the interval $[t_0, t - \lambda)$, $\lambda > 0$. Similarly Player II makes his choice at time t based on his choices on $[t_0, t]$ and on Player I's choices over $[t_0, t-\mu)$, $\mu > 0$. We say that Player I has an information lag λ and that Player II has an information lag μ.

We now formally define the game. To keep matters simple we discuss games of fixed duration with payoff $g(x)$.

A strategy Γ^λ for Player I is a choice of a sequence $\pi = \{\pi_n\}$ of partitions of $[t_0, T]$ and a choice of a sequence of maps $\Gamma^\lambda = \{\Gamma^\lambda_n\}$, where the maps Γ^λ_n are to be defined presently. We require that $\|\pi_n\| \le A/n$ for some constant A independent of n and satisfying $A > T_1 - T_0$. Let $t_0 < t_1 < \cdots < t_p = T_1$, denote the parition points of π_n, where $p = p(n)$ depends on n. Each map Γ^λ_n is a collection of maps $\Gamma^\lambda_{n,1}, \ldots, \Gamma^\lambda_{n,p}$ defined as follows. The map $\Gamma^\lambda_{n,1}$ selects an element in $\mathcal{Y}[t_0, t_1]$. For $2 \le j \le p$, if $t_{j-1} - \lambda \le t_0$, then $\Gamma^\lambda_{n,j}$ is a map from $\mathcal{Y}[t_0, t_{j-1})$ to $\mathcal{Y}[t_{j-1}, t_j)$; if $t_{j-1} - \lambda > t_0$, $\Gamma^\lambda_{n,j}$ is a map from $\mathcal{Y}[t_0, t_{j-1}) \times \mathcal{Z}[t_0, t_{j-1} - \lambda)$ to $\mathcal{Y}[t_{j-1}, t_j)$.

Similarly, a strategy Δ^μ for Player II is a choice of sequence of partitions $\bar{\pi} = \{\pi_n\}$ of $[t_0, T]$ satisfying $\|\bar{\pi}_n\| \le \bar{A}/n$, $\bar{A} > T_1 - T_0$, and a choice of sequence of maps $\Delta^\mu = \{\Delta^\mu_n\}$. Each Δ^n_μ is a collection of maps $\Delta^\mu_{n,1}, \ldots, \Delta^\mu_{n,q}$ where $q = q(n)$ as follows. Let $\bar{\pi}_n$ have partition points $t_0 = s_0 < s_1 < s_2 < \cdots < s_q = T_1$. Then $\Delta^\mu_{n,1}$ selects a function v in $\mathcal{Z}[s_0, s_1)$. For $2 \le j \le q$, if $s_{j-1} - \mu \le s_0$ then $\Delta^\mu_{n,j}$ is a map from $\mathcal{Z}[s_0, s_{j-1})$ to $\mathcal{Z}[s_{j-1}, s_j)$; if $s_{j-1} - \mu > s_0$, then $\Delta^\mu_{n,j}$ is a map from $\mathcal{Y}[s_0, s_{j-1} - \mu) \times \mathcal{Z}[s_0, s_{j-1})$ to $\mathcal{Z}[s_{j-1}, s_j)$. The notions of nth-stage trajectory, motion, payoff, value, and saddle point are as in games without lag.

In general, games with lags do not have values. The following simple example can be shown to be a game with lag that does not have a value. Let the dynamics be

$$\frac{dx}{dt} = u + v \quad 0 \le t \le 1$$
$$x(0) = 0$$

with x real, $|u|, |v| \le 1$.

The payoff function is $g(x(1)) = x^2(1)$. If Player I has an information lag $\lambda \geq 0$ and Player II has an information lag $\mu > 0$, then it can be shown that for λ, $\mu < 1$ the lower value is zero and the upper value is greater than or equal to $\mu^2/4$.

Let $W^-_{\lambda,\mu}(t_0, x_0)$ denote the lower value of a game with lag λ for Player I and lag μ for Player II. Let $W^+_{\lambda,\mu}(t_0, x_0)$ denote the corresponding upper value. The principal result concerning games of fixed duration with lags is the following.

Theorem 8.1. *Let Assumption I hold and let the Isaacs condition* (1.5) *hold. Then there exists a non-negative function* η *defined in* $[0, \theta)$ *such that* $\eta(\theta) \to 0$ *as* $\theta \to 0$ *and such that*

$$0 \leq W^+_{\lambda,\mu}(t_0, x_0) - W^-_{\lambda,\mu}(t_0, x_0) \leq \eta(\lambda) + \eta(\mu).$$

Although this result would be expected on intuitive grounds, the proof is quite delicate. For details of the proof and the arguments used to show that the example does not have value, see [19]. Also see [19] for a discussion of games of generalized pursuit and evasion and games of survival with lags.

9. State Constrained Problems

Another class of games that occurs in applications are games in which $x(t)$ the state of game at time t is required to lie in a preassigned closed set $\mathcal{H}(t)$. For a treatment of this problem see [14], [15].

References

[1] L.D. Berkovitz, The existence of value and saddle point in games of fixed duration, *SIAM J. Control*, **23** (1985), 172–196.

[2] L.D. Berkovitz, Differential games of generalized pursuit and evasion, *SIAM J. Control*, **24** (1986), 361–373.

[3] L.D. Berkovitz, Differential games without the Isaacs condition, in *Recent Advances in Communication and Control Theory*, Optimization Software, (R.E. Kalman, G.I. Marchuk, A.E. Roberts, and A.J. Viterbi, eds.), New York, 1987, pp. 308–336.

[4] L.D. Berkovitz, Characterizations of the values of differential games, *Appl. Math. Optim.*, **17** (1988), 177–183.

[5] L.D. Berkovitz, Differential games of survival, *J. Math. Anal. Appl.*, **129** (1988), 493–504.

[6] L.D. Berkovitz, Correction to the existence of value and saddle point in games of fixed duration, *SIAM J. Control*, **26** (1988), 740–742.

[7] M.G. Crandall, L.C. Evans, and P.L. Lions, Some properties of viscosity solutions of Hamilton-Jacobi equations, *Trans. Amer. Math. Soc.*, **282** (1984), 487–502.

[8] M.G. Crandall and P.L. Lions, Remarks on the existence and uniqueness of unbounded viscosity solutions of Hamilton-Jacobi equations, *Illinois J. Math.*, **31** (1987), 665–687.
[9] R.J. Elliott and N.J. Kalton, The existence of value in differential games, *Mem. Amer. Math. Soc.*, **126** (1972).
[10] W.H. Fleming, The convergence problem for differential games, *J. Math. Anal. Appl.*, **3** (1961), 102–116.
[11] W.H. Fleming, The convergence problem for differential games II, *Annals of Mathematics Study No. 52*, Princeton University Press, Princeton, NJ (1964), 195–210.
[12] A. Friedman, *Differential Games*, Wiley-Interscience, New York, 1971.
[13] A. Friedman, *Differential Games*, CBMS Regional Conference Series in Mathematics No. 18, American Mathematical Society, Providence, 1974.
[14] K. Haji-Ghassemi, On differential games of fixed duration with phase coordinate restrictions on one player, *SIAM J. Control Optim.*, **28**, no. 3 (1990), 624–652.
[15] K. Haji-Ghassemi, Differential games of fixed duration with state constraints, *J. Optimization Theory Appl.*, **68** (1991), 513–538.
[16] R. Isaacs, *Differential Games*, John Wiley and Sons, New York, London, Sydney, 1965.
[17] N.N. Krasovskii and A.I. Subbotin, *Positional Differential Games*, Nauka, Moscow, 1974.
[18] N.N. Krasovskii and A.I. Subbotin, *Game Theoretical Control Problems*, Springer-Verlag, New York, 1988.
[19] X. Qian, *Topics in Differential Games*, Thesis, Department of Mathematics, Purdue University, West Lafayette, IN, August 1992.
[20] J. Warga, *Optimal Control of Differential and Functional Equations*, Academic Press, New York, 1972.
[21] P. Varaiya and J. Lin, Existence of saddle points in differential games, *SIAM J. Control*, **7** (1969), 153–163.
[22] J. Yong, On the Isaacs equation of differential games of fixed duration, *J. Optim. Theory Appl.*, **50** (1986), 359–364.

Leonard D. Berkovitz
Department of Mathematics
Purdue University
West Lafayette, IN 47907

H^∞-Optimal Control of Singularly Perturbed Systems with Sampled-State Measurement*

Zigang Pan and Tamer Başar

1 Introduction

One of the important recent developments in control theory has been the recognition of the close relationship that exists between H^∞-optimal control problems, (originally formulated in the frequency domain [1] [2], and then extended to state space formulations [3] [4] [5] [6] [7] [8]) and a class of linear-quadratic differential games [9] [10] [11] [12] [13], which has not only led to simpler derivations of existing results on the former, but also enabled us to develop worst-case (H^∞-optimal) controllers under various information patterns, such as (in addition to perfect and imperfect state measurements) delayed state and sampled state measurements [14] [15]. An up-to-date coverage of this relationship and the derivation of H^∞-optimal controllers under different information patterns can be found in the recent book [16], which also contains an extensive list of references on the topic.

It is by now well-known that in both finite- and infinite-horizon formulations, under sampled-data state measurements, one can come arbitrarily close to the H^∞-optimal performance level by designing a dynamic controller, of the same order as that of the plant (which may also include the dynamics of the disturbance, if any), whose construction involves the solution of a parametrized differential or algebraic Riccati equation, subject to the condition that the system can be stabilized to attain the performance level on each sampling interval. The problem we address in this paper is the possible order reduction of this controller when the plant exhibits open-loop time-scale separation. In state space, such systems are commonly modeled using the mathematical framework of singular perturbations, with a small parameter, say ϵ, determining the degree of separation between the "slow" and "fast" modes of the system [17]. In this framework, our objective may be rephrased as one of obtaining "approximate" controllers which do not depend on the singular perturbation parameter ϵ, and proving that these approximate controllers can be used "reliably" on the original system when $\epsilon > 0$ is sufficiently small.

*Research supported in part by the National Science Foundation under Grant ECS 91-13153, and in part by the U.S. Department of Energy under Grant DE-FG-02-88-ER-13939.

In an earlier paper, [18], we have already initiated a study on this problem when the controller has access to full state information, where we use the framework of differential games. One of our results in this context has been that the H^∞-optimal performance does not show continuity at $\epsilon = 0$, in the sense that the performance of the original system as $\epsilon \to 0$ is not necessarily the same as the H^∞-optimal performance of the reduced-order (slow) system, even if the fast subsystem is stable. We have actually proven that the former is upper bounded by the maximum of the H^∞-optimal performances of the "slow" and "fast" subsystems, appropriately defined. We have also constructed composite controllers (from parametrized solutions of slow and fast games), independently of ϵ, which guarantee a desired achievable performance level for the full-order plant when ϵ is sufficiently small. In another paper, [19], we studied the same problem for the case when full state measurements are not available, which is the well-known four-block problem. In this case the composite controller exhibits a performance loss due to the lack of knowledge of the small singular perturbation parameter, which was not the case under perfect state measurements.

In the present paper, we develop counterparts of the results of [18] for the sampled-data measurements case. Some of our conclusions are similar in spirit to those of [18] (though details of proofs are much more involved), while some others are quite different. For a brief preview of our main results, let us introduce the quantity $\gamma^*_{SD}(\epsilon)$ to denote the H^∞-optimum performance of the full-order system under sampled-data state measurements, and the quantities γ_{SDs} and γ_{of} to denote the H^∞-optimum performances of (appropriately defined) reduced slow subsystems under sampled-data state measurements and fast subsystems under open-loop state measurements, respectively. We first show that $\gamma^*_{SD}(\epsilon = 0)$ is bounded above by $\max\{\gamma_{SDs}, \gamma_{of}\}$. We then obtain a "slow" controller, designed based on the slow subsystem only (though with the knowledge of γ_{of}) and independent of the singular perturbation parameter, under which the associated differential game has a bounded upper value, to achieve any desired achievable H^∞-performance level for the full-order problem.

The problem of designing controllers for singularly perturbed systems subject to unknown disturbances has been studied before in the literature, notably in papers [20], [21], [22], where the objective has been to obtain composite controllers that guarantee stability of the overall (possibly nonlinear) system. The main approach of the authors in these papers has involved the construction of appropriate Lyapunov functions, in terms of which a class of stabilizing controllers has been characterized. However, no optimality properties have been associated with these controllers, which is our main concern in this paper. Yet another paper that deals with uncertain (linear) systems which exhibit time-scale separation is [23], which obtains a two-frequency-scale decomposition for H^∞-disk problems, but does not address the issue of optimal controller design. There are a number of papers that have studied the problem of sampled-data H^∞-optimal control for regular systems; see, for example, [14] [15] [24] [25].

The balance of the present paper is organized as follows. In the next section (Section 2) we formulate the singularly perturbed H^∞-optimal control problem with sampled-data state measurements, and identify the associated linear-quadratic differential game. We also provide in that section the solution to the full-order problem, for both finite and infinite horizons. Then, we identify appropriate slow and fast subsystems, and define the associated differential games. In Section 3, we construct the "slow" sampled-data controller, and study its optimality as well as its robustness to order reduction. Section 4 presents some numerical results to illustrate the theory, and Section 5 provides a discussion on some immediate extensions. The paper ends with three Appendices, which provide the details of some of the derivations given in the main body of the paper.

2 Problem Formulation

The system under consideration, with slow and fast dynamics, is described in the standard "singularly perturbed" form by

$$\left\{ \begin{array}{rcll} \dot{x}_1 & = & A_{11}(t)x_1 + A_{12}(t)x_2 + B_1(t)u + D_1(t)w; & x_1(0) = 0 \\ \epsilon\dot{x}_2 & = & A_{21}(t)x_1 + A_{22}(t)x_2 + B_2(t)u + D_2(t)w; & x_2(0) = 0 \end{array} \right. \tag{1}$$

where $x' := (x_1', x_2')$ is the n-dimensional state vector, with x_1 of dimension n_1 and x_2 of dimension $n_2 := n - n_1$; u is the control input, and w is the disturbance, each belonging to appropriate ($\mathcal{L}^2$) Hilbert spaces $\mathcal{H}_x$, $\mathcal{H}_u$ and $\mathcal{H}_w$, respectively, defined on the time interval $[0, t_f]$. The control input u is generated by a sampled-data control policy μ_{SD}, according to

$$u(t) = \mu_{SD}(t, x(t_k), x(t_{k-1}), \ldots, x(t_1), x_0) \quad t_k \le t < t_{k+1} \tag{2}$$

where $0 < t_1 < \cdots t_K = t_f$ are the sampling times, $\mu(\cdot)$ is piecewise continuous in t and measurable in $x_0, x(t_1), \ldots, x(t_K)$. Let us denote the class of all these controllers by $\mathcal{M}_{SD}$. With this system, we associate the standard quadratic performance index

$$\begin{aligned} L(u,w) &= |x(t_f)|^2_{Q_f} + \int_0^{t_f} (|x(t)|^2_{Q(t)} + |u(t)|^2)\, dt \\ &\equiv |x(t_f)|^2_{Q_f} + \|x\|^2_Q + \|u\|^2; \quad Q_f \ge 0, \quad Q(.) \ge 0 \end{aligned} \tag{3}$$

where Q_f will show dependence on $\epsilon > 0$, as to be clarified later. Let us also introduce the notation $J(\mu_{SD}, w)$ to denote $L(u, w)$, with u given by (2). The H^∞-optimal control problem is the minimization of the quantity

$$\sup_{w \in \mathcal{H}_w} \{J(\mu_{SD}, w)\}^{1/2} / \|w\| \tag{4}$$

over all permissible controllers μ_{SD}, or if a minimum does not exist to find a controller μ_{SD} that will ensure a performance within a given neighborhood of the infimum of (4). Let us denote this infimum by $\gamma_{SD}^*(\epsilon)$, i. e.

$$\inf_{\mu_{SD}\in\mathcal{M}_{SD}} \sup_{w\in\mathcal{H}_w} \{J(\mu_{SD},w)\}^{1/2}/\|w\| = \gamma_{SD}^*(\epsilon) \tag{5}$$

where we explicitly show the dependence of γ_{SD}^* on the singular perturbation parameter $\epsilon > 0$.

For each $\epsilon > 0$, we can associate a soft-constrained linear-quadratic differential game with this worst-case design problem (see [16]), which has the cost function

$$L_\gamma(u,w) = L(u,w) - \gamma^2\|w\|^2 \tag{6}$$

The performance level $\gamma_{SD}^*(\epsilon)$ in (5) is then the "smallest" value of $\gamma \geq 0$ under which the differential game with state equation (1) and cost function (6) has a bounded upper value, when u is chosen according to (2). We also know that for every fixed $\epsilon > 0$, and for each $\gamma > \gamma_{SD}^*(\epsilon)$, this differential game admits a saddle-point solution, with the saddle-point controller μ_{SD}^* being a linear feedback law that depends only on the most recent sampled value of the state (see [16], chapter 4).

Even though the problem formulated above has been solved completely for every $\epsilon > 0$, the computation of $\gamma_{SD}^*(\epsilon)$ and a corresponding H^∞-optimal or suboptimal controller for small values of $\epsilon > 0$ present serious difficulties, due to numerical stiffness. To remedy this, we pose in this paper the question of whether $\gamma_{SD}^*(\epsilon)$ and the H^∞-optimal controller can be determined for small value of $\epsilon > 0$ by solving well-behaved ϵ-independent smaller-order problems, as in the case of the singularly perturbed linear-quadratic regulator problem ([26]). Another question of interest is whether the fast subsystem dynamics can be neglected completely in the design of such controllers, and if so under what conditions.

Before studying these questions, we first present the solution to the full-order problem, under four basic assumptions:

Assumption 1 Q_f and $Q(.)$ in (3) are partitioned as

$$Q_f = \begin{bmatrix} Q_{f11} & \epsilon Q_{f12} \\ \epsilon Q_{f21} & \epsilon Q_{f22} \end{bmatrix}; \qquad Q(t) = \begin{bmatrix} Q_{11}(t) & Q_{12}(t) \\ Q_{21}(t) & Q_{22}(t) \end{bmatrix}$$

where in each case the 11-block is of dimension $n_1 \times n_1$, and the 22-block of dimension $n_2 \times n_2$.

Assumption 2 $A_{ij}(t)$, $Q_{ij}(t)$, $B_i(t)$, $D_i(t)$ $(i = 1,2 \;\; j = 1,2)$ are continuously differentiable, and $\|A_{ij}\|$, $\|Q_{ij}\|$, $\|B_i\|$, $\|D_i\|$, $\|\dot{A}_{ij}\|$, $\|\dot{Q}_{ij}\|$, $\|\dot{B}_i\|$, $\|\dot{D}_i\|$ $(i = 1,2 \;\; j = 1,2)$ are bounded on $t \in [0,t_f]$.

Assumption 3 $A_{22}(t)$ and $Q_{22}(t)$ are invertible for all $t \in [0,t_f]$.

Assumption 4 $T_m := \min_{1\le k\le K} t_k - t_{k-1}$ is positive and ϵ-free.

Let us further introduce the notation $A(t)$ and $A_\epsilon(t)$, to denote the partitioned matrices

$$A(t) := \begin{bmatrix} A_{11}(t) & A_{12}(t) \\ A_{21}(t) & A_{22}(t) \end{bmatrix}; \qquad A_\epsilon(t) := \begin{bmatrix} A_{11}(t) & A_{12}(t) \\ \frac{1}{\epsilon}A_{21}(t) & \frac{1}{\epsilon}A_{22}(t) \end{bmatrix}$$

where we take $\epsilon > 0$. Similarly, we introduce the partitioned matrices

$$B(t) := \begin{bmatrix} B_1(t) \\ B_2(t) \end{bmatrix}; \qquad B_\epsilon(t) := \begin{bmatrix} B_1(t) \\ \frac{1}{\epsilon}B_2(t) \end{bmatrix}$$

$$D(t) := \begin{bmatrix} D_1(t) \\ D_2(t) \end{bmatrix}; \qquad D_\epsilon(t) := \begin{bmatrix} D_1(t) \\ \frac{1}{\epsilon}D_2(t) \end{bmatrix}$$

and define for each $\epsilon > 0$,

$$S(t;\gamma) = B(t)B'(t) - \frac{1}{\gamma^2}D(t)D'(t)$$

$$S_\epsilon(t;\gamma) = B_\epsilon(t)B_\epsilon'(t) - \frac{1}{\gamma^2}D_\epsilon(t)D_\epsilon'(t)$$

with the ij-th block of $S(t;\gamma)$ denoted by $S_{ij}(t;\gamma)$, $i,j = 1,2$.

We know from the existing theory on linear-quadratic differential games (see [27] Chapter 4 pp. 105, [16]) that for each $\epsilon > 0$, there exists a $\hat\gamma_{SD}(\epsilon) \ge 0$ such that for all $\gamma > \hat\gamma_{SD}(\epsilon)$ the zero-sum differential game described by (1) and (6) has a bounded upper value (which in this case is equal to *zero*, because the initial states are zero), as well as a saddle point. For all such γ, the following generalized Riccati differential equations do not have conjugate points in the given time intervals:

$$\dot{\tilde Z} + A_\epsilon'\tilde Z + \tilde Z A_\epsilon - \tilde Z S_\epsilon \tilde Z + Q = 0; \qquad \tilde Z(t_f) = Q_f \quad t \in [0, t_f] \tag{7}$$

$$\begin{aligned} &\dot{\tilde P}_k + A_\epsilon'\tilde P_k + \tilde P_k A_\epsilon + \tilde P_k \tfrac{1}{\gamma^2} D_\epsilon D_\epsilon' \tilde P_k + Q = 0 \\ &\tilde P_k(t_k) = \tilde Z(t_k) \quad t \in [t_{k-1}, t_k], \quad k = K, K-1, \dots, 1 \end{aligned} \tag{8}$$

and a saddle-point controller is given by

$$\begin{aligned} u_{SD}^*(t) = \mu_{SD}^*(t, x(t)) &= -B_\epsilon'\tilde Z(t)\Phi_\epsilon(t, t_k)x(t_k) \\ t_k \le t < t_{k+1}, &\quad k = 0, 1, \dots, K \end{aligned} \tag{9}$$

where Φ_ϵ is the state transition function associated with the matrix

$$\tilde F(t) := A_\epsilon - [B_\epsilon B_\epsilon' - \frac{1}{\gamma^2} D_\epsilon D_\epsilon']\tilde Z(t) \tag{10}$$

For $\gamma < \hat\gamma_{SD}(\epsilon)$, on the other hand, at least one of the generalized Riccati equations (7)–(8) has a conjugate point in the corresponding open interval and the game has infinite upper value under sampled-data information. The level $\hat{\gamma}_{SD}(\epsilon)$ is indeed the H^∞-optimal performance level (5).

Since GRDEs of the type above (7) and (8) have been studied extensively in a recent paper [18], we will apply those results directly without further study of the full-order GRDEs here.

To obtain the ϵ-free solution, we now introduce the slow and fast subsystems as in [18].

The slow subsystems and the associated soft-constrained games

To obtain the slow dynamics associated with (1), we let $\epsilon = 0$ and solve for x_2 (to be denoted $\bar{x}_2$) in terms of $x_1 =: x_s$, $u =: u_s$, $w =: w_s$, and under the working Assumption 3:

$$\bar{x}_2 = -A_{22}^{-1}(A_{21}x_s + B_2u_s + D_2w_s) \tag{11}$$

We also introduce the transformation (12) given below (as in [18], [19])

$$\begin{aligned}\hat{w}_s &= (\gamma^2 I - D_2' A_{22}'^{-1} Q_{22} A_{22}^{-1} D_2)^{1/2}[w_s - \frac{1}{\gamma^2} D_2'(A_{22}Q_{22}^{-1}A_{22}' \\ &\quad -\frac{1}{\gamma^2} D_2 D_2')^{-1}(B_2u_s + (A_{21} - A_{22}Q_{22}^{-1}Q_{21})x_s)]\end{aligned} \tag{12}$$

which is valid under the condition

$$\gamma^2 I - D_2' A_{22}'^{-1} Q_{22} A_{22}^{-1} D_2 > 0 \tag{13}$$

Note that the dependence of $\hat{w}_s$ on u_s is allowed here since we are interested in the upper value of the game. Using this transformation in (1), (11) and (6), we arrive at the following LQ structure:

$$\dot{x}_s = A^\square x_s + B^\square u_s + D^\square \hat{w}_s; \quad x_s(0) = 0 \tag{14}$$

$$L_{\gamma s} = |x_s(t_f)|^2_{Q_{f11}} + \int_0^{t_f} (|x_s|^2_{Q^\square} + 2x_s' P^\square u_s + |u_s|^2_{R^\square} - |\hat{w}_s|^2)\, dt \tag{15}$$

where

$$\begin{aligned}A^\square(\gamma) &:= A_{11} - A_{12}Q_{22}^{-1}Q_{21} - (A_{12}Q_{22}^{-1}A_{22}' - \frac{1}{\gamma^2}D_1D_2')(A_{22}Q_{22}^{-1}A_{22}' \\ &\quad -\frac{1}{\gamma^2}D_2D_2')^{-1}(A_{21} - A_{22}Q_{22}^{-1}Q_{21})\end{aligned} \tag{16}$$

$$\begin{aligned}B^\square(\gamma) &:= B_1 - (A_{12}Q_{22}^{-1}A_{22}' - \frac{1}{\gamma^2}D_1D_2')(A_{22}Q_{22}^{-1}A_{22}' \\ &\quad -\frac{1}{\gamma^2}D_2D_2')^{-1}B_2\end{aligned} \tag{17}$$

$$\begin{aligned}
D^\Box(\gamma) &:= (D_1 - A_{12}A_{22}^{-1}D_2)(\gamma^2 I - D_2' A_{22}'^{-1} Q_{22} A_{22}^{-1} D_2)^{-1/2} && (18)\\
Q^\Box(\gamma) &:= Q_{11} - Q_{12}Q_{22}^{-1}Q_{21} + (A_{21}' - Q_{12}Q_{22}^{-1}A_{22}')(A_{22}Q_{22}^{-1}A_{22}' \\
&\quad -\frac{1}{\gamma^2}D_2D_2')^{-1}(A_{21} - A_{22}Q_{22}^{-1}Q_{21}) && (19)\\
P^\Box(\gamma) &:= (A_{21}' - Q_{12}Q_{22}^{-1}A_{22}')(A_{22}Q_{22}^{-1}A_{22}' - \frac{1}{\gamma^2}D_2D_2')^{-1}B_2 && (20)\\
R^\Box(\gamma) &:= I + B_2'(A_{22}Q_{22}^{-1}A_{22}' - \frac{1}{\gamma^2}D_2D_2')^{-1}B_2 && (21)
\end{aligned}$$

The above is the equivalence form of the representation in [18] pp.35–36.

The GRDE associated with the above transformed game is[1]

$$\dot{Z}_s + \hat{A}_0' Z_s + Z_s \hat{A}_0 - Z_s \hat{S}_0 Z_s + \hat{Q} = 0; \qquad Z_s(t_f) = Q_{f11} \tag{22}$$

where we have introduced

$$\hat{A}_0 := A^\Box - B^\Box R^{\Box -1} P^{\Box\prime} \qquad \hat{S}_0 := B^\Box R^{\Box -1} B^{\Box\prime} - D^\Box D^{\Box\prime}$$
$$\hat{Q} := Q^\Box - P^\Box R^{\Box -1} P^{\Box\prime}$$

It is shown in [18] pp.36–37 that $\hat{A}$, $\hat{S}_0$ and $\hat{Q}$ can be rewritten as

$$\begin{aligned}
\hat{A}_0 &= A_{11} - A_{12}Q_{22}^{-1}Q_{21} - (S_{12} + A_{12}Q_{22}^{-1}A_{22}')(S_{22} + A_{22}Q_{22}^{-1}A_{22}')^{-1} \\
&\quad \cdot(A_{21} - A_{22}Q_{22}^{-1}Q_{21}) && (23)\\
\hat{S}_0 &= S_{11} + A_{12}Q_{22}^{-1}A_{12}' - (S_{12} + A_{12}Q_{22}^{-1}A_{22}')(S_{22} + A_{22}Q_{22}^{-1}A_{22}')^{-1} \\
&\quad \cdot(S_{21} + A_{22}Q_{22}^{-1}A_{12}') && (24)\\
\hat{Q} &= Q_{11} - Q_{12}Q_{22}^{-1}Q_{21} + (A_{21}' - Q_{12}Q_{22}^{-1}A_{22}')(S_{22} + A_{22}Q_{22}^{-1}A_{22}')^{-1} \\
&\quad \cdot(A_{21} - A_{22}Q_{22}^{-1}Q_{21}) && (25)
\end{aligned}$$

Now, we introduce the following set:

$$\hat{\Gamma}_s := \{\gamma' > 0 : \forall\gamma \geq \gamma',\ \text{(13) is satisfied and (22) has a bounded nonnegative definite solution, } Z_{s\gamma} \text{ over } [0, t_f]\} \tag{26}$$

and further define

$$\hat{\gamma}_s := \inf\{\gamma \in \hat{\Gamma}_s\} \tag{27}$$

On each of the sampling interval, the system is controlled by an open-loop control law. The "open-loop" GRDE associated with the kth sampling interval, where $k = 1, 2, \ldots, K$, is

$$\dot{P}_{sk} + A^{\Box\prime}P_{sk} + P_{sk}A^\Box + P_{sk}D^\Box D^{\Box\prime}P_{sk} + Q^\Box = 0; \quad P_{sk}(t_k) = Z_{s\gamma}(t_k) \tag{28}$$

We introduce the following sets, for $k = K, K-1, \ldots, 1$:

$$\hat{\Gamma}_{sk} := \{\gamma' \in \hat{\Gamma}_s : \forall\gamma \geq \gamma',\ \text{(13) is satisfied and (28) admits a bounded nonnegative definite solution } P_{s\gamma k} \text{ over } [t_{k-1}, t_k]\}$$

[1] Under condition (13), $Q^\Box > 0$.

and define

$$\gamma_{sk} := \inf\{\gamma \in \hat{\Gamma}_{sk}\}, \quad k = K, K-1, \ldots, 1 \tag{29}$$

Now, we further define

$$\gamma_{SDs} := \max\{\gamma_{sK}, \ldots, \gamma_{s1}\} \tag{30}$$

We will see this γ_{SDs} plays an important role in our problem.

The fast subsystem and the associated soft-constrained game

Let $x_f := x_2 - \bar{x}_2$, $u_f := u - u_s$, $w_f := w - w_s$ and $\tau = \frac{t'-t}{\epsilon}$ where we take t to be frozen, and t' to vary on the same scale as t. We define the fast subsystem and the associated cost (as in the standard regulator problem; see [26]) by

$$\frac{d}{d\tau}x_f^t = A_{22}(t)x_f^t + B_2(t)u_f^t + D_2(t)w_f^t; \quad x_f^t(0) = x_f(t) \tag{31}$$

$$L_{\gamma f}^t = \int_0^\infty (|x_f^t|^2_{Q_{22}(t)} + |u_f^t|^2 - \gamma^2|w_f^t|^2)\, d\tau \tag{32}$$

The "feedback" GARE associated with this infinite-horizon game, for each t, is

$$A_{22}'(t)Z_f + Z_f A_{22}(t) + Q_{22}(t) - Z_f S_{22}(t) Z_f = 0 \tag{33}$$

and the "open-loop" GARE is

$$A_{22}'(t)Z_{of} + Z_{of}A_{22}(t) + Q_{22}(t) + Z_{of}\frac{1}{\gamma^2}D_2(t)D_2(t)'Z_{of} = 0 \tag{34}$$

We now let γ_f^t denote the minimax disturbance attenuation bound for the H^∞-optimal control problem defined by (31)–(32) under closed-loop information, and γ_{of}^t denote the same under open-loop information,[2] where we must have $\gamma_{of}^t \geq \gamma_f^t$. For every $\gamma > \gamma_f^t$, let $Z_{f\gamma}(t)$ be the minimal positive definite solution for (33) (and, for every $\gamma > \gamma_{of}^t$, let $Z_{of\gamma}(t)$ be the minimal positive solution to (34)). We now define

$$\gamma_f := \sup_{t\in[0,t_f]} \gamma_f^t; \quad \gamma_{of} := \sup_{t\in[0,t_f]} \gamma_{of}^t$$

Then, for every $\gamma > \gamma_{of}$, the GAREs (33) and (34) both admit positive definite solutions for all $t \in [0, t_f]$. Let

$$\bar{\gamma}_{SD} := \max\{\gamma_{of}, \gamma_{SDs}\} \tag{35}$$

We will shortly see that this value plays an important role in our problem.

[2]To ensure that $\gamma_{of}^t < \infty$, it will be sufficient to take $A_{22}(t)$ to be Hurwitz.

The infinite-horizon case

We now turn to the infinite-horizon case (i. e., as $t_f \to \infty$, as well as when $t_f = \infty$). We take A, B, D, Q to be time-invariant, and $Q_f = 0$. Furthermore, we assume that (A_ϵ, B_ϵ) is controllable, and (A_ϵ, Q) is observable. Because of the nature of the infinite-horizon game, the GRDEs in (8) will be identical, except for the length of the corresponding sampling interval. We let

$$T := \sup_{k=0,1,\ldots} t_{k+1} - t_k \tag{36}$$

Then, for each $\epsilon > 0$, such that $\exists \hat{\gamma}_{SD\infty} > 0$, $\forall \gamma > \hat{\gamma}_{SD\infty}$, the following GARE and GRDE have positive definite solutions:

$$A_\epsilon' \tilde{Z}_\infty + \tilde{Z}_\infty A_\epsilon - \tilde{Z}_\infty S_\epsilon \tilde{Z}_\infty + Q = 0 \tag{37}$$

$$\dot{\tilde{P}} + A_\epsilon' \tilde{P} + \tilde{P} A_\epsilon + \tilde{P} \frac{1}{\gamma^2} D_\epsilon D_\epsilon' \tilde{P} + Q = 0; \quad \tilde{P}(T) = \tilde{Z}_\infty, \quad t \in [0, T] \tag{38}$$

and hence the infinite-horizon soft-constrained game has a finite value. The upper value is achieved by the controller

$$u_{SD}^*(t) = \mu_{SD}^\infty(t, x) = -B_\epsilon' \tilde{Z}_\infty(\epsilon) \Phi(t, t_k) x(t_k), \quad t_k \le t < t_{k+1} \tag{39}$$

where Φ is the state transition function associated with the matrix

$$\tilde{F}_\infty := A_\epsilon - (B_\epsilon B_\epsilon' - \frac{1}{\gamma^2} D_\epsilon D_\epsilon') \tilde{Z}_\infty \tag{40}$$

which is Hurwitz for all $\gamma > \hat{\gamma}_{SD\infty}$. For $\gamma < \hat{\gamma}_{SD\infty}$, either the GARE does not have a nonnegative definite solution or the GRDE has at least one conjugate point in $[0, T]$, and the game has infinite upper value. The level $\hat{\gamma}_{SD\infty}$ is again the one defined in (5).

Again, since GARE and GRDE of the type above (37) and (38) have been extensively studied in [18], we will not further discuss their direct derivation here.

As in the finite-horizon case, we study the derivation of (approximate) ϵ-free solutions, by solving the slow and fast subgames separately. By following steps similar to those in the finite-horizon case, we first decompose the system into slow and fast subsystems. The slow game is described by

$$\dot{x}_s = A^\square x_s + B^\square u_s + D^\square \hat{w}_s; \quad x_s(0) = 0 \tag{41}$$

$$L_{\gamma s} = \int_0^\infty (|x_s|^2_{Q^\square} + 2 x_s' P^\square u_s + |u_s|^2_{R^\square} - |\hat{w}_s|^2)\, dt \tag{42}$$

under the condition (13) and the transformation (12), where $A^\square$, $B^\square$, $D^\square$, $Q^\square$, $P^\square$ and $R^\square$ are as defined before, with the only difference being that they are now time-invariant. The associated GARE is

$$\hat{A}_0' \tilde{Z}_s + \tilde{Z}_s \hat{A}_0 - \tilde{Z}_s \hat{S}_0 \tilde{Z}_s + \hat{Q} = 0 \tag{43}$$

where $\hat{A}_0$, $\hat{S}_0$ and $\hat{Q}$ are defined as before, but now they are time-invariant.

Now, we introduce the set

$$\hat{\Gamma}_{s\infty} := \{\gamma' > 0 : \forall\gamma \geq \gamma',\ (13) \text{ is satisfied and } (43) \text{ has a minimal positive definite solution, } Z_{s\gamma}, \text{ and } \hat{A}_0 - \hat{S}_0 Z_{s\gamma} \text{ is Hurwitz.}\} \quad (44)$$

and further define[3]

$$\hat{\gamma}_{s\infty} := \inf\{\gamma \in \hat{\Gamma}_{s\infty}\} \quad (45)$$

For every $\gamma > \hat{\gamma}_{s\infty}$, let $Z_{s\gamma}$ be the minimal positive definite solution to (43).

To study the system behavior in between the sampling intervals, we consider only the sampling interval with the longest duration, T.

The GARE associated with it is:

$$\dot{P}_{sT} + A^{\square\prime} P_{sT} + P_{sT} A^{\square} + P_{sT} D^{\square} D^{\square\prime} P_{sT} + Q^{\square} = 0; \quad P_{sT}(T) = Z_{s\gamma} \quad (46)$$

Similar to the finite-horizon case, we introduce the following set:

$$\hat{\Gamma}_{sT} := \{\gamma' \in \hat{\Gamma}_{s\infty} : \forall\gamma \geq \gamma',\ (13) \text{ is satisfied and } (46) \text{ has a bounded nonnegative definite solution } P_{s\gamma T} \text{ over } [0,T]\}$$

and define

$$\gamma_{SDs\infty} := \inf\{\gamma \in \hat{\Gamma}_{sT}\}. \quad (47)$$

The fast part of the system is the same as in the finite-horizon case, where the coefficient matrices are now constants. The fast game is described by (31)–(32), and the GARE is the same as (33), and the "open-loop" GARE is the same as (34). We will use $\gamma_{f\infty}$ to denote the minimax disturbance attenuation bound (and $\gamma_{of\infty}$ for the open-loop case), and let $Z_{f\gamma}$ denote the minimal positive definite solution to (33) (and $Z_{of\gamma}$ for the open-loop case). We define, similar to (35),

$$\bar{\gamma}_{SD\infty} := \max\{\gamma_{of\infty}, \gamma_{SDs\infty}\} \quad (48)$$

This quantity will also play an important role in our analysis in the next section.

3 Main Results

Before we present the main results, we first introduce some useful matrices, defined in [18]:

$$U^{\cdot} \equiv U_1 Z_{s\gamma} + U_2$$

[3] Again, for $\hat{\gamma}_{s\infty}$ to be finite, it will be sufficient to have the pair $(A^{\square}, B^{\square})$ controllable at $\gamma = \infty$ (see [16]), which is equivalent to having the pair $A_{11} - A_{12}A_{22}^{-1}A_{21}, B_1 - A_{12}A_{22}^{-1}B_2)$ controllable.

$$
\begin{aligned}
V &\equiv V_1 Z_{s\gamma} + V_2 \\
U_1 &:= Q_{22}^{-1} A_{12}' - Q_{22}^{-1} A_{22}' (S_{22} + A_{22} Q_{22}^{-1} A_{22}')^{-1} (S_{21} + A_{22} Q_{22}^{-1} A_{12}') \\
U_2 &:= Q_{22}^{-1} Q_{21} + Q_{22}^{-1} A_{22}' (S_{22} + A_{22} Q_{22}^{-1} A_{22}')^{-1} (A_{21} - A_{22} Q_{22}^{-1} Q_{21}) \\
V_1 &:= -(S_{22} + A_{22} Q_{22}^{-1} A_{22}')^{-1} (S_{21} + A_{22} Q_{22}^{-1} A_{12}') \\
V_2 &:= (S_{22} + A_{22} Q_{22}^{-1} A_{22}')^{-1} (A_{21} - A_{22} Q_{22}^{-1} Q_{21})
\end{aligned}
$$

We now introduce the ϵ-free sampled-data controller we will use in the full-order system to achieve a desired performance level:

$$
u_{SD}(t) = \mu_{SD}(t, x_{1\,k-1}) = -(B_1' Z_{s\gamma} + B_2' V) x_1^*(t), \quad t \in [t_{k-1}, t_k] \tag{49}
$$

where x_1^* is the solution to

$$
\dot{x}_1^* = (\hat{A}_0 - \hat{S}_0 Z_{s\gamma}) x_1^*, \quad x_1^*(t_{k-1}) = x_{1\,k-1} \quad t \in [t_{k-1}, t_k] \tag{50}
$$

We introduce the maximizing disturbance corresponding to this control as

$$
w_{SD}(t) = \frac{1}{\gamma^2}(D_1' Z_{s\gamma} + D_2' V) x_1^*(t), \quad t \in [t_{k-1}, t_k] \quad k = K, \ldots, 1 \tag{51}
$$

The corresponding trajectory for x_2, obtained by setting $\epsilon = 0$ and letting u and w in (11) be the ones defined in (49) and (51), is

$$
x_2^* = -U x_1^* \tag{52}
$$

It is not difficult to see that, on each sampling interval $[t_{k-1}, t_k]$ and under the initial condition $x_1^*(t_{k-1}) = x_{1\,k-1}$, the x_1^* generated by (50) is actually the optimal trajectory for the slow subgame under closed-loop information (the terminal cost term in the cost function being $|x_s|^2_{Z_{s\gamma}(t_k)}$) with the optimal control and corresponding maximizing disturbance being u_{SD} and w_{SD}, respectively, (actually, they are particular representations of the optimal control and corresponding disturbance along the optimal trajectory), and x_2^* also coincides with $\bar{x}_2$, introduced in (11), generated by x_1^*, u_{SD} and w_{SD} (see Appendix A). Let x^* denote $(x_1^{*\prime}\ x_2^{*\prime})'$.

The finite-horizon case

Theorem 1 *For the singularly perturbed system (1)–(6), let Assumptions 1–4 be satisfied, $A_{22}(t)$ be Hurwitz for each $t \in [0, t_f]$, and the following condition hold:*

$Q_{f22} \le Z_{f\gamma}(t_f)$, where $Z_{f\gamma}(t_f)$ is the solution to (33) at $t = t_f$ with γ fixed.

Then,

1. *$\gamma_{SD}^*(\epsilon) \le \bar{\gamma}_{SD}$, asymptotically as $\epsilon \to 0$, where $\bar{\gamma}_{SD}$, as defined in (35), is finite.*

2. $\forall \gamma > \bar{\gamma}_{SD}$, $\exists \epsilon_\gamma > 0$ *such that* $\forall \epsilon \in [0, \epsilon_\gamma)$, *the set of equations (7)–(8) admits nonnegative definite solutions, and consequently, the full-order game (1)–(6) has a finite value.*

3. $\forall \gamma > \bar{\gamma}_{SD}$, *if we apply the controller defined in (49),* μ_{SD}, *to the system, then* $\exists \epsilon'_\gamma > 0$ *such that* $\forall \epsilon \in [0, \epsilon'_\gamma)$, *the disturbance attenuation level* γ *is attained for the full-order system.*

Proof We first note that under Assumption 3, and with A_{22} being Hurwitz for every fixed $t \in [0, t_f]$, the fast subsystem is open-loop stable and hence $\gamma_{of} < \infty$. Then we have $\bar{\gamma}_{SD}$ finite. Fix $\gamma > \bar{\gamma}_{SD}$,[4] by Theorem 2 in [18],[5] (7) does not have a conjugate point on $[0, t_f]$ for small ϵ, and the solution can be approximated by

$$\tilde{Z} = \begin{bmatrix} Z_{s\gamma}(t) + O(\epsilon) & \epsilon(Z'_c + Z_{cb}(\tau)') + O(\epsilon^2) \\ \epsilon(Z_c + Z_{cb}(\tau)) + O(\epsilon^2) & \epsilon(Z_{f\gamma}(t) + Z_{fb}(\tau)) + O(\epsilon^2) \end{bmatrix} \tag{53}$$

for all $t \in [0, t_f]$, where $Z_c := Z_{f\gamma}(t)U + V$, $Z_{cb}(\tau)$ and $Z_{fb}(\tau)$ are the boundary layer terms and as $\tau \to -\infty$ they converge to 0 exponentially in the τ time scale.

On each sampling interval $[t_{k-1}, t_k]$, we have $Z_{of\gamma}(t_k) \geq Z_{f\gamma}(t_k)$ and $\gamma > \max\{\gamma_{sk}, \gamma_{of}\}$, and hence each "open-loop" GRDE in (8) does not have a conjugate point on its corresponding time interval for small ϵ (see [18] Theorem 2). Thus, results 1) and 2) are proved.

To prove 3), we will use a dynamic programming argument to show that the control u_{SD} actually achieves the γ performance level, i. e.,

$$\sup_{w \in \mathcal{H}_w} L_\gamma(u_{SD}, w) = 0.$$

First, we write out the dynamic programming equation, adapted to the sampling intervals over which u_{SD} was defined:

$$\begin{aligned} &\sup_{w_{[0,t_f]}} L_\gamma(u_{SD}, w) \\ &= \sup_{w_{[0,t_1]}} \{ \int_0^{t_1} (|x|_Q^2 + |u_{SD}|^2 - \gamma^2 |w|^2)\, dt + \sup_{w_{[t_1,t_2]}} \{ \cdots \\ &\quad + \sup_{w_{[t_{K-1},t_f]}} \{ |x(t_f)|_{Q_f}^2 + \int_{t_{K-1}}^{t_f} (|x|_Q^2 + |u_{SD}|^2 - \gamma^2 |w|^2)\, dt \} \cdots \} \end{aligned}$$

In this connection let us introduce

$$L_K(w) = |x(t_f)|_{Q_f}^2 + \int_{t_{K-1}}^{t_f} (|x(t)|_{Q(t)}^2 + |u_{SD}|^2 - \gamma^2 |w|^2)\, dt \tag{54}$$

---

[4] Since $\gamma > \gamma_{of}$, then the condition (13) is satisfied by Lemma 2 in [18].
[5] As discussed in [18], $\hat{\gamma}_s \geq \gamma_s$, where γ_s is defined in [18] pp.11.

$$\begin{aligned} L_K^* &= \sup_{w[t_{K-1},t_f]} L_K(w) && (55) \\ L_i(w) &= L_{i+1}^* + \int_{t_{i-1}}^{t_i} (|x(t)|^2_{Q(t)} + |u_{SD}|^2 - \gamma^2 |w|^2)\,dt && (56) \\ L_i^* &= \sup_{w[t_{i-1},t_i]} L_i(w)\ ; \quad i = K-1, \ldots, 1 && (57) \end{aligned}$$

If we can show that L_1^* is bounded for small ϵ (by linearity of the problem, it must be zero), then it follows that the control achieves the γ bound.

We first consider stage K. Let $x(t_{K-1}) = x_{K-1}$. Introduce a new state variable:

$$\tilde{x} := x - x^* \tag{58}$$

It is shown in Appendix B that $\tilde{x}$ satisfies the singularly perturbed differential equation

$$\begin{bmatrix} \dot{\tilde{x}}_1 \\ \epsilon \dot{\tilde{x}}_2 \end{bmatrix} = A(t)\tilde{x} + D(t)w - D(t)w_{SD}(t) - \begin{bmatrix} 0 \\ O(\epsilon) \end{bmatrix} x_1^* \tag{59}$$

Furthermore, we can rewrite the cost $L_K(w)$ as

$$\begin{aligned} L_K(w) &= |x(t_f)|^2_{Q_f} + \int_{t_{K-1}}^{t_f} (|x(t)|^2_{Q(t)} + |u_{SD}(t)|^2 - \gamma^2 |w(t)|^2)\,dt \\ &= |x^*(t_f)|^2_{Q_f} + 2\tilde{x}'(t_f) Q_f x^*(t_f) + |\tilde{x}(t_f)|^2_{Q_f} + \int_{t_{K-1}}^{t_f} (|\tilde{x}(t)|^2_{Q(t)} \\ &\quad + 2\tilde{x}'(t) Q(t) x^*(t) + |x^*(t)|^2_{Q(t)} + |u_{SD}(t)|^2 - \gamma^2 |w(t)|^2 \\ &\quad + \gamma^2 |w_{SD}(t)|^2 - \gamma^2 |w_{SD}(t)|^2)\,dt \end{aligned}$$

Since x^*, u_{SD} and w_{SD} coincide with the solution of the slow subsystem under perfect state measurements, we have that

$$\begin{aligned} |x_1^*(t_{K-1})|^2_{Z_{s\gamma}(t_{K-1})} &= |x_1^*(t_f)|^2_{Q_{f11}} + \int_{t_{K-1}}^{t_f} (|x^*(t)|^2_{Q(t)} \\ &\quad + |u_{SD}(t)|^2 - \gamma^2 |w_{SD}(t)|^2)\,dt \end{aligned} \tag{60}$$

Now, using the structure of Q_f and a "completion of squares" argument, we arrive at the following expression:[6]

$$\begin{aligned} L_K^* &= |x_{1\,K-1}|^2_{Z_{s\gamma}(t_{K-1}) + O(\epsilon)} + |\tilde{x}(t_{K-1})|^2_{P_\epsilon(t_{K-1})} \\ &\quad + 2Y_\epsilon'(t_{K-1})\tilde{x}(t_{K-1}) + m(t_{K-1}) \end{aligned}$$

where $P_\epsilon(t)$, $Y_\epsilon(t)$ and $m(t)$ are the solutions to

$$\dot{P}_\epsilon + A_\epsilon' P_\epsilon + P_\epsilon A_\epsilon + P_\epsilon \tfrac{1}{\gamma^2} D_\epsilon D_\epsilon' P_\epsilon + Q = 0; \quad P_\epsilon(t_f) = Q_f \tag{61}$$

[6] The term $|x_{1\,K-1}|^2_{O(\epsilon)}$ is due to $O(\epsilon)$ terms in the terminal cost term $|x^*(t_f)|^2_{Q_f}$.

$$\dot{Y}_\epsilon + (A'_\epsilon + P_\epsilon \tfrac{1}{\gamma^2} D_\epsilon D'_\epsilon) Y_\epsilon - P_\epsilon D_\epsilon w_{SD} - P_\epsilon \begin{bmatrix} 0 \\ O(1) \end{bmatrix} x_1^* + Qx^* = 0 \tag{62}$$
$$Y_\epsilon(t_f) = Q_f x^*(t_f)$$

$$\dot{m} = -\tfrac{1}{\gamma^2} |Y_\epsilon - \begin{bmatrix} Z_{s\gamma} \\ \epsilon V \end{bmatrix} x_1^*|^2_{D_\epsilon D'_\epsilon} + 2Y'_\epsilon \begin{bmatrix} 0 \\ O(1) \end{bmatrix} x_1^*; \quad m(t_f) = 0 \tag{63}$$

on the time interval $[t_{K-1}, t_f]$. Since $\gamma > \max\{\gamma_{of}, \gamma_{sK}\}$, by Theorem 2 in [18], P_ϵ exists for small ϵ and can be approximated by

$$P_\epsilon = \begin{bmatrix} P_{s\gamma}(t) + O(\epsilon) & \epsilon(P'_c(t) + P'_{cb}(\tau)) + O(\epsilon^2) \\ \epsilon(P_c(t) + P_{cb}(\tau)) + O(\epsilon^2) & \epsilon(Z_{of\gamma}(t) + P_{fb}(\tau)) + O(\epsilon^2) \end{bmatrix} \tag{64}$$

where $\tau := \frac{t-t_f}{\epsilon}$, $P_{cb}(\tau)$ and $P_{fb}(\tau)$ are the boundary correction terms and they converge to zero exponentially in τ time scale.

Next, we introduce $Z_\epsilon(t)$, which is the solution to the following GRDE:

$$\dot{Z}_\epsilon + \hat{A}'_\epsilon Z_\epsilon + Z_\epsilon \hat{A}_\epsilon + Z_\epsilon \frac{1}{\gamma^2} D_\epsilon D'_\epsilon Z_\epsilon + \hat{Q}_w = 0; \quad Z_\epsilon(t_f) = Q_f \quad t \in [0, t_f] \tag{65}$$

where

$$\hat{A}_\epsilon := A_\epsilon - B_\epsilon B' \begin{bmatrix} Z_{s\gamma} & 0 \\ V & 0 \end{bmatrix} \tag{66}$$

$$\hat{Q}_w := Q + \begin{bmatrix} Z_{s\gamma} & V' \\ 0 & 0 \end{bmatrix} BB' \begin{bmatrix} Z_{s\gamma} & 0 \\ V & 0 \end{bmatrix} \tag{67}$$

We can see that this GRDE is exactly the same equation as the GRDE we got in the perfect state measurements case when we apply the slow controller in the full-order system. So, by Theorem 2 in [18], the Equation (65) has a nonnegative-definite solution on $[0, t_f]$ for small ϵ, which can be approximated by

$$Z_\epsilon = \begin{bmatrix} Z_{s\gamma}(t) + O(\epsilon) & \epsilon(Z'_{oc} + Z_{cb}(\tau)') + O(\epsilon^2) \\ \epsilon(Z_{oc} + Z_{cb}(\tau)) + O(\epsilon^2) & \epsilon(Z_{of\gamma}(t) + Z_{fb}(\tau)) + O(\epsilon^2) \end{bmatrix} \tag{68}$$

where $Z_{oc} := Z_{of\gamma}U + V$, the $Z_{cb}(\tau)$ and $Z_{fb}(\tau)$ are still the boundary layer correction terms and they converge to zero exponentially in the τ time scale.

Next we state the following identity, which will be used shortly:

$$\begin{aligned} Z_\epsilon x^* - \begin{bmatrix} Z_{s\gamma} \\ \epsilon V \end{bmatrix} x_1^* &= (Z_\epsilon - \begin{bmatrix} Z_{s\gamma} & 0 \\ \epsilon V & 0 \end{bmatrix}) x^* \\ &= \begin{bmatrix} O(\epsilon) & O(\epsilon) \\ \epsilon(Z_{of\gamma}U + Z_{cb}) + O(\epsilon^2) & \epsilon(Z_{of\gamma} + Z_{fb}) + O(\epsilon^2) \end{bmatrix} x^* \\ &= \begin{bmatrix} O(\epsilon) \\ \epsilon(Z_{cb} - Z_{fb}U) + O(\epsilon^2) \end{bmatrix} x_1^* \end{aligned} \tag{69}$$

Start with the following structure (where the partition is the same as x), which defines the variables Y_1 and Y_2:

$$Y_\epsilon - Z_\epsilon x^* := \begin{bmatrix} Y_1 \\ \epsilon Y_2 \end{bmatrix} \tag{70}$$

and note that $(Y_1', Y_2')'$ thus defined satisfies the following differential equation (the detailed derivation can be found in Appendix C):

$$\begin{aligned} \begin{bmatrix} \dot{Y}_1 \\ \epsilon\dot{Y}_2 \end{bmatrix} &= -\begin{bmatrix} F_{11} & F_{12} \\ F_{21} & F_{22} \end{bmatrix}\begin{bmatrix} Y_1 \\ Y_2 \end{bmatrix} - \begin{bmatrix} O(\epsilon) \\ O(\epsilon) \end{bmatrix} x_1^* \\ &\quad + \begin{bmatrix} O(1) \\ O(\epsilon) \end{bmatrix}(Z_{cb} - Z_{fb}U)x_1^*; \quad \begin{bmatrix} Y_1(t_f) \\ Y_2(t_f) \end{bmatrix} = 0 \end{aligned} \tag{71}$$

where

$$\begin{aligned} F_{11} &:= A_{11}' + P_{s\gamma}\frac{1}{\gamma^2}D_1D_1' + (P_c' + P_{cb}')\frac{1}{\gamma^2}D_2D_1' + O(\epsilon) \\ F_{12} &:= A_{21}' + P_{s\gamma}\frac{1}{\gamma^2}D_1D_2' + (P_c' + P_{cb}')\frac{1}{\gamma^2}D_2D_2' + O(\epsilon) \\ F_{21} &:= A_{12}' + (Z_{of\gamma} + P_{fb})\frac{1}{\gamma^2}D_2D_1' + O(\epsilon) \\ F_{22} &:= A_{22}' + (Z_{of\gamma} + P_{fb})\frac{1}{\gamma^2}D_2D_2' + O(\epsilon) \end{aligned}$$

This is a singularly perturbed system. Let t_a be the time by which the boundary layer terms Z_{cb}, Z_{fb}, P_{cb} and P_{fb} becomes of $O(\epsilon)$. Obviously, $t_a - t_f = O(\epsilon \ln \epsilon)$. Since $A_{22}' + Z_{of\gamma}\frac{1}{\gamma^2}D_2D_2'$ is Hurwitz for every fixed t and its eigenvalues are uniformly bounded away from 0 by $-c$ for some $c \in \mathcal{R}^+$ on $[0, t_f]$ (where c is ϵ-free), we have $A_{22}' + (Z_{of\gamma} + P_{fb})\frac{1}{\gamma^2}D_2D_2' + O(\epsilon)$ also Hurwitz, with its eigenvalues are bounded away from 0 by $-c/2$ uniformly after a boundary layer $[t_b, t_f]$, where now, $t_b - t_f = O(\epsilon)$. We choose ϵ small enough to make $t_{K-1} < t_a < t_b$ (under Assumption 4). By the approximation results in singular perturbation theory (see [17] Chapter 5 Theorem 6.1), in $[t_{K-1}, t_b]$, the state transition function for Y_1 and Y_2 is of $O(1)$.

In $[t_b, t_f]$, the growth of Y_1 and Y_2 can be bounded by

$$\begin{aligned} \begin{bmatrix} Y_1 \\ Y_2 \end{bmatrix} &\leq \int_0^{O(\epsilon)} \exp(\frac{1}{\epsilon}O(1)t)\, O(1)\, dt \\ &= O(\epsilon) \end{aligned}$$

In $[t_a, t_b]$, the growth of Y_1 and Y_2 can be bounded by

$$\begin{aligned} \begin{bmatrix} Y_1 \\ Y_2 \end{bmatrix} &\leq \int_{O(\epsilon)}^{O(\epsilon \ln \epsilon)} O(1) \cdot O(1)\, dt + O(\epsilon) \cdot O(1) \\ &= o(\epsilon^{1/2}) \end{aligned}$$

Then, in $[t_{K-1}, t_a]$, the growth can be bounded by

$$\begin{aligned} \begin{bmatrix} Y_1 \\ Y_2 \end{bmatrix} &\leq \int_{t_{K-1}}^{t_a} O(1)\cdot O(\epsilon)\, dt + o(\epsilon^{1/2})\cdot O(1) \\ &= o(\epsilon^{1/2}) \end{aligned}$$

Hence, we actually have

$$\begin{bmatrix} Y_1(t) \\ Y_2(t) \end{bmatrix} = o(\epsilon^{1/2}); \quad \forall t \in [t_{K-1}, t_f]$$

Then, we can further deduce that

$$Y_\epsilon = \begin{bmatrix} Z_{s\gamma} \\ \epsilon(V + Z_{cb} - Z_{fb}U) \end{bmatrix} x_1^* + \begin{bmatrix} o(\epsilon^{1/2}) \\ o(\epsilon^{3/2}) \end{bmatrix} \tag{72}$$

Now, it is clear that

$$\begin{aligned} m(t_{K-1}) &= \int_{t_{K-1}}^{t_f} \dot{m}\, dt \\ &\leq \int_{t_{K-1}}^{t_a} O(\epsilon)\, dt + \int_{t_a}^{t_f} (O(\epsilon) + b(\tau))\, dt \\ &= O(\epsilon) \end{aligned}$$

where $b(\tau)$ is some term that is exponentially decaying in the τ time scale.

Since the terminal condition and driving terms in (62) are linear functions of $x_{1\,K-1}$, Y_ϵ is also a linear function of $x_{1\,K-1}$ and consequently $\dot{m}(t)$ and $m(t_{K-1})$ are quadratic functions of it. Thus,

$$m(t_{K-1}) = x'_{1\,K-1} O(\epsilon) x_{1\,K-1}$$

Since $x^*(t_{K-1}) = [x'_{1\,K-1}, -x'_{1\,K-1}U'(t_{K-1})]'$ and $x_1(t_{K-1}) = x_{1\,K-1}$, we have

$$\tilde{x}(t_{K-1}) = \begin{bmatrix} 0 \\ x_{2\,K-1} + U(t_{K-1})x_{1\,K-1} \end{bmatrix}$$

Consequently,

$$\begin{aligned} L_K^* &= |x_{1\,K-1}|^2_{Z_{s\gamma}(t_{K-1})+O(\epsilon)} + |\tilde{x}(t_{K-1})|^2_{P_\epsilon(t_{K-1})} + 2Y_\epsilon'(t_{K-1})\tilde{x}(t_{K-1}) \\ &= |x_{1\,K-1}|^2_{Z_{s\gamma}(t_{K-1})+O(\epsilon)} + |x_{2\,K-1} + U(t_{K-1})x_{1\,K-1}|^2_{\epsilon Z_{of\gamma}(t_{K-1})+O(\epsilon^2)} \\ &\quad + 2x'_{1\,K-1}(\epsilon V(t_{K-1}) + o(\epsilon^{3/2}))'(x_{2\,K-1} + U(t_{K-1})x_{1\,K-1}) \\ &= x'_{K-1} \begin{bmatrix} Z_{s\gamma} + O(\epsilon) & \epsilon Z'_{oc} + o(\epsilon^{3/2}) \\ \epsilon Z_{oc} + o(\epsilon^{3/2}) & \epsilon Z_{of\gamma}(t) + O(\epsilon^2) \end{bmatrix}\Bigg|_{t=t_{K-1}} x_{K-1} \end{aligned} \tag{73}$$

Now, we consider the next stage, $K-1$. The arguments used for the analysis of stage K are equally valid here, with the only difference being that we now have to watch for the effect of the additional terminal cost perturbation term, which is

$$x'_{K-1}\begin{bmatrix} O(\epsilon) & o(\epsilon^{3/2}) \\ o(\epsilon^{3/2}) & O(\epsilon^2) \end{bmatrix} x_{K-1},$$

The solution to (61), P_ϵ, will still exist, and will have the same form as in (64) for small ϵ (since the solution of ϵ-free lower order problems depend continuously on the terminal condition). Due to the perturbation term, the terminal conditions for Y_1 and Y_2 will not be zero as in stage K, but they will be $o(\epsilon^{1/2})$. Note that, in this stage, there is no longer a boundary layer term for Z_ϵ. Thus, it is easy to see that Y_1 and Y_2 will still be of order $o(\epsilon^{1/2})$ in $[t_{K-2}, t_{K-1}]$, and consequently the approximation in (72) still holds.

Also, the perturbation term will not affect the approximation for the terms containing only x^* because the state transition function for x^* is of $O(1)$. Then, L^*_{K-1} will be of the same approximation form as L^*_K but the evaluation time is at t_{K-2}.

Since there is only a finite number of sampling intervals, we can find $\epsilon'_\gamma > 0$ such that $\forall \epsilon \in [0, \epsilon'_\gamma)$, L^*_1 is bounded (and L^*_1 have the approximation form as in (73)). This completes the proof of 3). □

The infinite-horizon case

To guarantee that $\bar{\gamma}_{SD\infty}$ is finite (although the theorem to be presented below is true also when it is infinite), we make two additional assumptions:

Assumption 5 $(A_{11} - A_{12}A_{22}^{-1}A_{21}, Q_{11} - Q_{12}Q_{22}^{-1}Q_{21})$ is observable.

Assumption 6 $(A_{11} - A_{12}A_{22}^{-1}A_{21}, B_1 - A_{12}A_{22}^{-1}B_2)$ is controllable and A_{22} is Hurwitz.

Theorem 2 *Consider the singularly perturbed system (1)–(6), with $t_f = \infty$, $Q_f = 0$ and A, B, D, Q time-invariant. If Assumptions 1–6 hold, then*

1. *$\gamma^*_{SD\infty}(\epsilon) \to \bar{\gamma}_{SD\infty}$ asymptotically as $\epsilon \to 0^+$, where $\bar{\gamma}_{SD\infty}$, defined in (48), is finite.*
2. *$\forall \gamma > \bar{\gamma}_{SD\infty}$, $\exists \epsilon_\gamma > 0$ such that $\forall \epsilon \in [0, \epsilon_\gamma)$, equation (37) admits a positive definite solution and (38) admits a nonnegative definite solution over the interval $[0, T]$, and consequently, the game has a finite upper value.*
3. *$\forall \gamma > \bar{\gamma}_{SD\infty}$, if we apply the controller μ_{SD} defined in (49) to the system, then $\exists \epsilon'_\gamma > 0$ such that, $\forall \epsilon \in [0, \epsilon'_\gamma)$, the disturbance attenuation level γ is attained for the full-order system.*

Proof We first note that under the assumptions stated, the slow game has a saddle-point solution for sufficiently large values of γ, since at $\gamma = \infty$ the problem becomes a "regular" LQR problems admitting stabilizing optimal controller. This shows that $\hat{\gamma}_{s\infty}$ is finite. Also, since the fast subsystem is open-loop stable, $\gamma_{of\infty} < \infty$. These implies that $\bar{\gamma}_{SD\infty}$ is finite. Now fix $\gamma > \bar{\gamma}_{SD\infty}$. By arguments similar to those used in the finite-horizon case, we can easily deduce that 2) is true, and $\limsup_{\epsilon \to 0^+} \gamma^*_{SD\infty}(\epsilon) \leq \bar{\gamma}_{SD\infty}$. For any $\gamma < \bar{\gamma}_{SD\infty}$, by the Theorem 2 of [28] and its proof, either the GARE (37) doesn't admit any nonnegative definite solutions or the GRDE (38) has at least one conjugate point in the interval $[0, T]$. Hence, 1) follows.

For 3), we first note here that the controller is the same as in the finite-horizon case. Since we have already shown validity of the result when t_f is finite, it will be sufficient here to find a $\epsilon'_\gamma > 0$ that works (uniformly) when $t_f \to \infty$. Toward this end, we first note that since $\hat{A}_0 - \hat{S}_0 Z_{s\gamma}$ is Hurwitz, we can find a constant matrix $\Delta > 0$ such that

$$(\hat{A}_0 - \hat{S}_0 Z_{s\gamma})'\Delta + \Delta(\hat{A}_0 - \hat{S}_0 Z_{s\gamma}) < 0.$$

Then, we have

$$|x_1^*(T)|^2_\Delta < \alpha |x_1^*(0)|^2_\Delta; \quad \text{for some } \alpha \in (0,1)$$

Consider a sample interval of length T (which we may as well assume to be $[0, T]$ due to the time-invariant nature), with the "cost to go" being

$$x(T)' \begin{bmatrix} Z_{s\gamma} + \delta\Delta & \epsilon(U'Z_{of\gamma} + V') \\ \epsilon(Z_{of\gamma}U + V) & \epsilon(Z_{of\gamma} + \delta I_{n_2}) \end{bmatrix} x(T) := |x(T)|^2_\Xi \tag{74}$$

where δ is some small positive real number. We introduce the cost function to be

$$L_T(w) := |x(T)|^2_\Xi + \int_0^T (|x(t)|^2_{Q(t)} + |u_{SD}|^2 - \gamma^2 |w|^2)\, dt$$

By following a line of reasoning similar to that in the finite-horizon case, we first introduce the new set of state variables $\tilde{x} := x - x^*$. By Appendix B, $\tilde{x}$ satisfies the following dynamics:

$$\begin{bmatrix} \dot{\tilde{x}}_1 \\ \epsilon\dot{\tilde{x}}_2 \end{bmatrix} = A(t)\tilde{x} + D(t)w - D(t)w_{SD}(t) - \begin{bmatrix} 0 \\ \epsilon U(\hat{A}_0 - \hat{S}_0 Z_{s\gamma}) \end{bmatrix} x_1^*$$

Then, we can rewrite the L_T into the following form:

$$\begin{aligned} L_T(w) \;&=\; |x^*(T)|^2_\Xi + 2\tilde{x}'(T)\Xi x^*(T) + |\tilde{x}(T)|^2_\Xi + \int_0^T (|\tilde{x}(t)|^2_{Q(t)} \\ &\quad + 2\tilde{x}'(t)Q(t)x^*(t) + |x^*(t)|^2_{Q(t)} + |u_{SD}(t)|^2 - \gamma^2|w(t)|^2 \\ &\quad + \gamma^2|w_{SD}(t)|^2 - \gamma^2|w_{SD}(t)|^2)\, dt \end{aligned}$$

By following the same arguments as in the finite-horizon case, we deduce that those terms containing only x^* in L_T can be approximated by

$$\begin{aligned} & |x_1^*(0)|^2_{Z_{s\gamma}+O(\epsilon)+O(\epsilon\delta)} + |x_1^*(T)|^2_{\delta\Delta} \\ \leq\ & |x_1^*(0)|^2_{Z_{s\gamma}+O(\epsilon)+O(\epsilon\delta)} + \alpha|x_1^*(0)|^2_{\delta\Delta} \\ =\ & |x_1^*(0)|^2_{Z_{s\gamma}+\alpha\delta\Delta+O(\epsilon)+O(\epsilon\delta)} \end{aligned}$$

By using a "completion of squares" argument, we can compute that

$$L_K^* = |x_1(0)|^2_{Z_{s\gamma}+\alpha\delta\Delta+O(\epsilon)+O(\epsilon\delta)} + |\tilde{x}(0)|^2_{P_\epsilon(0)} + 2Y_\epsilon'(0)\tilde{x}(0) + m(0)$$

where $P_\epsilon(t)$, $Y_\epsilon(t)$ and $m(t)$ are again the solutions to

$$\dot{P}_\epsilon + A_\epsilon' P_\epsilon + P_\epsilon A_\epsilon + P_\epsilon \tfrac{1}{\gamma^2} D_\epsilon D_\epsilon' P_\epsilon + Q = 0; \qquad P_\epsilon(T) = \Xi \tag{75}$$

$$\begin{aligned} & \dot{Y}_\epsilon + (A_\epsilon' + P_\epsilon \tfrac{1}{\gamma^2} D_\epsilon D_\epsilon')Y_\epsilon - P_\epsilon D_\epsilon w_{SD} - P_\epsilon \begin{bmatrix} 0 \\ O(1) \end{bmatrix} x_1^* + Qx^* = 0 \\ & Y_\epsilon(T) = \Xi x^*(T) \end{aligned} \tag{76}$$

$$\dot{m} = -\tfrac{1}{\gamma^2}|Y_\epsilon - \begin{bmatrix} Z_{s\gamma} \\ \epsilon V \end{bmatrix} x_1^*|^2_{D_\epsilon D_\epsilon'} + 2Y_\epsilon' \begin{bmatrix} 0 \\ O(1) \end{bmatrix} x_1^*; \quad m(T) = 0 \tag{77}$$

on the time interval $[0, T]$. Since $\gamma > \max\{\gamma_{SDs\infty}, \gamma_{of\infty}\}$ and the solution to the ϵ-free lower-order GRDE depends continuously on terminal condition, we can choose δ small enough (but still ϵ-free) to make P_ϵ exist for small ϵ. Then P_ϵ will have the same approximation form as in (64).

Let Z_ϵ be the solution to the following GARE:

$$\hat{A}_\epsilon' Z_\epsilon + Z_\epsilon \hat{A}_\epsilon + Z_\epsilon \frac{1}{\gamma^2} D_\epsilon D_\epsilon' Z_\epsilon + \hat{Q}_w = 0 \tag{78}$$

where $\hat{A}_\epsilon$ and $\hat{Q}_w$ are defined in (66) and (67). By the same argument as in finite-horizon case, Z_ϵ exists, and it can be approximated by

$$Z_\epsilon = \begin{bmatrix} Z_{s\gamma}(t) + O(\epsilon) & \epsilon(U'Z_{of\gamma} + V') + O(\epsilon^2) \\ \epsilon(Z_{of\gamma}U + V) + O(\epsilon^2) & \epsilon Z_{of\gamma}(t) + O(\epsilon^2) \end{bmatrix} \tag{79}$$

Note here Z_ϵ does not have any boundary layer terms. Thus, we have

$$Z_\epsilon x^* - \begin{bmatrix} Z_{s\gamma} \\ \epsilon V \end{bmatrix} x_1^* \;=\; O(\epsilon)$$

Next we define Y_1 and Y_2 as in the finite-horizon case (see (70)), and follow a similar reasoning, to show that

$$\begin{bmatrix} Y_1(t) \\ Y_2(t) \end{bmatrix} = O(\epsilon) + O(\delta); \quad \forall t \in [0, T]$$

Consequently, we have the following:

$$\begin{aligned} Y_\epsilon &= \begin{bmatrix} Z_{s\gamma} \\ \epsilon V \end{bmatrix} x_1^* + \begin{bmatrix} O(\epsilon)+O(\delta) \\ O(\epsilon^2)+O(\epsilon\delta) \end{bmatrix} \\ m(0) &= O(\epsilon^2)+O(\epsilon\delta)+O(\delta^2) \end{aligned}$$

Hence, by choosing ϵ small enough so that the boundary terms of P_ϵ die out at $t=0$, we arrive at

$$\begin{aligned} L_T^* &= x'(0)\begin{bmatrix} Z_{s\gamma}+\alpha\delta\Delta+O(\epsilon)+O(\delta^2) & \epsilon Z_{oc}'+O(\epsilon^2)+O(\epsilon\delta) \\ \epsilon Z_{oc}+O(\epsilon^2)+O(\epsilon\delta) & \epsilon Z_{of\gamma}+O(\epsilon^2) \end{bmatrix} x(0) \\ &:= |x(0)|^2_{\Xi} \end{aligned}$$

Now, we first choose a small enough $\delta_0>0$ so that $O(\delta^2)\ll O(\delta)$; then we find a small enough $\epsilon_\gamma'>0$ so that $O(\epsilon)\ll O(\delta)$, $O(\epsilon\delta)\ll O(\delta)$ and $O(\epsilon^2)\ll O(\epsilon\delta)$. Then, $\forall\epsilon\in(0,\epsilon_\gamma']$, $\Xi>\tilde{\Xi}$.

Noticing the following two facts:

1. We can obtain the above results for every $T'\in[T_m,T]$, the quantity δ_0 can be chosen uniformly on $[T_m,T]$, and the dependence of ϵ_γ' on $T'\in[T_m,T]$ is continuous.

2. If the terminal cost is decreased, the maximal cost for a given controller is not increased.

Thus, we deduce that, we can find a $\epsilon_\gamma'>0$ that works for all $T'\in[T_m,T]$. Then, $\forall\epsilon\in(0,\epsilon_\gamma']$, the infinite horizon game (1)–(6) has a bounded upper value , which is bounded by Ξ. □

4 Examples

We present here some numerical results for the infinite-horizon case with uniform sampling. As stressed earlier, the two quantities $\gamma_{SDs\infty}$ and $\gamma_{of\infty}$ play important roles in the computation of an approximate value for $\gamma^*_{SD\infty}(\epsilon)$. In the examples below, we will also compare the effectiveness of the full-order optimal sampled-data controller with that of the "slow" sampled-data controller introduced in (49) and (50) by plotting the system response (under these controllers) to various disturbance inputs.

Example 1 Consider the system

$$\begin{bmatrix} \dot{x}_1 \\ \epsilon\dot{x}_2 \end{bmatrix} = \begin{bmatrix} 2 & 1 \\ -1 & -2 \end{bmatrix}\begin{bmatrix} x_1 \\ x_2 \end{bmatrix} + \begin{bmatrix} 2 \\ 2 \end{bmatrix} u + \begin{bmatrix} 1 \\ 3 \end{bmatrix} w \tag{80}$$

$$L_\gamma = \int_0^\infty (2x_1^2+2x_1x_2+3x_2^2+|u|^2-\gamma^2|w|^2)\,dt \tag{81}$$

where we take the uniform sampling rate to be

$$T = 0.1$$

Then, the two critical performance levels are:

$$\gamma_{of\infty} = 2.5981; \quad \gamma_{SDs\infty} = 2.8574,$$

which show that $\bar{\gamma}_{SD\infty} = 2.8574$. We can also compute the minimax disturbance attenuation level $\gamma^*_{SD\infty}(\epsilon)$ of system (80)–(81) for different values of ϵ:

$\gamma^*_{SD\infty}(\epsilon)$	1.3311	2.0883	2.8127	2.8536
ϵ	1	0.1	0.01	0.001

We can see that as $\epsilon \to 0$, $\gamma^*_{SD\infty}(\epsilon) \to \bar{\gamma}_{SD\infty}$. Now, we choose $\gamma = 3 > \bar{\gamma}_{SD\infty}$, and design the "slow" sampled-data controller based on this value of γ:

$$Z_{s\gamma} = 2.1379; \quad \mu_{SD} = -1.1618 x_1^*$$

where

$$\dot{x}_1^* \;\; = \;\; -2.6890 x_1^*; \quad x_1^*(t_k) = x_1(t_k) \quad t \in [t_k, t_{k+1}) \quad k = 0, 1, \ldots$$

Now, fix $\epsilon = 0.001$, and design the optimal sampled-data controller $\mu^*_{SD\infty}$ for the full-order system:

$$\mu^*_{SD\infty} = -3.3917\hat{x}_1 - 1.0697\hat{x}_2$$

where

$$\begin{bmatrix} \dot{\hat{x}}_1 \\ \dot{\hat{x}}_2 \end{bmatrix} = \begin{bmatrix} -4.6930 & -9.6106 \\ -7512.2 & -3604.3 \end{bmatrix} \begin{bmatrix} \hat{x}_1 \\ \hat{x}_2 \end{bmatrix}; \quad \hat{x}_i(t_k) = x_i(t_k)$$
$$t \in [t_k, t_{k+1}) \quad k = 0, 1, \ldots$$

In the next three figures, we show some simulation results on the response of the system to three different disturbance inputs, under the μ_{SD} and $\mu^*_{SD\infty}$ given above, with $\epsilon = 0.001$. The disturbances are plotted in Figures 1(a), 2(a) and 3(a), which are constant, exponentially decaying, and sinusoidal, respectively; Figures 1(b) and (c), 2(b) and (c), 3(b) and (c), respectively, show plots of the corresponding full-order sampled-data control inputs and two components of the state responses to the controllers. Figures 1(d) and (e), 2(d) and (e), 3(d) and (e), respectively, do the same for the "slow" sampled-data controller. Note that the full-order controller results in a better state response, but a larger and more oscillatory control signal. ◇

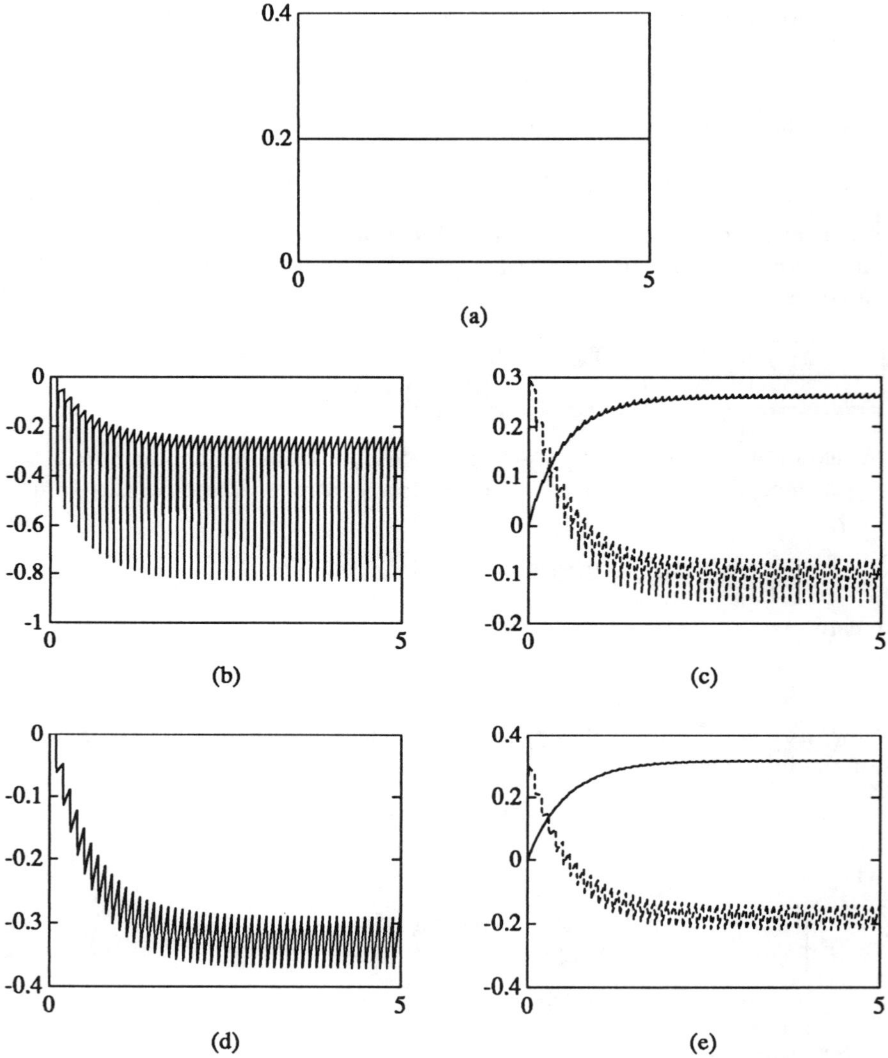

Figure 1: Input/output signals for Example 1: Constant disturbance (a) disturbance input; (b) full-order controller; (c) state responses to full-order controller; (d) slow controller; (e) state responses to slow controller

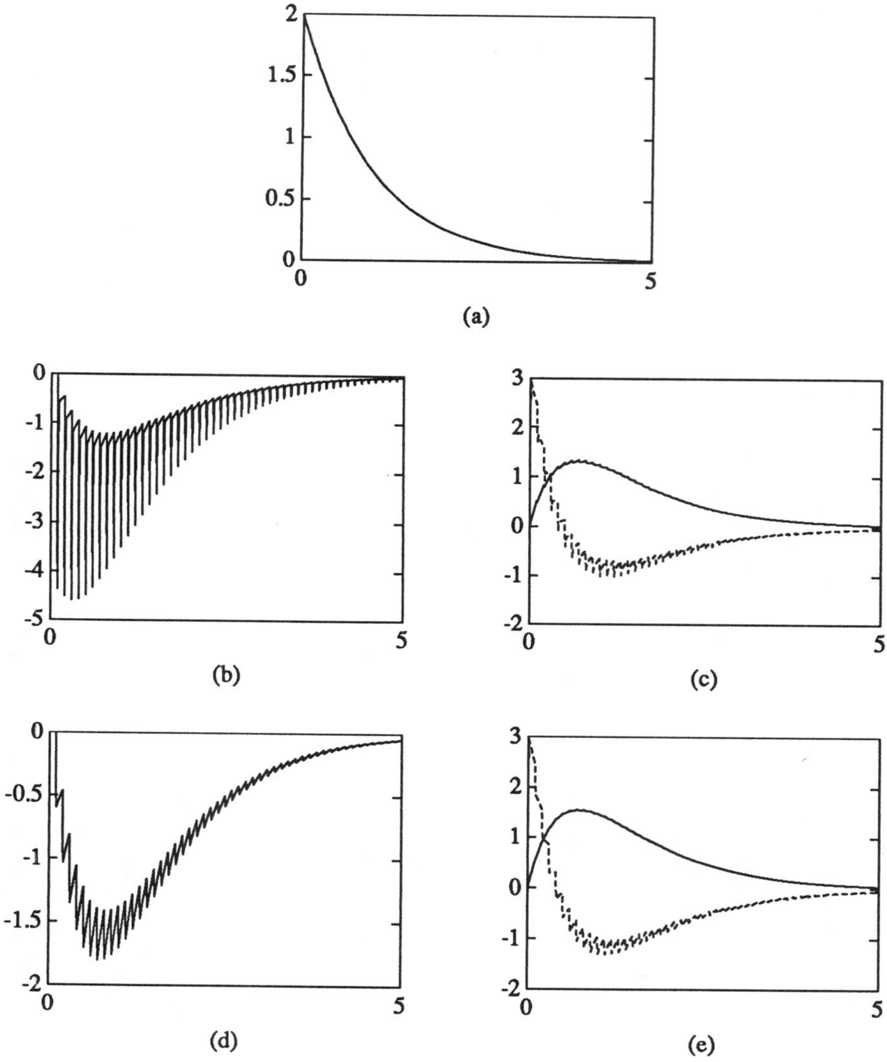

Figure 2: Input/output signals for Example 1: Exponentially decaying disturbance
(a) disturbance input; (b) full-order controller; (c) state responses to full-order controller; (d) slow controller; (e) state responses to slow controller

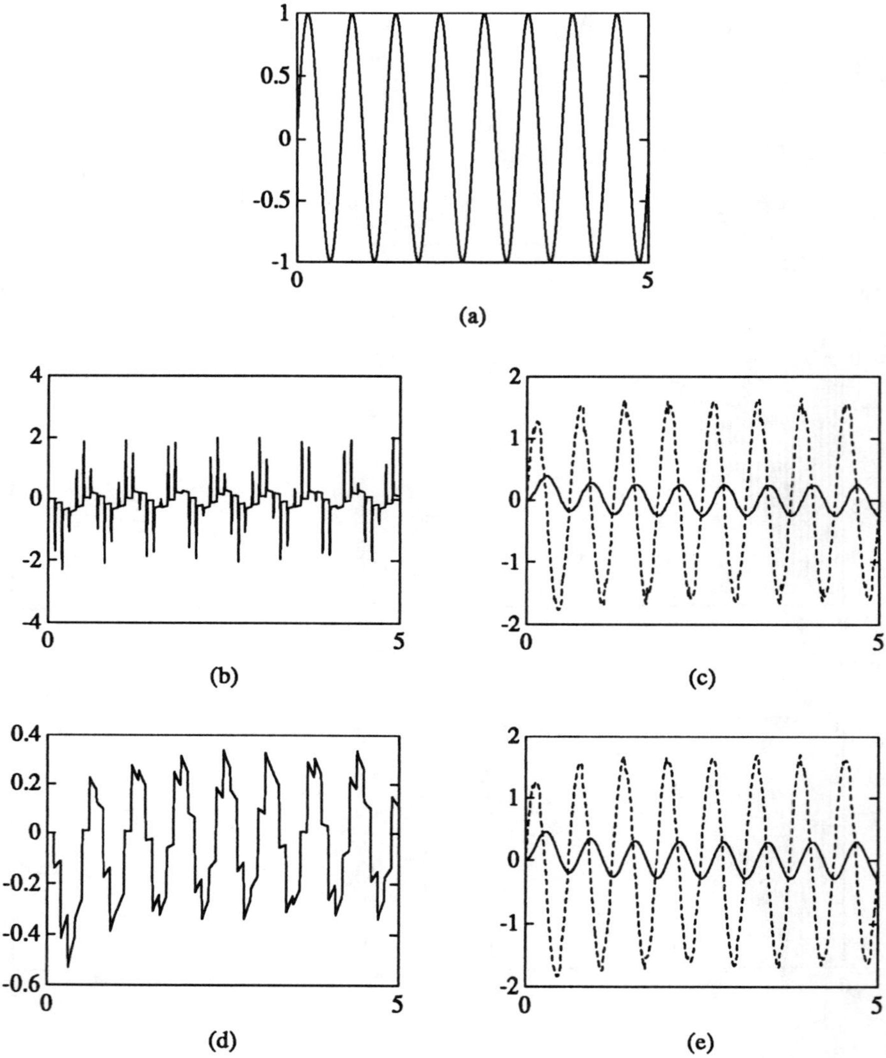

Figure 3: Input/output signals for Example 1: Sinusoidal disturbance (a) disturbance input; (b) full-order controller; (c) state responses to full-order controller; (d) slow controller; (e) state responses to slow controller

Example 2 Consider the system

$$\begin{bmatrix} \dot{x}_1 \\ \epsilon\dot{x}_2 \end{bmatrix} = \begin{bmatrix} 2 & 1 \\ 2 & -2 \end{bmatrix}\begin{bmatrix} x_1 \\ x_2 \end{bmatrix} + \begin{bmatrix} 2 \\ 1 \end{bmatrix} u + \begin{bmatrix} 1 \\ 1 \end{bmatrix} w \qquad (82)$$

$$L_\gamma = \int_0^\infty (2x_1^2 + 2x_1x_2 + x_2^2 + |u|^2 - \gamma^2|w|^2)\, dt \qquad (83)$$

where the uniform sampling rate is again taken to be $T = 0.1$. The two critical performance levels are:

$$\gamma_{of\infty} = 0.49941; \quad \gamma_{SDs\infty} = 1.0084.$$

Note that this depicts a different feature than Example 1, since $\gamma_{s\infty} > \gamma_{of\infty}$, whereas in Example 1 it was the opposite. Again in this case, we have $\bar{\gamma}_{SD\infty} = \gamma_{SDs\infty} = 1.0084$. We can also compute the minimax disturbance attenuation level $\gamma^*_{SD\infty}(\epsilon)$ of the system (82)–(83) for different values of ϵ,

$\gamma^*_{SD\infty}(\epsilon)$	0.75253	0.92789	1.0001	1.0080
ϵ	1	0.1	0.01	0.001

Note that as $\epsilon \to 0$, $\gamma^*_{SD\infty}(\epsilon) \to \bar{\gamma}_{SD\infty}$. Now, we choose $\gamma = 1.01 > \bar{\gamma}_{SD\infty}$, and design the "slow" sampled-data controller accordingly:

$$Z_{s\gamma} = 1.7836; \quad \mu_{SD} = -4.9973x_1^*$$

where

$$\dot{x}_1^* = -4.7678x_1^*; \quad x_1^*(t_k) = x_1(t_k) \quad t \in [t_k, t_{k+1}) \quad k = 0, 1, \ldots$$

Now, fix $\epsilon = 0.001$, and design the optimal sampled-data controller $\mu^*_{SD\infty}$ for the full-order system

$$\mu^*_{SD\infty} = -4.9746\hat{x}_1 - 0.25303\hat{x}_2$$

where

$$\begin{bmatrix} \dot{\hat{x}}_1 \\ \dot{\hat{x}}_2 \end{bmatrix} = \begin{bmatrix} -4.8210 & -0.74061 \\ 153.60 & -2006.4 \end{bmatrix}\begin{bmatrix} \hat{x}_1 \\ \hat{x}_2 \end{bmatrix}; \quad \hat{x}_i(t_k) = x_i(t_k)$$
$$t \in [t_k, t_{k+1}) \quad k = 0, 1, \ldots$$

We again present some simulation results, given in Figures 4, 5 and 6. This time the "slow" controller results in a slightly larger and more oscillatory control signal, worse response for the fast state variable, but a better response for the slow state variable. This is mainly due to the fact that the fast subsystem is less sensitive to noise than the slow subsystem. ◇

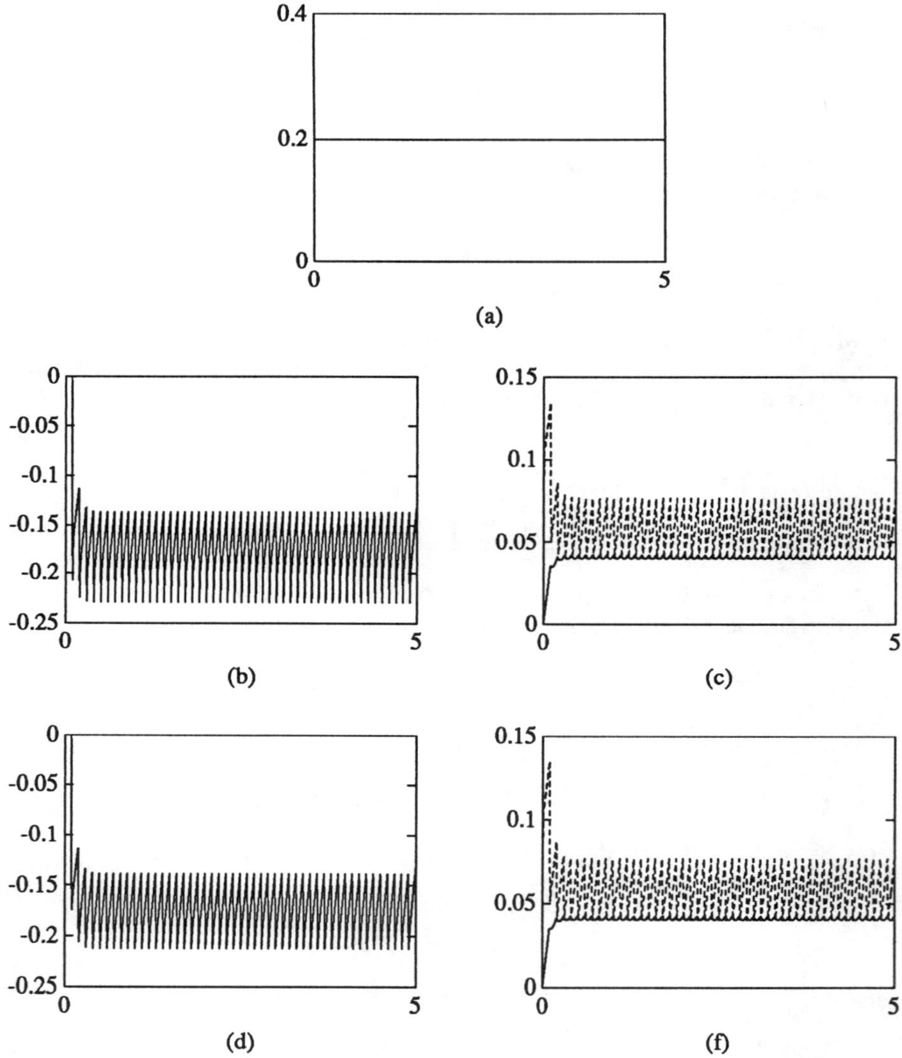

Figure 4: Input/output signals for Example 2: Constant disturbance (a) disturbance input; (b) full-order controller; (c) state responses to full-order controller; (d) slow controller; (e) state responses to slow controller

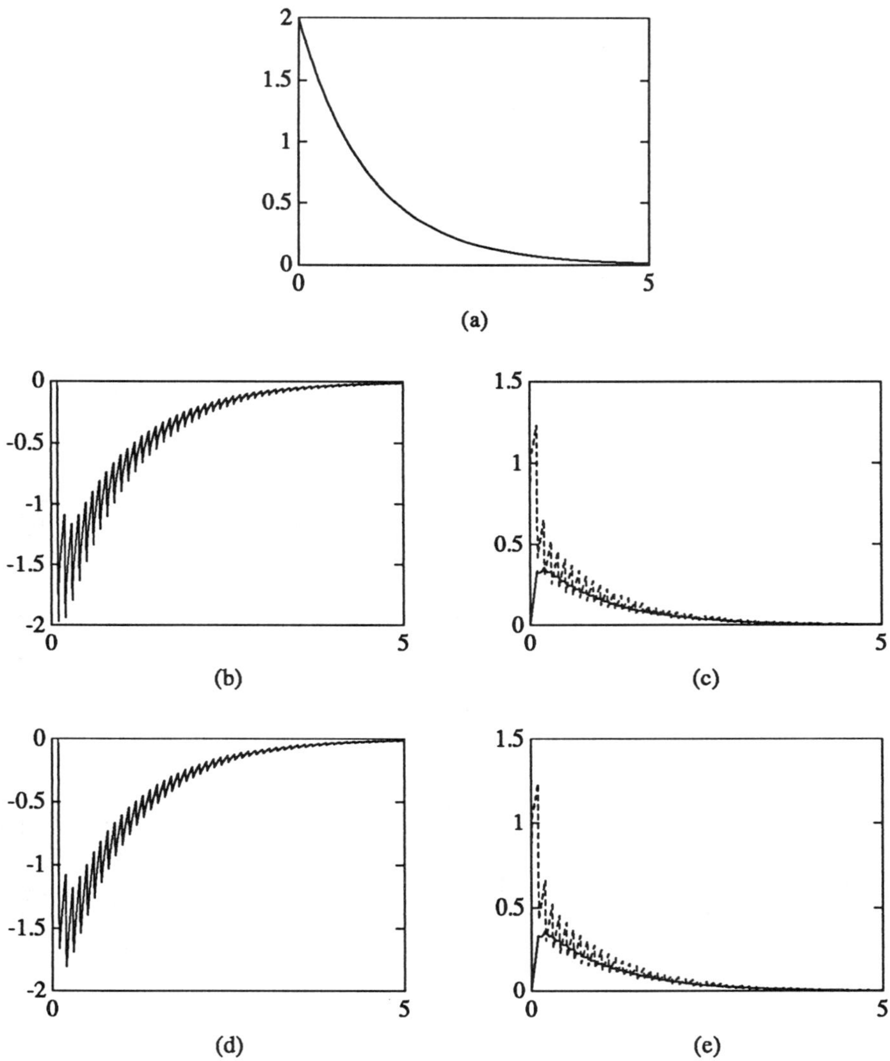

Figure 5: Input/output signals for Example 2: Exponentially decaying disturbance
(a) disturbance input; (b) full-order controller; (c) state responses to full-order controller; (d) slow controller; (e) state responses to slow controller

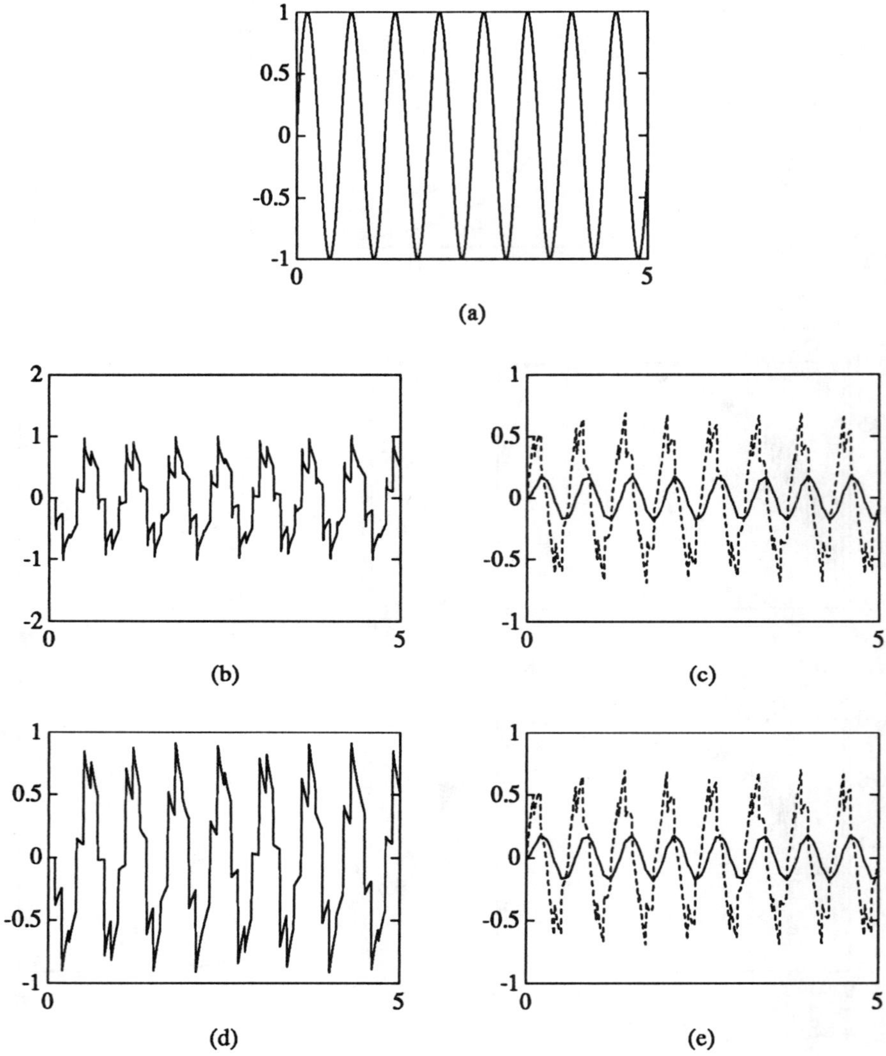

Figure 6: Input/output signals for Example 2: Sinusoidal disturbance (a) disturbance input; (b) full-order controller; (c) state responses to full-order controller; (d) slow controller; (e) state responses to slow controller

5 Conclusion

In this paper, we have provided a complete analysis of the singularly perturbed H^∞-optimal control problem with sampled-data state measurements in both finite and infinite horizons, by relating it to a class of singularly perturbed differential games. One of the main results of the paper is that one can totally neglect the fast dynamics, and still achieve a performance very close to that achieved using full-order controllers, which is quite different from the results obtained in the perfect and imperfect (nonsampled) state measurements cases. We construct a "slow" controller, based on the slow subsystem only and independent of the singular perturbation parameter, which can achieve any desired achievable performance level for the full-order system. Several numerical examples are included, which illustrate the theoretical results.

One immediate, but not trivial, extension of these results would be to the imperfect sampled-data measurements case, so as to obtain the counterparts of the results of [19] in the sampled-data measurements case. Another extension would be to obtain the counterparts of the results here for multi-time scale systems, which, however, seems to be rather routine (though cumbersome) not requiring any new conceptual developments. Another problem of interest would be the singularly perturbed version of the problem studied in [15], where we have the additional restriction that control action remains constant on each sampling interval. Again techniques developed here seem to be of direct relevance to solve this singularly perturbed digital control problem.

Appendix

A

Over a typical sampling interval $[t_{k-1}, t_k]$, consider the following game kernel for the slow subsystem:

$$L_{ck} = |x_s(t_f)|^2_{Z_{s\gamma}(t_k)} + \int_{t_{k-1}}^{t_k} (|x_s|^2_{Q^\square} + 2x_s' P^\square u_s + |u_s|^2_{R^\square} - |\hat{w}_s|^2)\, dt$$

It is easy to see, from the discussion in Section 3 of [18] about the SF information pattern for the slow subsystem, that u_{SD} given by (49) is the minimizing control input that achieves the upper value of the game under open-loop information, w_{SD} given by (51) is the corresponding maximizing disturbance under open-loop information, and x_1^* is the corresponding optimal trajectory.

With u_{SD}, w_{SD}, x_1^* given as stated, we can now compute $\bar{x}_2$:

$$\bar{x}_2 \;=\; -A_{22}^{-1}(A_{21}x_1^* + B_2 u_{SD} + D_2 w_{SD})$$

$$\begin{aligned}
&= -A_{22}^{-1}(A_{21} - B_2(B_1'Z_{s\gamma} + B_2'V) + \frac{1}{\gamma^2}D_2(D_1'Z_{s\gamma} + D_2'V))x_1^* \\
&= -A_{22}^{-1}(A_{21} - S_{21}Z_{s\gamma} - S_{22}V)x_1^* = -A_{22}^{-1}(A_{22}U)x_1^* = -Ux_1^*
\end{aligned}$$

B

We first obtain an expression for $\dot{\tilde{x}}_1$:

$$\begin{aligned}
\dot{\tilde{x}}_1 &= \dot{x}_1 - \dot{x}_1^* \\
&= A_{11}x_1 + A_{12}x_2 + B_1u_{SD} + D_1w - (\hat{A}_0 - \hat{S}_0Z_{s\gamma})x_1^* \\
&= A_{11}x_1 + A_{12}x_2 + B_1u_{SD} + D_1w - (A_{11} - S_{11}Z_{s\gamma} - S_{12}V \\
&\qquad -A_{12}U)x_1^* \\
&= A_{11}\tilde{x}_1 + A_{12}\tilde{x}_2 + D_1w - D_1w_{SD}
\end{aligned}$$

Similarly, for $\epsilon\dot{\tilde{x}}_2$:

$$\begin{aligned}
\epsilon\dot{\tilde{x}}_2 &= \epsilon(\dot{x}_2 - \dot{x}_2^*) \\
&= A_{21}x_1 + A_{22}x_2 + B_2u_{SD} + D_2w - \epsilon(\dot{U} + U(\hat{A}_0 - \hat{S}_0Z_{s\gamma}))x_1^* \\
&= A_{21}\tilde{x}_1 + A_{22}\tilde{x}_2 + B_2u_{SD} + D_2w + A_{21}x_1^* + A_{22}x_2^* - \epsilon(\dot{U} \\
&\qquad +U(\hat{A}_0 - \hat{S}_0Z_{s\gamma}))x_1^* \\
&= A_{21}\tilde{x}_1 + A_{22}\tilde{x}_2 + D_2w - D_2w_{SD} - \epsilon(\dot{U} + U(\hat{A}_0 - \hat{S}_0Z_{s\gamma}))x_1^*
\end{aligned}$$

Thus, we have

$$\begin{bmatrix} \dot{\tilde{x}}_1 \\ \epsilon\dot{\tilde{x}}_2 \end{bmatrix} = A(t)\tilde{x} + D(t)w - D(t)w_{SD}(t) - \begin{bmatrix} 0 \\ \epsilon(\dot{U} + U(\hat{A}_0 - \hat{S}_0Z_{s\gamma})) \end{bmatrix} x_1^*$$

Note that if the system is time-invariant then $\dot{U} = 0$.

C

The verification of (71) involves some substitutions and straightforward manipulations, as given below:

$$\begin{aligned}
\begin{bmatrix} \dot{Y}_1 \\ \epsilon\dot{Y}_2 \end{bmatrix} &= -(A_\epsilon' + P_\epsilon\frac{1}{\gamma^2}D_\epsilon D_\epsilon')\begin{bmatrix} Y_1 \\ \epsilon Y_2 \end{bmatrix} - (A_\epsilon' + P_\epsilon\frac{1}{\gamma^2}D_\epsilon D_\epsilon')Z_\epsilon x^* \\
&\quad +P_\epsilon D_\epsilon w_{SD} + P_\epsilon\begin{bmatrix} 0 \\ O(1) \end{bmatrix} x_1^* - Qx^* + (\hat{A}_\epsilon'Z_\epsilon + Z_\epsilon\hat{A}_\epsilon \\
&\quad +Z_\epsilon\frac{1}{\gamma^2}D_\epsilon D_\epsilon'Z_\epsilon + \hat{Q}_w)x^* - Z_\epsilon\begin{bmatrix} \hat{A}_0 - \hat{S}_0Z_{s\gamma} \\ O(1) \end{bmatrix} x_1^* \\
&= -(A' + P_\epsilon\frac{1}{\gamma^2}D_\epsilon D')\begin{bmatrix} Y_1 \\ Y_2 \end{bmatrix} - P_\epsilon\frac{1}{\gamma^2}D_\epsilon D_\epsilon'(Z_\epsilon x^* - \begin{bmatrix} Z_{s\gamma} \\ \epsilon V \end{bmatrix} x_1^*)
\end{aligned}$$

$$
\begin{aligned}
& +P_\epsilon \begin{bmatrix} 0 \\ O(1) \end{bmatrix} x_1^* - \begin{bmatrix} Z_{s\gamma} & V' \\ 0 & 0 \end{bmatrix} BB'_\epsilon (Z_\epsilon x^* - \begin{bmatrix} Z_{s\gamma} \\ \epsilon V \end{bmatrix} x_1^*) \\
& +Z_\epsilon (\hat{A}_\epsilon + \frac{1}{\gamma^2} D_\epsilon D'_\epsilon Z_\epsilon - \begin{bmatrix} \hat{A}_0 - \hat{S}_0 Z_{s\gamma} & 0 \\ O(1) & 0 \end{bmatrix}) x^* \\
= \quad & -(A' + P_\epsilon \frac{1}{\gamma^2} D_\epsilon D') \begin{bmatrix} Y_1 \\ Y_2 \end{bmatrix} + P_\epsilon \begin{bmatrix} 0 \\ O(1) \end{bmatrix} x_1^* - (P_\epsilon \frac{1}{\gamma^2} D_\epsilon D' \\
& + \begin{bmatrix} Z_{s\gamma} B_1 + V' B_2 \\ 0 \end{bmatrix} B') \begin{bmatrix} O(\epsilon) \\ Z_{cb} - Z_{fb} U + O(\epsilon) \end{bmatrix} x_1^* \\
& + Z_\epsilon \begin{bmatrix} \frac{1}{\gamma^2} D_1 D'_2 (Z_{cb} - Z_{fb} U) + O(\epsilon) \\ \frac{1}{\epsilon} (\frac{1}{\gamma^2} D_2 D'_2 (Z_{cb} - Z_{fb} U) + O(\epsilon)) \end{bmatrix} x_1^*
\end{aligned}
$$

The above can be seen to be equivalent to (71), by writing out the boundary layer differential equations for Z_{fb} and P_{fb} (from (65) and (61)), and showing that they are actually identical.

Zigang Pan and Tamer Başar
Decision and Control Laboratory
Coordinated Science Laboratory and the
Department of Electrical and Computer Engineering
University of Illinois
1308 West Main Street
Urbana, IL 61801, USA

REFERENCES

[1] G. Zames, "Feedback and optimal sensitivity: Model reference transformation, multiplicative seminorms and approximate inverses," *IEEE Trans. Automat. Control*, vol. AC-26, pp. 301–320, 1981.

[2] B. A. Francis, *A Course in H_∞ Control Theory*, vol. 88 of *Lecture Notes in Control and Information Sciences*. New York, NY: Springer-Verlag, 1987.

[3] J. Doyle, K. Glover, P. Khargonekar, and B. Francis, "State-space solutions to standard H_2 and H_∞ control problems," *IEEE Trans. Automat. Control*, vol. AC-34, no. 8, pp. 831–847, 1989.

[4] K. Glover and J. C. Doyle, "State-space formulae for all stabilizing controllers that satisfy an H^∞-norm bound and relations to risk sensitivity," *Systems and Control Letters*, vol. 11, pp. 167–172, 1988.

[5] P. P. Khargonekar, I. R. Petersen, and M. A. Rotea, "H_∞-optimal control with state-feedback," *IEEE Trans. Automat. Control*, vol. AC-33, no. 8, pp. 786–788, 1988.

[6] P. P. Khargonekar, K. N. Nagpal, and K. R. Poolla, "H^∞ control with transients," *SIAM Journal on Control and Optimization*, vol. 29, pp. 1373–1393, November 1991.

[7] R. Ravi, K. M. Nagpal, and P. P. Khargonekar, "H_∞ control for linear time varying systems: a state-space approach," *SIAM Journal on Control and Optimization*, vol. 29, pp. 1394–1413, November 1991.

[8] A. A. Stoorvogel, *The H_∞ Control Problem: A State Space Approach.* Prentice Hall, 1992.

[9] T. Başar, "A dynamic games approach to controller design: Disturbance rejection in discrete time," in *Proc. IEEE 29th Conf. on Decision and Control*, (Tampa, FL), pp. 407–414, December 13-15, 1989. Also appeared in the *IEEE Transactions on Automatic Control*, vol. 36, no. 8, pp. 936–952, August 1991.

[10] T. Başar, "Game theory and H^∞-optimal control: The continuous-time case," in *Differential Games: Developments in Modelling and Computation*, Springer-Verlag, August 1991. R. P. Hamalainen and H. K. Ehtamo, eds. Lecture Notes in Control and Information Sciences, vol. 156, pp. 171–186.

[11] K. Uchida and M. Fujita, "On the central controller: Characterizations via differential games and LEQG control problems," *Systems and Control Letters*, vol. 13, no. 1, pp. 9–13, 1989.

[12] D. Limebeer, B. Anderson, P. Khargonekar, and M. Green, "A game theoretic approach to H_∞ control for time varying systems," in *Proceedings of the International Symposium on the Mathematical Theory of Networks and Systems*, (Amsterdam, The Netherlands), 1989.

[13] I. Rhee and J. L. Speyer, "A game theoretic controller and its relationship to H^∞ and linear-exponential Gaussian synthesis," in *Proc. IEEE 29th Conf. on Decision and Control*, (Tampa, FL), pp. 909–915, December 13-15, 1989.

[14] T. Başar, "Optimum H^∞ designs under sampled state measurements," *Systems and Control Letters*, vol. 16, pp. 399–409, June 1991.

[15] H. T. Toivonen, "Sampled-data H_∞ optimal control of time-varying systems," *Automatica*, vol. 28, no. 4, pp. 823–826, 1992.

[16] T. Başar and P. Bernhard, *H^∞-Optimal Control and Related Minimax Design Problems: A Dynamic Game Approach.* Boston, MA: Birkhäuser, 1991.

[17] P. V. Kokotović, H. K. Khalil, and J. O'Reilly, *Singular Perturbation Methods in Control: Analysis and Design.* New York, NY: Academic Press, 1986.

[18] Z. Pan and T. Başar, "H^∞-optimal control for singularly perturbed systems. Part I: Perfect state measurements," *Automatica*, vol. 29, March 1993.

[19] Z. Pan and T. Başar, "H^∞-optimal control for singularly perturbed systems. Part II: Imperfect state measurements," in *Proceedings of the 31st IEEE Conference on Decision and Control*, (Tucson, AZ), pp. 943–948, December 1992.

[20] F. Garofalo and G. Leitmann, "A composite controller ensuring ultimate boundedness for a class of singularly perturbed uncertain systems," *Dynamics and Stability of Systems*, vol. 3, no. 3 & 4, pp. 135–145, 1988.

[21] M. Corless, G. Leitmann, and E. P. Ryan, "Control of uncertain systems with neglected dynamics," in *Deterministic Control of Uncertain Systems* (A. S. I. Zinober, ed.), ch. 12, pp. 252–268, London: IEE Press, 1990.

[22] F. Garofalo and G. Leitmann, "Nonlinear composite control of a class of nominally linear singularly perturbed uncertain systems," in *Deterministic Control of Uncertain Systems* (A. S. I. Zinober, ed.), ch. 13, pp. 269–288, London: IEE Press, 1990.

[23] D. W. Luse and J. A. Ball, "Frequency-scale decomposition of H^∞-disk problems," *SIAM Journal on Control and Optimization*, vol. 27, pp. 814–835, July 1989.

[24] T. Chen and B. Francis, "On the L^2 induced norm of a sampled-data system," *Systems and Control Letters*, vol. 15, pp. 211–219, 1990.

[25] B. Bamieh and J. B. Pearson, "A general framework for linear periodic systems with application to H^∞ sampled-data control," *IEEE Trans. Automat. Control*, vol. 37, pp. 418–435, April 1992.

[26] J. H. Chow and P. V. Kokotović, "A decomposition of near-optimum regulators for systems with slow and fast modes," *IEEE Trans. Automat. Control*, vol. AC-21, pp. 701–705, October 1976.

[27] T. Başar and G. J. Olsder, *Dynamic Noncooperative Game Theory.* London/New York: Academic Press, 1982. Second printing 1989.

[28] Z. Pan and T. Başar, "A tight bound for the H^∞ performance of singularly perturbed systems," CSL Report, University of Illinois, Urbana, IL, March 1992.

New Results on Nonlinear H_∞-Control via Measurement Feedback*

Alberto Isidori

1 Introduction

In the last few years, the solution of the H_∞ (sub)optimal control problem via state-space methods was developed by several authors (see, e.g. the prize-winning paper [1], the theses [2] [3] and the recent paper [4]). In the state-space formulation, the problem of minimizing the H_∞ norm (or, equivalently, the L_2 gain) of a closed loop system is viewed as a two-person, zero sum, differential game and, thus, the existence of the desired controller can be related to the existence of a solution of the algebraic Riccati equations arising in linear quadratic differential game theory (see, e.g. [5], [6] and [7]).

The advantage of the state-space approach and of the related game-theoretic interpretation is that the analysis of the H_∞ control problem for a linear system (namely, the characterization of necessary and sufficient conditions for the existence of a solution, the construction of one particular solution and the description of the set of all possible solutions) can be rendered exclusively dependent on *time domain* arguments, thus laying the basis for appropriate extension to broader classes of systems.

The analysis of the "nonlinear equivalent" of the H_∞ control problem has been addressed in a number of recent papers (see [8], [9], [10], [11]). In this domain, the Riccati equation of the linear quadratic differential game is replaced by the (more general) Hamilton-Jacobi-Isaacs equation of a nonlinear differential game. As a matter of fact, it is rather straightforward to realize, as pointed out in [7], that the existence of a positive definite solution of an appropriate Isaacs equation is a sufficient condition for the existence of a *full-information* feedback law providing internal stability and disturbance attenuation, in the sense of the L_2 gain [9], [10]. In addition, it has also been shown that, under suitable hypotheses, the existence of such a solution is a necessary condition for the solvability of the problem [11].

More recently (see [12], [14]) the design of a control law based on *measurement* feedback was also investigated. In particular, in [12], it was shown that the problem in question is solvable if, in addition to the Isaacs

*Work supported in part by MURST, by NSF under grant ECS-9208306 and by AFOSR under grant 91-0266

equation arising in the design of a full-information feedback law, a suitable additional Hamilton-Jacobi equation, associated with the design of an output injection gain, has a positive definite solution. In this paper, the results of [12] are extended in a number of ways: in particular, we suppress some technical simplifying hypotheses and we improve the procedure for the design of the output injection gain, by showing how this can be related to the solution of a "dual" Isaacs equation.

2 Notations and problem statement

Consider a nonlinear system modeled by equations of the form

$$\begin{aligned} \dot{x} &= f(x) + g_1(x)w + g_2(x)u \\ z &= h_1(x) + k_{12}(x)u \\ y &= h_2(x) + k_{21}(x)w \end{aligned} \tag{1}$$

The first equation of this system describes a plant with *state* x, defined on a neighborhood X of the origin in $I\!R^n$ with *control input* $u \in I\!R^m$ and subject to a set of *exogenous input* variables $w \in I\!R^r$ which includes *disturbances* (to be rejected) and/or *references* (to be tracked). The second equation defines a *penalty* variable $z \in I\!R^s$, which may include a *tracking error*, for instance the difference between the actual plant output and its desired reference behavior, expressed as a function of some of the exogenous variables w, as well as a *cost* of the input u needed to achieve the prescribed control goal. The third equation defines a set of *measured variables* $y \in I\!R^p$, which also are functions of the state plant x and the exogenous input w.

The mappings $f(x), g_1(x), g_2(x), h_1(x), h_2(x), k_{12}(x), k_{21}(x)$ are smooth (i.e. C^∞) mappings defined in a neighborhood of the origin in $I\!R^n$. We assume also that $f(0) = 0$, $h_1(0) = 0$ and $h_2(0) = 0$.

The control action to (1) is to be provided by a *controller*, which processes the measured variable y and generates the appropriate control input u, and is modeled by equations of the form

$$\begin{aligned} \dot{\xi} &= \eta(\xi, y) \\ u &= \theta(\xi, y), \end{aligned} \tag{2}$$

in which ξ is defined on a neighborhood Ξ of the origin in $I\!R^\nu$ and $\eta : \Xi \times I\!R^p \to I\!R^\nu$, $\theta : \Xi \times I\!R^p \to I\!R^m$ are C^k functions (for some $k \geq 1$), satisfying $\eta(0,0) = 0$ and $\theta(0,0) = 0$.

The purpose of the control is twofold: *to achieve closed loop stability* and *to attenuate the influence* of the exogenous input w on the penalty variable z. A controller which locally asymptotically stabilizes the equilibrium $(x, \xi) = (0,0)$ of the closed loop system is said to be an *admissible controller*. The requirement of disturbance attenuation, in the present paper, is characterized by means of the so-called L_2 gain of a system, which is defined as follows. Consider a nonlinear system Σ described by equations of the form

$$\dot{x} = f(x) + g(x)u$$
$$y = h(x)$$

with $x \in \mathbb{R}^n$, $u \in \mathbb{R}^m$ and $y \in \mathbb{R}^p$. Let U_ρ denote the open ball in $\mathbb{R}^n$ of radius ρ centered at $x = 0$ and $\mathcal{U}_\delta$ the set of all piecewise continuous functions $u : \mathbb{R} \to \mathbb{R}^m$ satisfying $\|u(t)\| < \delta$ for all $t \in \mathbb{R}$. Suppose $x = 0$ is a locally asymptotically stable equilibrium of Σ. Then, for each $\epsilon > 0$ there exists $\delta(\epsilon) > 0$ such that for every input $u(\cdot) \in \mathcal{U}_{\delta(\epsilon)}$ the state response $x(\cdot)$ of Σ from the initial state $x(0) = 0$ exists and satisfies $x(t) \in U_\epsilon$ for all $t \geq 0$. Σ is said to have L_2 gain less than or equal to γ if for each $u(\cdot) \in \mathcal{U}_{\delta(\epsilon)}$ the response $y(\cdot)$ of Σ to $u(\cdot)$ from the initial state $x(0) = 0$ satisfies

$$\int_0^t \|y(s)\|^2 ds \leq \gamma^2 \int_0^t \|u(s)\|^2 ds.$$

for all $t > 0$. In what follows, by *problem of local disturbance attenuation with internal stability* we mean the problem of finding an admissible controller yielding a closed loop system with an L_2 gain which less than or equal to a prescribed number γ.

3 Disturbance attenuation via state feedback

In this Section, we present sufficient conditions for the solution of the problem of local disturbance attenuation via state feedback. To this purpose, we assume that the parameters which charaterize the plant (1) satisfy the following hypotheses.

Assumption A1 The matrix

$$R_2(x) = k_{12}^{\mathrm{T}}(x)k_{12}(x)$$

is nonsingular (and thus positive definite) for each x.$\triangleleft$

Assumption A2 Any bounded trajectory $x(t)$ of the system

$$\dot{x}(t) = f(x(t)) + g_2(x(t))u(t)$$

satisfying

$$h_1(x(t)) + k_{12}(x(t))u(t) = 0$$

for all $t \geq 0$, is such that $\lim_{t\to\infty} x(t) = 0$.$\triangleleft$

Remark. Assumptions A1 and A2 are the nonlinear versions of some standard assumptions usually considered in the case of linear systems (see e.g. [4]). In particular, it is possible to show that, if the system (1) is a linear system, i.e. a system described by equations of the form

$$\begin{aligned} \dot{x} &= Ax + B_1 w + B_2 u \\ z &= C_1 x + D_{12} u \\ y &= C_2 x + D_{21} w, \end{aligned} \tag{3}$$

assumption A2 corresponds to the assumption that the matrix

$$\begin{pmatrix} A - j\omega I & B_2 \\ C_1 & D_{12} \end{pmatrix}$$

has independent columns for each $\omega \in I\!R$.◁

The following result, which is a simple extension of the main result of [9], shows how the problem of local disturbance attenuation can be solved via state feedback.

Theorem 1 *Consider the system (1) and suppose assumptions* A1 *and* A2 *are satisfied. Suppose there exists a smooth positive definite function $V(x)$, locally defined in a neighborhood of the origin in $I\!R^n$ and vanishing at $x = 0$, which satisfies the Hamilton-Jacobi-Isaacs inequality*

$$V_x f(x) + h_1^{\mathrm{T}}(x) h_1(x) + \gamma^2 \alpha_1^{\mathrm{T}}(x)\alpha_1(x) - \alpha_2^{\mathrm{T}}(x) R_2(x) \alpha_2(x) \le 0 \tag{4}$$

where

$$\begin{aligned} \alpha_1(x) &= \frac{1}{2\gamma^2} g_1^{\mathrm{T}}(x) V_x^{\mathrm{T}} \\ \alpha_2(x) &= -R_2^{-1}(x)(\frac{1}{2} g_2^{\mathrm{T}}(x) V_x^{\mathrm{T}} + k_{12}^{\mathrm{T}}(x) h_1(x)). \end{aligned} \tag{5}$$

Then, the problem of local disturbance attenuation with internal stability is solved by the state feedback

$$u = \alpha_2(x).$$

Proof. The quantity

$$H(x, w, u) = V_x(f(x) + g_1(x) w + g_2(x) u) + \|h_1(x) + k_{12}(x) u\|^2 - \gamma^2 \|w\|^2$$

is a polynomial of second degree in (w, u). Moreover, by construction,

$$\begin{aligned} \left(\frac{\partial H}{\partial w}\right)_{\substack{w=\alpha_1(x)\\ u=\alpha_2(x)}} &= V_x g_1(x) - 2\gamma^2 \alpha_1^{\mathrm{T}}(x) = 0 \\ \left(\frac{\partial H}{\partial u}\right)_{\substack{w=\alpha_1(x)\\ u=\alpha_2(x)}} &= V_x g_2(x) + 2 h_1^{\mathrm{T}}(x) k_{12}(x) + 2\alpha_2^{\mathrm{T}}(x) R_2(x) = 0. \end{aligned}$$

The inequality (4) implies

$$H(x, \alpha_1(x), \alpha_2(x)) \le 0.$$

and therefore

$$H(x,u,w) \leq \|u-\alpha_2(x)\|^2_{R_2(x)} - \gamma^2\|w-\alpha_1(x)\|^2$$

or, what is the same,

$$\begin{aligned} &V_x(f(x)+g_1(x)w+g_2(x)u)+\|h_1(x)+k_{12}(x)u\|^2-\gamma^2\|w\|^2 \\ &\leq \|u-\alpha_2(x)\|^2_{R_2(x)} - \gamma^2\|w-\alpha_1(x)\|^2. \end{aligned} \tag{6}$$

In particular, the choice $u=\alpha_2(x)$ yields

$$V_x(f(x)+g_1(x)w+g_2(x)\alpha_2(x))+\|h_1(x)+k_{12}(x)\alpha_2(x)\|^2-\gamma^2\|w\|^2 \leq 0. \tag{7}$$

Then, along the trajectories of the closed loop system with $w=0$ we have

$$\frac{dV(x(t))}{dt} \leq -\|h_1(x)+k_{12}(x)\alpha_2(x)\|^2 \leq 0.$$

This proves that the equilibrium $x=0$ of the closed loop system is stable. To prove asymptotic stability, observe that any trajectory $x(\cdot)$ such that $V(x(t))=0$ for all $t \geq 0$ is necessarily a trajectory of

$$\dot{x} = f(x)+g_2(x)\alpha_2(x)$$

such that $x(t)$ is bounded and

$$h_1(x(t))+k_{12}(x(t))\alpha_2(x(t))=0$$

for all $t \geq 0$. Thus, by hypothesis A2 it is concluded that $\lim_{t\to\infty} x(t)=0$.

To prove that the L_2 gain is less than or equal to γ, it suffices to integrate (7) from 0 to t along the trajectories of the closed loop system. ◁

Remark. It is well known (see e.g. [7]) that the problem of rendering the inequality

$$\int_0^\infty \|z(t)\|^2 dt \leq \gamma^2 \int_0^\infty \|w(t)\|^2 dt$$

satisfied for each input $w(\cdot) \in \mathcal{L}_2[0,\infty)$ can be cast as a *two players, zero sum, differential game*

$$\dot{x} = f(x)+g_1(x)w+g_2(x)u$$

of infinite duration $[0,\infty)$, fixed initial condition $x(0)=0$ and value functional

$$J(u,w) = \int_0^\infty L(x(s),w(s),u(s))ds$$

where $L(x,w,u)$ is given by

$$L(x,w,u) = \|h_1(x)+k_{12}(x)u\|^2-\gamma^2\|w\|^2.$$

The *minimizing* player controls u and the *maximizing* player controls w (the utimate goal of the minimizing player being - in the present setting of the problem of disturbance attenuation - to render $J(u, w)$ nonpositive for all w).

In this setup, it is easily realized that if $V(x)$ is a solution of the Hamilton-Jacobi-Isaacs *equation*

$$V_x f(x) + h_1^{\mathrm{T}}(x)h_1(x) + \gamma^2\alpha_1^{\mathrm{T}}(x)\alpha_1(x) - \alpha_2^{\mathrm{T}}(x)R_2(x)\alpha_2(x) = 0,$$

the strategy

$$u_*(t) = \alpha_2(x(t))$$

is the best strategy for the minimizing player, while the strategy

$$w_*(t) = \alpha_1(x(t))$$

is the best strategy for the maximizing player. Since in the present case w is a disturbance, the latter can be interpreted as the *worst* possible disturbance affecting the system.◁

4 Disturbance attenuation via measurement feedback

Preliminaries

If the state x of the plant is not available for measurement, it is reasonable to expect that its actual value might be replaced, in the expression of the feedback law

$$u = \alpha_2(x),$$

by a suitable estimate ξ provided by some auxiliary dynamics.

As suggested in [12], we generate this estimate ξ by means of an *exact copy* of the dynamics of x, *corrected* by a term proportional to the error induced on the measured variable y, that is a dynamics of the form

$$\dot{\xi} = f(\xi) + g_1(\xi)w + g_2(\xi)u + G(\xi)(y - h_2(\xi) - k_{21}(\xi)w)$$

where $G(\xi)$ is a matrix to be determined.

Moreover, since the dynamics defined in this way requires the explicit knowledge of the disturbance w, which in general is not available for feedback, following again [12] we replace the actual value of the latter by the *worst* possible one, which, as seen in the previous section, has the expression

$$w = \alpha_1(x).$$

Changing x into ξ in the expressions of the feedback law and of the worst disturbance, this yields a dynamical feedback law of the form

$$\begin{aligned}\dot{\xi} &= f(\xi) + g_1(\xi)\alpha_1(\xi) + g_2(\xi)\alpha_2(\xi) + G(\xi)(y - h_2(\xi) - k_{21}(\xi)\alpha_1(\xi))\\ u &= \alpha_2(\xi).\end{aligned}$$

For the sake of notational simplicity, we set

$$\begin{aligned}\widetilde{f}(\xi) &= f(\xi) + g_1(\xi)\alpha_1(\xi)\\ \widetilde{h}_2(\xi) &= h_2(\xi) + k_{21}(\xi)\alpha_1(\xi),\end{aligned}$$

in which case the dynamic feedback law introduced above reduces to the simple form

$$\begin{aligned}\dot{\xi} &= \widetilde{f}(\xi) + g_2(\xi)\alpha_2(\xi) + G(\xi)(y - \widetilde{h}_2(\xi))\\ u &= \alpha_2(\xi).\end{aligned} \tag{8}$$

In what follows, we will show that, under suitable hypotheses, a control law of this type solves the problem of disturbance attenuation via measurement feedback. More precisely we will show that, if $G(\xi)$ is appropriately chosen, the corresponding closed loop system, which has the form

$$\begin{aligned}\dot{x} &= f(x) + g_1(x)w + g_2(x)\alpha_2(\xi)\\ \dot{\xi} &= \widetilde{f}(\xi) + g_2(\xi)\alpha_2(\xi) + G(\xi)(h_2(x) + k_{21}(x)w - \widetilde{h}_2(\xi))\\ z &= h_1(x) + k_{12}(x)\alpha_2(\xi),\end{aligned}$$

is asymptotically stable and has an L_2 gain which is less than or equal to γ.

To see how disturbance attenuation can be achieved, recall (see (6)) that the function $V(x)$ is such that the following inequality

$$\begin{aligned}&\frac{dV(x(t))}{dt} + \|z(t)\|^2 - \gamma^2\|w(t)\|^2\\ &\quad\le \|\alpha_2(\xi(t)) - \alpha_2(x(t))\|^2_{R_2(x(t))} - \gamma^2\|w - \alpha_1(x(t))\|^2\end{aligned}$$

holds along any trajectory $(x(\cdot), \xi(\cdot))$ of the closed loop system. In addition, suppose there exists a real-valued function $W(x,\xi)$, locally defined in a neighborhood of the origin in $\mathbb{R}^n \times \mathbb{R}^n$ and such that $W(x,\xi) > 0$ for $x \ne \xi$, satisfying the inequality

$$\frac{dW(x(t),\xi(t))}{dt} + \|\alpha_2(\xi(t)) - \alpha_2(x(t))\|^2_{R_2(x(t))} - \gamma^2\|w - \alpha_1(x(t))\|^2 \le 0 \tag{9}$$

along any trajectory $(x(\cdot), \xi(\cdot))$ of the closed loop system. If this is the case, then

$$\frac{dV(x(t))}{dt} + \frac{dW(x(t),\xi(t))}{dt} + \|z(t)\|^2 - \gamma^2\|w(t)\|^2 \le 0,$$

from which it is easily concluded that, if the closed loop system is asymptotically stable, its L_2 gain is less than or equal to γ. Deferring for a moment the issue of asymptotic stability, we observe that the inequality (9) can be enforced via a suitable Hamilton-Jacobi inequality for $W(x,\xi)$. In fact, it suffices to put the equations describing the dynamics of the closed loop system in the form

$$\dot{x}^e = f^e(x^e) + g^e(x^e)r \tag{10}$$

where $r = w - \alpha_1(x)$, $x^e = col(x,\xi)$,

$$f^e(x^e) = \begin{pmatrix} \widetilde{f}(x) + g_2(x)\alpha_2(\xi) \\ \widetilde{f}(\xi) + g_2(\xi)\alpha_2(\xi) + G(\xi)(\widetilde{h}_2(x) - \widetilde{h}_2(\xi)) \end{pmatrix},$$

$$g^e(x^e) = \begin{pmatrix} g_1(x) \\ G(\xi)k_{21}(x) \end{pmatrix},$$

to set

$$h^e(x^e) = \alpha_2(\xi) - \alpha_2(x),$$

to observe that the function

$$H^e(x^e,r) = W_{x^e}(f_e(x^e) + g_e(x^e)r) + \|h_e(x^e)\|^2_{R_2(x)} - \gamma^2\|r\|^2$$

is a polynomial of second degree in r and to impose that its *maximum*, achieved at

$$r = \frac{1}{2\gamma^2}W_{x^e}g_e^{\mathrm{T}}(x^e) \tag{11}$$

is non positive. This yields a Hamilton-Jacobi inequality for $W(x,\xi)$, which has the form

$$W_{x^e}f^e(x^e) + h^{e\mathrm{T}}(x^e)R_2(x)h^e(x^e) + \frac{1}{4\gamma^2}W_{x^e}g^e(x^e)g^{e\mathrm{T}}(x^e)W_{x^e}^{\mathrm{T}} \leq 0.$$

and the latter indeed implies (9).

On the other hand, the issue of stability can be handled in the following simple way. From the relations derived above, we see that the positive definite function

$$U(x,\xi) = V(x) + W(x,\xi)$$

satisfies

$$\frac{dU(x(t),\xi(t))}{dt} \leq -\|h_1(x) + k_{12}(x)\alpha_2(\xi)\|^2$$

along any trajectory $(x(\cdot),\xi(\cdot))$ of the closed loop system with $w(\cdot) = 0$. This proves that the equilibrium $(x,\xi) = (0,0)$ of the closed loop system is stable. To prove asymptotic stability we use again Assumption A2 and, in addition, we postulate that the output injection $-G(\xi)\widetilde{h}_2(\xi)$ stabilizes $\dot{x} = \widetilde{f}(\xi)$. This yields the result indicated in the following statement, which extends the main result of [12].

Lemma 1 *Consider the system (1) and suppose assumptions* A1 *and* A2 *are satisfied. Suppose there exists a smooth positive definite function $V(x)$, locally defined in a neighborhood of the origin in $I\!R^n$ and vanishing* *at $x = 0$, which satisfies the Hamilton-Jacobi-Isaacs inequality (4). Suppose there exists a smooth real-valued function $W(x^e)$, locally defined in a neighborhood of the origin in $\mathbb{R}^{2n}$, vanishing at $x^e = 0$ and such that $W(x^e) > 0$ for $x^e \neq col(x, x)$, and a $n \times p$ matrix of smooth functions $G(\xi)$, locally defined in a neighborhood of the origin in $\mathbb{R}^n$, such that*

(i) *the Hamilton-Jacobi inequality*

$$W_{x^e} f^e(x^e) + h^{eT}(x^e) R_2(x) h^e(x^e) + \frac{1}{4\gamma^2} W_{x^e} g^e(x^e) g^{eT}(x^e) W_{x^e}^T \leq 0 \quad (12)$$

holds, and

(ii) *the equilibrium $\xi = 0$ of the system*

$$\dot{\xi} = \widetilde{f}(\xi) - G(\xi)\widetilde{h}_2(\xi) \quad (13)$$

is locally asymptotically stable.

Then, the problem of local disturbance attenuation with internal stability is solved by the output feedback (8).

Proof. We only need to complete the proof that the closed loop system is locally asymptotically stable. To this end, observe that any trajectory satisfying

$$\frac{dU(x(t), \xi(t))}{dt} = 0$$

is necessarily a trajectory of

$$\dot{x} = f(x) + g_2(x)\alpha_2(\xi)$$

such that $x(t)$ is bounded and

$$h_1(x(t)) + k_{12}(x(t))\alpha_2(\xi(t)) = 0$$

for all $t \geq 0$. As a consequence, by hypothesis A2, $\lim_{t\to\infty} x(t) = 0$. Moreover, since the matrix $k_{12}(x)$ has linearly independed columns by hypothesis A1, it can also be concluded that $\lim_{t\to\infty} \alpha_2(\xi(t)) = 0$. Thus, the ω-limit set of such a trajectory is a subset of

$$M = \{(x, \xi) : x = 0, \alpha_2(\xi) = 0\}.$$

Any initial condition on this ω-limit set yields a trajectory in which

$$x(t) = 0$$

for all $t \geq 0$, while $\xi(t)$ is a trajectory of

$$\dot{\xi} = \widetilde{f}(\xi) - G(\xi)\widetilde{h}_2(\xi)$$

Thus, hypothesis (ii) implies $\lim_{t\to\infty}\xi(t) = 0$ and by the invariance principle we can conclude local asymptotic stability. ◃

A separation property

The Hamilton-Jacobi inequality (12) introduced in the previous subsection depends on the choice of gain matrix $G(\xi)$ which characterizes the dynamic feedback law (8). Thus, from a practical point of wiew, the approach to the solution of the problem of disturbance attenuation described in Lemma 1 has the disadvantage of requiring to solve the inequality (12) for $W(x^e)$ and $G(\xi)$ simultaneously. Another disadvantage of the approach presented in the previous Lemma is that the number of independent variables in the Hamilton-Jacobi inequality (12), associated with the augmented system (10), is twice as much as the number of independent variables in the Hamilton-Jacobi-Isaacs inequality (4). In this subsection, we address both these issues. In particular, we show that the existence of $W(x^e)$ and $G(\xi)$ satisfying the conditions (i) and (ii) of Lemma 1 is implied by the existence of a solution of a Hamilton-Jacobi-Isaacs inequality which has only n independent variables and does not involve the gain matrix $G(\xi)$.

Lemma 2 *Suppose $V(x)$ is a smooth solution of (4). Let $Q(x)$ be a positive definite smooth function, locally defined in a neighborhood of $x = 0$ and vanishing at $x = 0$, and set*

$$\begin{aligned} S(x) &= Q_x(\tilde{f}(x) - G(x)\tilde{h}_2(x)) + \alpha_2^{\mathrm{T}}(x)R_2(x)\alpha_2(x) \\ &+ \frac{1}{4\gamma^2}Q_x(g_1(x) - G(x)k_{21}(x))(g_1(x) - G(x)k_{21}(x))^{\mathrm{T}}Q_x^{\mathrm{T}}. \end{aligned} \tag{14}$$

Suppose $S(x) < 0$ for each nonzero x and its Hessian matrix is nonsingular at $x = 0$. Then the function

$$W(x^e) = Q(x - \xi)$$

satisfies conditions (i) *and* (ii) *of Lemma 1*

Proof. Set $W(x^e) = Q(x - \xi)$ and let $K(x, \xi)$ denote the function

$$K(x,\xi) = W_{x^e}f^e(x^e) + h^{e\mathrm{T}}(x^e)R_2(x)h^e(x^e) + \frac{1}{4\gamma^2}W_{x^e}g^e(x^e)g^{e\mathrm{T}}(x^e)W_{x^e}^{\mathrm{T}}.$$

To prove that $W(x^e) = Q(x - \xi)$ satisfies (i) is to prove that $K(x, \xi)$ is nonpositive. To this end, set

$$e = x - \xi$$

and

$$H(e,\xi) = [K(x,\xi)]_{x=\xi+e}.$$

Then, a simple calculation shows that

$$H(0,\xi)=0, \qquad \left[\frac{\partial H(e,\xi)}{\partial e}\right]_{e=0}=0,$$

so that $H(e,\xi)$ can be expressed as

$$H(e,\xi)=e^{\mathrm{T}}R(e,\xi)e$$

for some continuous matrix $R(e,\xi)$. Moreover, it is also easy to check that

$$R(0,0)=\left[\frac{\partial^2 S(x)}{\partial x^2}\right]_{x=0}.$$

Since the latter matrix is by assumption negative definite, the function $H(e,\xi)$ is nonpositive in a neighborhood of $(e,\xi)=(0,0)$ and (i) follows.

To prove condition (ii), observe that the hypothesis $S(x)<0$ implies

$$Q_x(\tilde{f}(x)-G(x)\tilde{h}_2(x))<0.$$

Thus, the positive definite function $Q(\xi)$ is a Lyapunov function for (13), and the equilibrium $\xi=0$ of the latter is locally asymptotically stable. $\triangleleft$

The new Hamilton-Jacobi inequality introduced in this Lemma has only n independent variables, but still needs to be solved simultaneously for a positive definite function $Q(x)$ and for the gain matrix $G(x)$. Fortunately, however, these two undeterminates can be separated, under the following additional hypothesis on the plant (1).

Assumption A3 The matrix

$$R_1(x)=k_{21}(x)k_{21}^{\mathrm{T}}(x)$$

is nonsingular (and thus positive definite) for each x.$\triangleleft$

In fact, a standard "completion of the squares" argument shows that the function $S(x)$ introduced in the previous Lemma is such that

$$\begin{aligned} S(x) \;\geq\; & Q_x(\tilde{f}(x)-g_1(x)k_{21}^{\mathrm{T}}(x)R_1^{-1}(x)\tilde{h}_2(x)) \\ + \; & \alpha_2^{\mathrm{T}}(x)R_2(x)\alpha_2(x)-\gamma^2\tilde{h}_2^{\mathrm{T}}(x)R_1^{-1}(x)\tilde{h}_2(x) \\ + \; & \frac{1}{4\gamma^2}Q_x g_1(x)[I-k_{21}^{\mathrm{T}}(x)R_1^{-1}(x)k_{21}(x)]g_1^{\mathrm{T}}(x)Q_x^{\mathrm{T}} \end{aligned} \tag{15}$$

with the equality holding if and only if

$$Q_xG(x)=(2\gamma^2\tilde{h}_2^{\mathrm{T}}(x)+Q_xg_1(x)k_{21}^{\mathrm{T}}(x))R_1^{-1}(x). \tag{16}$$

Thus, in order to have $S(x)<0$, it suffices to assume that the right-hand side of (15), which does not contain $G(\xi)$, is negative for each nonzero

x and then to choose $G(x)$ as indicated by (16). In this way one is led to consider a new Hamilton-Jacobi-Isaacs inequality for $Q(x)$ which, assumes the form

$$\begin{aligned} & Q_x(\widetilde{f}(x) - g_1(x)k_{21}^{\mathrm{T}}(x)R_1^{-1}(x)\widetilde{h}_2(x)) \\ + \quad & \alpha_2^{\mathrm{T}}(x)R_2(x)\alpha_2(x) - \gamma^2\widetilde{h}_2^{\mathrm{T}}(x)R_1^{-1}(x)\widetilde{h}_2(x) \\ + \quad & \frac{1}{4\gamma^2}Q_x g_1(x)[I - k_{21}^{\mathrm{T}}(x)R_1^{-1}(x)k_{21}(x)]g_1^{\mathrm{T}}(x)Q_x^{\mathrm{T}} < 0 \end{aligned} \tag{17}$$

Note also that a matrix $G(x)$, of smooth functions of x, which renders the identity (16) satisfied exists if and only if the function $Q(x)$ is such that

$$Q_x L(x) = \widetilde{h}_2^{\mathrm{T}}(x)$$

for some $n \times p$ matrix $L(x)$ of smooth functions of x, in which case $G(x)$ is indeed given by

$$G(x) = (2\gamma^2 L(x) + g_1(x)k_{21}^{\mathrm{T}}(x))R_1^{-1}(x). \tag{18}$$

We obtain in this way the result summarized in the following statement (see also [15]).

Theorem 2 *Consider the system (1) and suppose assumptions* A1, A2 *and* A3 *are satisfied.*

Suppose there exists a smooth positive definite function $V(x)$, locally defined in a neighborhood of $x = 0$ and vanishing at $x = 0$, which satisfies the inequality (4).

Suppose there exists a smoth positive definite function $Q(x)$, locally defined in a neighborhood of $x = 0$ and vanishing at $x = 0$, which satisfies the inequality (17), such that the Hessian matrix of the left-hand side of (17) is nonsingular and such that $Q_x L(x) = \widetilde{h}_2^{\mathrm{T}}(x)$ for some $n \times p$ matrix $L(x)$ of smooth functions of x.

Then, the problem of local disturbance attenuation with internal stability is solved by the output feedback (8), with $\alpha_1(\xi), \alpha_2(\xi)$ and $G(\xi)$ chosen as in (5) and (18).

Remark. The inequality (17) can be viewed as a "dual" Hamilton-Jacobi-Isaacs inequality, arising in a differential game associated with the determination of the output injection gain. In the case of a linear system, this inequality reduces to a Riccati inequality corresponding to the equation for Z_∞ considered in [1] (see also [6]). $\triangleleft$

Remark. Note that the fulfillment of the (strict) Hamilton-Jacobi-Isaacs inequality associated with the design of the output injection gain is required to be completely determined by its second order terms. This stronger characterization, which has no counterpart the design of the state feedback law, is not surprising at all. The situation, in fact, is somewhat similar to a situation already seen in the analysis of the problem of local stabilization, via smooth dynamic output feedback, of a system of the form

$$\begin{aligned}\dot{x} &= f(x) + g(x)u \\ y &= h(x),\end{aligned}$$

for which stabilizability via output feedback is implied by smooth stabilizability via state feedback and detectability in the first approximation. ◁

5 Conclusions

In this paper we have shown how the problem of disturbance attenuation via measurement feedback can be solved for an affine nonlinear system. The proposed solution relies upon the existence of positive definite smooth solutions of a pair of Hamilton-Jacobi inequalities, which are used to characterize a state feedback control and, respectively, an output injection gain.

The author wishes to thank T.Basar and A.Haurie for numerous interesting discussions.

REFERENCES

[1] J.C. Doyle, K. Glover, P.P. Khargonekar, and B.A. Francis, State space solutions to standard H_2 and H_∞ control problems, *IEEE Trans. Autom. Control*, **AC-34**:831–846, 1989.

[2] A.A. Stoorvogel, *The H_∞ control problem: a state space approach*, PhD thesis, Technical University Eindhoven, 1990.

[3] C. Scherer, *The Riccati inequality and state-space H_∞-optimal control*, PhD thesis, University of Würzburg, 1990.

[4] I.R. Petersen, B.D.O. Anderson and E.A. Jonckheere, A first principles solution to the nonsingular H_∞ control problem, *Int. J. of Robust and Nonlinear Control*, **1**:153–170, 1991.

[5] E.F. Mageirou and Y.C. Ho, Decentralized stabilization via game theoretic methods, *Automatica*, **13**:888–896, 1977.

[6] G. Tadmor, Worst case design in time domain, *Math. Control, Signals and Systems*, **3**:301–324, 1990.

[7] T. Basar and P. Bernhard, *H_∞-optimal control and related Minimax problems*, Birkhauser, 1990.

[8] J.A. Ball and J.W. Helton, H_∞ optimal control for nonlinear plants: connection with differential games, In *Proc. of 28th Conf. Decision and Control*, pages 956–962, Tampa, FL, December 1989.

[9] A.J. Van der Schaft, A state-space approach to nonlinear H_∞ control, *Syst. and Contr. Lett.*, **16**:1–8, 1991.

[10] A. Isidori, Feedback control of nonlinear systems, In *Proc. of 1st European Control Conf.*, pages 1001-1012, Grenoble, France, July 1991.

[11] A.J. Van der Schaft, L_2-gain analysis of nonlinear systems and nonlinear H_∞ control, *IEEE Trans. Autom. Control*, **AC-37**:770–784, 1992.

[12] A. Isidori and A. Astolfi, Nonlinear H_∞ control via measurement feedback, *J. Math. Systems, Estimation and Control*, **2**:31-44, 1992.

[13] A. Isidori and A. Astolfi, Disturbance attenuation and H_∞ control via measurement feedback in nonlinear systems, *IEEE Trans. Autom. Control*, **AC-37**:1283–1293, 1992.

[14] J. Ball, J.W. Helton, and M.L. Walker, A variational approach to nonlinear H_∞ control, Tech. memorandum, University of California at San Diego, August 1991.

[15] A. Isidori, Dissipation inequalities in nonlinear H_∞ control, *Proc. 31th Conf. Decis. Contr.*, pages 3265-3270, Tucson, AZ, 1992.

Alberto Isidori
Departement of Computer and Systems Science
Università di Roma "La Sapienza"
00184 Rome, Italy

and

Department of Systems Sciences and Mathematics,
Washington University
St. Louis, MO 63130, USA

Reentry Trajectory Optimization under Atmospheric Uncertainty as a Differential Game*

Michael H. Breitner and H. Joseph Pesch

Dedicated to Professor Dr. Dr. h. c. Roland Bulirsch
on the occasion of his 60th birthday.

Abstract. If solutions of optimal control problems are to be realized in practical applications, one has to take into account the influence of unpredictable disturbances. If lower and upper bounds for the disturbances are known, one can investigate the so-called worst case. This worst case can be formulated as a two-person differential game and its solution in closed form provides the optimal feedback controller against all possible disturbances. Following this approach, realistically modelled optimal control problems lead in general to nonseparable, non zero-sum differential games where inequality constraints have to be taken into account. A maximum cross-range reentry of a space-shuttle in the presence of uncertain air density fluctuations serves as an example. For the treatment of terminal conditions and path constraints in which controls of both players are involved, the responsibility of the players for obeying these constraints is investigated, and a new transformation technique is used. The open-loop representation of the optimal feedback controller is computed along various saddle-point trajectories representing the worst case. The computational method is based on the numerical solution of the multipoint boundary-value problem which arises from the necessary conditions for a saddle-point solution of the game. Extensive numerical results are presented for a space-shuttle reentry under a dynamic pressure and an aerodynamic heating constraint. It is outlined, that the proposed method allows not only the analysis of the worst case, but is also the first step towards the construction of the optimal feedback controller via a successive computation of the open-loop representation or via a neural network training.

Key Words. Constrained differential games, unknown disturbances, space-shuttle reentry, air density fluctuations, optimal control problems.

---

*This research was partly supported by the German National Science Foundation (Research Center on "Applied Optimization and Control") and by the Bavarian Science Foundation (Bavarian Consortium on High Performance Scientific Computing).

1 Introduction

The purpose of this paper is to develop the theoretical basis and a numerical method for the computation of the optimal feedback controller for the following rather general optimal control problem with unknown disturbances:
Let the dynamic system and the associated boundary conditions be given in the form

$$\dot{z}(t) = f(z(t), u(t), w(t)) \ , \tag{1}$$
$$z(t_0) = z_0 \ , \quad ||\psi(z(t_f))||_p \leq \varepsilon \tag{2}$$

where $z(t)$ denotes the state vector, $u(t)$ the control vector and $w(t)$ the unknown disturbance vector. The derivative with respect to time t, $t \in [t_0, t_f]$, is denoted by a dot. The initial time is t_0 and the final time t_f which can be specified or unspecified. It is assumed that the disturbance vector $w(t)$ lies within a known bounded set $\Omega(t, z(t))$ for all $t \in [t_0, t_f]$. The initial state is prescribed by the vector z_0, and the terminal conditions are given by an inequality constraint. In order to have a differentiable norm, we require p to be an even integer. The accuracy by which the terminal conditions are to be satisfied can be regulated by the parameter ε. Furthermore, constraints might occur in the form of so-called control variable inequality or mixed constraints,

$$C(z(t), u(t), w(t)) \leq 0 \ . \tag{3}$$

We now search for control functions and their associated solutions of the differential equations so that the boundary conditions as well as the inequality constraints are satisfied. If a higher accuracy for the terminal conditions than ε can be enforced against all possible disturbances w, a performance index

$$I[u] = \phi(t_f, z(t_f)) \tag{4}$$

to be minimized, can be additionally taken into account.

The optimal open-loop solution $\tilde{z}^*(t)$ of the related undisturbed optimal control problem is usually computed first. Next, the problem (1)–(4) is linearized around $\tilde{z}^*(t)$. Then, a feedback controller tracking $\tilde{z}^*(t)$ against the unkown disturbances must be constructed, for example by using Lyapunov theory [13], H^∞-control theory [1], or the theory of linear differential games [3]. On the other hand, the construction of a feedback controller using neighboring optimal trajectories is also possible; see [14] and [12]. However, all these feedback controllers are only suboptimal for the disturbed optimal control problem (1)–(4).

In the present paper, we show that the problem of constructing the optimal feedback controller for the disturbed optimal control problem, can be formulated as a two-person differential game. For this purpose, the theory of separable inequality-constrained zero-sum games [4] is generalized to nonseparable games. The application of the maximinimum principle allows the computation of the optimal open-loop strategies and yields a criterion to verify whether the game is zero-sum. In addition, a method for the treatment of terminal conditions and path constraints, in which even controls of both players are involved, is described. This method also includes an investigation of the responsibility of the players for those constraints. The computation of the optimal controls, or more precisely of the open-loop representations of the optimal strategies, is based on the numerical computation of a solution of the multipoint boundary-value problem which arises from the necessary conditions for optimality. By this method, the optimal feedback strategies against all possible disturbances are obtained along the so-called saddle-point trajectories. An extensive insight is gained into the structure of the optimal solution for the worst case. Moreover, a benchmark for the assessment of various suboptimal feedback controllers can be obtained.

A realistically modelled space-shuttle reentry problem where the cross-range is to be maximized and where a dynamic pressure and an aerodynamic heating constraint are taken into account, serves as a real-life example to apply the theory and the numerical method. The unpredictable fluctuations of the air density, for which lower and upper bounds are known, are the unknown disturbances. These disturbances are of great influence on the constraints and the performance index. The optimal feedback controller for the worst possible variations of the air density and the associated reentry trajectories are computed and presented for that space-shuttle reentry problem.

2 Optimal Control Problem for a Space-Shuttle Reentry

Assuming no planet rotation and oblateness, no winds and a stationary atmosphere, a point mass vehicle and a constant drag polar, the equations of motion for a space-shuttle-type vehicle in a flight path oriented coordinate system can be written as

$$\dot{h} = v \sin\gamma , \tag{5}$$

$$\dot{\Lambda} = \frac{v}{R+h} \cos\gamma \sin\chi , \tag{6}$$

$$\dot{\Theta} = \frac{v}{R+h} \frac{\cos\gamma \cos\chi}{\cos\Lambda} , \tag{7}$$

$$\dot{v} = -C_L^n \frac{S\, v^2 \rho}{2\, m} \sigma - C_{D_0} \frac{S\, v^2 \rho}{2m} \sigma - g \sin\gamma , \tag{8}$$

$$\dot{\chi} = C_L \sin\mu \frac{S\, v \rho}{2\, m \cos\gamma} \sigma - \frac{v}{R+h} \cos\gamma \cos\chi \tan\Lambda , \tag{9}$$

$$\dot{\gamma} = C_L \cos\mu \frac{S\,v\,\rho}{2\,m}\sigma + \left(\frac{v}{R+h} - \frac{g}{v}\right)\cos\gamma\,. \tag{10}$$

Here h denotes the altitude above the earth's surface, Λ the cross-range angle, Θ the down-range angle, v the velocity, χ the heading angle, and γ the flight path angle. Besides these six state variables there are two control variables, the lift coefficient C_L ($C_L \geq 0$) and the bank angle μ ($-90[\text{deg}] \leq \mu \leq 90[\text{deg}]$). Moreover, R is the earth's radius and $g(h)$ the gravity acceleration; S, m, C_{D_0}, and n are the aerodynamic reference area, the mass, the zero-drag coefficient, and the exponent of C_L in the induced drag of the shuttle, respectively. The air density $\rho\,\sigma$ consists of a prescribed air density model $\rho(h)$, e.g., $\rho = \rho_0 \exp(-\beta\,h)$, and a variable deviation factor σ to describe the density fluctuations. For more details of the space-shuttle model, see [12].

Table 1: Initial and final conditions.

	$t = 0$	$t = t_f$
h [km]	95.0	30.0
Λ [deg]	0.00	
Θ [deg]		3.0
v [km s^{-1}]	7.850	1.116
χ [deg]	0.00	
γ [deg]	-1.25	-2.70

The initial conditions at the reentry into the atmosphere and the final conditions at the unspecified terminal time t_f are given in Table 1 . The initial conditions correspond to a reentry out off the equatorial plane with 97% of the total energy for a stable circular orbit in an altitude of 242 km. The terminal conditions for h, v and γ enable the entry into the final quasi-steady glide for a landing at an appropriate landing site.

For a realistic description of a reentry maneuver, the limitation of the dynamic pressure and of the aerodynamic heating at a critical stagnation point is of utmost importance. Both state constraints are of the first order and are transformed into the single zeroth-order state constraint

$$C_L \leq C_{L,\max}(h, v\,\sqrt[6]{\sigma}; \vartheta_{\max})\;; \tag{11}$$

see [7]. The complicated function $C_{L,\max}$ can be found in [12]. The parameter $\vartheta_{\max}$ indicates the maximum permitted skin temperature at the critical stagnation point. For the given data, only the heating constraint becomes active. The maximum dynamic pressure $q_{\max} = 40\,000$ [N/m^2] is not reached in the problems investigated here; see Figs. 6 and 10. Therefore, the term $v\,\sqrt[6]{\sigma}$ occurs in Eq. (11), because the temperature is proportional to $v^3\,\sqrt{\rho\,\sigma}$.

Since the terminal conditions for h, v and γ can be met within a higher than the prescribed accuracy against all possible disturbances for technical reasons, it is possible to use the remaining degree of freedom to maximize the cross-range. This enables to reach landing sites in southern France. Thus, we have to minimize the performance index

$$I[C_L, \mu, \sigma] := -\Lambda(t_f) \ . \tag{12}$$

In order to define the preceding optimal control problem completely, σ must be known as a function of the time and/or the state variables. However, only lower and upper bounds depending on the altitude are available in reality,

$$\sigma_{\min}(h) \leq \sigma \leq \sigma_{\max}(h) \ . \tag{13}$$

Some values of ρ, $\sigma_{\min}$ and $\sigma_{\max}$ are given in Table 2. Note that the dependence of the air density on season, day time, and locus is assumed to be already considered in the density model $\rho(h)$. The density model used for the reentry problem is based on data of Refs. [18] and [16].

Table 2: Air density and bounds for the density deviation.

h [km]	ρ [g m^{-3}]	$\sigma_{\max}$ [1]	$\sigma_{\min}$ [1]
100	.000391	1.106	.894
90	.00347	1.133	.867
80	.0219	1.148	.852
70	.0974	1.148	.852
60	.343	1.133	.867
50	1.14	1.106	.894
40	4.33	1.078	.922
30	19.1	1.050	.950

3 Differential Game Approach for a Space-Shuttle Reentry under Uncertain Air Density

By means of the necessary conditions for optimality (see, e.g., [5]), a multipoint boundary-value problem is obtained that must be satisfied by the solution of the optimal control problem. This boundary-value problem can be solved by the multiple shooting method; see, e.g., [17] and [6]. Note that the numerical solution of the boundary-value problem is only a candidate for the optimal solution of the control problem. The numerical solution for the space-shuttle reentry problem of section 2 with $\sigma \equiv 1$ was first given in [8].

In this section we investigate first the optimal control problem with the control constraint (11) omitted. The optimal solution for the control unconstrained case is shown in Fig. 1 (thin line).

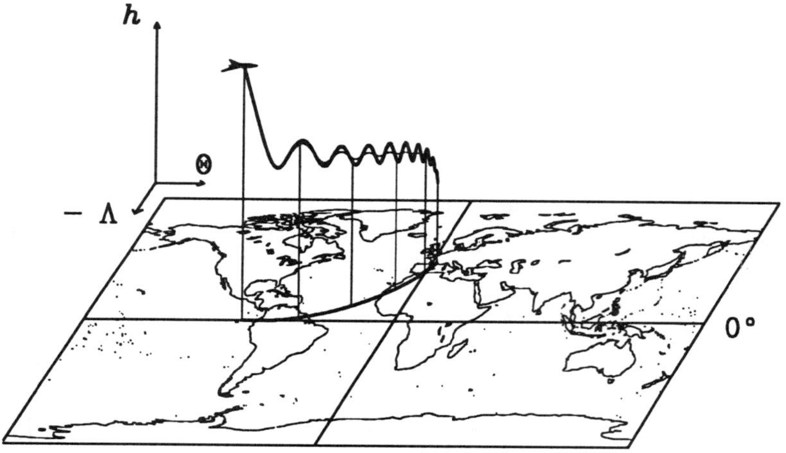

Figure 1: Optimal reentry trajectories with $\sigma \equiv 1$ (thin line) and $\sigma_{\min} \leq \sigma^* \leq \sigma_{\max}$ (thick line); control unconstrained cases.

In order to take into account fluctuations of the air density in the optimization process, the unknown disturbance factor σ is now considered as a control variable of an opposite player, say, the disturber D, in a two-person differential game. Here, complete information of the actual state variables and the dynamic system is available for both players, the disturber D and the optimizer O. Contrary to this, we assume that the disturber D has an information advantage with respect to the control variables, i. e. , he knows the actual value of the control variables choosen by the optimizer O. Recall, that we want to investigate the worst case here.

The optimizer O must be able to achieve the final conditions of Table 1 against all admissible disturbances at least approximately. This controllability proposition can be satisfied for the problem under investigation if, for example, the terminal conditions of Table 1 are replaced by the terminal inequality

$$F := \left\| \left(\frac{v(t_f) - v_f}{v_f}, \frac{\gamma(t_f) - \gamma_f}{\gamma_f}, \frac{h(t_f) - h_f}{h_f} \right)^{\mathsf{T}} \right\|_2 \leq \varepsilon \, . \tag{14}$$

We call the resulting problem the modified optimal control problem. If the optimizer O can enforce penetration into the terminal set defined by (14), the terminal time t_f of the game is reached when the optimizer O can no longer keep the state within this hyperset. Of course, this special situation is induced here by the monotonous increase of the cross-range angle Λ. We want to emphasize here that a primary objective of O is to reach the terminal set against all admissible disturbance strategies of D. We say, O is responsible for reaching the terminal set. The minimization of the performance index is only the secondary objective of the player O.

The first step for the treatment of the disturbed optimal control problem is a homotopy in the parameter ε from $\varepsilon = 0$ to, say, $\varepsilon = 0.01$. This value guarantees that the deviations from the prescribed terminal conditions given in Table 1 stay below 1%. The differential game is now mounted on this modified optimal control problem.

To compute the optimal strategies, we first consider the Hamiltonian of the problem,

$$H(z, \lambda, C_L, \mu, \sigma) := \lambda^\top f(z, C_L, \mu, \sigma) \tag{15}$$

with z denoting the state vector and f the right-hand sides of the dynamic equations (5)–(10). The vector λ denotes the gradient of the performance index I with respect to the state variables. For details see, e.g., [4]. The Hamiltonian can be rewritten in the form

$$H(z, \lambda, C_L, \mu, \sigma) = C_1(z)\, A(z, \lambda, C_L, \mu)\, \sigma + C_2(z, \lambda)\ . \tag{16}$$

Note that this Hamiltonian coincides with the Hamiltonian of the modified optimal control problem for $\sigma \equiv 1$. Since the present differential game is nonseparable, the optimal open-loop strategies for the two players can be computed by Isaacs' maximinimum principle (see [10] and [11]). To calculate the optimal strategies C_L^*, μ^* and σ^*, we first compute

$$\tilde{\sigma}(z, \lambda, C_L, \mu) := \arg \max_\sigma H(z, \lambda, C_L, \mu, \sigma)\ , \tag{17}$$

$$(C_L^*(z, \lambda), \mu^*(z, \lambda)) := \arg \min_{C_L,\, \mu} H(z, \lambda, C_L, \mu, \tilde{\sigma}(z, \lambda, C_L, \mu))\ , \tag{18}$$

and then

$$\sigma^*(z, \lambda) := \arg \max_\sigma H(z, \lambda, C_L^*(z, \lambda), \mu^*(z, \lambda), \sigma)\ . \tag{19}$$

Following this procedure, we obtain for the optimal open-loop strategies,

$$\sin \mu^* = -\frac{\lambda_\chi / \cos\gamma}{\sqrt{(\lambda_\chi / \cos\gamma)^2 + \lambda_\gamma^2}}\ , \tag{20}$$

$$\cos \mu^* = -\frac{\lambda_\gamma}{\sqrt{(\lambda_\chi / \cos\gamma)^2 + \lambda_\gamma^2}}\ , \tag{21}$$

$$C_L^* = \left(-\frac{\sqrt{(\lambda_\chi / \cos\gamma)^2 + \lambda_\gamma^2}}{v\,\lambda_v\,n} \right)^{\dfrac{1}{n-1}} , \tag{22}$$

$$\sigma^* = \begin{cases} \sigma_{\min} & \text{if} \quad A < 0 , \\ \text{undefined} & \text{if} \quad A = 0 , \\ \sigma_{\max} & \text{if} \quad A > 0 . \end{cases} \tag{23}$$

Since both players optimize the same performance index, the minimax condition

$$\min_{C_L,\,\mu}\ \max_{\sigma} H(z, \lambda, C_L, \mu, \sigma) = \max_{\sigma}\ \min_{C_L,\,\mu} H(z, \lambda, C_L, \mu, \sigma) = 0 , \tag{24}$$

if holding for all z and λ in a neighborhood of an optimal trajectory, is necessary and sufficient for the game to be zero-sum. This condition is always satisfied for the reentry problem in discussion. Moreover, the optimal open-loop strategies C_L^* and μ^* coincide here with the optimal control variables of the modified optimal control problem. A formulation of the game with an information advantage for O, i. e. with $\min\max H$ instead of $\max\min H$, leads to the same open-loop strategies for O and D. The calculations in this section are based on the assumption, which is reasonable, that σ^* is of bang-bang type. The case of singular subarcs, along which there holds $A \equiv 0$, is not considered here.

Since the performance index I is known on the terminal hypermanifold defined by the equality sign in (14), the terminal conditions for λ are obtained by differentiating the performance index with respect to the state variables. Hence, we have the following nonlinear terminal conditions for the adjoint variables,

$$H(z, \lambda, C_L^*, \mu^*, \sigma^*)\big|_{t=t_f} = 0 , \tag{25}$$

$$\lambda_\Lambda = -1 , \tag{26}$$

$$\lambda_v = \lambda_h \frac{h_f^2}{h(t_f) - h_f} \frac{v(t_f) - v_f}{v_f^2} , \tag{27}$$

$$\lambda_\chi = 0 , \tag{28}$$

$$\lambda_\gamma = \lambda_h \frac{h_f^2}{h(t_f) - h_f} \frac{\gamma(t_f) - \gamma_f}{\gamma_f^2} . \tag{29}$$

The boundary-value problem is completed by the differential equation for λ,

$$\dot{\lambda} = -\frac{\partial}{\partial z} H(z, \lambda, C_L^*(z, \lambda), \mu^*(z, \lambda), \sigma^*(z, \lambda)) . \tag{30}$$

For details see, e.g., Refs. [2] and [4]. Note that the differential equation for the down-range angle is decoupled.

For every saddle-point trajectory $z^*(t)$ related to the secondary objective minimization of I the compliance with the primary objectives of the optimizer O has to be verified a posteriori. In detail, for all $z \in z^*(t)$ the precondition

$$\max_{\sigma} \min_{C_L,\mu} \left(\min_{t\in[0,\infty[} F(t) \right) \leq \varepsilon \tag{31}$$

has to be checked with an auxiliary two-person differential game. However, in the reentry problem in consideration the precondition (31) is satisfied for technical reasons.

The optimal solution of the differential game can now be interpreted as the optimal solution of the modified optimal control problem if a most unfavourable course of the air density is assumed. Furthermore, the close relationship between these two problems enables the following procedure. Since $C_1 > 0$, the switching function A in Eq. (16) along the optimal solution of the modified control problem indicates the bang-bang switching structure of the optimal open-loop strategy σ^* for the disturber D to be expected if the bounds $\sigma_{\min}$ and $\sigma_{\max}$ are slightly freed from one. Therefore, the numerical solution of the multipoint boundary-value problem associated with the modified optimal control problem can serve as an appropriate starting trajectory for the homotopy in the two parameters $\sigma_{\min}$ and $\sigma_{\max}$. After the first homotopy step, the solution of the multipoint boundary-value problem, now associated with the differential game, is a candidate for the saddle-point trajectory of the game. At the end of the homotopy, $\sigma_{\min}(h)$ and $\sigma_{\max}(h)$ reach the functions the values of which are given in Table 2. The numerical solutions of the various multipoint boundary-value problems have been obtained by the multiple shooting algorithm given in [9].

The history of the function $C_1 A$ is shown in Fig. 2. The magnitude of the function $C_1 A$ is a measure for the influence of the disturbance on the performance index. The optimal open-loop strategies σ^*, C_L^*, and μ^* are shown in Figs. 3–5. In Fig. 1, the reentry trajectories for both the modified optimal control problem (thin line) and the disturbed optimal control problem (thick line) are compared. The flight oscillations are typical for maximum-range flights of gliders (pumping). The same trajectories are shown in an altitude-velocity diagram in Fig. 6. This diagram indicates that the dynamic pressure constraint is nowhere violated. In addition, the loss of the initial total energy E_0 during the reentry can be seen. The skin temperature ϑ due to the aerodynamic heating for both problems is given in Fig. 7. The discontinuities are caused by the discontinuities in σ^*; compare Eq. (11). Note that the maximum temperature of about 1522 [°C] occurs at the first peak in the undisturbed problem, but of the magnitude of about 1701 [°C] at the fourth peak in the disturbed problem. In the next section the maximum temperature $\vartheta_{\max}$ is decreased to 1400 [°C].

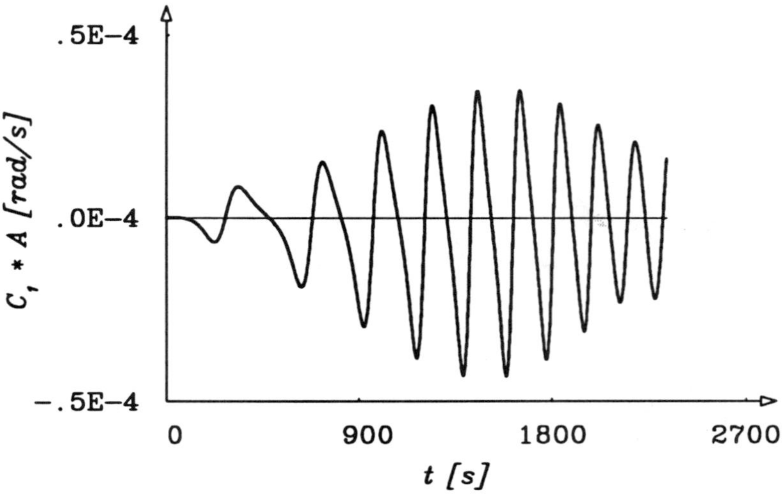

Figure 2: Switching function for σ^*; disturbed unconstrained optimal control problem.

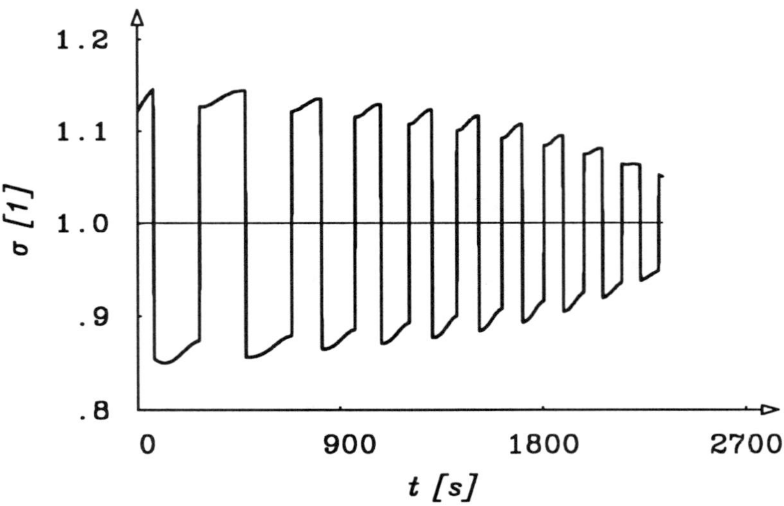

Figure 3: Optimal open-loop strategy σ^*; disturbed unconstrained optimal control problem.

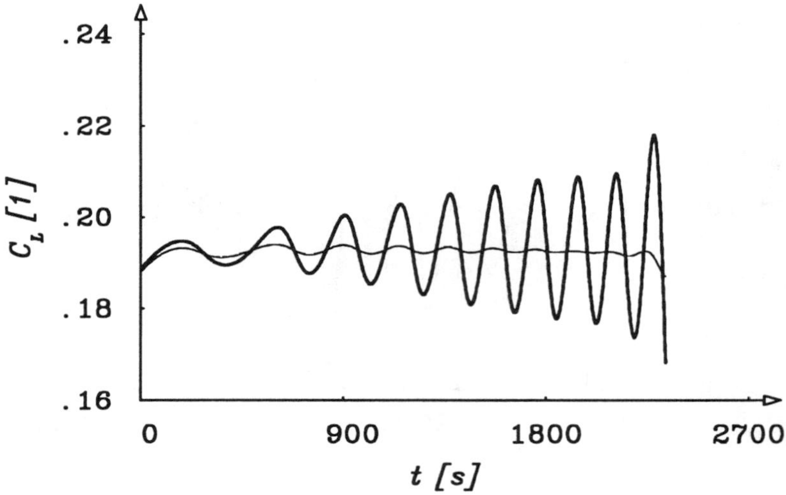

Figure 4: Optimal open-loop strategy C_L^* (thick line) and optimal control C_L^* (thin line); unconstrained problems.

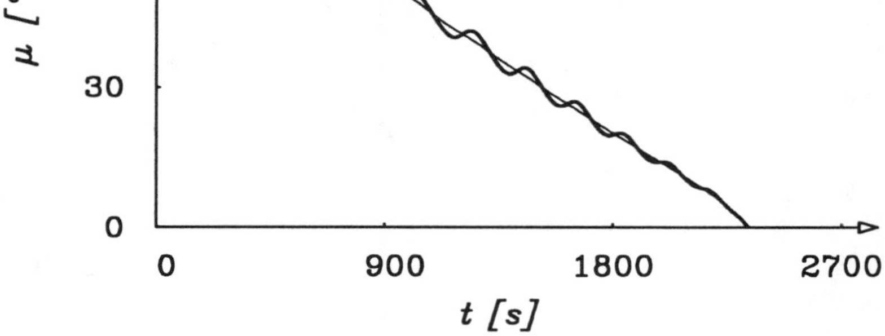

Figure 5: Optimal open-loop strategy μ^* (thick line) and optimal control μ^* (thin line); unconstrained problems.

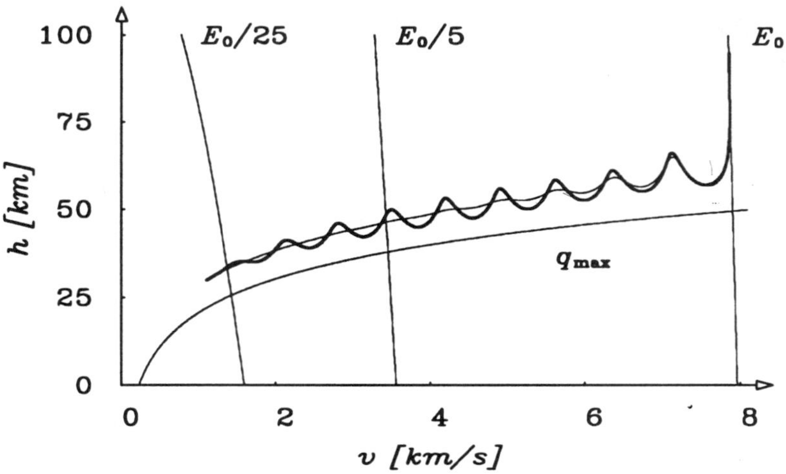

Figure 6: Optimal trajectories for the disturbed optimal control problem (thick line) and for the modified optimal control problem (thin line); unconstrained cases.

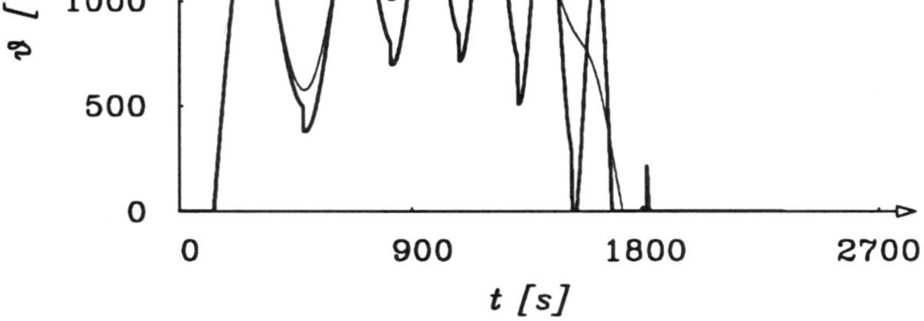

Figure 7: Temperature ϑ for the disturbed optimal control problem (thick line) and the modified optimal control problem (thin line); unconstrained cases.

4 Differential Game for the Temperature-Constrained Disturbed Optimal Control Problem

The compliance with the temperature constraint (11) is now the second primary objective of the optimizer O, i.e., O is also responsible for the strict observance of this constraint. Therefore, we must transform the inequality constraint (11) such that this constraint can be satisfied at any time against all admissible air density fluctuations σ. Hence, there always must hold

$$C_L \leq \min_{\sigma} C_{L,\max}(h, v\ \sqrt[6]{\sigma}; \vartheta_{\max}) \ . \tag{32}$$

Because of the special function $C_{L,\max}$, the minimum is always taken at $\sigma = \sigma_{\max}$. This modified constraint is called active if there exists an admissible disturbance σ such that the equality sign holds in Eq. (32). In this case, the subarcs are called constrained and the optimizer O must choose the optimal open-loop strategy C_L^* as

$$C_L^* = C_{L,\max}(h, v\ \sqrt[6]{\sigma_{\max}}; \vartheta_{\max}) \ . \tag{33}$$

Note that, on constrained subarcs, the term H_{C_L} no longer vanishes in the differential equations for the adjoint variables (30).

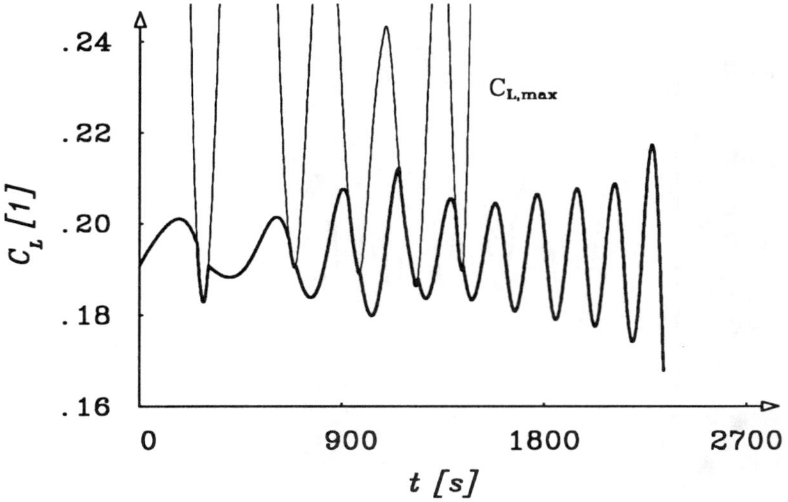

Figure 8: Optimal open-loop strategy C_L^* (thick line) and $C_{L,\max}$ for $\vartheta_{\max} = 1400$ [°C] (thin line); constrained case.

By a second homotopy, the temperature constraint is now introduced and tightened step by step until $\vartheta_{\max} = 1400$ [°C] is reached. As the first result, the history of the optimal lift coefficient is shown in Fig. 8. Five constrained subarcs occur.

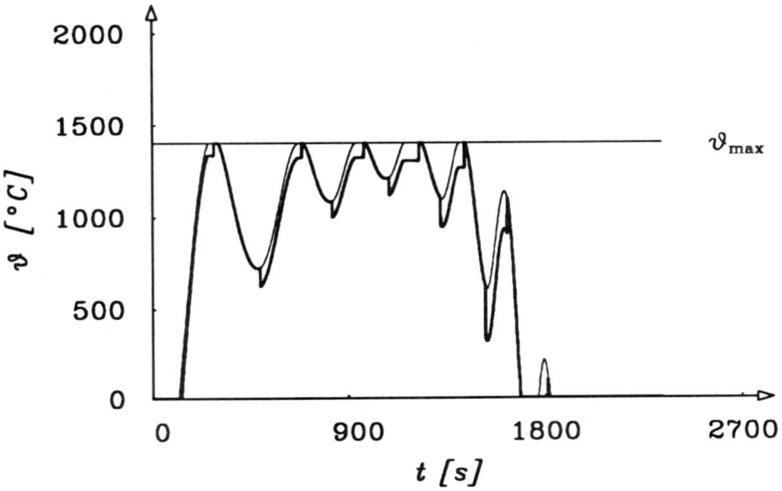

Figure 9: Temperature ϑ for the disturbed optimal control problem (thick line) and temperature for $\sigma \equiv \sigma_{\max}$ (thin line); constrained case.

The history of the temperature is shown in Fig. 9. The thin line indicates the history of the temperature which is the worst possible one for the optimizer O. Note that, along the constrained subarcs, i.e., where the transformed inequality (32) is active, the temperature ϑ is not everywhere equal to $\vartheta_{\max}$. The discontinuities in ϑ result from the bang-bang structure of the optimal open-loop strategy σ^*. Finally, the optimal trajectories for the disturbed optimal control problems are depicted in Fig. 10.

5 Conclusions

Since the worst case of the air density fluctuations is extremly unlikely, the actual trajectory will deviate from the precalculated saddle-point trajectory immediately after the start of the reentry maneuver. Nevertheless it is not only possible to analyze the worst case by the method presented in this paper. We can benefit from this method in two more ways. At first, many saddle-point trajectories filling up the relevant part of the state space can be computed. They can serve as a basis for the training of a neural network to approximate the optimal feedback controller. Secondly neighboring boundary-value problems can be solved successively. Their solutions are connected with the saddle-point trajectories for the remaining flight and enable tests with realistic air density distributions. These tests yield benchmarks for suboptimal feedback controllers. Moreover, the method is expected to become capable of real-time applications if one takes advantage of the inherent parallel structure of the multiple shooting method. For deterministic optimal control problems, first suggestions for a real-time-capable multiple shooting method have been already developed; see [15].

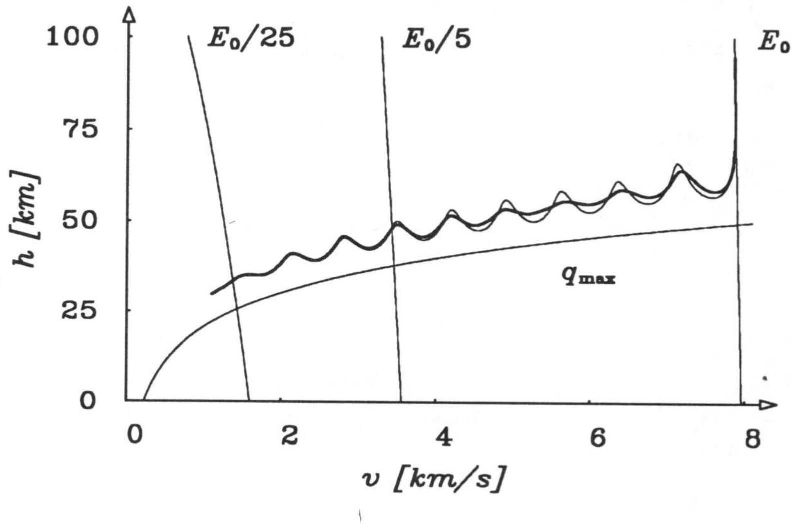

Figure 10: Optimal trajectories for the disturbed optimal control problems: constrained case (thick line) and unconstrained case (thin line).

Acknowledgement

We are greatly indebted to Prof. Dr. Josef Shinar from the Technion, Israel Institute of Technology, for fruitful discussions and helpful comments.

Address

Michael H. Breitner and Dr. H. Josef Pesch

Technische Universität München
Mathematisches Institut

Postfach 20 24 20

D–8000 München 2, Germany

E-mail: Michael.Breitner@Mathematik.TU-Muenchen.dbp.de

REFERENCES

[1] Başar, T. and Bernhard, P.: *H^∞-Optimal Control and Related Minimax Design Problems*, Birkhäuser, Boston, Massachusetts, 1991.

[2] Başar, T. and Olsder, G. J.: *Dynamic Noncooperative Game Theory*, Academic Press, London, Great Britain, 1982.

[3] Botkin, N. D., Kein, V. M., Patsko, V. S., and Turova, V. L.: *Aircraft Landing Control in the Presence of Windshear*, Problems of Control and Information Theory, Vol. 18, pp. 223–235, 1989.

[4] Breitner, M. H., Pesch, H. J., and Grimm, W.: *Complex Differential Games of Pursuit-Evasion Type with State Constraints, Part 1: Necessary Conditions for Optimal Open-Loop Strategies, Part 2: Numerical Computation of Optimal Open-Loop Strategies*, to appear in Journal of Optimization Theory and Applications, Vol. 78, No. 3, 1993.

[5] Bryson, A. E. and Ho, Y. C.: *Applied Optimal Control*, Hemisphere, New York, New York, 1975.

[6] Bulirsch, R., Montrone, F., and Pesch, H. J.: *Abort Landing in the Presence of a Windshear as a Minimax Optimal Control Problem, Part 1: Necessary Conditions, Part 2: Multiple Shooting and Homotopy*, Journal of Optimization Theory and Applications, Vol. 70, pp. 1–23, 223–254, 1991.

[7] Dickmanns, E. D.: *Maximum Range Three-Dimensional Lifting Planetary Entry*, NASA, Marshall Space Flight Center, Alabama, Technical Report No. R-M-199, 1972.

[8] Dickmanns, E. D. and Pesch, H. J.: *Influence of a Reradiative Heating Constraint on Lifting Entry Trajectories for Maximum Lateral Range*, Proceedings of the 11th International Symposium on Space Technology and Science, Tokyo, Japan, pp. 241–246, 1975.

[9] Hiltmann, P.: *Numerische Lösung von Mehrpunkt-Randwertproblemen und Aufgaben der optimalen Steuerung mit Steuerfunktionen über endlichdimensionalen Räumen*, Doctoral Thesis, Mathematisches Institut, Technische Universität München, München, Germany, 1990.

[10] Isaacs, R.: *Games of Pursuit*, Paper No. P-257, RAND Corporation, Santa Monica, California, 1951.

[11] Isaacs, R.: *Differential Games*, 3rd Printing, Krieger, New York, New York, 1975.

[12] Kugelmann, B. and Pesch, H. J.: *A New General Guidance Method in Constrained Optimal Control, Part 1: The Numerical Method, Part*

2: Application to Space Shuttle Guidance, Journal of Optimization Theory and Applications, Vol. 67, pp. 421–435, 437–446, 1990.

[13] Leitmann, G.: *Deterministic Control of Uncertain Systems Via a Constructive Use of Lyapunov Stability Theory*, Lecture Notes in Control and Information Sciences 143, Springer, Berlin, Germany, 1989.

[14] Pesch, H. J.: *Real-time Computation of Feedback Controls for Constrained Optimal Control Problems, Part 1: Neighbouring Extremals, Part 2: A Correction Method Based on Multiple Shooting*, Optimal Control Applications and Methods, Vol. 10, pp. 129–145, 147–171, 1989.

[15] Pesch, H. J.: *Optimal and Nearly Optimal Guidance by Multiple Shooting*, Mécanique Spatiale - Space Dynamics, Proceedings of the International Symposium, Toulouse, France, 1989, Edited by Centre National d'Etudes Spatiales, Cepadues Editions, Toulouse, France, pp. 761–771, 1990.

[16] Pichler, H.: *Dynamik der Atmosphäre*, Bibliographisches Institut, Mannheim, Germany, 1986.

[17] Stoer, J. and Bulirsch, R.: *Introduction to Numerical Analysis*, Springer, 2nd edition, New York, New York, 1993.

[18] Valley, L.: *Handbook of Geophysics and Space Environments*, McGraw-Hill, New York, New York, 1983.

Michael H. Breitner and H. Josef Pesch
Technische Universität München
Mathematisches Institut
Postfach 20 24 20
D-8000 München 2, Germany

PART II

Zero-sum differential games: Pursuit-evasion games and numerical schemes

Fully Discrete Schemes for the Value Function of Pursuit-Evasion Games

M. Bardi, M. Falcone and P. Soravia

Abstract. We consider the classical pursuit-evasion problem and an approximation scheme based on Dynamic Programming. We prove the convergence of the scheme to the value function of the game by using some recent results and methods of the theory of viscosity solutions to the Isaacs equations. The more restrictive assumption is the continuity of the value function, but we can eliminate it when dealing with control problems with a single player. We test the algorithm on two simple examples with explicit solution.

0. Introduction

We are given a dynamical system controlled by two players

$$\dot{y} = f(y, a, b) , \qquad t > 0 , \tag{0.1}$$

where $y(t) \in \mathbb{R}^M$, $a(t) \in A$, $b(t) \in B$, A and B are given compact sets, and f is at least continuous. We are also given a closed target set $\mathcal{T} \subset \mathbb{R}^M$, and we consider the generalized pursuit-evasion game where the first player "a" seeks to minimize the time taken by the system to reach $\mathcal{T}$, while the second player "b" wants to maximize it. Let $T(x)$ be the lower value function in the sense of Varaiya, Roxin, Elliott and Kalton [EK] of this game, where x is the position of the system at initial time $t = 0$. In our previous papers [BS1, BS2] we have shown that the exponential (Kružkov) transform of T, that is

$$v(x) \equiv \begin{cases} 1 - e^{-T(x)} & \text{if } T(x) < +\infty , \\ 1 & \text{if } T(x) = +\infty , \end{cases} \tag{0.2}$$

is a viscosity solution of the Hamilton-Jacobi-Isaacs equation

$$v(x) + \min_{b \in B} \max_{a \in A} \{-f(x, a, b) \cdot Dv(x) - 1\} = 0 \quad \text{in } \mathbb{R}^M \backslash \mathcal{T} \equiv \mathcal{T}^c , \tag{0.3}$$

and that, if v is continuous, it is the unique bounded viscosity solution of (0.3) satisfying the natural Dirichlet boundary condition

$$v(x) = 0 \qquad \text{for } x \in \partial\mathcal{T} . \tag{0.4}$$

We also gave sufficient conditions for the continuity of T and v, and showed that under the Isaacs' condition (the min and max in (0.3) commute) T coincides with the upper value and therefore it is the value of the game. In [BS3] and [S] we proved the coincidence of this value with Fleming's and Friedman's values respectively.

In this paper we give a discrete approximation scheme for the boundary value problem (0.3), (0.4) and prove the uniform convergence of approximate solutions to $v(x)$ as the discretization steps go to 0. Our scheme is an adaptation to games of the one of Falcone [F] for infinite horizon control problems and of the scheme for the minimum time problem studied in [BF2]. The computation of an approximate solution is got by finding a fixed point of a monotone contraction in $[0,1]^N$, where N is the number of nodes of the mesh. The cost of the fixed point algorithm is drastically reduced by an acceleration technique based on the monotonicity property of the contraction operator (see [F] and [CDF] for details). Our fully discrete version of (0.3), (0.4) (i.e. (HJD) in the next section) is obtained by discretizing in space the Isaacs equation for the discrete-time game we studied in [BS3]. Therefore it has a game theoretic interpretation and it gives approximate feedback controls on the nodes (but we do not try to prove any precise convergence result for them here). A similar discrete-time discrete-space game was studied by Laporte, Nicolas and Bernhard [LNB].

We remark that our scheme is designed to fit also problems with constraints on the state of the system $y(t)$, but here we do not give a convergence result covering this situation as well.

Our convergence result is new even for control problems with a single player and in this case we can improve it to treat problems with discontinuous value function v.

After the foundation of the theory of viscosity solution, the approximation of Hamilton-Jacobi equations was studied by Capuzzo Dolcetta [CD], Crandall–Lions [CL], Souganidis [So], Capuzzo Dolcetta–Ishii [CDI], Gonzalez–Rofman [GR] and the authors [F], [BF1], [FF] (see also the survey paper of Capuzzo Dolcetta and Falcone [CDF]).

Some recent papers dealing with explicit applications to differential games, besides [BS3], are Barles–Souganidis [BSo] on stochastic games, Gonzalez and Tidball [GT], [T] on deterministic games with stopping times (see also the references therein), and Alziary [A2] on the classical problem of a lion pursuing a slower antelope in a bounded and closed domain.

The paper is organized as follows. Section 1 describes the discrete problem, two convergence theorems, and outlines the methods used for the proofs. Section 2 contains the more technical proofs. In Section 3 we present some numerical results for simple 1-dimensional examples.

1. Algorithm and Convergence Results

First of all we describe the discrete version of the boundary value problem (0.3), (0.4). The idea is to build first a discrete approximation on a polyhedron adding appropriate discrete boundary conditions and then extend the solution over $\mathbb{R}^M$ by linear interpolation.

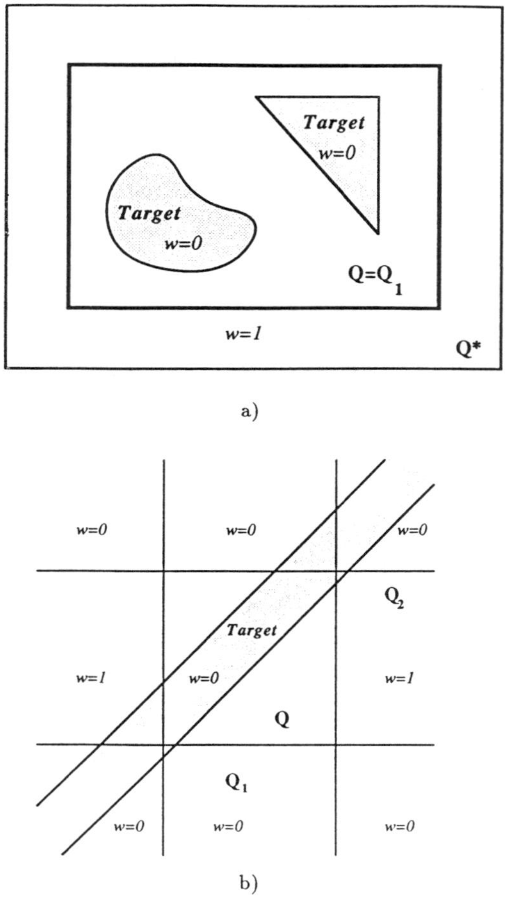

Figure 1. The role of Q_1 and Q_2; (a) bounded target, (b) unbounded target.

We call a polyhedron P *"discretized with step k"* if we are given a finite family of simplices $\{S_j\}$ such that $P = \cup_j S_j$, $int(S_i) \cap int(S_j) = \emptyset$ for $i \neq j$, $\max_j diam(S_j) = k$. We denote by x_i the vertices of the simplices of the tringulation and we write each $x \in P$ in a unique way as a convex combination of the vertices of a simplex S_j containing x, that is

$$x = \sum_{i=1}^{N} \lambda_i x_i \quad \text{with } \lambda_i \geq 0, \quad \sum_{i=1}^{N} \lambda_i = 1, \quad \lambda_i = 0 \ \text{ if } x_i \notin S_j.$$

Let us fix now a discretized polyhedron Q (this will be the one on which we want to determine the approximate solution). For $h > 0$ fixed we define a new polyhedron

$$Q^* \equiv \{x \in \mathbb{R}^M : \|x - y\|_\infty \leq hM_f, \quad y \in Q\}$$

where $\|\cdot\|_\infty$ denotes the maximum norm in $\mathbb{R}^M$ and M_f is such that $|f(x_i, a, b)| \leq M_f$ for all $x_i \in Q$ and all a, b. We discretize Q^* by adding N^* nodes to the previous triangulation of Q and we observe that, by definition, the points $z_i(a,b) \equiv x_i + hf(x_i, a, b) \in Q^*$ for all $x_i \in Q$ and all a, b. Let the nodes of the triangulation be numbered in a way such that $x_i \in Q$ for $i \in I \equiv \{1, \ldots, N\}$ and $x_i \in Q^* \backslash Q$ for $i = N+1, \ldots, N+N^*$. We fix also two sets Q_1 and Q_2 such that $Q = Q_1 \cap Q_2$ (see the comments after (HJD) about the role of these sets in the discretization). We define the map $F : \mathbb{R}^{N+N^*} \to \mathbb{R}^{N+N^*}$ componentwise as follows:

$$F_i(V) \equiv \begin{cases} \gamma \max\limits_{b} \min\limits_{a} P_i(a,b,V) + 1 - \gamma & \text{if } i \in I_{in}, \\ 1 & \text{if } i \in I_{out2}, \\ 0 & \text{if } i \in I_{\mathcal{T}} \cup I_{out1}, \end{cases}$$

where $\gamma \equiv e^{-h}$, $P_i(a,b,V) \equiv \sum_{j=1}^{N+N^*} \lambda_{ij}(a,b) V_j$ and $z_i(a,b) \equiv x_i + hf(x_i, a, b) \in Q^*$

$$\sum_{j=1}^{N+N^*} \lambda_{ij}(a,b) x_j = z_i(a,b) \ , \quad \lambda_{ij} \in [0,1], \quad \sum_{j=1}^{N+N^*} \lambda_{ij} = 1 \ ,$$

$$\begin{aligned} &I_{in} \equiv \{i : x_i \in Q \backslash \mathcal{T}\} \ , \quad I_{\mathcal{T}} \equiv \{i : x_i \in \mathcal{T} \cap Q\} \ , \\ &I_{out1} \equiv \{i : x_i \notin Q_2\} \ , \quad I_{out2} \equiv \{i : x_i \in Q_2 \backslash Q\} \ , \\ &Q = Q_1 \cap Q_2 \ . \end{aligned}$$

Notice that, by definition, the last N^* components of the map F are fixed so that we can identify F and its restriction to $\mathbb{R}^N$ at least on the set $\mathcal{V} \equiv \{V \in \mathbb{R}^{N+N^*} : V_i = 1 \text{ if } i \in I_{out2} \text{ and } V_i = 0 \text{ if } i \in I_{\mathcal{T}} \cup I_{out1}\}$. This is why the definitions of F_i for $i \in I_{\mathcal{T}} \cup I_{out1}$ and $i \in I_{out2}$ should be interpreted as discrete boundary conditions.

Proposition. (i) $F : [0,1]^N \to [0,1]^N$; (ii) *F is monotone with respect to the partial order: $U \geq V$ if $U_i \geq V_i$ for all i*; (iii) *F is a contraction in $[0,1]^N$ with respect to the norm $\|V\|_\infty = \max_{i=1,\ldots,N} |V_i|$.*

We defer the proof of the Proposition to Section 2. The unique fixed point of F is our approximate solution on the nodes of the grid belonging to Q. To have an approximate solution defined everywhere we can extend the triangulation over $\mathbb{R}^M$ and define by linear interpolation an affine function on each simplex, that is, we define $w : \mathbb{R}^M \to [0,1]$ as the unique function such that

$$(HJD)\begin{cases} w(x) = \sum_j \lambda_j w(x_j) & \text{if } x = \sum_j \lambda_j x_j \ , \\ w(x_i) = \gamma \max\limits_b \min\limits_a w(x_i & \\ +hf(x_i,a,b)) + 1 - \gamma & \text{if } x_i \in Q\backslash \mathcal{T} \ , \\ w(x_i) = 1 & \text{if } x_i \in Q_2\backslash Q \\ w(x_i) = 0 & \text{if } x_i \in (\mathcal{T} \cap Q) \cup (\mathbb{R}^M \backslash Q_2) \ . \end{cases}$$

To understand better the discrete version (HJD) of the boundary value problem (0.3), (0.4), the reader should first give a look at our paper [BS3], where we replaced the continuous-time system (0.1) by a discrete-time system (with time step $h > 0$) and we studied the associated discrete Isaacs equation

$$(HJ_h)\begin{cases} v_h(x) = \gamma \max\limits_b \min\limits_a v_h(x + hf(x,a,b)) + 1 - \gamma & \text{in } x_i \in \mathbb{R}^M \backslash \mathcal{T} \ , \\ v_h(x) = 0 & \text{on } \mathcal{T} \end{cases}$$

In (HJD) we are making two further approximation steps with respect to (HJ$_h$): *(i)* the discretization of the state variable x by restricting ourselves to the nodes x_i of the triangulation, *(ii)* the cut-off of the possibly unbounded set $\mathbb{R}^N \setminus \mathcal{T}$ by restricting ourselves to the compact set Q. To motivate the introduction of two additional sets, Q_1 and Q_2, we remark that if $Q_2 = \mathbb{R}^M$, $Q_1 = Q$, the condition $w(x_i) = 1$ off Q_1 means that the first player loses the game if the system exits Q before reaching the target (see Figure 1a). This is a natural approximation for the case of the single player [BF2] or for bounded targets, but it is not realistic for true pursuit-evasion games. In the classical problem where the first player (the "pursuer") wants to catch the second player (the "evader") in minimum time, we have $M = 2L$, the pursuer controls the first L state variables, the evader controls the other variables and the target is either the "diagonal" (i.e. $\{(x,x) : x \in \mathbb{R}^L\}$) or a neighbourood of the diagonal. In this case, it is natural to take $Q_1 = K \times \mathbb{R}^L$, $Q_2 = \mathbb{R}^L \times K$, where K is a compact set in $\mathbb{R}^L$, so each player loses the game if his state variables exit K first (see Figure 1b).

To study the consistency of our scheme we need the

Definition. A sequence (Q^n, h_n) of discretized polyhedrons Q^n and real numbers $h_n > 0$, is "admissible" if $Q^{n+1} \supseteq Q^n$ for all n, $\cup_n Q_n = \mathbb{R}^M$, $h_n \searrow 0$ and $\frac{k_n}{h_n} \to 0$ as $n \to \infty$. (From now on $k_n > 0$ will be the "discretization step" of the polyhedron Q^n).

Let w_n be the unique solution of (HJD) with $Q = Q^n$, $h = h_n$. Our main result is the following

Theorem 1. *Assume* $|f(x,a,b) - f(y,a,b)| \le L|x-y|$ *for all* x, y, a, b; $|f(x,a,b)| \le M_f$ *for all* $x \in \partial\mathcal{T}$ *and all* a, b; $\mathcal{T}$ *is the closure of an open set with Lipschitz boundary;* (Q^n, h_n) *is an admissible sequence; there is a bounded continuous viscosity solution* v *of* (0.3), (0.4). *Then* w_n *converge to* v *as* $n \to \infty$, *uniformly on any compact set of* $\mathbb{R}^M$.

The strategy of proof follows our previous papers [BF1] and [BS3] and is based on the "weak limits" introduced by Barles and Perthame [BP]:

$$\overline{v}(x) \equiv \limsup_{(y,n)\to(x,\infty)} w_n(y), \quad \underline{v}(x) \equiv \liminf_{(y,n)\to(x,\infty)} w_n(y)\ .$$

The key property of $\overline{v}$ and $\underline{v}$ is given by the following

Lemma. $\overline{v}$ *and* $\underline{v}$ *are, respectively, a subsolution and a supersolution of* (0.3) *in the viscosity sense coupled with the boundary condition*

$$v = 0 \quad or \quad v + \min_b \max_a \{-f(x,a,b)\cdot Dv - 1\} = 0 \quad on \ \partial\mathcal{T}, \tag{1.1}$$

to be understood in the viscosity sense.

We defer the proof of this lemma to the next section. By applying the Lemma, we can easily complete the proof of Theorem 1 by means of a comparison result for viscosity sub and supersolution of the boundary value problem (0.3), (1.1) in [BS4]. In fact the regularity of f and $\mathcal{T}$ and the continuity of v allow to compare the upper semicontinuous subsolution $\overline{v}$ with v and get $\overline{v} \le v$ in $\mathcal{T}^c$, and similarly $v \le \underline{v}$. Then $\underline{v} = \overline{v}$, and it is easy to check that this equality implies the uniform convergence of w_n to v on compact sets.

We refer the reader to [BS3] for more details on this strategy of proof and for some possible variants. For instance the continuity assumption on v can be replaced by a suitable uniform estimate on w_n around $\partial\mathcal{T}$. Moreover it is not hard to show that $-\log(1-w_n)$ converge to the lower capture time T uniformly on compact subsets of $\{x : T(x) < +\infty\}$. Note also that the choice of Q_1 and Q_2 and the "discrete boundary conditions" on I_{out1} and I_{out2} do not influence the limit as $n \to \infty$ because $\cup_n Q^n = \mathbb{R}^M$.

Remark. The same convergence results can be obtained for a different algorithm based on a discretization of Q and the following definition of $F : \mathbb{R}^N \to \mathbb{R}^N$,

$$F_i(V) \equiv \begin{cases} \gamma \max\limits_b \min\limits_a P_i(a,b,V) + 1 - \gamma & \text{if } i \in I_{in}, \\ 0 & \text{if } i \in I_{\mathcal{T}}, \end{cases}$$

where γ, I_{in}, $I_{\mathcal{T}}$ have the same meaning as before,

$$P_i(a,b,V) \equiv \begin{cases} \sum_{j=1}^{N} \lambda_{ij}(a,b)V_j & \text{if } z_i(a,b) \in Q, \\ 1 & \text{if } z_i(a,b) \in Q_2 \backslash Q, \\ 0 & \text{if } z_i(a,b) \in \mathbb{R}^M \backslash Q_2. \end{cases}$$

and

$$\sum_{j=1}^{N} \lambda_{ij}(a,b)x_j = z_i(a,b) \ , \quad \lambda_{ij} \in [0,1], \quad \sum_{j=1}^{N} \lambda_{ij} = 1 \ .$$

The advantage of this formulation is that no interpolation is made to compute the values at nodes exiting Q under the action of the flow f. This choice is more efficient from the computational point of view but the definition of the corresponding problem (HJD) is more tricky. □

It would be desirable for many applications to know the convergence of the scheme also for games with barriers, that is, when v is discontinuous. At the moment we can do this only for control problems with a single player by means of a result of one of the authors and Staicu [BSt]. We define

$$\hat{v}(x) = \inf_{a(\cdot)} \{1 - e^{-t} | y_x(t, a(\cdot)) \in \operatorname{int}(\mathcal{T})\} \ ,$$

where the controls $a(\cdot) : [0, +\infty[\to A$ are measurable functions and y_x denotes the solution of (0.1) (with b constant!) such that $y_x(0) = x$. Note that $\hat{v}$ is the exponential transform of the minimum time necessary to enter the interior of $\mathcal{T}$, and not just $\mathcal{T}$.

Theorem 2. *Assume $|f(x,a) - f(y,a)| \le L|x-y|$ for all x, y, a; $\mathcal{T}$ is the closure of an open set; (Q^n, h_n) is an admissible sequence. Then w_n converge to $\hat{v}$ as $n \to \infty$ uniformly on any compact set where $\hat{v}$ is continuous.*

Proof. In [BSt] it is proved that $\hat{v}$ is a "complete solution" of the boundary value problem (0.3) plus the boundary conditions

$$\begin{aligned} &\hat{v} \ge 0 && \text{on } \partial\mathcal{T} \\ &\hat{v} \le 0 && \text{or } \hat{v} + \min_a \{-f(x,a) \cdot D\hat{v} - 1\} \le 0 \quad \text{on } \partial\mathcal{T} \ , \end{aligned}$$

which means that its upper semicontinuous envelope $\hat{v}^*$ is greater than or equal to any subsolution, e.g. $\overline{v}$, and its lower semicontinuous envelope $\hat{v}_*$ is smaller than or equal to any supersolution, e.g. $\underline{v}$. Therefore

$$\hat{v}_* \le \underline{v} \le \overline{v} \le \hat{v}^* \ ,$$

and all inequalities are equalities at points where $\hat{v}^* = \hat{v}_*$, that is, where $\hat{v}$ is continuous. □

An easy consequence of Theorem 2 is the uniform convergence of $-\log(1-w_n)$ to the minimal time function T on any compact set where T is finite and continuous.

We end this section with some remarks on the possible developments of the results presented here. First of all we can get estimate of the rate of convergence of the scheme, by using mainly viscosity solutions methods as in [CL], [CDI], [F], [A2], [BSa].

It is also interesting to study differential games with state constraints, where the first player has to keep the system in a given set Q_1 and the second player in Q_2. Our discretized problem (HJD) is designed to fit this situation as well, at least if $Q = Q_1 \cap Q_2$ is a polyhedron. If we pass to the limit as $h \searrow 0$, $\frac{k}{h} \to 0$, and Q^n is kept fixed, we get that $\overline{v}$ and $\underline{v}$ are, respectively, sub and supersolution of (0.3) in $Q \backslash \mathcal{T}$, of (1.1) on $\partial\mathcal{T}$ and of

$$\overline{v}(x) + \min_{b \in B} \max_{a \in A}\{-f(x,a,b) \cdot D\overline{v}(x) - 1\} \le 0 \quad \text{on } \partial Q \cap Q_2 \,,$$

$$\underline{v}(x) + \min_{b \in B} \max_{a \in A}\{-f(x,a,b) \cdot D\underline{v}(x) - 1\} \ge 0 \quad \text{on } \partial Q \cap Q_1 \,.$$

To prove the convergence of the scheme as above we need a suitable comparison theorem for this new boundary value problem: this can be obtained by refining the arguments of [BS4] and [A1] and it will be done in a future paper.

2. Proofs

Proof of the Proposition.

(i) If $V_j \in [0,1]$ for all j then $P_i(a,b,V) \in [0,1]$ for all a, b, i, and therefore $0 < 1-\gamma \le F_i(V) \le 1$ for all $i \in I_{in}$.

(ii) If $V \le W$, $P_i(a,b,V) \le P_i(a,b,W)$ for all a, b because $\lambda_{ij} \ge 0$. Then $F_i(V) \le F_i(W)$ for all $i \in I_{in}$.

(iii) We fix V, W, i, $\overline{b}$ such that

$$\max_b \min_a P_i(a,b,V) = \min_a P_i(a,\overline{b},V) \,,$$

and $\overline{a}$ such that

$$\min_a P_i(a,b,W) = P_i(\overline{a},\overline{b},W) \,.$$

For $i \in I_{in}$ we have

$$\begin{aligned} F_i(V) - F_i(W) &\le \gamma P_i(\overline{a},\overline{b},V) - \gamma P_i(\overline{a},\overline{b},W) \\ &= \gamma \sum_{i,j} \lambda_{ij}(\overline{a},\overline{b})(V_j - W_j) \le \gamma \|V-W\|_\infty \,. \end{aligned}$$

□

Proof of the Lemma. We prove only the statement about $\overline{v}$ because the proof for $\underline{v}$ is similar. Let

$$H(x,p) \equiv \min_a \max_b \{-f(x,a,b) \cdot p - 1\} ,$$

and $\Omega \equiv \mathcal{T}^c = \mathbb{R}^M \backslash \mathcal{T}$. We recall that $\overline{v}$ is a viscosity subsolution of (0.3) and (1.1) if and only if, for all $\varphi \in C^1(\overline{\Omega})$, at a local strict maximum point y for $\overline{v} - \varphi$ we have

$$\overline{v}(y) + H(y, D\varphi(y)) \leq 0 \quad \text{if } y \in \Omega ,$$

and the same inequality holds for every $y \in \partial\Omega$ such that $\overline{v}(y) > 0$.

Let x^n be such that

$$\max_{cl(Q^n \cap \Omega \cap B(y,1))} (w_n - \varphi) = (w_n - \varphi)(x^n),$$

where cl indicates the closure and $B(y,r) = \{x : |x-y| < r\}$. By lemma $A.1$ of [BP]

$$x^n \to y , \qquad w_n(x^n) \to \overline{v}(y) \qquad \text{as } n \to \infty .$$

We can choose $\varepsilon > 0$ such that $B(y,\varepsilon) \subseteq \Omega$ if $y \in \Omega$, $\varphi(y) > \varepsilon$ in $B(y,\varepsilon)$ if $y \in \partial\Omega$, and $B(y,\varepsilon) \subseteq Q^n$ for n large enough. Moreover, let us take an n sufficiently large so that the following holds:

$$\begin{cases} x^n \in B\left(y, \frac{\varepsilon}{3}\right), & |h_n f(x^n,a,b)| \leq \varepsilon/3, \\ (2+C)k_n \leq \varepsilon/3, & \varphi(x^n) - w_n(x^n) < \varepsilon , \end{cases} \tag{2.1}$$

where $C \equiv \sup_{B(y,1)} |D\varphi|$.

The following calculations are made for n fixed, so we drop the subscript in w_n, h_n, γ_n and the superscript in x^n. First of all we note that $w(x) > 0$ since $\varepsilon > \varphi(x) - w(x) > \varepsilon - w(x)$. Thus, if $x = \Sigma_i \lambda_i x_i$ there exists a node x_i such that $\lambda_i \neq 0$ and $w(x_i) > 0$, and this implies

$$w(x_i) = \max_b \min_a \gamma w(x_i + hf(x_i,a,b)) + 1 - \gamma .$$

Then for any $\rho > 0$ there exists b_i such that

$$w(x_i) - \gamma w(x_i + hf(x_i,a,b_i)) - (1-\gamma) < \rho h , \quad \forall a . \tag{2.2}$$

Let us prove first that

$$w(x) - \varphi(x) \geq w(x_i + hf(x_i,a,b_i)) - \varphi(x_i + hf(x_i,a,b_i)) - Ck . \tag{2.3}$$

If $z_i \equiv x_i + hf(x_i,a,b_i) \in \overline{\Omega}$ this follows from (2.1) which implies $z_i \in B(y,1)$. Let $z_i \notin \overline{\Omega}$ and $z_i = \Sigma \mu_j x_j$, we have the following two cases:

(i) $\mu_j \neq 0$ and $x_j \in \overline{\Omega}$, which implies

$$w(x) - \varphi(x) \geq w(x_j) - \varphi(x_j)$$

(ii) $\mu_j \neq 0$ and $x_j \notin \overline{\Omega}$, which implies

$$\varphi(x_j) - w(x_j) = \varphi(x_j) > \varepsilon > \varphi(x) - w(x) .$$

Then we conclude that

$$w(x) - \varphi(x) \geq \Sigma\mu_j(w(x_j) - \varphi(x_j)) \geq w(z_i) - \varphi(z_i) - Ck .$$

Let us prove now that

$$|w(x) - w(x_i)| \leq Ck . \tag{2.4}$$

By construction, there exists a simplex S_j such that $x \in S_j$ and $|x - x_i| \leq k$ for any $x_i \in S_j$. Let $X \equiv int(S_j)$, since we can assume $x \in$ rel int$X$ and $(w-\varphi)|_X$ has a maximum at x, we have

$$\|D_X w\|_{L^\infty(Q)} = |D_X w(x)| = |D_X \varphi(x)| \leq C,$$

where D_X denotes the (constant) gradient of w on X, and then (2.4) holds.
Starting from (2.2) and applying (2.3), (2.4) for any $a \in A$ we have

$$\begin{aligned}\rho h &> (1-\gamma)w(x) + \gamma(w(x) - w(x_i + hf(x_i, a, b_i)) - (1-\gamma) - Ck \\ &\geq (1-\gamma)w(x) + \gamma(\varphi(x) - \varphi(x_i + hf(x_i, a, b_i)) - (1-\gamma) - (1+\gamma)Ck \\ &\geq (1-\gamma)w(x) + \gamma(\varphi(x) - \varphi(x + hf(x, a, b_i)) - (1-\gamma) \\ &\qquad - (1 + 2\gamma + \gamma hL)Ck.\end{aligned}$$

Since ρ is arbitrary we get, by making again explicit the dependence on n of the variables,

$$\begin{aligned}0 \geq \frac{1-\gamma_n}{h_n} w_n(x) + \min_b \max_a \left\{ \gamma_n \frac{\varphi(x^n) - \varphi(x^n + h_n f(x^n, a, b))}{h_n} - \frac{1-\gamma_n}{h_n} \right\} \\ - C\frac{k_n}{h_n}(1 + 2\gamma_n + \gamma_n h_n L) .\end{aligned}$$

The result is obtained passing to the limit on n for admissible sequences, since by definition $\gamma_n \equiv e^{-h_n}$ and for such sequences $\frac{k_n}{h_n} \to 0$.
□

3. Two Simple Examples

In this section we present two examples of 1-dimensional pursuit-evasion games for which the solution can be derived exactly.

Test 1. We consider the following dynamics for the players

$$\begin{aligned}\dot{x} &= a(x-1)(x+1)v_1\\ \dot{y} &= b(y-1)(y+1)v_2\end{aligned} \tag{3.1}$$

where x, y are the (1-dimensional) state variable respectively of the pursuer and the quarry and v_1, v_2 are two positive real parameters representing their relative velocities. We choose $A = B \equiv [-1,1]$.

We study this game in $Q \equiv [-1,1]^2$ which is invariant with respect to the trajectories and we set $\mathcal{T} \equiv \{(x,y) : x = y\}$. Starting the game from (x_0, y_0), $x_0 < y_0$ the optimal strategies for the quarry and the pursuer will be $b^* \equiv -1$ and $a^* \equiv -1$, since this correspond for both to the motion toward 1 at the maximum speed.

We can solve explicitly (3.1) obtaining

$$x^*(t) = \frac{1 - k_1^2 e^{-2v_1 t}}{1 + k_1 e^{-2v_1 t}} \tag{3.2}$$

$$y^*(t) = \frac{1 - k_2^2 e^{-2v_2 t}}{1 + k_2 e^{-2v_2 t}} \tag{3.3}$$

where $k_1 \equiv (|x_0 - 1|/|x_0 + 1|)^{1/2}$ and $k_2 \equiv (|y_0 - 1|)/|y_0 + 1|)^{1/2}$.

Setting $x^*(t) = y^*(t)$ we can compute the value function of the game. When $v_1 \le v_2$ that equation implies $x_0 = y_0$, then the game has no solution and $T(x) = +\infty$ off the diagonal of Q except for $(x_0, y_0) \in \partial Q \cap (\{y = -1\} \cup \{y = 1\})$ where the game has always a solution.

For $v_1 > v_2$ the capture time is

$$t^* = \frac{1}{2(v_1 - v_2)} \ln \left(\frac{k_1}{k_2}\right)^2$$

and by the Kružkov transform we get the solution of (0.3), (0.4)

$$v(x_0, y_0) = 1 - \left(\frac{k_1}{k_2}\right)^{\frac{1}{v_2 - v_1}}$$

valid on $\{(x,y) : x \le y\} \cap Q$. For $x_0 > y_0$, repeating the same argument we get

$$v(x_0, y_0) = 1 - \left(\frac{k_2}{k_1}\right)^{\frac{1}{v_2 - v_1}} .$$

Figures 2 and 3 show the numerical results obtained by applying the algorithm on a grid of 1849 nodes ($k = 0.04$) and for $h = 0.01$.

Notice that the approximate solution of Figure 2 does not have jumps (whereas the exact solution does) due to the smoothing effect of the linear interpolation which, in particular, makes the values at nodes close to the

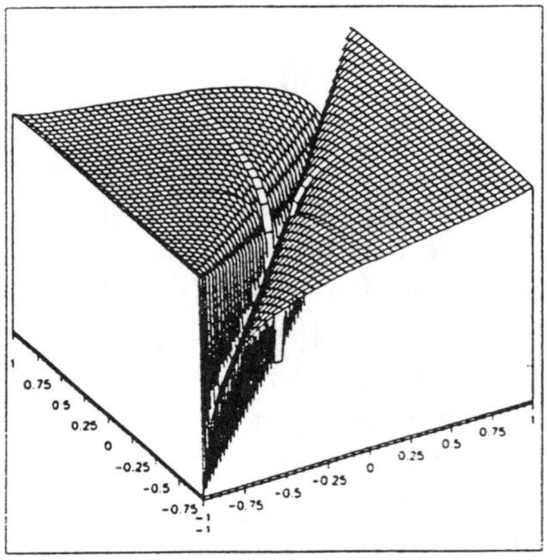

a)

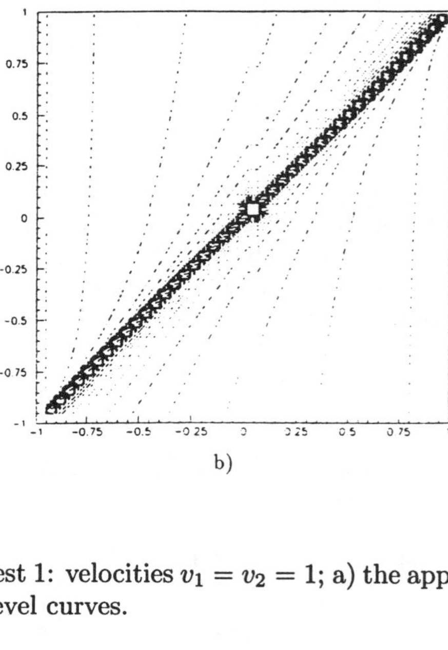

b)

Figure 2. Test 1: velocities $v_1 = v_2 = 1$; a) the approximate value function $v(x)$, b) its level curves.

diagonal be close to 0. Anyway the slope is very high near the diagonal as the level curves show.

For $v_1 > v_2$, the exact solution is continuous and the approximation is accurate (Figure 3), infact the maximum error on the nodes in $int(Q)$ is 0.05962329.

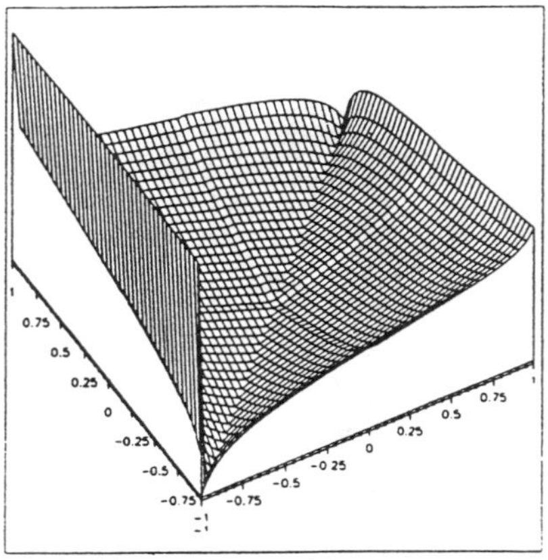

a)

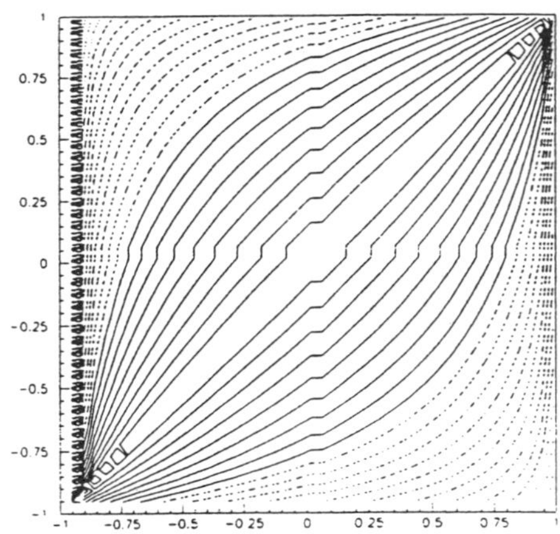

b)

Figure 3. Test 1: velocities $v_1 = 5$, $v_2 = 1$; a) the approximate value function $v(x)$, b) its level curves.

Test 2. Let the dynamics of the pursuer and the quarry be given respectively by

$$\dot{x} = ax(x-1)(x+1)v_1$$
$$\dot{y} = by(y-1)(y+1)v_2$$

where a, b, v_1, v_2 have the same meaning as in the previous test and $A = B \equiv [-1,1]$. We study the game in the square $Q \equiv [-1,1]^2$ (which is always invariant for the dynamics) and set $\mathcal{T} \equiv \{(x,y) : x = y\}$.

We divide the square into four regions by the lines $x = 0$ and $y = 0$. In $R_2 \equiv [-1,0] \times [0,1]$ and $R_4 = [0,1] \times [-1,0]$ the game has no solution and $T(x) \equiv +\infty$. In the remaining two squares R_1 and R_3 the game has a solution depending on the values of v_1 and v_2.

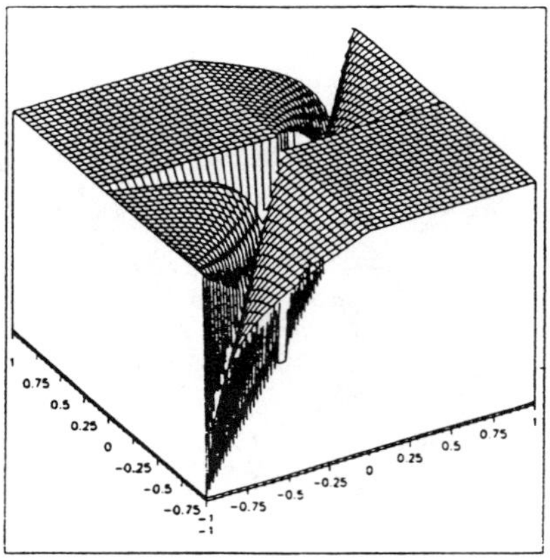

a)

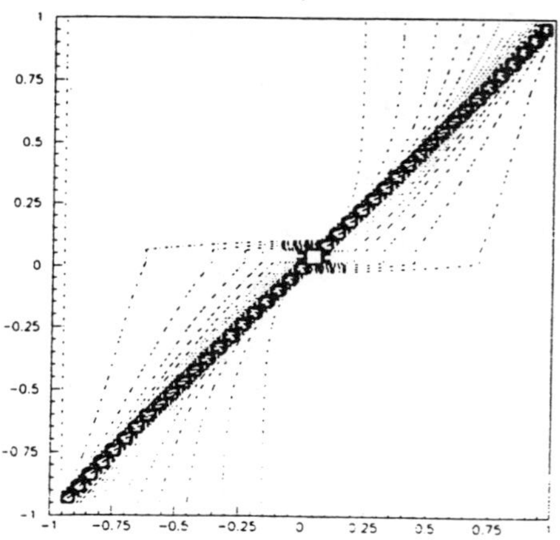

b)

Figure 4. Test 2: velocities $v_1 = v_2 = 1$; a) the approximate value function $v(x)$, b) its level curves.

As in Test 1 we can compute the exact solution, for example, for a starting point $(x_0, y_0) \in R_1$, $x_0 < y_0$, assuming $v_1 > v_2$. This solution is

$$v(x_0, y_0) = 1 - \left(\frac{k_1}{k_2}\right)^{\frac{1}{v_2 - v_1}}$$

where

$$k_1 = (\sqrt{|x_0^2 - 1|})/x_0 \ , \qquad k_2 \equiv (\sqrt{|y_0^2 - 1|})/y_0 \ .$$

Figures 4 and 5 show the numerical results obtained by applying the algorithm on a grid of 1849 nodes ($k = 0.04$) and $h = 0.01$.

Notice that the detection of R_2 and R_4 is quite accurate (they corresponds to the region where the solution is flat since $v \equiv 1$ there). Again there is a smoothing effect which rounds up the discontinuities of the solution (Figure 4), but in R_1 and R_3 the approximation is accurate when $v_1 > v_2$ and the solution is continuous (Figure 5). The maximum error on the nodes belonging to the interior of the regions $R_i, i = 1, \ldots, 4$, is 0.05754636.

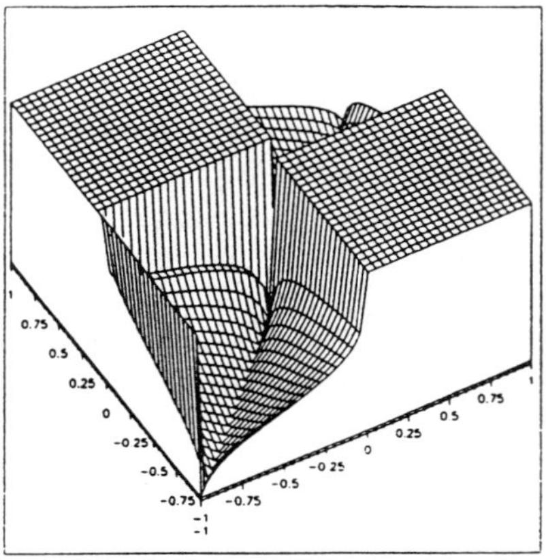

a)

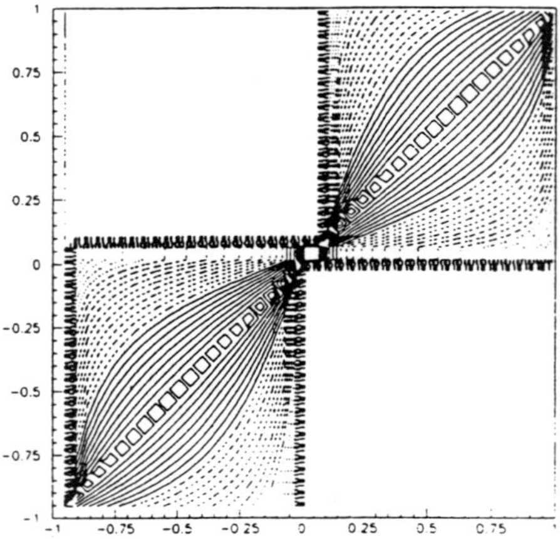

b)

Figure 5. Test 2: velocities $v_1 = 5$, $v_2 = 1$; a) the approximate value function $v(x)$, b) its level curves.

Acknowledgments. We wish to thank P. Lanucara for his contribution to the development of the code that produced the numerical results.

References

[A1] B. Alziary de Roquefort, Jeux différentiels et approximation numérique de fonctions valeur, 1re partie: étude théorique, *RAIRO Math. Model. Numer. Anal.*, **25** (1991), 517–533.

[A2] B. Alziary de Roquefort, Jeux différentiels et approximation numérique de fonctions valuer, 2e partie: étude numérique, *RAIRO Math. Model. Numer. Anal.*, **25** (1991), 535–560.

[BF1] M. Bardi, M. Falcone, An approximation scheme for the minimum time function, *SIAM J. Control Optim.*, **28** (1990), 950–965.

[BF2] M. Bardi, M. Falcone, Numerical solution of the minimum time problem, in preparation.

[BP] G. Barles, B. Perthame, Discontinuous solutions of deterministic optimal stopping time problems, *RAIRO Math. Model. Numer. Anal.*, **21** (1987), 557–579.

[BSa] M. Bardi, C. Sartori, Convergence results for Hamilton-Jacobi-Bellman equations in variable domains, *Differential Integral Equations*, **5** (1992), 805–816.

[BS1] M. Bardi, P. Soravia, A PDE framework for games of pursuit-evasion type, in *Differential Games and Applications*, T. Basar & P. Bernhard eds., pp. 62–71, Lecture Notes in Control and Information Sciences 144, Springer-Verlag, 1989.

[BS2] M. Bardi, P. Soravia, Hamilton-Jacobi equations with singular boundary conditions on a free boundary and applications to differential games, *Trans. Amer. Math. Soc.*, **325** (1991), 205–229.

[BS3] M. Bardi, P. Soravia, Approximation of differential games of pursuit-evasion by discrete-time games, in *Differential Games - Developments in Modeling and Computation*, R.P. Hamalainen & H.K. Ethamo eds., pp. 131–143, Lecture Notes in Control and Information Sciences 156, Springer Verlag, 1991.

[BS4] M. Bardi, P. Soravia, A comparison result for Hamilton-Jacobi equations and applications to some differential games lacking controllability, *Funkcial. Ekvac.*, to appear.

[BSo] G. Barles, P.E. Souganidis, Convergence of approximation schemes for fully nonlinear second order equations, *Asymptotic Anal.*, **4** (1991), 271–283.

[BSt] M. Bardi, V. Staicu, The Bellman equation for time-optimal control of non-controllable nonlinear systems, to appear in *Acta Appl. Math.*, **31**(1993), 201–223.

[CD] I. Capuzzo Dolcetta, On a discrete approximation of the Hamilton –Jacobi equation of dynamic programming, *Appl. Math. Optim.*, **10** (1983), 367–377.

[CDF] I. Capuzzo Dolcetta, M. Falcone, Discrete dynamic programming and viscosity solutions of the Bellman equation, *Ann. Inst. Henry Poincaré Anal. Nonlin.*, **6** (supplement) (1989), 161–184.

[CDI] I. Capuzzo Dolcetta, H. Ishii, Approximate solutions of the Bellman equation of deterministic control theory, *Appl. Math. Optim.*, **11** (1984), 161–181.

[CL] M.C. Crandall, P.L. Lions, Two approximations of solutions of Hamilton-Jacobi equations, *Math. Comp.*, **43** (1984), 1–19.

[EK] R.J. Elliott, N.J. Kalton, The existence of value in differential games, *Mem. Amer. Math. Soc.*, **126** (1972).

[F] M. Falcone, A numerical approach to the infinite horizon problem of deterministic control theory, *Appl. Math. Optim.*, **15** (1987), 1–13.

[FF1] M. Falcone, R. Ferretti, High-order approximations for viscosity solutions of Hamiltonian-Jacobi-Bellman equations, to appear in *Nonlinear Variational Problems and Partial Differential Equations, vol. III,* A. Marino and M.K.V. Murthy eds., Pitman Research Series in Mathematics, Longman.

[FF2] M. Falcone, R. Ferretti, Discrete time high–order schemes for viscosity solutions of Hamilton–Jacobi–Bellman equations, 1992, Numerische Mathematik, to appear.

[GR] R.L.V. Gonzalez, E. Rofman, On deterministic control problems: an approximation procedure for the optimal cost, part 1 and 2, *SIAM J. Control Optim.*, **23**(1985), 242–285.

[GT] R.L.V. Gonzalez, M.M. Tidball, Fast solution of discrete Isaacs' inequalities, *Rapport de Recherche INRIA*, 1167 (1990).

[LNB] V. Laporte, J.M. Nicolas, P. Bernhard, About the resolution of discrete pursuit games and its application to naval warfare, in *Differential Games - Developments in modeling and computation,* R.P. Hamalainen & H.K. Ethamo eds, pp. 151–163, Lecture Notes in Control and Information Sciences 156, Springer Verlag, 1991.

[S] P. Soravia, The concept of value in differential games of survival and viscosity solutions of Hamilton-Jacobi equations, *Differential Integral Equations*, **5** (1992), 1049–1068.

[So] P.E. Souganidis, Approximation schemes for viscosity solutions of Hamilton-Jacobi equations, *J. Differential Equations*, **57** (1985), 1–43.

[T] M.M. Tidball, Sobre la resolucion numerica de la ecuacion de Hamilton-Jacobi-Bellman, Thesis, Universidad Nacional de Rosario (1991).

M. Bardi and P. Soravia
Dipartimento di Matematica Pura e Applicata
Università di Padova, Via Belzoni 7
35131 Padova, Italy

M. Falcone
Dipartimento di Matematica
Università di Roma, "La Sapienza"
P. Aldo Moro 2, 00185 Roma, Italy

Zero Sum Differential Games With Stopping Times: Some Results About its Numerical Resolution

Mabel M. Tidball and R.L.V. González

Abstract. In this work we consider a zero sum differential game problem with stopping times. We discretize the associated Isaacs equation and we obtain an approximate solution that converges to the value function of the game when the parameter of discretization tends to zero. We give also an estimate of the error of discretization. The discrete solution of the problem is the fixed point of a contractive operator and we present an accelerated algorithm to find it. We prove that this algorithm converges to the value function in a finite number of steps.

Introduction

In this work we study a differential game problem with stopping times. For this special game, the value function V is totally characterized by means of a system of inequalities associated to the Isaacs equation of the problem (see [9], [18]). This system of inequalities defines a bilateral variational inequality (BVI). Many important properties of this BVI can be obtained when it is transformed in a system of unilateral quasi-variational inequalities (see [9] and [18]).

To obtain numerical solutions we discretize these inequalities via finite element techniques, through a suitable extension of the methodology presented in [10]. Our discrete solution V^k converges to the value V when the parameter of discretization k tends to zero. We give an estimate of the rate of convergence; the obtained estimate, of type $\sqrt{k}$, is optimal as a consequence of the results presented in [15], [16].

To find V^k, it is necessary to solve a fixed point problem associated to a contractive operator. In order to eliminate the inefficiency of the usual fixed point algorithm (see [2]), we develop an accelerated ad-hoc method. The developed procedure, which has been specially designed for this particular game, consists of an extension of the methodology presented in [11] for optimal control problems. We prove that the accelerated algorithm converges to the discrete solution in a finite number of steps. The proof is based on the use of a system of quasi-variational inequalities associated with the bilateral inequality. Several stopping rules for the internal loop of the algorithm are presented. They allow us to extend the original acceleration procedure to problems where some of the original hypothesis are not verified (see [11]). Numerical examples and comparative results are shown in Section 5. Extensions to the general case of general differential games are considered in [12].

1. Description of the Problem

In this differential game problem, the decision variables used by the players are the stopping times for the uncontrolled system. The evolution of the system is described by the ordinary differential equation:

$$\frac{d}{\xi}dt = g(\xi) \tag{1}$$

$$\xi(0) = x, \qquad x \in \Omega \subset R^n$$

where Ω is an open bounded domain. It is assumed that for every $t \geq 0$, the state of the system $\xi(t) \in \Omega$. As the game is of zero sum, the goal of the first player is to maximize the payoff J and the goal of the second player is to minimize it. Denoting τ_1 and τ_2 the stopping times chosen by players 1 and 2 respectively, J takes the form:

$$\begin{aligned} J(\tau_1, \tau_2) = &\int_0^{\tau_1 \wedge \tau_2} f(\xi(s))e^{-\lambda s} ds \\ &+ \psi(\xi(\tau_1))e^{-\lambda\tau_1}\chi_{\tau_1<\tau_2} + \psi_2(\xi(\tau_2))e^{-\lambda\tau_2}\chi_{\tau_2\geq\tau_1} \end{aligned} \tag{2}$$

$\lambda > 0$ is the discount rate and $f(\xi)$ is the instantaneous cost. ψ_1, ψ_2 are the final costs and they verify:

$$\psi_2(x) \geq \psi_1(x) + \mu \qquad \forall\, x \in \Omega,\ \mu > 0. \tag{3}$$

This inequality does not represent a real restriction as we can see in [21], but it plays a very important role in the proof of the existence of the value function. For this game, the lower and upper values $\underline{V}$ and $\overline{V}$ are respectively defined by:

$$\underline{V}(x) = \sup_{\tau_1} \inf_{\tau_2} J(x, \tau_1, \tau_2), \tag{4}$$

$$\overline{V}(x) = \inf_{\tau_2} \sup_{\tau_1} J(x, \tau_1, \tau_2).$$

If (3) is verified, it can be proved that the game has a unique value (see [21]):

$$V(x) = \underline{V}(x) = \overline{V}(x) \tag{5}$$

This function continuously depends on the initial position x (see [8]). In terms of the value function of the game, saddle strategies $(\overline{\tau_1}, \overline{\tau_2})$ can be found such that:

$$J(x, \overline{\tau_1}, \overline{\tau_2}) = V(x), \tag{6}$$

$$J(x, \tau_1, \overline{\tau_2}) \leq J(x, \overline{\tau_1}, \overline{\tau_2}) \leq J(x, \overline{\tau_1}, \tau_2) \qquad \forall\, \tau_1, \tau_2. \tag{7}$$

That is, the function J has a saddle point at $\overline{\tau_1}, \overline{\tau_2}$. These strategies are defined in the following way (see [1]):

$$\overline{\tau_i} = \inf\{s/V(\xi(s)) = \psi_i(\xi(s))\}, \qquad i = 1, 2. \tag{8}$$

with the usual convention $\inf(\emptyset) = +\infty$.

Thus, function V allows us to characterize this game problem and enables us to construct saddle strategies. In consequence, to solve this game we will study here how to identify V through a system of inequalities and we will present a numerical procedure to compute the discretization of that system.

2. Hamilton–Jacobi–Isaacs Inequality

Using the techniques applied in [3], [6], [8], [14], [15], [23], and with the following hypothesis (H): *g, f, ψ_i are Lipschitz continuous functions (with constants L_g, L_f, L_{ψ_i})*, the following theorems can be proved (related results can be seen in [20]).

Theorem 2.1. *If* (H) *holds, then V is a Hölder continuous function and the following inequality is verified, $\forall x, y \in \Omega$:*

$$|V(x) - V(y)| \leq C\|x - y\|^\gamma \tag{9}$$

with $\gamma = \lambda/L_g$, when $\lambda < L_g$, $\gamma = 1$ when $\gamma > L_g$ and $\gamma \in (0,1)$ when $\gamma = L_g$.

Theorem 2.2. *The value function V is the unique solution in the viscosity sense in Ω of the system:*

$$\psi_1(x) \leq V(x) \leq \psi_2(x) \qquad \forall\, x \in \Omega \tag{10}$$

$$\left(\frac{\partial V(x)}{\partial x}\right) g(x) + f(x) - \lambda V(x))(V(x) - \psi_2(x)) \geq 0 \tag{11}$$

$$\left(\frac{\partial V(x)}{\partial x}\right) g(x) + f(x) - \lambda V(x))(V(x) - \psi_1(x)) \geq 0 \tag{12}$$

Remark 2.1. The proof of these results is contained in [17]. By virtue of this result, the solution of the game problem is reduced to find the unique viscosity solution of (10)–(12). To compute the function V numerically, we solve (10)–(12) by discretization, finding an approximate solution V^k. In the following sections, we present the discretization procedure, results of convergence and an accelerated method to compute V^k.

3. The Discretized Problem

3.1. The discretization procedure

In this section we will introduce the set W^k, where the approximation V^k of V is going to be found. Functions $w^k \in W^k$ will satisfy properties related to (10), (11) and (12). The solution of the discretized problem will be the fixed point of a contractive operator.

Discretization of the domain Ω. We will identify the discretization with the parameter k, which also indicates the size of the discretization.

Let Ω be an open set of $\Re^n$ and S_j^k a family of regular triangulations of Ω, i.e., a finite sets of simplices (see [10], [15], [22]) that approaches Ω in the following sense: $\Omega_k = \cup_j S_j^k$ is a polyhedron of $\Re^n$ such that the following properties are satisfied:

$$\max_j(diam S_j^k) = k \tag{13}$$

- $\Omega_k \to \Omega$, when $k \to 0$, in the following sense:

$$\forall\, K \text{ compact contained in } \Omega,\ \exists \bar{k}(K)/K \subset \Omega_k \quad \forall\, k < \bar{k}(K) \tag{14}$$

- If d_i is the diameter of the simplex S_i^k, then: $\exists\ \gamma_1 > 0$ such that for every simplex S_i of Ω_k; there is a sphere with radius r_i inside the simplex such that:

$$r_i \geq \gamma_1 d_i \tag{15}$$

  and there also exists γ_2, independent of the discretization such that

$$\gamma_2 \geq \frac{k}{d_i} \tag{16}$$

- It is assumed that for any $x_p \in \partial\Omega_k$, the point $x_p + \eta g(x_p) \in \Omega_k$ for η small enough (this is the discrete equivalence of the condition of invariance of Ω).

Remark 3.1. This condition holds in particular if for the continuous system, the field g satisfies: $\langle g(x), \nu(x)\rangle < 0, \forall x \in \partial\Omega$, where ν is the outer normal and $\langle .,. \rangle$ denotes the inner product in $\Re^n$.

Discretization of the functional space. The space $W^{1,\infty}(\Omega)$ is approximated by the set W^k of functions $w^k : \Omega_k \to \Re$, where

- w^k are continuous functions in Ω_k
- $\frac{\partial w^k}{\partial x}$ are constant inside each simplex of Ω_k

that is, w^k are linear finite elements defined on Ω_k. We denote with $Z_p = x_p$, $i = 1, \dots, N_k$ the set of nodes (vertices of simplex) of Ω_k, where N_k is the number of vertices of Ω_k. Then, the elements w^k are completely characterized by the values $w^k(x_p)$, $p = 1, \dots, N_k$. For simplicity of notation, below we will use the abridged expression w_p or $w(p)$ for $w^k(x_p)$, which defines an isomorphism $W^k \longleftrightarrow \Re^{N_k}$.

Discretization of Isaacs inequalities. We define the discretized operator L^k:

$$L^k V^k(x_p) = \frac{\partial V^k}{\partial x_g}(x_p) \cdot \|g(x_p)\| + f(x_p) - \lambda V^k(x_p), \tag{17}$$

where

$$\frac{\partial V^k}{\partial x_g}(x_p)\|g(x_p)\| = \frac{V^k(a_p) - V^k(x_p)}{\|a_p - x_p\|} \cdot \|g(x_p)\| \quad \text{if } g(x_p) \neq 0,$$

$$\frac{\partial V^k}{\partial x_g}(x_p)\|g(x_p)\| = 0 \quad \text{if } g(x_p) = 0$$

and

$$a_p = x_p + g(x_p)\overline{h}_p,$$

where

$$\overline{h}_p = \max\{s > 0 / x_p + sg(x_p) \in S_j^k, x_p \in S_j^k\}.$$

We define the discretization of inequalities (10), (11) and (12) in the following way, $\forall x_p \in Z_k$:

$$\psi_1(x_p) \leq V^k(x_p) \leq \psi_2(x_p) \tag{18}$$

$$V^k(x_p) > \psi_1(x_p) \Rightarrow L^k V^k(x_p) \geq 0 \tag{19}$$

$$V^k(x_p) < \psi_2(x_p) \Rightarrow L^k V^k(x_p) \leq 0. \tag{20}$$

3.2. The discrete problem

The discrete problem takes the following form: Problem P^k

Find an element $V^k \in W^k$, which verifies the relations (18)–(20)

Taking into account the isomorphism $W^k \longleftrightarrow \Re^{N_k}$ defined above, it is easy to prove that V^k is given by the unique fixed point of the following operator $M : \Re^{N_k} \rightarrow \Re^{N_k}$:

$$(Mw)_p = \min(\max((Bw)_p + \gamma_p, \psi_1(x_p)), \psi_2(x_p)) \tag{21}$$

where B is a $N_k \times N_k$ matrix, defined by: $B_{ij} = \beta_k(x_i) \times \alpha_{ij}$; α_{ij} are non-negative numbers such that

$$a_p = x_p + g(x_p)\overline{h}_p = \sum_{j=1}^{N_k} \alpha_{pj} x_j$$

and

$$\beta_k(x_p) = \begin{cases} 0 & \text{if } g(x_p) = 0 \\ \frac{\|g(x_p)\|}{\|g(x_p)\| + \lambda\|a_p - x_p\|} & \text{if } g(x_p) \neq 0 \end{cases} \tag{22}$$

$$\gamma(x_p) = \begin{cases} \frac{f(x_p)}{\lambda} & \text{if } g(x_p) = 0 \\ \frac{f(x_p)\|a_p - x_p\|}{\|g(x_p)\| + \lambda\|a_p - x_p\|} & \text{if } g(x_p) \neq 0 \end{cases} \tag{23}$$

Remark 3.2. From (21) it is clear that

$$(Mw)_p = P_{[\psi_1(x_p), \psi_2(x_p)]}((Bw)_p + \gamma_p)$$

where $P_I(x)$ denotes the projection of x on the interval I.

Remark 3.3. From (22) it results that operator M is contractive with a contractive constant α, being:

$$\alpha = \left(1 + \frac{\lambda\gamma_1}{M_g\gamma_2} k\right), \quad M_g = \max\left\{\|g(x)\|, x \in \Omega\right\}. \tag{24}$$

Theorem 3.1. *There exists a unique solution of* (P^k).

The proof of the results of this paragraph can be seen in [17], [23].

3.3. Convergence of the discrete solution

The following rate of convergence is valid (see [17], [23] for a detailed proof).

Theorem 3.2. *If assumptions* (H) *and* (9) *are satisfied, we have:*

$$\|V - V^k\| \leq Ck^{\gamma/2} \tag{25}$$

with $\gamma = \frac{\lambda}{L_g}$, *when* $\lambda < L_g$, $\gamma = 1$ *when* $\gamma > L_g$ *and* $\gamma \in (0,1)$ *when* $\gamma = L_g$.

Remark 3.4. The rate (25) is optimal as it can be easily proved using the techniques and counter-examples introduced in [15].

Remark 3.5. The proof of Theorem 3.2 is based essentially on the discrete maximum principle (DMP) satisfied by the discretization scheme L^k (see [4], [5], [10], for a discussion and applications of DMP). This result of convergence is obtained by means of regularization and interpolation techniques. We introduce a smooth function V_ρ^k which approximately satisfies the discretized conditions (18)-(20) and we obtain (using the DMP), an estimate for the difference $\|V_\rho^k - V^k\|$.

4. The Accelerated Algorithm

4.1. Description of the accelerated procedure

To solve (21) we can use the usual fixed point algorithm:

Algorithm A0

Step 0: Set $w^0 \in W^k$, set $\nu = 0$.
Step 1: Set $w^{\nu+1} = Mw^\nu$.
Step 2: If $\|w^\nu - w^{\nu+1}\| < \varepsilon$, Stop ($V^k \approx w^\nu$); if not, set $\nu = \nu + 1$ and go to Step 1.

As it is well known, this algorithm is very inefficient when the contraction factor α tends to 1 (see [2], [19]). To have an efficient method, we use the accelerated algorithm presented in the following lines. This algorithm is based on the method introduced in [11].

Auxiliary definitions:

- Given $I(j)$, $j = 1, \ldots, N_k$, a family of indices taking values in $\{0, 1, 2\}$ we define the affine operator M_I

$$M_I : \Re^{N_k} \to \Re^{N_k},$$

in the form:

$$(M_I v)(j) = \begin{cases} \psi_1(j) & \text{if } I(j) = 1 \\ \psi_2(j) & \text{if } I(j) = 2 \\ (Bv + \gamma)(j) & \text{if } I(j) = 0 \end{cases} \tag{26}$$

- We define:

$$I(j, w) = \begin{cases} 0 & \text{if } (Mw)(j) \in (\psi_1(j), \psi_2(j)) \\ 1 & \text{if } (Mw)(j) = \psi_1(j) \\ 2 & \text{if } (Mw)(j) = \psi_2(j). \end{cases}$$

- Given the sequence $\{w^{\mu,\nu}, \mu \geq 0, \nu \geq 0\}$, we say that $R(w^{\mu,0}, \ldots, w^{\mu,\nu})$ is a **"stopping rule"** if it takes only the values 0 or 1.

We will assume that the following condition is satisfied:

$$\forall\ (\mu, w^{\mu,0})\ \exists\ \overline{\nu}(\mu, w^{\mu,0})\ \ /\ \ R(w^{\mu,0}, \dots, w^{\mu,\overline{\nu}(\mu)}) = 1$$

where $\{w^{\mu,\nu},\ \nu = 0, 1, \dots\}$ is the sequence generated by the iteration $w^{\mu,\nu+1} = Mw^{\mu,\nu}$, with initial condition $w^{\mu,0}$.

This stopping rule will be used in the following algorithm to switch from the usual Picards iteration to a policy iteration step.

Algorithm A1

Step 0: Set $w^{0,0} \in \Re^{N_k}$, set $\mu = 0$, $\nu = 0$ and begin the procedure.
Step 1: Compute $w^{\mu,\nu+1} = Mw^{\mu,\nu}$.
Step 2: If $w^{\mu,\nu+1} = w^{\mu,\nu}$, then stop (the discrete solution is $V^k = w^{\mu,\nu}$; else, go to Step 3.
Step 3: If $R(w^{\mu,0}, \dots, w^{\mu,\nu}) = 1$, then set $I(j) = I(j, w^{\mu,\nu})$, $\forall\ j = 1, \dots, n$ and go to Step 4; else, set $\nu = \nu + 1$ and go to Step 1.
Step 4: Compute $w^{\mu+1}$ as the fixed point of operator M_I, set $w^{\mu+1,0} = w^{\mu+1}$, $\nu = 0$, $\mu = \mu + 1$ and go to Step 1.

Remark 4.1. It is clear from (26), that M_I is a contractive operator. The fixed point of operator M_I is in fact obtained solving the linear system $w = M_I w$.

Remark 4.2. Algorithm A1 is a modification of the policy iteration method currently used to solve optimal control problems on Markov chains [19]; it also inherits the property of convergence in a finite number of steps.

4.2. Convergence of algorithm A1

Theorem 4.1. *If condition* (27) *is satisfied, then algorithm A1 converges in a finite number of steps to the fixed point V^k of the operator M.*

To prove this theorem, we need to introduce some definitions and properties:

Definitions. Each $w \in \Re^{N_k}$ can be associated with pairs of elements (w_1, w_2), $w_i \in \Re^{N_k}$, $i = 1, 2$ such that $w_2 - w_1 = w$. So we define:

- $$J_w = \{(w_1, w_2)/w_2 - w_1 = w\} \tag{28}$$

- $$\begin{cases} \hat{\psi}_2 = \psi_2, & \hat{\psi}_1 = -\psi_1 \\ \hat{\gamma}_2 = \frac{\gamma}{2}, & \hat{\gamma}_1 = -\frac{\gamma}{2} \\ \hat{i} = 1, & \text{if } i = 2 \\ \hat{i} = 2, & \text{if } i = 1 \end{cases} \tag{29}$$

Note. This definition corresponds to the transformation (studied in [9], [18]) of a bilateral inequality in a system of unilateral inequalities. We define the operator $\hat{M} : \Re^{2N_k} \to \Re^{2N_k}$ such that:

$$\left(\hat{M}(w_1, w_2)\right)_i (j) = \min\{(Bw_{\hat{i}} + \hat{\gamma}_{\hat{i}})(j) + \hat{\psi}_i(j), (Bw_i + \hat{\gamma}_i)(j)\} \quad (30)$$

Properties of the operator $\hat{M}$. Let $w \in \Re^{2N_k}$ and $(w_1, w_2) \in J_w$, then:

- $\left(\hat{M}(w_1, w_2)\right)_i (j)$ is the maximum subsolution of the system of inequalities

$$\begin{cases} u_i \leq u_{\hat{i}} + \hat{\psi}_i & i = 1, 2 \\ u_i \leq Bw_i + \hat{\gamma}_i \end{cases}$$

- $$\left(\hat{M}(w_1, w_2)\right)_2 - \left(\hat{M}(w_1, w_2)\right)_1 = Mw \quad \forall\, (w_1, w_2) \in J_w. \quad (31)$$

- $$\text{If } V^k = MV^k \Rightarrow \hat{M}(J_{V^k}) \subset J_{V^k} \quad (32)$$

- $\hat{M}$ is monotone, that is:

$$\hat{w}_i \geq w_i \Rightarrow \left(\hat{M}(\hat{w}_1, \hat{w}_2)\right)_i \geq \left(\hat{M}(w_1, w_2)\right)_i, \quad i = 1, 2 \quad (33)$$

Remark 4.3. In this frame of transformed pairs $(w_1, w_2) \in \Re^{2N_k}$, the concept of fixed point is replaced by the concept of maximum subsolution.

Definition: Let $\hat{M}_I : \Re^{2N_k} \to \Re^{2N_k}$ be such that $(v_1, v_2) \to \big((\hat{M}_I(v_1, v_2))_1, (\hat{M}_I(v_1, v_2))_2\big)$, where:

$$\left(\hat{M}_I(v_1, v_2)\right)_i (j) = \begin{cases} (Bv_i + \hat{\gamma}_i)(j) & \text{if } I(j) \neq i \\ ((Bv_{\hat{i}} + \hat{\gamma}_{\hat{i}} + \hat{\psi}_i)(j) & \text{if } I(j) = 1 \end{cases} \quad (34)$$

The proof of following properties are almost self-evident. For sake of simplicity we will omit them, see [13], [17] for a detailed proof.

Properties of the operator $\hat{M}_I$

- $$\left(\hat{M}_I(v_1, v_2)\right)_2 - \left(\hat{M}_I(v_1, v_2)\right)_1 = (M_I(v)) \quad \forall\, (v_1, v_2) \in J_v \quad (35)$$

- $\hat{M}_I$ is contractive (therefore it has a unique fixed point).

(36)

- $\hat{M}_I$ is monotone:

$$\hat{v}_i \geq v_i \Rightarrow \left(\hat{M}_I(\hat{v}_1, \hat{v}_2)\right)_i \geq \left(\hat{M}_I(v_1, v_2)\right)_i \tag{37}$$

- if $\hat{w}_i$ is the fixed point of $\hat{M}_I$ then

$$\hat{w} = \hat{w}_2 - \hat{w}_1 \tag{38}$$

  is the fixed point of the operator M_I
- if $\hat{w}$ is the fixed point of the operator M_I then, there exists a unique $(\hat{w}_1, \hat{w}_2)$ such that

$$\hat{w} = \hat{w}_2 - \hat{w}_1 \tag{39}$$

  with $\hat{w}_i$ the fixed point of $\hat{M}_I$.

In order to prove the convergence of algorithm A1, we will associate to each $w^{\mu,\nu}$ a pair $(w_1^{\mu,\nu}, w_2^{\mu,\nu}) \in J_{w^{\mu,\nu}}$

We will prove convergence for sequences $w_i^{\mu,\nu}$, $i = 1, 2$ and from here, Theorem 4.1 will be immediately proved. We will denote by $(\rightarrow)$ such association. So

1. $$w^{0,0} \in \Re^{N_k} \rightarrow (w^{0,0})_i \in \Re^{N_k} \text{ such that } (w_1^{0,0}, w_2^{0,0}) \in J_{w^{0,0}} \tag{40}$$

2. $$w^{\mu,\nu+1} \rightarrow \hat{M}(w_1^{\mu,\nu}, w_2^{\mu,\nu}) \tag{41}$$

3. $\hat{w}^{\mu+1}$ (the fixed point of the operator M_I) $\rightarrow (\hat{w}^{\mu+1})_i$ (the fixed point of the operator $\hat{M}_I$) (42)

4. $$w^{\mu,0} \rightarrow (\hat{w}^{\mu})_i. \tag{43}$$

By virtue of (31), (35), (38) and (39) the following diagrams, which synthesize the above association, are commutative:

$$\begin{array}{ll} w \in \Re^{N_k} \longleftrightarrow (w_1, w_2) \in J_w & v \in \Re^{N_k} \longleftrightarrow (v_1, v_2) \in J_v \\ Mw \in \Re^{N_k} \longleftrightarrow \hat{M}(w_1, w_2) & M_I v \in \Re^{N_k} \longleftrightarrow \hat{M}_I(v_1, v_2) \end{array}$$

Properties of the sequences $(w^{\mu,\nu})_i$. The following properties are direct consequences of definitions (40)–(43).

- $(w^{\mu,\nu+1})_i = \min\left\{(Bw_{\hat{i}}^{\mu,\nu} + \hat{\gamma}_{\hat{i}}) + \hat{\psi}_i, (Bw_i^{\mu,\nu} + \hat{\gamma}_i)\right\}$, $\nu \geq 0$ (by (30)). (44)

- $$(\hat{w}^{\mu+1})_i = \begin{cases} (B\hat{w}_i^{\mu+1} + \hat{\gamma}_i) & \text{if } i \neq 1 \\ (B\hat{w}_{\hat{i}}^{\mu+1} + \hat{\gamma}_{\hat{i}}) + \hat{\psi}_i & \text{if } i = 1, \quad \text{(see (34))}. \end{cases} \tag{45}$$
- $(\hat{w}^{\mu+1})_i$ can be obtained by the following convergent iteration (see (36)):
$$(v^{\zeta+1})_i = \hat{M}_I(v_1^\zeta, v_2^\zeta) \tag{46}$$
with initial condition $(v^0)_i = (w^{\mu,\overline{\nu}(\mu)})_i$.

Proposition 4.1. *$(w^{\mu,\nu})_i$ is a nonincreasing sequence for $\mu \geq 1$, $\nu \geq 0$, in the following sense:*

$$(w^{\mu,\nu+1})_i \leq (w^{\mu,\nu})_i \tag{47}$$

$$(w^{\mu+1,0})_i \leq (w^{\mu,\overline{\nu}(0)})_i \tag{48}$$

Proof. In fact, by definition of $w^{\mu,0}$, we have for all $\mu \geq 1$: $(w^{\mu,0})_i = (\hat{w}^\mu)_i$. Given μ, we know by algorithm A1 that the following equality holds:

$$(\hat{w}^\mu)_i = \begin{cases} (B(\hat{w}^\mu)_i + \hat{\gamma}_i) & \text{if } i \neq I(w^{\mu-1,\overline{\nu}(\mu-1)}) \\ (B(\hat{w}^\mu)_{\hat{i}} + \hat{\gamma}_{\hat{i}} + \hat{\psi}_i) & \text{if } i = I(w^{\mu-1,\overline{\nu}(\mu-1)}) \end{cases}$$

With $\overline{\nu}(\mu)$ we mean the value of ν where the test of Step 3 is satisfied, therefore

$$(w^{\mu,1})_i = \left(\hat{M}(w_1^{\mu,0}, w_2^{\mu,0})\right)_i = \min\left\{B(w^{\mu,0})_{\hat{i}} + \hat{\gamma}_{\hat{i}} + \hat{\psi}_i, B(w^{\mu,0})_i + \hat{\gamma}_i\right\}$$
$$= \min\left\{B(\hat{w}^\mu)_{\hat{i}} + \hat{\gamma}_{\hat{i}} + \hat{\psi}_i, B(\hat{w}^\mu)_i + \hat{\gamma}_i\right\} \leq (\hat{w}^\mu)_i = (w^{\mu,0})_i$$

Repeating Steps 1 to 3 both sequences $(w^{\mu,\nu})_i$ are obtained, according to the following law:

$$(w^{\mu,\nu+1})_i = \left(\hat{M}(w_1^{\mu,\nu}, w_2^{\mu,\nu})\right)_i.$$

Then, by the monotony of the operator $\hat{M}$ (see (33)), we obtain that:

$$(w^{\mu,\nu+1})_i \leq (w^{\mu,\nu})_i, \quad \forall\, \nu \leq \overline{\nu}(\mu), \quad \forall\, \mu \geq 1$$

and consequently (47) is proved. By definition $(w^{\mu+1,0})_i = (\hat{w}^{\mu+1})_i$. In Step 4 the element $\hat{w}^{\mu+1}$ is generated as:

$$(\hat{w}^{\mu+1})_i = \left(\hat{M}_I(\hat{w}_1^{\mu+1}, \hat{w}_2^{\mu+1})\right)_i.$$

Since $(\hat{w}^{\mu+1})_i$ is the fixed point of the operator $\left(\hat{M}_I(w_1, w_2)\right)_i$, see (36) and (42), we have:

$$(\hat{w}^{\mu+1})_i = \lim_{\zeta\to\infty} v^\zeta.$$

where $(v^{\varsigma+1})_i = \left(\hat{M}_I(\hat{v}_1^\varsigma, \hat{v}_2^\varsigma)\right)_i$, with initial condition $(v^0)_i = (w^{\mu,\overline{\nu}(\mu)})_i$. But $(w^{\mu,\overline{\nu}(\mu)})_i = \left(\hat{M}(w_1^{\mu,\overline{\nu}(\mu)-1}, w_2^{\mu,\overline{\nu}(\mu)-1})\right)_i$, moreover, by (18)

$$\left(\hat{M}_I(w_1, w_2)\right)_i = \left(\hat{M}(w_1, w_2)\right)_i \quad \forall\, (w_1, w_2) \in J_w,\ I = I(w).$$

Then

$$\begin{aligned}&\left(\hat{M}(w_1^{\mu,\overline{\nu}(\mu)-1}, w_2^{\mu,\overline{\nu}(\mu)-1})\right)_i \\ &\quad = \left(\hat{M}_I(w_1^{\mu,\overline{\nu}(\mu)-1}, w_2^{\mu,\overline{\nu}(\mu)-1})\right)_i, \quad I = I(w^{\mu,\overline{\nu}(\mu)-1})\end{aligned}$$

therefore, by (47)

$$(v^0)_i = (w^{\mu,\overline{\nu}(\mu)})_i = \left(\hat{M}_I(w_1^{\mu,\overline{\nu}(\mu)-1}, w_2^{\mu,\overline{\nu}(\mu)-1})\right)_i \le (w^{\mu,\overline{\nu}(\mu)-1})_i$$

so

$$(v^1)_i = \left(\hat{M}_I(v_1^0, v_2^0)\right)_i \le \left(\hat{M}_I(w_1^{\mu,\overline{\nu}(\mu)-1}, w_2^{\mu,\overline{\nu}(\mu)-1})\right)_i = (v^0)_i.$$

From the monotony of $\hat{M}_I$ it follows that $(v^\varsigma)_i$ is not increasing and consequently:

$$(\hat{w}^{\mu+1})_i = \lim_{\varsigma\to\infty} (v^\varsigma)_i \le (v^0)_i = (w^{\mu,\overline{\nu}(\mu)})_i.$$

and this proves (48). □

Proof of Theorem 4.1. By condition (44) the test of Step 3 is satisfied after a finite number of iterations (on ν) leaving the [1,2,3] loop and generating a new element $\hat{w}^{\mu+1}$. Therefore the theorem will be proved if that sequence is finite. By Proposition 4.1 $(\hat{w}^\mu)_i$ is nonincreasing. Considering that there is only a finite number of possibilities for $I(j, \hat{w}^\mu)$, only a finite number of different values for $(\hat{w}^\mu)_i$ may be generated. Therefore there exists one $\overline{\mu}$ such that $(\hat{w}^{\overline{\mu}})_i = (\hat{w}^{\overline{\mu}+1})_i$. Since

$$(\hat{w}^{\overline{\mu}+1})_i \le (w^{\overline{\mu},\nu+1})_i \le (w^{\overline{\mu},\nu})_i \le (w^{\overline{\mu},0})_i = (\hat{w}^{\overline{\mu}})_i$$

we have $(w^{\overline{\mu},1})_i = (w^{\overline{\mu},0})_i$, then by virtue of (48) $w^{\overline{\mu},1} = w^{\overline{\mu},0}$. Hence, the test in Step 2 is satisfied for the pair of indexes ($\mu = \overline{\mu}, \nu = 0$) and therefore the fixed point of operator M is found. □

4.3. Analysis of different stopping rules

- **Rule R1:** Let $\Gamma : N_0 \to N$, where $N_0 = N \cup \{0\}$; we define:

$$R1(w^{\mu,0}, \ldots, w^{\mu,\nu+1}) = \begin{cases} 1 & \text{if } \nu \ge \Gamma(\mu) \\ 0 & \text{otherwise.} \end{cases}$$

It is obvious for this rule that condition (27) is satisfied.

- **Rule R2:**

$$R2(w^{\mu,0},\dots,w^{\mu,\nu+\hat{p}}) = \begin{cases} 1 & \text{if } I(j,w^{\mu,\nu}) = I(j,w^{\mu,\nu+p}),\ \forall\, j,\ 1 \le p \le \hat{p} \\ 0 & \text{otherwise.} \end{cases}$$

Using this rule, Algorithm A1 would be modified in the following way:

Description of the modified accelerated algorithm. Algorithm A1

Step 0: Set $\mu = 0$, $\nu = 0$, $w^{\mu,\nu} \in \Re^{N_k}$, and $\hat{p} \in N$ arbitrary.
Step 1: Set $p = 0$.
Step 2: Compute $w^{\mu,\nu+1} = Mw^{\mu,\nu}$.
Step 3: Set $\hat{I}(j) = I(j, w^{\mu,\nu})$, $\forall\, j$.
Step 4: If $w^{\mu,\nu+1} = w^{\mu,\nu}$, then stop (the discrete solution is $V^k = w^{\mu,\nu}$); else go to Step 5.
Step 5: If $\nu \ge 1$ and $I(\cdot, w^{\mu,\nu}) = I(\cdot, w^{\mu,\nu-1})$ then set $p = p + 1$, and go to Step 6; else set $p = 0$ $\nu = \nu + 1$, and go to Step 2.
Step 6: If $p \ge \hat{p}$ then compute $\hat{w}^{\mu+1}$ as the fixed point of the operator $M_{\hat{I}}$, set $w^{\mu+1,0} = \hat{w}^{\mu+1}$, $\nu = 0$, $\mu = \mu + 1$ and go to Step 1; else, set $\nu = \nu + 1$ and go Step 2.

Proposition 4.2. *If the following hypotheses hold:*

$$V^k(J) = \psi_2(j) \Rightarrow (BV^k + \gamma)(j) > \psi_2(j) \tag{49}$$
$$V^k(j) = \psi_1(j) \Rightarrow (BV^k + \gamma)(j) < \psi_1(j) \tag{50}$$

then condition (27) *is satisfied for rule R2.*

Proof. We have to prove that for all μ, the test of Step 6 is satisfied after a finite number of repetitions of the $[1,\dots,5]$-loop allowing the generation of a new element $\hat{w}^{\mu+1}$. In fact, by contradiction we assume that $\tilde{\mu}$ exists such that the test of Step 6 is never satisfied. Consequently the algorithm would generate an infinite sequence $w^{\tilde{\mu},\nu}$. This sequence, by definition of Step 2, is identical to the one generated by algorithm A0. In that case, we know that the sequence converges to V^k.

For each index j one of the three following conditions must be verified:

$$\text{i)}\quad I(j, V^k) = 0$$
$$\text{ii)}\quad I(j, V^k) = 1$$
$$\text{iii)}\quad I(j, V^k) = 2$$

In case i) we have $V^k(j) \in (\psi_1(j), \psi_2(j))$, there exists ν_0 such that

$$\forall\, \nu \ge \nu_0,\quad w^{\tilde{\mu},\nu}(j) \in (\psi_1(j), \psi_2(j))$$

and

$$I(j, w^{\tilde{\mu},\nu_0+p}) = 0, \quad \forall\, p \geq 0.$$

In case ii) $V^k(j) = \psi_1(j)$, and we have:

$$\lim_{\nu}(Bw^{\tilde{\mu},\nu} + \gamma)(j) = (BV^k + \gamma)(j) < \psi_1(j)$$

then for all $\nu \geq \nu_1$

$$(Bw^{\tilde{\mu},\nu} + \gamma)(j) < \psi_1(j)$$

so for all $\nu \geq \nu_1$, we have $I(j, w^{\tilde{\mu},\nu}) = 1$.

Case iii) is completely analogous to case ii).

Then after a finite number of iterations, $I(\cdot, w)$ remains constant, and the test of Step 6 is verified; this allows the generation of a new element $\hat{w}^{\mu+1}$. □

- **Rule R3:**

To avoid the restrictive condition used in Proposition 4.2, let $\hat{p} \in N$, $\varepsilon_\nu = K\|w^{\mu,\nu} - w^{\mu,\nu-1}\|$, $K \geq 2\alpha/(1-\alpha)$, given by (24).

We define:

$$(M_I w^{\mu,\nu})(j) = \begin{cases} \psi_1(j) & \text{if } I(j) = 1 \\ \psi_2(j) & \text{if } I(j) = 2 \\ (Bw^{\mu,\nu} + \gamma)(j) & \text{if } I(j) = 0 \end{cases}$$

We introduce the following notation:

$$I_{\varepsilon_\nu}(j, w^{\mu,\nu}) = \{i \,/\, \exists\, \vec{I},\ I(j) = i, \text{ with } \left|w^{\mu,\nu+1}(j) - (M_I w^{\mu,\nu})(j)\right| < \varepsilon_\nu\} \tag{51}$$

and we define:

$$R3(w^{\mu,0}, \ldots, w^{\mu,\nu+\hat{p}}) = \begin{cases} 1 & \text{if } I_{\varepsilon_\nu}(j, w^{\mu,\nu}) = I_{\varepsilon_\nu}(j, w^{\mu,\nu+p})\ \forall\, 1 \leq j \leq n \\ 0 & \text{otherwise.} \end{cases}$$

Proposition 4.3. *Condition* (27) *is satisfied by Rule 3.*

Proof. In the same way as in the proof of Proposition 4.2, we have to prove that it is not possible to obtain an infinite sequence

$$\{w^{\mu,\nu},\ \nu = 1, \ldots\}, \quad \text{such that } R3(w^{\mu,0}, \ldots, w^{\mu,\nu}) = 0. \tag{52}$$

Assume an infinite sequence for some $\hat{\mu}$ is generated, then it converges to V^k. But if the hypotheses of Proposition 4.2 hold, then

$$\psi_2(j) = V^k(j) \Rightarrow (BV^k + \gamma)(j) > \psi_2(j),$$
$$\psi_1(j) = V^k(j) \Rightarrow (BV^k + \gamma)(j) < \psi_1(j).$$

In the same way as in the proof of Proposition 4.2 we have that for each j, only one of the following conditions must be true:

i) If $I(j, V^k) = 0$, then $I_{\varepsilon_\nu}(j, w^{\mu,\nu+p}) = \{0\}$, $\forall\, \nu > \nu_0$, $p \geq 0$.
ii) If $I(j, V^k) = 1$, then $I_{\varepsilon_\nu}(j, w^{\mu,\nu+p}) = \{1\}$, $\forall\, \nu > \nu_1$, $p \geq 0$.
iii) Similarly for $I(j, V^k) = 2$.

It remains to prove that (27) is valid for all those cases where the hypotheses of Proposition 4.2 are not satisfied. That is, when

$$\psi_2(j) = V^k(j) \text{ and } (BV^k + \gamma)(j) = \psi_2(j),$$
$$\psi_1(j) = V^k(j) \text{ and } (BV^k + \gamma)(j) = \psi_1(j).$$

We will only prove it for the first case, (it is analogous for the second). By continuity of the operator M, for all $\nu \geq \nu_2$, $I_{\varepsilon_\nu}(j, w^{\mu,\nu}) \subset \{0, 2\}$.

Let us see that $2 \in I_{\varepsilon_\nu}(j, w^{\mu,\nu})\ \forall\, \nu \geq \nu_2$. M is a contraction, in fact by (24) $\exists\, \alpha \in (0,1)$ such that

$$\|Mw^{\mu,\nu-1} - MV^k\| \leq \alpha \|w^{\mu,\nu-1} - \overline{w}\|.$$

By the triangular property:

$$\|w^{\mu,\nu} - V^k\| \leq \alpha \|w^{\mu,\nu-1} - V^k\| + \alpha \|w^{\mu,\nu-1} - w^{\mu,\nu}\|$$

from this inequality:

$$(1 - \alpha)\|w^{\mu,\nu} - V^k\| \leq \alpha \|w^{\mu,\nu-1} - w^{\mu,\nu}\| \tag{53}$$

then

$$\begin{aligned} |w^{\mu,\nu}(j) - \psi_2(j)| &= \left|w^{\mu,\nu}(j) - V^k(j)\right| \\ &\leq \|w^{\mu,\nu} - V^k\| \\ &\leq \frac{\alpha}{(1-\alpha)}\|w^{\mu,\nu-1} - w^{\mu,\nu}\| < K\|w^{\mu,\nu-1} - w^{\mu,\nu}\| \end{aligned} \tag{54}$$

and consequently, by virtue of (51), $2 \in I_{\varepsilon_\nu}(j, w^{\mu,\nu})\ \forall\, \nu \geq \nu_2$.

Let us show now that $0 \in I_{\varepsilon_\nu}(j, w^{\mu,\nu})\ \forall\, \nu \geq \nu_2$.

$$\begin{aligned} \|Bw^{\mu,\nu} + \gamma - w^{\mu,\nu+1}\| &= \|B(w^{\mu,\nu} - V^k) + V^k - w^{\mu,\nu+1}\| \\ &\leq \alpha\|w^{\mu,\nu} - V^k\| + \|w^{\mu,\nu+1} - V^k\| \leq 2\alpha \|w^{\mu,\nu} - V^k\| \end{aligned}$$

From this inequality and (53):

$$\begin{aligned} \left|(Bw^{\mu,\nu} + \gamma)(j) - w^{\mu,\nu+1}(j)\right| &\leq \frac{2\alpha^2}{(1-\alpha)}\|w^{\mu,\nu-1} - w^{\mu,\nu}\| \\ &\leq K\|w^{\mu,\nu-1} - w^{\mu,\nu}\| \end{aligned} \tag{55}$$

and consequently by (51), $0 \in I_{\varepsilon_\nu}(j, w^{\mu,\nu}) \ \forall\, \nu \geq \nu_2$.

Thus, we have proved that for every possible case there exists $\hat{\nu}$, such that the families of indices $I_{\varepsilon_\nu}(j, w^{\mu,\nu})$ satisfy $R3(w^{\mu,0}, \ldots, w^{\mu,\hat{\nu}+\hat{p}}) = 1$, for $\nu \geq \hat{\nu}$. This proves the impossibility of constructing sequence (52). □

• **Rule R4:** Let $\bar{\varepsilon} > 0$, $r \in (0,1]$ and $\varepsilon_\nu = \bar{\varepsilon}\|w^{\mu,\nu} - w^{\mu,\nu-1}\|^r$. We define:

$$R4(w^{\mu,0}, \ldots, w^{\mu,\nu+1}) = \begin{cases} 1 & \text{if } \|w^{\mu,\nu+1} - w^{\mu,\nu}\| \leq \varepsilon_\nu \\ 0 & \text{in other case.} \end{cases}$$

It is obvious that condition (27) is satisfied $\forall\, \nu \geq \overline{\nu}$, $\overline{\nu}$ being the first index such that

$$\|w^{\mu,\overline{\nu}} - w^{\mu,\overline{\nu}-1}\| < \left(\frac{\bar{\varepsilon}}{\alpha}\right)^{1/(1-r)}.$$

• **Rule R5:** Let $\hat{p} \in N$, $\varepsilon_\nu = \|w^{\mu,\nu} - w^{\mu,\overline{\nu}-1}\|^r$, $r \in (0,1)$,

$$\begin{aligned} \varepsilon_\nu(j, w^{\mu,\nu}) = \{&i/\exists\, \vec{I} \text{ satisfying } I(j) = i \\ &\text{and } |w^{\mu,\nu+1}(j) - (M_I w^{\mu,\nu})(j)| < \varepsilon_\nu\}. \end{aligned}$$

So we define:

$$\begin{aligned} &R5(w^{\mu,0}, \ldots, w^{\mu,\nu+\hat{p}}) \\ &\quad = \begin{cases} 1 & \text{if } I_{\varepsilon_\nu}(j, w^{\mu,\nu}) = I_{\varepsilon_\nu}(j, w^{\mu,\nu+p}) \text{ with } 1 \leq p \leq \hat{p}, \\ 0 & \text{otherwise.} \end{cases} \end{aligned}$$

As in the previous cases we prove that it is not possible to obtain an infinite sequence:

$$\{w^{\mu,\nu},\ \nu = 1, \ldots\}, \ \text{ satisfying } \ R5(w^{\mu,o}, \ldots, w^{\mu,\nu}) = 0.$$

The proof combines the arguments used for the case of rule 3 and rule 4.

5. Numerical Examples

In the following tables we will show some results concerning computing times and number of iterations of the usual fixed point Algorithm A0, and accelerated algorithm A1 implemented with rule R3. Table 1 corresponds to a problem where $N_k = 36$ and Table 2 to other one where $N_k = 64$.

The problem's data:

$$\begin{aligned} &\Omega = [-1, 1] \times [0, 2] \\ &g_1(x_1, x_2) = (x_2 - 1) - 6.7(\rho - \rho_0)x_1, \\ &g_2(x_1, x_2) = (-x_1 - 6.7(\rho - \rho_0)(x_2 - 1) \end{aligned}$$

where

$\rho_0 = \sqrt{0.5}, \quad \rho = \sqrt{(x_1)^2 + (x_2 - 1)^2}$

$f(x_1, x_2) = 0.007\cos(33\, x_1) + 0.0067\sin(44\, x_2)$

$\psi_1(x_1, x_2) = -1 + 4(\rho - \rho_0), \quad \psi_2(x_1, x_2) = 1 + 4(\rho - \rho_0)$

λ is the discount rate and ε is the parameter used in algorithm A0.

Table 1. Comparison between Algorithms A0 and A1

ε	10^{-3}	10^{-6}	10^{-3}	10^{-6}
λ	A1 A0	A1 A0	A1 A0	A1 A0
1.0	19″ 1′14″	19″ 35″	5 22	5 51
0.4	21″ 2′56″	21″ 1′23″	6 57	6 124
0.2	25″ 5′52″	25″ 2′53″	8 123	8 253

Table 2. Comparison between Algorithms A0 and A1

	times		iterations	
ε	A1	A0	A1	A0
10^{-3}	17″	1′8″	5	86
10^{-6}	17″	1′38″	5	125
10^{-7}	17″	2′8″	5	165
10^{-9}	17″	2′37″	5	205

6. Conclusions

In this work we have studied a differential game problem with stopping times. This problem is equivalent to the analysis of a bilateral Isaacs inequality. To have a numerical approximation, we define a discretization procedure that reduces the game problem to finding the fixed point of a contractive operator. We have also obtained an explicit estimate of the rate of convergence for the discretization procedure.

To eliminate the very well known inefficiency of the usual fixed point algorithm, we have developed an accelerated ad-hoc method. Our numerical procedure consists of a suitable extension of the methodology presented in [10].

The algorithm combines Picards and policy iteration methods. Basically, it consists on the resolution of a finite number of suitable chosen systems of linear equations. The choice of terminating the usual Picard iteration is given by the use of various stopping rules, some of which have been analyzed in Section 4. These rules generalize the "control repetitions" rule used in [11] and allow the resolution of those problems where the restrictive conditions requested in [11] to assure the convergence of the algorithm are not satisfied.

The use of these rules, and extensions of the algorithms presented here and in [11], are considered in [12] to find the solution of more general fixed point problems; extensions and improvements of this methodology can be found in [7].

References

[1] A. Bensoussan and J.L. Lions, *Applications des Inéquations Variationnelles en Contrôle Stochastique*, Dunod, Paris, 1978.

[2] S.A. Belbas and I.D. Mayergoyz, Applications of fixed-point methods to discrete variational and quasi-variational inequalities, *Numerische Mathematik*, vol. 51, (1987), 631–654.

[3] I. Dolcetta Capuzzo and H. Ishii, Approximate solution of the Bellman equation of deterministic control theory, *Appl. Math. Optim.*, vol. 11, (1984), 161–181.

[4] P.G. Ciarlet, Discrete maximum principle for finite-difference operators, *Aequations Math.*, vol. 4, (1970), 338–352.

[5] P.G. Ciarlet and P.A. Raviart, Maximum principle and uniform convergence for the finite element method, *Computer Methods in Applied Mechanics and Engineering*, vol. 2, (1973), 17–31.

[6] M.G. Crandall, L.C. Evans, and P.L. Lions, Some properties of viscosity solutions of Hamilton–Jacobi equations, *Trans. Amer. Math. Soc.*, vol. 282, (1984), 487–502.

[7] S. Di Marco, Técnicas de descomposición-agregación en el tratamiento de la inecuación bilátera de Isaacs, *Mecánica Computacional*, vol. 12, AMCA, (1991), 509–518.

[8] A. Friedman, *Differential Games*, Wiley-Interscience, New York, 1971.

[9] M.G. Garroni and M.A. Vivaldi, Bilateral inequalities and implicit unilateral system of the non-variational type, *Manuscripta Mathematica*, 33, (1980), 177–215.

[10] R.L.V. Gonzalez and E. Rofman, On deterministic control problems: an approximation procedure for the optimal cost, Part I and II, *SIAM Journal on Control and Optimization*, vol. 23, (1985), 242–285.

[11] R.L.V. Gonzalez and C. Sagastizabal, Un algorithme pour la résolution rapide d'équations discrètes de Hamilton–Jacobi–Bellman, *Comptes Rendus Acad. Sc. Paris*, Serie I, Tome 311, (1990), 45–50.

[12] R.L.V. Gonzalez and M.M. Tidball, Fast solution of general nonlinear fixed point problems. In *System Modelling and Optimization*, Proceedings of the 15th IFIP Conference on System Modelling and Optimization, Zurich, Switzerland, September 2–6, 1991. *Lecture Notes in Control and Information Sciences*, vol. 180, pp. 35–44, Springer Verlag, New York, 1992.

[13] R.L.V. Gonzalez and M.M. Tidball, Fast solution of discrete Isaacs inequalities, *Rapport de Recherche*, Nr. 1167, INRIA, 1990.

[14] R.L.V. Gonzalez and M.M. Tidball, On a discrete time approximation of the Hamilton-Jacobi equation of dynamic programming, *Rapport de Recherche*, Nr. 1375, INRIA, 1991.

[15] R.L.V. Gonzalez and M.M. Tidball, On the rate of convergence of fully discrete solutions of Hamilton–Jacobi equations, *Rapport de Recherche*, Nr. 1376, INRIA, 1991.

[16] R.L.V. Gonzalez and M.M. Tidball, Sur l'ordre de convergence des solutions discrétisées en temps et en espace de l'équation de Hamilton–Jacobi, *Comptes Rendus Acad. Sci.*, Paris, Tomo 314, Série I, pp. 479–482, 1992.

[17] R.L.V. Gonzalez and M.M. Tidball, Numerical solution of zero sum differential games with stopping times. Unpublished paper.

[18] O. Nakoulima, Etude d'une inéquation variationnelle bilaterale et d'un système d'inéquations quasi-variationnelles unilaterales associé, Thèse 3r. Cycle, Université de Bordeaux I, Bordeaux, 1977.

[19] S.M. Ross, *Applied Probability Models with Optimization Applications*, Holden-Day, San Francisco, 1970.

[20] P.E. Souganidis, Max-min representation and product formulas for the viscosity solution of Hamilton–Jacobi equations with application to differential games, *Nonlinear Analysis, Theory, Methods and Applications*, vol. 9, no. 3, (1985), 217–257.

[21] L. Stettner, Zero-sum Markov games with stopping and impulsive strategies, *Appl. Math. Optim.*, vol. 9, (1982), 1–24.

[22] G. Strang and G. Fix, *An Analysis of the Finite Element Method*, Prentice-Hall, Englewood Cliffs, New York, 1973.

[23] M.M. Tidball, Sobre la resolución numérica de las ecuaciones de Hamilton–Jacobi–Bellman, Thesis, Universidad Nacional de Rosario, 1991.

Mabel M. Tidball
Departamento de Matemática
Facultad de Ciencias Exactas
Ingeniería y Agrimensura

R.L.V. González
Universidad Nacional de Rosario
Avda. Pellegrini 250
2000 Rosario, Argentina

Singular Paths in Differential Games with Simple Motion

Arik A. Melikyan

Abstract. The generalized viscosity solution of the basic (Bellman–Isaacs) equation in dynamic game theory is considered [1]. An assumptiom is made that the solution (game value) is nonsmooth on some hypersurface. Local necessary conditions in the form of equalities and inequalities are obtained for viscosity solution under different assumptions about the behaviour of the optimal paths in the vicinity of singular surface. The singular surfaces are investigated, which contain (as equivocal or focal ones) or not (dispersal one) the optimal singular paths [2,3]. Using the equality-conditions and singular charactristics technique [4] the equations of motion for the former surfaces are obtained. These equations appear to be not of Hamiltonian type but more general. Some results of this paper are obtained using other approaches in [4,5]; their connection with the properties of viscosity solutions is stated here.

1. Viscosity Solution of the Basic Equation

For many problems in optimal control and differential games the basic equation has the form:

$$\begin{gathered} H(x, u(x), p(x)) = 0, \quad x \in \Omega, \;\; u(x) = w(x), \;\; x \in M \subset \partial\Omega \\ p = \partial u / \partial x = u_x \end{gathered} \tag{1}$$

where u is the game value, $x \in \mathbf{R}^N$–the state vector. In the neighborhood of a singularity the problem (1) is considered in generalized sense [1].

A continuous function $u\colon \Omega \to \mathbf{R}^1$ is called viscosity solution of the boundary value problem (1) if for every test-function $\varphi(x) \in C^1(\Omega)$, such that local minimum (maximum) of $u(x) - \varphi(x)$ is attained at $x_0 \in \Omega$, the inequality holds:

$$H\big(x_0, u(x_0), \varphi_x(x_0)\big) \leq 0 \quad \big(H(x_0, u(x_0), \varphi_x(x_0)) \geq 0\big) \tag{2}$$

Here Ω is an open set in $\mathbf{R}^N$, Hamiltonian H is a continuous function, $H\colon \Omega \times \mathbf{R}^1 \times \mathbf{R}^N \to \mathbf{R}^1, w(x) \in C(\partial\Omega)$. This is one of the possible definitions of the viscosity solutions [1].

For some cases the existence and uniqueness of viscosity solutions are proved, and their connection with generalized optimality conditions in differential games is found [1,6].

The formulation in (1), (2) corresponds, for example, to an autonomous differential game with terminal set $M \subset \mathbf{R}^N \backslash \Omega$. Local considerations of this paper can be applied to nonautonomous problem as well if one of the components of x is treated as the time.

2. Simplest Singularity

Let the solution $u(x)$ of the problem (1), (2) exist and in a neighborhood D of a point $x^* \in \Omega$ have the properties: $u_i(x) \in C^2(D_i)$, $i = 0, 1$. Here D_i are open half-neighborhoods of x^* separated by smooth hypersurface Γ; $D = D_0 + \Gamma + D_1$; $u_i(x)$ is the restriction of $u(x)$ on D_i. Let the gradients $p_i(x) = u_{ix}(x)$ have continuous extensions up to Γ and nonzero vector $p_1(x) - p_0(x)$, $x \in \Gamma$, be directed from D_1 to D_0, i.e. if the functions $u_i(x)$ have smooth extensions on D, then $u(x) = \min[u_0(x), u_1(x)]$, $x \in D$. Thus, in domains D_i the solution $u(x)$ satisfies the equality (1) and on the surface Γ condition (2) must hold.

In the neighborhoods $N_i \subset \mathbf{R}^{2N+1}$ of the points $(x^*, u(x^*), p_i(x^*))$ the Hamiltonian $H(x, u, p)$ is supposed either to be twice differentiable or to have one of the forms: $H = \max[H^+, H^-]$ or $H = \min[H^+, H^-]$ with $H^\pm(x, u, p) \in C^2(N_i)$.

In optimal control and game problems Hamiltonians are constructed as min- or minmax-functions and usually do not depend upon Bellman function u:

$$H(x,p) = \min_\alpha \langle p, f(x,\alpha)\rangle, \quad \alpha \in A$$

$$H(x,p) = \min_\alpha \max_\beta \langle p, f(x,\alpha,\beta)\rangle, \quad \beta \in B.$$

The above functions H^+, H^- are some extensions in (x,p)-space of certain branches of H. The very existence of such an extension is an additional assumption.

3. Necessary Conditions for Viscosity Solution

Assumptions about the functions u, H made above lead to several consequences.

Lemma 1. *There exists no test-function $\varphi(x) \in C^1(D)$, such that local minimum of $u - \varphi$ is attained at $x \in \Gamma$.*

Lemma 2. *The functions $u_i(x)$ have smooth extensions in the whole neighborhood D. For any test-function $\varphi_\lambda(x) = \lambda u_1(x) + (1-\lambda)u_0(x)$, $0 < \lambda < 1$, maximum of $u - \varphi_\lambda$ is equal to zero and is attained at $x \in \Gamma$.*

Lemma 3. *For $x \in \Gamma$ and for all $\lambda \in [0,1]$ the following inequality holds:*

$$f(\lambda) \equiv H(x, u(x), p_\lambda(x)) \geq 0; \quad p_\lambda(x) = \varphi_{\lambda x}(x) = \lambda p_1(x) + (1-\lambda)p_0(x)$$

in the points of Hamiltonian's smoothness and at the ends of the interval $[0,1]$ *the function* $f(\lambda)$ *has the properties:*

$$f'(\lambda) = \langle H_p(x, u, p_\lambda), p_1 - p_0 \rangle$$

$$f(0) = 0, \ f'(+0) \geq 0; \ f(1) = 0, \ f'(1-0) \leq 0. \tag{3}$$

The proof of the Lemmas is rather simple. Lemma 1 means that only one inequality (2) has to be checked; Lemma 2 supplies test-functions for such a check. The conditions (3) give some geometric restrictions for the behaviour of the x-projection of the characteristics in the neighborhood of Γ, since the vector $p_1 - p_0$ is normal to the surface Γ and the characteristical system for the equation (1) is the following one:

$$\dot{x} = H_p, \ \dot{u} = \langle p, H_p \rangle, \ \dot{p} = -H_x - pH_u. \tag{4}$$

For smooth Hamiltonian inequalities (3) mean, that x-paths do not approach Γ from each side. So, some additional condition must hold, if the paths approach Γ from some side. Such a side, D_1 or/and D_0, will be called a singular side (of the singular surface Γ).

If the solution's singularity is of the form $u = \max[u_0, u_1]$, i.e. the vector $p_1 - p_0$ is directed to D_1, then the extrema in Lemmas 1, 2 and inequalities in Lemma 3 must be opposite.

4. Dispersal Surface

Suppose that the characteristics' projection leaves the surface Γ under nonzero angle, while $H \in C^2(N_i)$, $i = 0, 1$. Since the vector $p_1 - p_0$, $x \in \Gamma$, is directed from D_1 to D_0, the above assumptions mean:

$$\langle H_p(x, u, p_0), p_1 - p_0 \rangle > 0; \quad \langle H_p(x, u, p_1), p_1 - p_0 \rangle < 0.$$

These inequalities, in particular, guarantee the extension of the class C^2 for $u_i(x)$ in D as the solutions of $H(x, u_i(x), u_{ix}(x)) = 0$.

Thus, necessary conditions (3) are satisfied and none of sides of Γ is singular. But the check of the inequality $f(\lambda) \geq 0$ from Lemma 3 on the whole interval $\lambda \in [0,1]$ can reveal an edge of the dispersal surface Γ. This is not the case for the control problem with linear dynamics: $f(x, \alpha) = f_0(x) + \alpha f_1(x)$, $\alpha_0 \leq \alpha \leq \alpha_1$.

Lemma 4. *The function $f(\lambda)$ in linear control problem is piecewise linear, continuous and non-negative (Fig. 1).*

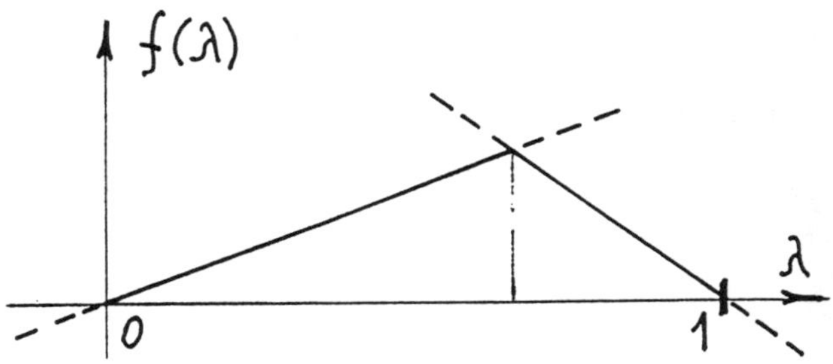

Figure 1

The example of other kind supplies the game problem of simple pursuit on a manifold [5].

The Hamiltonian of the problem and solution's branches are the followings ones:

$$H(x,p) = -\langle A(z)r,r\rangle^{1/2} + \nu\langle A(y)s,s\rangle^{1/2} + 1$$

$$u_0(x) = L^0(x)/(1-\nu), \quad u_1(x) = L^1(x)/(1-\nu)$$

$$z,y,r,s \in \mathbf{R}^n, \; x = (z,y) \in \mathbf{R}^{2n}, \; p = (r,s) \in \mathbf{R}^{2n}, \; N = 2n.$$

Here $A^{-1}(z)$ is the metric tensor of the manifold; $1, \nu$ are the velocities of motions of pursuer and evader, $\nu < 1$; $L^i(x)$ are the lengths of two different geodesic lines, connecting points z, y where pursuer and evader are situated; $u_i(x)$ are primary times of pursuit along the geodesics. The equalities $H(x, u_{ix}(x)) = 0$, $x \in D$, hold here due to eikonal equations [5]: $\langle A(z)L_z^i, L_z^i\rangle = 1$, $\langle A(y)L_y^i, L_y^i\rangle = 1$.

A part of the surface $\Gamma_0 = \{x \in \mathbf{R}^N;\; L^0(x) = L^1(x)\}$ of the equal pursuit times along two geodesics is a dispersal surface. The analysis of the function

$$f(\lambda) \equiv H\big(x, \lambda p_1 + (1-\lambda)p_0\big), \quad 0 \le \lambda \le 1, \; f(0) = f(1) = 0$$

shows that it has a local minimum at $\lambda = 1/2$ and two local maxima in the intervals $(0, 1/2)$, $(1/2, 1)$. Depending upon $x \in \Gamma_0$ the value of $f(1/2)$ may be positive as well as negative. Thus, dispersal surface is conditioned by:

$$\Gamma\colon L^0(x) = L^1(x); \quad H(x, (p_0 + p_1)/2) \ge 0. \tag{5}$$

For some manifolds the edge of Γ, defined by the equality $H(x, (p_0 + p_1)/2) = 0$ is nonempty [5]. This equality was found in [5] using other approach, A.I. Subbotin's conditions [6].

The function $f(\lambda)$, $\lambda \in [0,1]$, for different $x \in \Gamma_0$ is presented in Fig. 2. Here B_0 stands for the set of points $x \in \Gamma_0$, where x-projections of characteristics are tangent to Γ_0 from both sides: $\langle H_p(x,p_i), p_1 - p_0\rangle = 0$, $i = 0,1$, $x \in \Gamma_0$. One can see from the Fig. 2 that the part of Γ_0 lying between B and B_0 geometrically is still dispersal surface, but can't represent a viscosity solution.

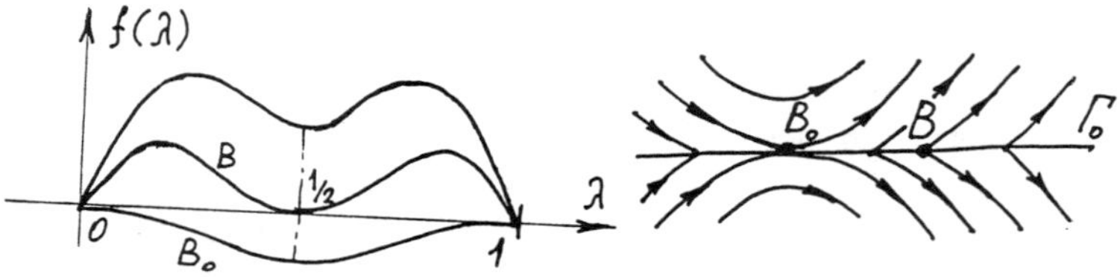

Figure 2

5. Equivocal Surface

Suppose now that the charactristics' projections approach (maybe, with tangency) from D_1 to Γ and leave Γ for D_0 under nonzero angles. Geometrically these conditions can be written as

$$\langle H_p(x,u,p_0), p_1 - p_0\rangle > 0,\ \langle H_p(x,u,p_1), p_1 - p_0\rangle \geq 0,\ x \in \Gamma. \quad (6)$$

Relations (6) must be in accord with the inequalities (3) of the Lemma 3. Left inequalities (3) and (6) are identical. Thus, the D_1 side of Γ is singular and certain conditions must hold on Γ from this side. We indicate two cases, when such a condition keeps in accord right inequalities as well. The first case is the problem with the Hamiltonian smooth in the neighborhood N_1.

Lemma 5. *Let $H \in C^2(N_1)$. Then the following tangency condition and inequality—necessary conditions for viscosity solution—hold:*

$$\begin{aligned} &f'(1) = \{F_1 H\} = \langle H_p(x, u(x), p_1(x)), p_1(x) - p_0(x)\rangle = 0,\\ &f''(1) = -\{\{F_1 H\} F_1\} \geq 0, \quad x \in \Gamma. \end{aligned} \quad (7)$$

The first statement of the Lemma follows immediately comparing (3) and (6). The function $F_1 = F_1(x, u_1)$ and the brackets $\{\ ,\ \}$ are defined below in (9), (10).

For the second case Hamiltonian in the domain N_1 has the form $H = \max[H^+, H^-]$, $H^\pm \in C^2(N_1)$, while $H^+(x, u_1(x), p_1(x)) = 0$, $x \in D_1$.

Lemma 6. *Let the projections of characteristics approach from D_1 to Γ without tangency (both inequalities in (6) are strict). Then the following necessary conditions for viscosity solution are true:*

$$\begin{aligned} &H^-(x, u(x), p_1(x)) = 0, \quad x \in \Gamma\\ &\langle H_p^-(x, u(x), p_1(x)), p_1(x) - p_0(x)\rangle \leq 0, \end{aligned} \quad (8)$$

Proof. The above assumptions mean that for some subinterval $[\lambda_1, 1] \subset [0, 1]$ the function f from Lemma 3 has the form: $f(\lambda) = \max[f^+(\lambda), f^-(\lambda)]$, while $f(1) = f^+(1)$. Let $\lambda^* \in [\lambda_1, 1]$ be a point where $f^+(\lambda^*) = f^-(\lambda^*)$. If we suppose that λ^* doesn't exist or $\lambda^* < 1$, then right inequality in (3) is violated. Hence, $\lambda^* = 1$ and the equality (8) holds. The inequality (8) means $f^{-\prime}(1) \leq 0$ and follows from (3).

Thus, if in D the equation $H^+ = 0$ holds, then H^- equals to zero on Γ. Note that representation of the type $H = \min[H^+, H^-]$ doesn't allow to obtain conditions (3). But if we suppose $u = \max[u_0,\ u_1]$, then this kind of singularity is compatible with the Hamiltonian of the form $H = \min[H^+, H^-]$.

The functions $f(\lambda)$ for the cases (7) and (8) are presented on Figs. 3 and 4.

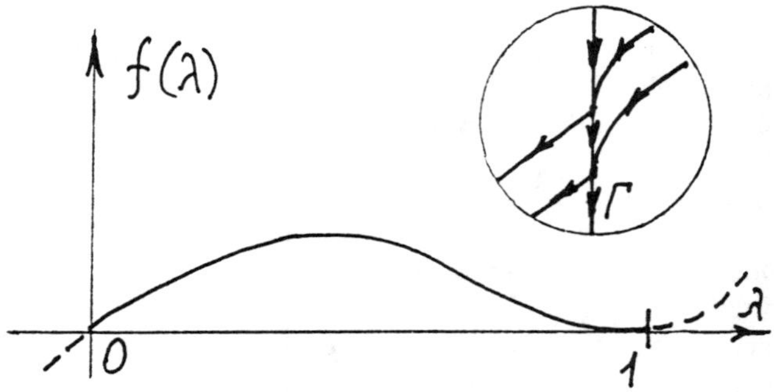

Figure 3

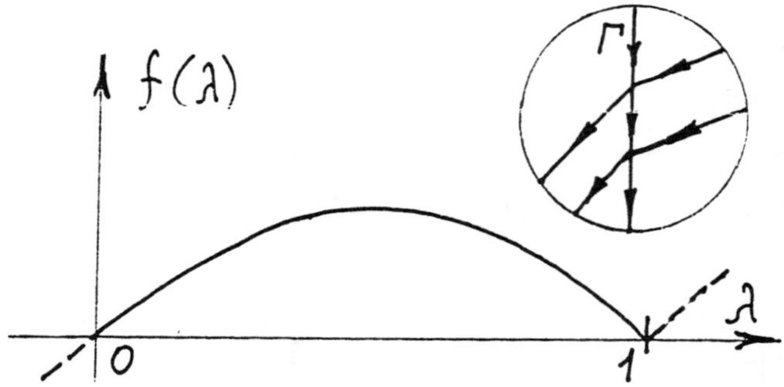

Figure 4

In addition to the equation (1) equality (7) or (8) gives one more condition on the surface Γ. Including the continuity condition of viscosity solution three equality conditions on the surface Γ hold:

$$\begin{aligned} &F_0(x, u_1, p_1) \equiv H = H^+ = 0, \quad F_1(x, u_1) \equiv u_1 - u_0(x) = 0, \\ &F_{-1}(x, u_1, p_1) = 0. \end{aligned} \tag{9}$$

Depending upon the case the function F_{-1} has one of the forms (7) or (8). Arguments for the functions $F_{\pm 1}$ are indicated under assumption that functions $u_0(x)$, $p_0(x)$ are known in advance, which is the case, if the inverse time technique is used [2].

The conditions (9) are very important for backward technique of construction of viscosity solution in the vicinity of the given singularity. Indeed, in inverse time the domain D_0 is first filled by the characteristics(x-parts of the system's (4) solutions). Since the Hamiltonian H is smooth in N_0, the characteristics do not indicate the existence of the singularity Γ. In direct time such an indication is the vanishing of H^- on Γ, if (4) is written in terms of H^+. Thus, for further extension into D_1 the surface Γ must be found using some special technique.

6. Focal Surface

Here optimal regular paths approach Γ from both sides, the latters being singular ones. Depending upon the smoothness of Hamiltonian the conditions of the form (7) or (8) are true for each side of Γ. So, three types of focal surfaces can exist: 1) regular paths contact Γ from both sides, Fig. 5; 2) paths contact Γ from one side, Fig. 6; 3) regular paths approach Γ under nonzero angle from both sides, Fig. 7. On Figs. 5–7 the corresponding functions $f(\lambda)$ are presented. The case 3) is called in [7] singular focal surface.

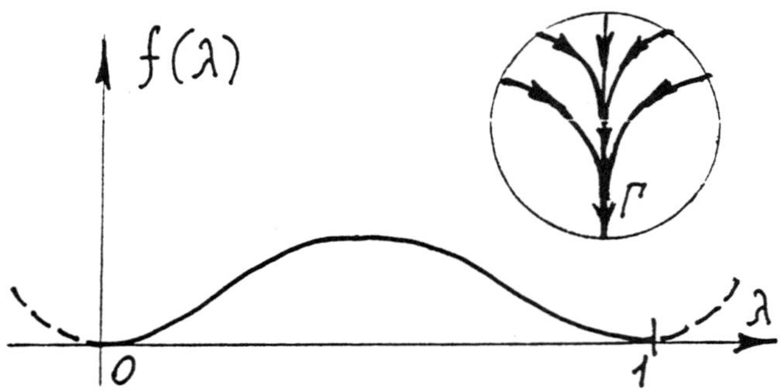

Figure 5

7. The Method of Singular Characteristics

One of the techniques for construction of Γ is the method of singular characteristics, developed in several papers of the author and described in [4]. The singular surface Γ is constructed using the system (4) based on singular Hamiltonian $H = H^\sigma$ of a special type, expressed in terms of functions (9):

$$\mu H^\sigma(x, u, p) = \{F_{-1}F_0\}F_1 + \{F_1F_{-1}\}F_0 + \{F_0F_1\}F_{-1}. \tag{10}$$

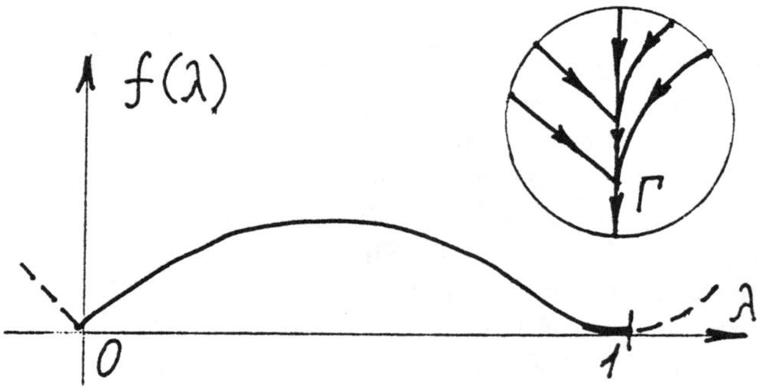

Figure 6

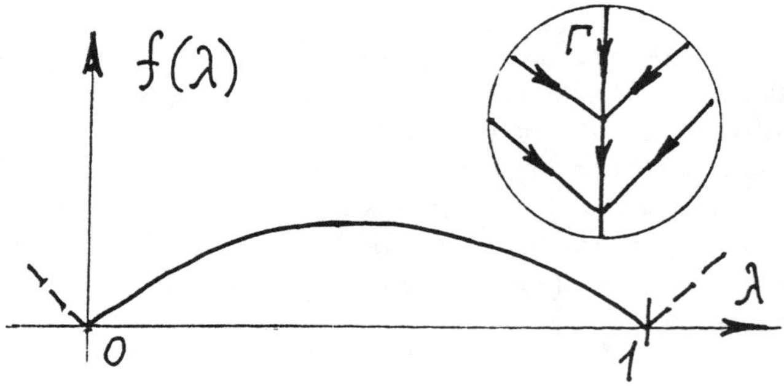

Figure 7

Here $\{FG\} = \langle F_x + pF_u, G_p\rangle - \langle G_x + pG_u, F_p\rangle$ are the Jacobi brackets; μ is nonzero homogenuity multiplier. The system (4) with Hamiltonian (10) must be restricted to the manifold

$$W = \{Q \in \mathbf{R}^{2N+1} : F_i(Q) = 0,\ |i| \leq 1\}, \quad Q = (x, u, p) \tag{11}$$

defined by the functions (9). This leads to the system

$$\begin{gathered}\dot{x} = \sum_i \lambda_i F_{ip},\ \dot{u} = \sum_i \lambda_i \langle p, F_{ip}\rangle,\ \dot{p} = -\sum_i \lambda_i (F_{ix} + pF_{iu}), |i| \leq 1 \\ \lambda_1 = \{F_{-1}F_0\},\ \lambda_0 = \{F_1 F_{-1}\},\ \lambda_{-1} = \{F_0 F_1\},\end{gathered} \tag{12}$$

The manifold W is invariant for the system (12), i.e. the functions F_i are first integrals of it.

Usually the Hamiltonian doesn't depend upon u for game problems, $H = H(x, p)$. Hence, the Hamiltonian part of the system of regular characteristics (4) is separated: $\dot{x} = H_p, \dot{p} = -H_x$. But the system of singular characteristics due to the presence of the function $F_1 \equiv u_1 - u_0(x)$, which determines the continuity condition, in the Hamiltonian (10), isn't already of a Hamiltonian type and has more general form (4). The second equation (4) again is separated due to linearity in u of the Hamiltonian (10).

For the integration of the system of singular characteristics (12) some initial data (Cauchy data) must be given on a manifold $B \subset \Gamma$, $\dim B = N-2$. The construction of the equivocal surface in the game of pursuit on a manifold is performed in [5]. The edge of dispersal surface (5) from Section 4 was taken as the set B with initial data: $p(x) = (p_0(x)+p_1(x))/2$, $x \in B$. The multiplier μ in (10) must be chosen in such a way which provides the optimal velocity $\dot{x} = H_p$ on Γ. In time optimal problem, for example, the following condition must hold: $\dot{u} = \langle p, H_p \rangle = -1$.

8. Equations for Equivocal Paths

For the equivocal surface-envelope, i.e. when the tangency condition (7) holds, we have $F_{-1} = \{F_0 F_1\} = \langle H_p, p_1 - p_0 \rangle$, and the system (12) can be written in terms of two functions H, F_1:

$$\dot{x} = H_p,\ \dot{u} = \langle p, H_p \rangle,\ \dot{p} = -H_x - pH_u - \frac{\{\{HF_1\}H\}}{\{\{F_1H\}F_1\}}(p - p_0(x)). \qquad (13)$$

Here indices by u_1, p_1 are omitted, $\mu = \{\{F_1 H\} F_1\}$ in (10). Note, that $u_0(x) \in C^2(D)$, since it can be extended to D as the solution of $H(x, u_0(x), p_0(x)) = 0$.

For the second case, when $F_{-1} = H^-$ the system (12) has the form:

$$\begin{aligned}
&\dot{x} = \lambda_+ H_p^+ + \lambda_- H_p^-,\ \dot{u} = \lambda_+ \langle p, H_p^+ \rangle + \lambda_- \langle p, H_p^- \rangle \\
&\dot{p} = -\lambda_+ (H_x^+ + pH_u^+) - \lambda_- (H_x^- + pH_u^-) - \frac{\{H^- H^+\}}{\mu}(p - p_0(x)) \qquad (14) \\
&\lambda_+ = \{F_1 H^-\}/\mu,\ \mu = \{F_1 H^-\} + \{H^+ F_1\},\ \lambda_+ + \lambda_- = 1.
\end{aligned}$$

9. Equations for Focal Paths

None of the functions $u_0(x)$, $u_1(x)$ can be considered to be known in advance during backward constructions. The surface Γ first must be found with the values of $u(x), p_0(x), p_1(x)$ on it. We formulate here the result only for the focal surface of the first type with the Hamiltonian $H = H(x,p)$.

Lemma 7. *Singular focal paths are governed by the following system of differential and finite equations, obtained on the base of* (13):

$$\begin{aligned}
\dot{x} &= H_p(x, p_0) = H_p(x, p_1) \\
\dot{p}_0 &= -H_x(x, p_0) - [(A_0 + r)/B_0](p_0 - q) \qquad (15) \\
\dot{p}_1 &= -H_x(x, p_1) - [(A_1 + r)/B_1](p_1 - q)
\end{aligned}$$

$$\Phi(x,p_0,p_1,q,r)=0, \quad \varphi(x,p_0,p_1,q,r)=0$$

Here $q=v_x(x)\in\mathbf{R}^N$ is the gradient of an auxiliary function $v(x)\in C^2(D)$, $v(x)=u(x)$, $x\in\Gamma$, and $\Phi\in\mathbf{R}^N$, A_i, B_i, r, $\varphi\in\mathbf{R}^1$:

$$\begin{aligned}
r &= \langle\dot q, H_p(x,p_0)\rangle = \langle\dot q, H_p(x,p_1)\rangle\\
\Phi &= d/dt[H_p(x,p_0)-H_p(x,p_1)]\\
\varphi &= d/dt[\langle p_0,H_p(x,p_0)\rangle - \langle p_1,H_p(x,p_1)\rangle]\\
A_i &= A_i(x,p_i,q)\equiv\{H\{F_1H\}\}-r\\
B_i &= B_i(x,p_i,q)\equiv\{F_1\{HF_1\}\}.
\end{aligned}$$

Thus, the complete state-vector of the system (15) is $(x,p_0,p_1,q,r)\in\mathbf{R}^{4N+1}$ and it includes differential system of the order $3N$ and $N+1$ finite equations. In principle q and r can be expressed from this finite equations as the functions of x, p_0, p_1 and substituted in differential equations.

The system (15) for second type focal surface can be constructed in similar way using (13), (14). The third type needs special consideration.

If for some symmetry or other reason the surface Γ can be supposed to be a hyperplane, say, $x_N=0$, then the system (15) is simplifyed. The auxilliary function v can be taken as (conditioned that $u_i(x)\in C^2(\Gamma)$):

$$v(x_1,\dots,x_N)=u_i(x_1,\dots,x_{N-1},0), q_N=0, \quad x\in D \tag{16}$$

$$q_k=p_k^0=p_k^1, \quad k=1,\dots,N-1, \ x\in\Gamma.$$

Thus, the normal to Γ vectors have only one non-zero component:

$$p_i-q=(0,\dots,0,p_N^i); p_i=(p_1^i,\dots,p_N^i), \quad i=0,1.$$

The system (15) is now reduced to the ODE system of the order $2(N-1)$ with respect to $x_k, p_k=p_k^i$, $k=1,\dots,N-1$, and one algebraic or differential equation with respect to p_N (the former having two roots p_N^0, p_N^1):

$$\begin{aligned}
\dot x_k &= H_k(x_k,0,p_k,p_N), \quad k=1,\dots,N-1\\
\dot x_N &= H_N(x_k,0,p_k,p_N)=0,\\
\dot p_k &= -H_{x_k}(x_k,0,p_k,p_N),\\
\dot p_N &= -H_{x_N}(x_k,0,p_k,p_N)-\{H_NH\}/H_{NN}.
\end{aligned} \tag{17}$$

Here H_k, H_N, H_{NN} stand for the first and second derivatives with respect to p_k, p_N. Parameter p_N can be found from the equation for x_N in (17) and substituted into equations for x_k, p_k. The resulting system of the order $2(N-1)$ is of a Hamiltonian type, but the complete system of the order $2N$ isn't such one. If one prefers to deal only with differential equations, then the equality $H_N=0$ should be differentiated with respect to time to obtain the last equation in (17). One can show this equation to coinside with the corresponding one in (15).

References

[1] P.-L. Lions and P.E. Souganidis, Differential games, optimal control and directional derivatives of viscosity solutions of Bellman's and Isaacs' equations. *SIAM Journal of Control and Optimization*, vol. 23, no. 4, (1985), 566-583.
[2] R. Isaacs, *Differential Games*, Wiley, New York, 1965.
[3] P. Bernhard, Singular surfaces in differential games: an introduction. In: *Differential Games and Applications*, Springer, Berlin, 1977, pp. 1-33.
[4] A. A. Melikyan,The method of characteristics for constructing singular paths and manifolds in optimal control and differential games. In: *Lecture Notes in Control and Information Sciences*, vol. 156, Springer-Verlag, Berlin, 1991, pp. 81-90.
[5] A. A. Melikyan and N.V. Ovakimyan, Differential games of simple pursuit and approach on the manifolds. Yerevan, Academy of Sciences of Armenian Republic, preprint, 1993.
[6] A.I. Subbotin, A generalization of the basic equation of the theory of differential games, *Soviet Math. Dokl.*, vol. 22, (1980), pp. 358-362.
[7] J.V. Breakwell and P. Bernhard, A simple game with a singular focal line, *JOTA*, vol. 64, no. 2, (1990), pp. 419-428.

Arik A. Melikyan
Institute for Problems in Mechanics
Russian Academy of Sciences
Prospect Vernadskogo, 101
117526, Moscow, Russia

The Circular Wall Pursuit

J. Lewin

Abstract. A simple game in a plane is used to demonstrate the phenomenon of a *Focal Surface*. A formulation in 4 dimensions is used to get an invariance relation that gives better insight to the nature of the solution of the 2 dimensional model.

1. Introduction

The classical *Wall Pursuit* problem was discussed already by R. Isaacs (see [1]). It treats a pursuit evasion case where the evader is confined to run along a straight wall while the pursuer moves in the half plane adjacent to the wall. The solution of this game is rather simple and it is characterized by two fields of *regular optimal trajectories* emanating backwards from the *Target Set* intersecting along a *Dispersal Surface.*

In *The Circular Wall Pursuit* the evader is confined to run along the circumference of a circle while the pursuer moves in it. In the following result we shall see that the solution is not as simple.

2. The Game

The Circular Wall Pursuit takes place inside a circle with a radius of magnitude 1. The evader, Player E, is confined to move on the circumference of the circle. His velocity has the magnitude of 1. The pursuer, Player P, moves in the circle with a speed of magnitude w[1] and is capable to turn instantaneously. The pursuer wants to capture the evader by reducing his distance from player E to be less than ρ as early as possible.

We shall construct the solution for this problem by using two formulations simultaneously. One formulation will be in a *fixed reference frames* yielding a 4-dimensional game model and the other will be in a *relative reference frame* which will turn to be a 2-dimensional model.

[1] When $w > 1$ *pursuer* is faster, and when $w < 1$ *pursuer* is slower than *evader.*

The Frames of Reference

The "Real" Frame. The origin of the reference frame is located at the center of the circular arena (at the point O in Figure 1). The position of player P at any instance is given by the two-dimensional vector r_p and the position of player E at any instance is given by the two-dimensional vector r_e.

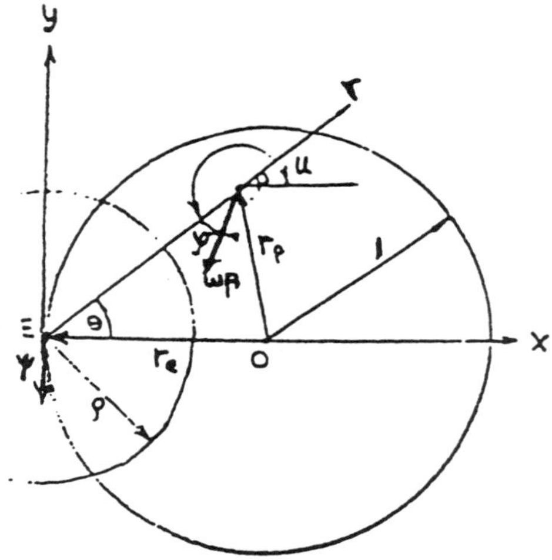

Figure 1. Frames of reference.

The state vector of the system is a 4-dimensional vector which has two 2-dimensional vectors as its components.

$$z = \begin{bmatrix} r_p \\ r_e \end{bmatrix}.$$

The Relative Frame. The origin of the relative reference frame is attached to player E and the $\theta = 0$ ray is confined to pass through the center of the arena (see Figure 1).

The state of the game is given by r, the distance between player P and player E, and θ, the angle of direction of the location of player P relative to the $\theta = 0$ ray (measured counter-clockwise).

2.1. The Game Model in the Real Frame

The Game Set

$$\mathcal{S} \equiv \{(r_p, r_e) \mid |r_p| \le 1,\ |r_e| = 1,\ |r_p - r_e| \ge \rho\}.$$

The Target Set

$$\mathcal{C} \equiv \{(r_p, r_e) \mid |r_p| \le 1,\ |r_e| = 1,\ |r_p - r_e| = \rho\}.$$

The Equations of Motion

$$\dot{r}_p = w\beta, \quad |\beta| = 1$$
$$\dot{r}_e = \psi e_\theta \times r_c, \quad |\psi| \le 1,\ |e_\theta| = 1,\ e_\theta, r_e = 0$$

(e_θ is a unit vector perpendicular to the arena in the sense that $\times$ is a *vector cross multiplication* and $e_\theta \times r_e$ is in the plane of the arena).

Player P chooses the direction of the unit vector β and player E determines the speed and sense of his motion along the circumference by choosing ψ.

The Outcome Functional

$$G(r_p, r_e) = 0, \quad L(r_p, r_e, \beta, \psi) = 1.$$

2.2. The Game Model in the Relative Frame

The Game Set

$$\mathcal{S} \equiv \{(r, \theta) \mid r > \rho,\ r^2 + 1 - 2r\cos\theta \le 1\}$$

or

$$\mathcal{S} \equiv \{(x, y) \mid (x+1)^2 + y^2 \le 1,\ x^2 + y^2 \ge \rho^2\}.$$

The Target Set

$$\mathcal{C} \equiv \{(r, \theta) \mid r = \rho,\ r^2 + 1 - 2r\cos\theta \le 1\}$$

or

$$\mathcal{C} \equiv \{(x, y) \mid (x+1)^2 + y^2 \le 1,\ x^2 + y^2 = \rho^2\}.$$

The Equations of Motion

$$\dot{r} = -w\cos\varphi + \psi\sin\theta$$
$$\dot{\theta} = -\frac{1}{r}w\sin\varphi + \left(\frac{\cos\theta}{r} - 1\right)\psi, \quad |\psi| \le 1.$$

Player P chooses φ and player E determines the speed and sense of his motion along the circumference by choosing ψ, or in *Cartesian coordinates*:

$$\dot{x} = -w\cos u$$
$$\dot{y} = -w\sin u + \psi, \quad |\psi| \le 1.$$

Here player P chooses u and player E determines the speed and sense of his motion along the circumference by choosing ψ.

The Outcome Functional

$$G(r,\theta)=0, \quad L(r,\theta,\varphi,\psi)=1.$$

3. Solution

We begin to solve the game problem in the *real* frame.

The Candidate Optimal Control Laws. The equations of motion are separable and ME1 is:

$$\min_{\beta} \max_{|\psi|\leq 1} \{w\beta \cdot J_{r_p} + \psi e_\theta \times r_e \cdot J_{r_e} + 1\}$$

the candidate optimal control laws are:

$$\begin{aligned}
&\dot{\psi} = \text{sgn}(e_\theta \times r_e \cdot J_{r_e}), \quad \text{for } e_\theta \times r_e \cdot J_{r_e} \neq 0 \\
&\dot{\psi} = \text{Any admissible } \psi, \quad \text{for } e_\theta \times r_e \cdot J_{r_e} = 0 \\
&\dot{\beta} \| - J_{r_p}
\end{aligned}$$

ME2 is:

$$-w|J_{r_p}| + |e_\theta \times r_e \cdot J_{r_e}| + 1 = 0.$$

3.1. The Adjoint Equations

$$\begin{aligned}
\overset{\circ}{J}_{r_p}(r_p, r_e) &= 0 \\
\overset{\circ}{J}_{r_e}(r_p, r_e) &= J_{r_e} \times c_\theta.
\end{aligned}$$

3.2. Invariance

When a system which can be modelled in two dimensions is represented in a model with four dimensions there are some redundancies which imply that there exist *invariance relations* between the main parameters of the model.

We observe that our system should not be affected by rotations of the arena hence the value function should be a function only of two parameters ϱ and ϑ where:

$$\begin{aligned}
\varrho &= |r_p - r_e| \\
\vartheta &= r_e \cdot (r_p - r_e)/|r_p - r_e| \cdot |r_e|
\end{aligned}$$

so

$$J(r_p, r_e) = J(\varrho(r_p, r_e), \vartheta(r_p, r_e)).$$

3.3. The Nonusuable Part

To find the *Nonusuable Part* we solve:

$$\min_{\varphi} \max_{\psi} \dot{r} > 0.$$

We obtain

For $w < 1$ The NUP is: $|\theta| > \sin^{-1} w$
For $w > 1$ The NUP is empty.

3.4. The Candidate Optimal Trajectories

We have that along regular segments of the optimal trajectories

$$J_{r_p}(r_p, r_e) = \text{const}$$

and that the pursuer runs on a straight line towards E_f, the final position of player E.

The *isochrone* τ in the *real frame* is a circle with a radius of magnitude $\rho + w\tau$ centered at E_f.

In the *relative frame* the τ *isochronic* is also a circle with a radius of magnitude $\rho + w\tau$.

The trajectory from ($x(0) = \rho \cos\theta$, $y(0) = \rho \sin\theta$) for points that belong to the *Usable Part*, is given by:

$$\begin{aligned} x &= 1 - \cos\tau + (\rho + w\tau)\cos(\tau + \theta) \\ y &= -\sin\tau + (\rho + w\tau)\sin(\tau + \theta). \end{aligned} \tag{1}$$

3.5. The Barrier

In the case $w < 1$ *A Barrier* may emanate from each of the BUP points $|\theta| = \sin^{-1} w$.

It is not difficult to show that the *Barrier Paths* are *Evolutes* to a circle of radius w centered at the center of the arena. We may use the geometric characterization of the *Evolute* to derive the following conclusions regarding the dependence of the properties of the *Barrier* on the parameters ρ and w.

A **Closed Barrier** exists for:

$$\sqrt{1 - w^2} - \rho > \frac{1}{2}\pi w - w \sin^{-1} w.$$

See the curve BC in Figure 2.

An **Open Barrier** exists for:

$$0 < \sqrt{1-w^2} - \rho < \frac{1}{2}\pi w - w \sin^{-1} w.$$

See the curve BO in Figure 3.

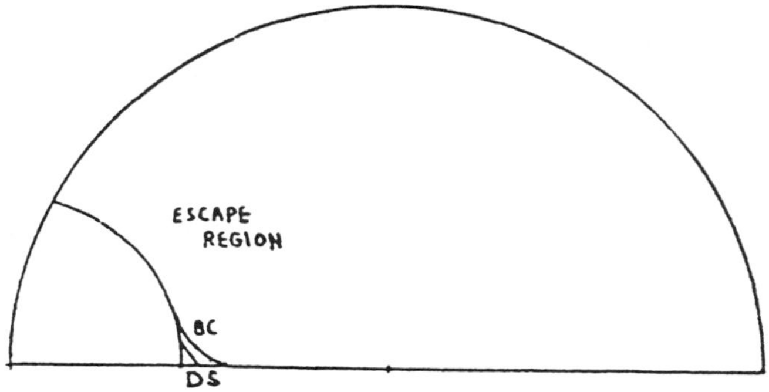

Figure 2. Closed barrier (region A: $w = 0.4$, $\rho = 0.45$).

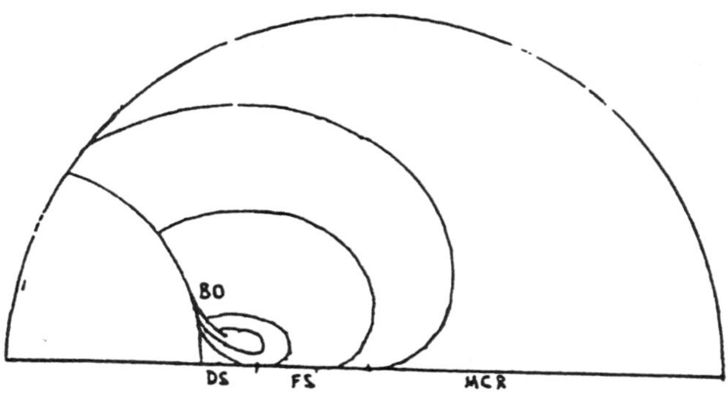

Figure 3. Open Barrier (region B: $w = 0.4$, $\rho = 0.54$).

No *Barrier* exists if $w > 1$ or for $w > 1$ if[2]

$$0 > \sqrt{1-w^2} - \rho.$$

See Figure 5.

---

[2] In this case the *Barrier Path* arrives at the BUP point from *outside* of the *Game Set*.

3.6. The Dispersal Surface

It is easy to see from equations (1) that at the point $(\rho, 0)$:

$$\frac{dy}{dx} = \frac{\rho - 1}{1 + w}. \tag{2}$$

So for $\rho > 1$ the trajectories emanating backwards from this point enclose a *void*. It is not difficult to realize that while in the *void* player P will run towards the center of the arena no matter what player E does. The *Isochrones* in the void are circular arcs centered at the center of the arena. We shall refer to the void as a region of *multiple choices* (for player E) and denote it in the figures as *MCR*.

For $\rho < 1$ we conclude from equation (2) that for θ smaller then some $\tilde{\theta}$ the trajectories may intersect the x-axis at a time $\tilde{\tau}(\theta)$:

$$\frac{1}{\sin(\theta + \tau)} = \frac{\rho + w\tilde{\tau}}{\sin \tilde{\tau}}. \tag{3}$$

Examining the $\dot{\theta}$ equation in the *relative* frame for $\theta = 0$ we conclude that it is a case of *Evader's Dispersal Surface* and it is player P who has to apply an *information nonadmissible control law* of the type $\varphi = \varphi(\psi)$ on it. (See the curve DS in Figure 2.) One particular trajectory arrives at the x-axis tangentially, this happens for $\theta = \tilde{\theta}$ where in addition to relation (3) the following is to be satisfied

$$-\cos\tilde{\tau} + (\rho + w\tilde{\tau})\cos(\tilde{\tau} + \tilde{\theta}) + \sin(\tilde{\tau} + \tilde{\theta}) = 0.$$

The *Dispersal Surface* cannot extend further from this point along the $\theta = 0$ axis.

We also notice that the first field of candidate optimal trajectories fails to cover the game set.

3.7. The Focal Surface

Next we examine the possibility to obtain another field of optimal trajectories assuming that a *singular arc* emanates from the edge of the DS and that tributaries arrive at it from both sides. We assume that it is player P who has to apply an *information nonadmissible control law* of the type $\varphi = \varphi(\psi)$ along the singular arc.

At the edge of the DS we have that $\dot{\theta} = 0$ on both arriving trajectories.

3.7.1. Conditions at the junction of the tributaries to the singular arc

Let the superscript (i) denote side-i of the *singular surface*. On the singular arc the problem:

$$\max_{\psi\in\{\psi\mid\frac{1}{r}w\sin\varphi+(\frac{1}{r}-1)\psi=0\}} H(r,\theta,\nabla J^{(i)},(r,\theta),\hat{\varphi}^{(i)}(\psi),\psi)$$

is solved by both $\psi^{(i)}$ and $\tilde{\psi}^{(i)}$.[3]

Here this problem is

$$\max_{\psi}\left\{-J_r^{(i)}(r,0)w\sqrt{1-[(1-r)\psi/w]^2}\right\}.$$

This problem has a unique solution (for ψ that provide that $\ddot{\theta}$ have the right sign for the paths that arrive from side-i)

$$\dot{\psi}^{(i)}=\tilde{\psi}^{(i)}.$$

The problem:

$$\min_{|\psi|\leq 1} H(r,\theta,\nabla J^{(i)}(r,\theta),\varphi,\dot{\psi}^{(i)})$$

has a unique solution so

$$\dot{\varphi}^{(i)}=\tilde{\varphi}^{(i)}(\dot{\psi}^{(i)})$$

and together:

$$\begin{aligned}\dot{\psi}^{(i)}&=\tilde{\psi}^{(i)}\\ \dot{\varphi}^{(i)}&=\tilde{\varphi}^{(j)}(\tilde{\psi}^{(i)}).\end{aligned}$$

The tributaries to the singular arc join tangentially and we have a case of a *Focal Surface*.

One can easily find the tributaries in the *real frame* and transform them to the *relative frame*.

Notice that the law $\varphi=\varphi(\psi)$ is here:

$$\sin\varphi(\psi)=(1-r)\psi/w.$$

At the center O of the arena where $r=1$ we have that $\varphi=0$ at both tributaries regardless of ψ. Here Player P do not use a law of the type $\varphi=\varphi(\psi)$ and the singular arc cannot be extended backward anymore. (See the curves FS in Figures 3 and 4.)

---

[3] The superscripts * and ˜ denote realizations of optimal strategies in the regular field adjacent to the singular arc and along the singular arc respectively.

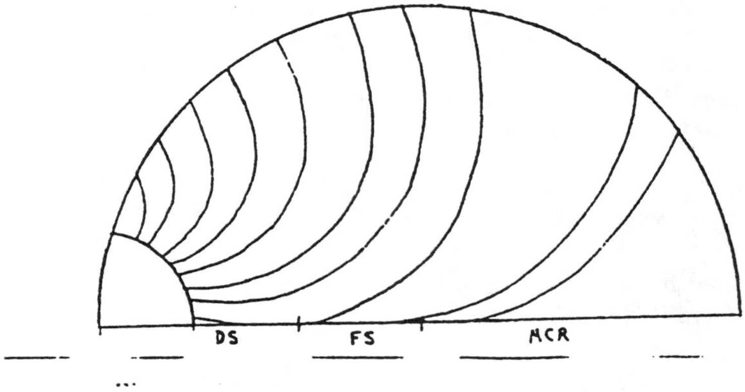

Figure 4. No barrier, (region C: $w = 1.6$, $\rho = 0.3$).

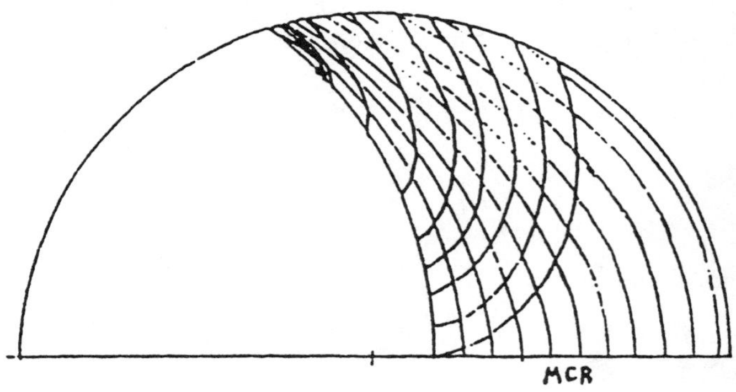

Figure 5. No barrier, (region D: $w = 0.8$, $\rho = 1.16$).

We still have a void in the game set uncovered by candidate optimal trajectories.

3.8. The Region of Multiple Choices

Recall that in the real frame we had:

$$J(r_p, r_e) = J(|r_p - r_e|,\ r_e(r_p - r_e)/|r_p - r_e| \cdot |r_e|).$$

If $\rho \leq 1$ then $r_p = 0$ belongs to the *Game Set* and:

$$J(0, r_e) = J(|r_e|) = \text{const.}$$

so

$$J_{r_e}(0, r_e) = 0$$

and we can deduce from the adjoint equations that along any regular optimal trajectory that emanate from the point O we have:

$$\dot{J}_{r_e}(r_p, r_e) = 0$$

and:

$$J_{r_p}(r_p, r_e) \parallel r_p, \quad J_{r_e}(r_p, r_e) = 0$$

and

$$\beta \parallel r_p, \quad \dot{\psi} = \text{any admissible } \psi.$$

So in the void player P runs toward the center while player E can choose any ψ. We have thus succeeded to cover the game set with candidate optimal trajectories that satisfy the necessary conditions and the solution is complete.

3.9. The Parameter Plane

We can summarize the features of the solution for various by partitioning a plane with coordinates ρ and w into regions (see Figure 6) and by describing the surfaces that occur in each region:

	DS	FS	BO	BC	MC
A	+	-	+	-	-
B	+	+	-	+	+
C	+	+	-	-	+
D	-	-	-	-	+

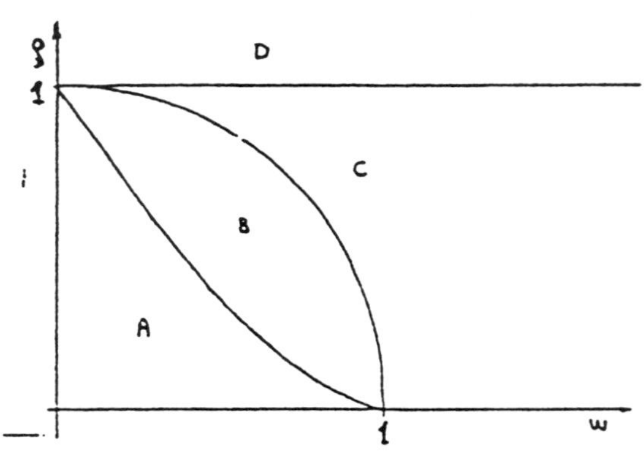

Figure 6. Regions in the parameter plane.

References

[1] Isaacs, R., *Differential Games*, Robert E. Krieger, New York, 1975.

[2] Bernhard, P., Singular surfaces in differential games, in *Differential Games and Applications*, P. Hagedorn, H.W. Knobloch, and G.J. Olsder, eds., Springer-Verlag, Berlin, 1977.

Joseph Lewin
Department of Aeronautics
Technion
Haifa, Israel

PART III

Mathematical programming techniques

Decomposition of Multi-Player Linear Programs

Richard Loulou, Gilles Savard, and Denis Lavigne

1. Problem Definition

We consider a set of n players, each of which produces and consumes a set of q commodities. Commodities produced by player i are consumed by that same player, and/or exported to players $j = 1, \ldots, n$ possibly with some loss factor along arc ij. We set for player i the following mathematical program

$$
(LP_i) \begin{cases} \min\limits_{x_i} c_i x_i \\ A_i x_i - \sum_j E_{ij}^{-1} s_{ij} + \sum_j s_{ji} \geq b_i \\ B_i x_i \geq d_i \\ s_{ij} \geq 0 \quad \text{all } j, \end{cases}
$$

where

x_i is the vector of activities of player i

c_i is the vector of unit costs of the activities

s_{ij} is the q dimensional vector of exportations from i to j (delivered to j)

E_{ij} is a $q \times q$ diagonal matrix (e_{ij}^{ℓ}) with e_{ij}^{ℓ} equal to the fraction of flow ℓ along ij that arrives at destination j

A_i defines the production constraints of player i

b_i is the vector of demands for the q commodities by player i

B_i, d_i define linear constraints involving only player i. These constraints will also be denoted $x_i \in L_i$ in the sequel.

Of course, the n linear programs (LP_i) are not well defined, since variables s_{ij} are common to several of them. If however we seek a global optimum of the sum of the n players' objectives, this can be achieved by solving the following global optimization problem (P)

$$(P)\begin{cases} \min_{x,s} \sum_i c_i x_i \\ A_i x_i - \sum_j E_{ij}^{-1} s_{ij} + \sum_j s_{ji} \geq b_i & \text{all } i \\ B_i x_i \geq d_i & \text{all } i \\ s_{ij} \geq 0 & \text{all } i, j. \end{cases}$$

The above formulation is suitable for a number of multi-regional problems where the regions can produce, import, and export a common set of commodities. In particular, it is a realistic model of power exchanges between neighbouring states, where each state's electricity is managed by a single utility that must satisfy the state's demand at minimum cost. In this example, there are q forms of electricity (perhaps depending on season, time of day, etc.), and e_{ij}^{ℓ} represents the energy "gain" along electric line (or gas pipeline) from i to j, with $e_{ij}^{\ell} \leq 1$.

It is implicitly assumed that (P) is too large to be conveniently solved as a single Linear Program, and so we will seek a global optimum of (P) through a decomposition approach. Of course, (P) may be viewed as a large linear program with a dual block diagonal structure, and considered a candidate for the classical Benders decomposition approach. Alternatively, its dual is solvable via Dantzig-Wolfe decomposition method.

However, we wish to consider (P) as the *composition* of the n individual (LP_i), and always keep in mind the seller-buyer dimension of the problem, so as to apply some ideas inspired from the classical economic equilibrium area. Our viewpoint is thus close to that adopted by [HW83]. It is well-known that a global optimum for (P) corresponds to a cooperative economic equilibrium between the n players exchanging the commodities at their marginal values (see [TJ71]). This viewpoint is somewhat formalized in theorem 1 below.

In the case where each player can be clearly identified as a pure seller or a pure buyer, the above problem resembles the well-known situation of the competitive equilibrium between several suppliers and several markets. The PIES algorithm (see [AH82]) solves such a situation when each market is described via an explicit demand function having certain properties. Our problem thus can be seen as a generalization of the PIES situation to the case where each demand function is not known explicitly, but rather as the implicit result of a linear program. The fact that such an implicit demand function is known to be non differentiable of course introduces difficulties which do not permit the application of a PIES like approach.

Alternatively, there may be situations where the pricing of exchanged commodities is subjected to taxes, etc., in which case there is no simple correspondance between the (regulated) economic equilibrium and a global optimization problem. [GM85] have generalized the PIES method to such equilibria, and we shall see that our approach is suitable both for the pure competitive case and the regulated case, thus also extending these authors work somewhat.

Some previous works by the authors and associates (see [LS91], and [HLS90]) have applied a similar approach as will be presented here (but in a much more ad hoc manner) to specific applications, notably in the planning of co-generation.

The following result, although not really new, is useful to state for it sets the stage for a description of our algorithms.

Let us first slightly transform (P) into the following equivalent form which we still denote (P) by a slight abuse of notation. Note that we have simply introduced the new variables s_i representing the net total outflow from region i.

$$(P)\begin{cases} \min_{x,s} \sum_i c_i x_i \\ A_i x_i - s_i \geq b_i \quad \text{all } i \\ B_i x_i \geq d_i \quad \text{all } i \\ s_i - \sum_j E_{ij}^{-1} s_{ij} + \sum_j s_{ji} = 0 \quad \text{all } i \\ s_{ij} \geq 0 \quad \text{all } i, j. \end{cases}$$

The dual (D) of (P) is

$$(D)\begin{cases} \max_{\lambda,\mu,\nu} \sum_i b_i \lambda_i + \sum_i d_i \mu_i \\ A_i^t \lambda_i + B_i^t \mu_i = c_i \quad \text{all } i \\ \nu_i - \lambda_i = 0 \quad \text{all } i \\ T\nu \leq 0 \\ \lambda_i, \mu_i \geq 0 \quad \text{all } i, \end{cases}$$

where T is an appropriate matrix.
We now state the following.

Theorem 1. *Let x^*, s^* be an optimal solution of (P) and λ_i^* be a vector of optimal dual variables for the ith group of demand constraints $A_i x_i - s_i \geq b_i$, then (s_i^*, λ_i^*) is a competitive equilibrium, i.e. (x_i^*, s_i^*) is an optimal solution (with λ_i^* fixed) of*

$$\begin{cases} \min_{x_i, s_i} c_i x_i - \lambda_i^* s_i \\ A_i x_i - s_i \geq b_i \\ B_i x_i \geq d_i. \end{cases}$$

In other words, using λ_i^* as prices achieves the decomposition of the global problem (P).

Proof. The proof can be derived by using the complementarity slackness property. □

In the following section, we motivate and describe the basic algorithm.

2. Decomposition as an Equilibrium Computation

Theorem 1 enables us to see the decomposition of (P) as the computation of equilibrium quantities (x_i, s_i) and prices λ_i, where λ_i are optimal dual variables of the demand constraints of the following LP program

$$(P_i)\begin{cases} \min\limits_{x_i} c_i x_i \\ A_i x_i \geq b_i + s_i \\ B_i x_i \geq d_i. \end{cases}$$

To facilitate the discussion, let us call $Z_i(s_i)$ the optimal value function of (P_i), parameterized by the export vector s_i. If $Z_i(\cdot)$ were known explicitly for all i, then the computation of an optimal solution for (P) would amount to a program of reasonable size, as follows

$$(Q)\begin{cases} \min\limits_{s_i, s_{ij}} \sum\limits_i Z_i(s_i) & \\ s_i - \sum\limits_j E_{ij}^{-1} s_{ij} + \sum\limits_j s_{ji} = 0 & \text{all } i \\ s_{ij} \geq 0 & \text{all } i, j. \end{cases}$$

Program (Q) may also be looked at as the computation of an economic equilibrium between n economic agents represented by their supply functions $Z_i(s_i)$, subject to marginal cost pricing. Note that $Z_i(s_i)$ is at the same time a supply function and a demand function, depending on the sign of s_i. If only one commodity is traded, it is well-known from LP theory that each $Z_i(s_i)$ is a convex, piecewise linear function.

For higher dimensions of the vectors s_i, it is still true that the functions $Z_i(s_i)$ are convex, but it is not easy, or even possible, to explicitly compute them. We therefore resort to an iterative procedure using local approximations of $Z_i(s_i)$ to compute tentative solutions of (Q). The general scheme goes as follows:

Generic Algorithm

Step 1: $k \leftarrow 0$; choose some $s_i(0)$.

Step 2: For each i, compute a closed form approximation of $Z_i(s_i)$ around the current value $s_i(k)$. Let $Z_i^k(s_i)$ be this approximation.

Step 3: Compute an optimal solution of program (Q_k) below

$$(Q_k)\begin{cases} \min_{s_i, s_{ij}} \sum_i Z_i^k(s_i) \\ s_i - \sum_j E_{ij}^{-1} s_{ij} + \sum_j s_{ji} = 0 \quad \text{all } i \\ s_{ij} \geq 0 \quad \text{all } i, j. \end{cases}$$

Let $s_i(k+1)$ be an optimal solution of (Q_k).
Step 4: if $s_i(k+1) = s_i(k)$ for all i, STOP; else let $k \leftarrow k+1$ and GOTO *step 2.*

Of course, the success of such an algorithm rests upon (i) finding a good approximation of $Z_i(s_i)$, and (ii) proving that the stopping criterion indeed yields an optimal solution. One of the simplest approximations of $Z_i(s_i)$ is the affine separable function, which at iteration k is equal to

$$Z_i^k(s_i) = Z_i(s_i(k)) + \lambda_i(k)(s_i - s_i(k))$$

where $\lambda_i(k)$ is the vector of optimal dual variables of (P_i) at point $s_i(k)$. This is easily obtained by solving all programs (P_i) with s_i fixed . We shall call the resulting algorithm the Generalized Cob Web (GCW) algorithm, by analogy with the famous cob web scheme used to compute an equilibrium between supply and demand functions. The analogy stems from the fact that, at each iteration, the only information passed is a price vector or a quantity vector (see [M87] for a discussion of the cob web approach).

The cob web algorithm does not perform well with linear programs and tends to exhibit oscillations (see [W81]). This is so because the information included in $Z_i^k(\cdot)$ is insufficient, especially when the current point $s_i(k)$ happens to be a degeneracy point of the program (P_i); in such a case, the dual variable is not valid in a neighbourhood of $s_i(k)$. The following section describes an improved local approximation scheme.

3. The Improved Generalized Cob Web Algorithms ($IGCW$)

A more elaborate approximation can be obtained by exploiting information given by a local parametric analysis of the q demand constraints of each $(P_i(k))$. The simplest parametric analysis on the ℓth demand constraint essentially provides a range of validity of the optimal dual variable $\lambda_i^\ell(k)$, assuming that all other demand constraints remain fixed; denote by $[s_{i-}^\ell(k), s_{i+}^\ell(k)]$ this range, with $s_i^\ell(k)$ lying in it. The same parametric analysis also provides the value $\lambda_{i-}^\ell(k)$ (respectively $\lambda_{i+}^\ell(k)$) of the ℓth dual variable that prevails when s_i^ℓ is just below $s_{i-}^\ell(k)$ (respectively just above $s_{i+}^\ell(k)$). Figure 1 shows a typical shape of the proposed approximation for $q = 1$, where p stands for λ.

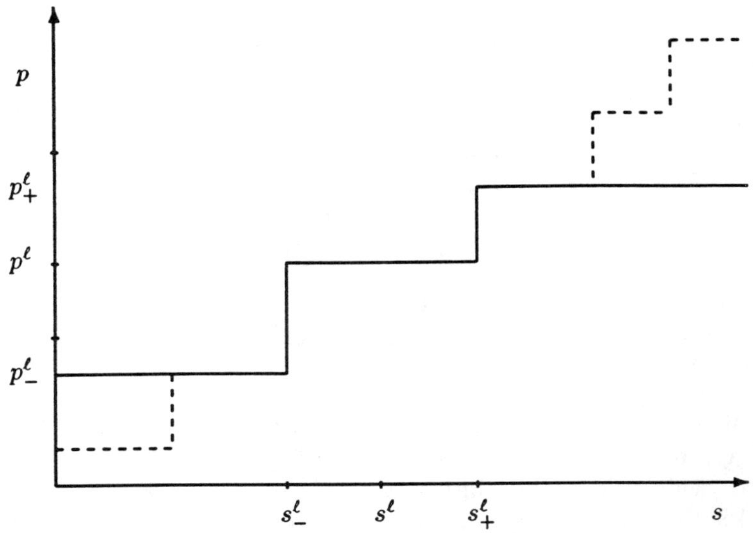

Figure 1. Approximation of the inverse supply function.

Using this approximation successively for each ℓ, program Q_k becomes a separable, convex, piece wise linear program, which is equivalent to a linear program with $3qn$ variables (in addition to the s_{ij} variables), obtained by replacing each s_i^ℓ by the sum of three variables $u_{1i}^\ell, u_{2i}^\ell, u_{3i}^\ell$. Therefore, the first version of an Improved Generalized Cob Web algorithm $IGCW_1$ may be written:

Algorithm $IGCW_1$

Step 1: $k \leftarrow 0$; choose some $s_i(0)$.

Step 2: For each $i = 1, \cdots, n$, solve (P_i) with $s_i = s_i(k)$; let $x_i(k)$ be an optimal solution, and $p_i^\ell(k)$ be the shadow price of the ℓth demand constraint of player i.

Step 3: Using local sensitivity analyses on each demand constraint of (P_i), compute for each $i = 1, \cdots, n$ and $\ell = 1, \cdots, q$, the range of $s_i^\ell(k)$ such that the current basis remains optimal for (P_i). Let $[s_{i-}^\ell(k), s_{i+}^\ell(k)]$ be this range, with $s_{i-}^\ell(k) \leq s_i^\ell(k) \leq s_{i+}^\ell(k)$. Furthermore the same parametric analysis provides the values $p_{i-}^\ell(k)$ and $p_{i+}^\ell(k)$, where:
$p_{i-}^\ell(k)$ is the shadow price when s_i^ℓ is just below $s_{i-}^\ell(k)$
$p_{i+}^\ell(k)$ is the shadow price when s_i^ℓ is just above $s_{i+}^\ell(k)$.

Step 4: Compute a tentative equilibrium between the n players, using the approximate supply/demand functions described at *step 3.* To do so, solve the following linear program:

$$\begin{aligned}\min \sum_i \sum_\ell \{p^\ell_{i-}(k)u^\ell_{i,1} + p^\ell_i(k)u^\ell_{i,2} + p^\ell_{i+}(k)u^\ell_{i,3}\}\\ \text{s.t. } (u_{i,1}+u_{i,2}+u_{i,3}) - \sum_j E^{-1}_{ij}s_{ij} + \sum_j s_{ji} = 0, \quad i=1,\ldots,n\\ (LEP_0) \qquad u_1 \le s_-(k)\\ u_2 \le s_+(k) - s_-(k)\\ u_1, u_2, u_3, s_{ij}, s_{ji} \ge 0.\end{aligned}$$

Let $s_i(k+1)$ be an optimal solution of (LEP_0).
Step 5: if $s_i(k+1) = s_i(k)$ for all i, STOP; else let $k \leftarrow k+1$ and GOTO *step 2.*

The model and algorithm of this section will be specialized to several commonly encountered cases in the following section.

Remark. In a small enough neighbourhood of current point $s_i(k)$, the epigraph of $Z^k_i(\cdot)$ is easily seen to be included in the epigraph of $Z_i(\cdot)$. Therefore, $Z^k_i(\cdot)$ is *locally* an inner approximation of $Z_i(\cdot)$. We thank an anonymous referee for mentioning this fact (see [G70] for a classification of approximations).

4. Asymmetric IGCW Algorithms

In our approach so far, we have treated the players symmetrically, and created a fictitious player who computes the tentative equilibrium whenever all real players provide him with approximations of their supply (demand) functions.

Although the symmetry of this scheme is a esthetically appealing, it may well not be the most efficient one: indeed, it should be intuitively clear that when computing a tentative equilibrium, one must use as few approximations as possible. This suggests that one player could be singled out and put in charge of computing the sequence of tentative equilibria, all other players having the role of providing him with their individual approximate demand/supply functions. In contrast, the algorithm described in section 3 gave an equal role to each player, and let a fictitious player in charge of computing the sequence of tentative equilibria.

The proposed alteration destroys the symmetry, but allows *one* demand/supply function to be described exactly, when computing tentative equilibria. This approach is indicated when the set of players is partitioned into one seller and several independent buyers, or one buyer and several independent sellers. We now describe some relevant cases.

a) The case of one seller and one buyer

We first tackle the case of two players, one supplier and one buyer, with the approach suggested above. This will then be easily extended to the case of one supplier and several buyers, or the case of one buyer and several suppliers. We denote the algorithm $IGCW_2$. In this description we choose the seller to be the player who computes the sequence of tentative equilibria. The algorithm corresponding to the other possibility is easily derived.

For two players, there is only one vector of exchange variables, which we denote s. For a well determined value of vector s, the two players linear programs are as follows:

$$
\begin{array}{llrll}
 & & \min \quad c_1 x & & \min \quad c_2 x \\
(P_1) & \text{s.t.} & A_1 x - E^{-1} s = b_1 & (P_2) & \text{s.t.} \ A_2 x + s = b_2 \\
 & & x \in L_1 & & x \in L_2 \\
 & & s \geq 0 & & s \geq 0
\end{array}
$$

Algorithm $IGCW_2$

Step 1: $k \leftarrow 0$; choose some $s(0)$.

Step 2: Solve (P_2) with $s = s(k)$; let $x_2(k)$ be an optimal solution and $p^\ell(k)$ be the shadow price of the ℓth demand constraint.

Step 3: Using a local parametric analysis of the demand constraint of (P_2), compute for each $\ell = 1, \ldots, q$, the range of $s^\ell(k)$ such that the current basis remains optimal for (P_2). Let $[s_+^\ell(k), s_-^\ell(k)]$ be this range, with $s_+^\ell(k) \leq s^\ell(k) \leq s_-^\ell(k)$. Furthermore the same parametric analysis provides the values $p_+^\ell(k)$ and $p_-^\ell(k)$, where:

$p_+^\ell(k)$ is the shadow price when s^ℓ is just below $s_+^\ell(k)$

$p_-^\ell(k)$ is the shadow price when s^ℓ is just above $s_-^\ell(k)$.

Step 4: Compute a tentative equilibrium between the seller's complete program and the approximation of the buyer's demand function described at *step 3*. To do so, solve the following linear program:

$$
\begin{aligned}
\min \ c_1 x - \sum_\ell \{p_+^\ell(k) u_1^\ell + p^\ell(k) u_2^\ell + p_-^\ell(k) u_3^\ell\} & \\
\text{s.t.} \quad A_1 x - E^{-1}(u_1 + u_2 + u_3) &= b_1 \\
u_1 &\leq s_+(k) \qquad (LEP_1) \\
u_2 &\leq s_-(k) - s_+(k) \\
u_1, u_2, u_3 &\geq 0 \\
x &\in L_1 .
\end{aligned}
$$

Let $s(k+1) = u_1 + u_2 + u_3$ be the equilibrium commodity vector just computed.

This step is illustrated by Figure 2 in the case $q = 1$.

Step 5: if $s(k+1) = s(k)$, STOP; else let $k \leftarrow k+1$ and GOTO *step 2.*

Note that we could have reversed the roles of the seller and the buyer in algorithm $ICGW_2$. In that case, the buyer would have been charged with computing the sequence of equilibria, using for that purpose the seller's supply function's approximation. The best choice of the player who will play the role of equilibrium computation remains open for the moment. We shall discuss this important question in section 5 when addressing the computational performance of the algorithms

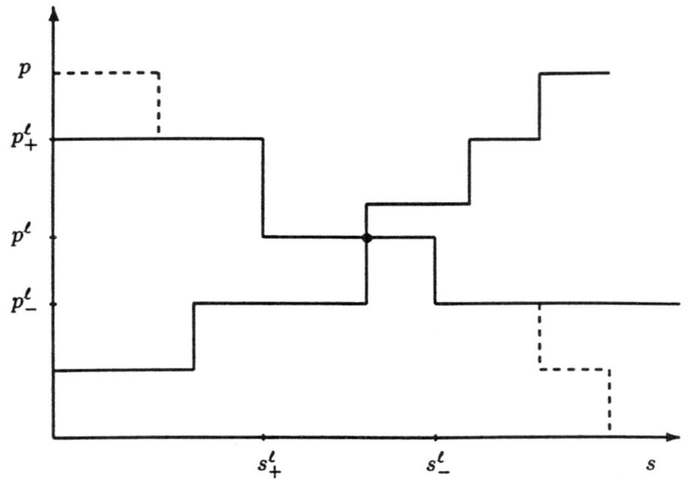

Figure 2. An approximate equilibrium.

b) The case of one seller and n buyers (or n sellers and one buyer)

The extension of algorithm $IGCW_2$ to a single seller and n buyers is quite simple: it suffices to perform steps 2 and 3 once for each buyer, and to modify step 4 to include the total costs of the n buyers.

Note that the case of n sellers and a single buyer can be treated via a simple interchange of the roles of seller and buyer in $IGCW_2$. In such a case, it is the buyer who computes the tentative equilibria, whereas each seller sends him his approximate supply function.

c) The case where exchanges occur in both directions

Consider again the case of two players, but where each player may buy *and* sell commodities from or to the other. In this situation, there are two vectors of non-negative variables, namely s_1 and s_2, representing the purchases by players 1 and 2 respectively. For well-determined values of the vectors s_1 and s_2, the two players' linear programs are:

$$\begin{array}{llll} & \min \quad c_1 x & & \min \quad c_2 x \\ (P_1) & \text{s.t. } A_1 x + s_1 - E^{-1} s_2 = b_1 & (P_2) & \text{s.t. } A_2 x - E^{-1} s_1 + s_2 = b_2 \\ & x \in L_1 & & x \in L_2 \\ & s_1, s_2 \geq 0 & & s_1, s_2 \geq 0 \end{array}$$

Referring to step 3 of $IGCW_2$, player 2 must now compute an approximation of a *hybrid demand/supply* function. In order to take into account the two possibilities (purchases and sales), and because of the loss factor matrix E, the step-wise approximation must now have one additional jump, corresponding to $s_1 = s_2 = 0$ and the magnitudes of the quantity and price ranges must be altered accordingly. Consequently, program (LEP_1) in step 4, now has four vectors of variables $u_1, \ldots, u_4$.

As before, the roles of players 1 and 2 can be reversed in the obvious way. Furthermore, this case can be easily extended to one player trading simultaneously with several others.

d) The 'heavy-duty' version of IGCW

Pushing the logic of $IGCW_2$ a little further, consider the case of one supplier and one buyer, but modify the algorithm in such a way that each player *in turn* plays the role of computing the tentative equilibrium. A word description of such a 'heavy-duty' algorithm, noted $HIGCW_2$, would be: first, the supplier provides the buyer with his approximate supply function, and the buyer computes a tentative equilibrium using the supplier's approximation plus his own exact demand function; secondly, the buyer provides the supplier with *his* approximate demand function, and the supplier in turn computes a tentative equilibrium, etc. One would expect this 'heavy-duty' approach to be at least as effective as the simpler asymmetric algorithm. Indeed, as will be shown in the following section, there are cases where the regular version of $IGCW_2$ works well or not at all, depending on which player is selected for the computation of the tentative equilibria at step 4: in such cases, the 'heavy-duty' version will be effective.

Comments. Note that the case of one seller and several buyers cannot be made 'heavy-duty', since it is not clear how the n buyers would be able to compute an equilibrium. Note also that the extension of $HIGCW_2$ to the case of bi-directional commodity exchanges is done in a straightforward manner by splitting the exchange vector s in two vectors s_1 and s_2, exactly as was done for the $IGCW_2$ algorithm.

e) Computation of regulated equilibria

We now turn to a new application of the above ideas, one that demonstrates an important advantage of our approach versus more classical decomposition methods. We treat the case when the buyers must purchase each commodity at a price equal to the marginal price plus taxes, tariffs, etc. Such situations are called regulated equilibria in [GM85]. These authors treat the case when the suppliers are described via a linear program, and the buyers via explicit, smooth demand functions. Therefore, our method extends theirs to the case when all players are described via linear programs or, in other words, when the demand functions are implicit, and non-smooth.

To compute a regulated equilibrium, the above algorithms require very few modifications. In fact, only step 4 is slightly altered as follows: replace the three shadow prices $p^{\ell}_{+}(k), p^{\ell}(k)$, and $p^{\ell}_{-}(k)$ by $T^{-1}(p^{\ell}_{+}(k)), T^{-1}(p^{\ell}(k))$, and $T^{-1}(p^{\ell}_{-}(k))$, respectively, where $T(p)$ is the price paid by buyer when marginal cost is p. Function $T(\cdot)$ is called the tariff function in [GM85]. We call $RIGCW_2$ the $IGCW_2$ algorithm modified with a tariff function.

The computational experiments that will be presented in section 6 are performed for cooperative as well as regulated equilibrium computation.

5. Demonstration Example

A very small example will serve three purposes: a) to demonstrate the main features of the algorithms, b) to exhibit a case when algorithm $IGCW_2$ works or fails, depending on which player is chosen for the computation of the tentative equilibria, and c) to show that even on such a small instance, the cob web algorithm may fail while $IGCW_2$ converges.

The example has two players and two commodities s^1, s^2 ; player 1 has only one decision variable x_1, and player 2 has two decision variables x_2, y_2 E is taken as the identity matrix; the two LP's are as follows:

$$
\begin{array}{rrl}
 & \min & 3x_1 \\
(P_1) & \text{s.t.} & x_1 - s^1 = 0 \\
 & & x_1 - s^2 = 0 \\
 & & s \geq 0
\end{array}
\qquad
\begin{array}{rrl}
 & \min & 2x_2 + 2y_2 \\
(P_2) & \text{s.t.} & x_2 + s^1 = 1 \\
 & & y_2 + s^2 = 1 \\
 & & s \geq 0.
\end{array}
$$

The above formulations have the following practical interpretation: player 1 is a supplier of electricity who manages one hydro-electric plant. The two commodities are winter and summer electricity, both of which can be produced by the same plant. There is no variable cost of production, only an yearly maintenance cost of 3, which is proportional to the plant capacity x_1. Player 2 is a single industrial customer requiring one unit of industrial heat in summer, and one in winter. Heat can be produced either with electricity purchased from the hydro supplier or with some other fuel such as gas, at a cost of 2 per unit. The global optimization problem represents the joint minimization of the hydroplant and the customer's annual costs.

Let us apply $IGCW_2$, with player 2 in charge of computing the equilibria.

Step 1: Let $s^1(0) = s^2(0) = 0$. *Step 2:* The optimal solution of (P_1) is $x_1 = 0$; the sensitivity analysis on each seasonal supply constraint yields a shadow price of 3, valid over $(0, +\infty)$.
Step 3: To compute a tentative equilibrium, player 2 solves the following LP

$$\begin{aligned} \min \quad & 2x_2 + 2y_2 + 3u_1^1 + 3u_1^2 \\ \text{s.t.} \quad & x_2 + u_1^1 = 1 \\ & y_2 + u_1^2 = 1 \\ & u_1^1, u_1^2 \geq 0 \end{aligned}$$

whose optimal solution is obviously $x_2 = y_2 = 1$, $u_1^1 = u_1^2 = 0$, so that $s^1(1) = s^2(1) = 0$, and the algorithm stops, having achieved a solution with a global cost of 4. In fact, the true global optimum is clearly $x_1 = 1, x_2 = y_2 = 0; s^1 = s^2 = 1$, with a global cost of 3.

The reason for the failure is that the separate sensitivity analyses of the two right hand sides of the (P_1) constraints have overestimated the cost of supplying one unit of s^1 *and* one unit of s^2 *simultaneously*. In other words, program (P_1) is degenerate at point $s^1 = s^2 = 0$.

It can be just as easily be verified that the same algorithm converges to the correct global optimum if the roles of the two players are simply reversed. Therefore, the application of the heavy-duty version of the algorithm also works, since the latter literally 'contains' both versions of the simple algorithm. This is the principal virtue of the heavy-duty version.

Finally, it is straightforward to see that the simple cob web algorithm fails to compute an equilibrium for our simple example, and that *the failure occurs irrespective of the way the cobweb is applied*: If the cobweb is applied in such a way that player 1 supplies its marginal cost, it stalls in exactly the same way as $ICWG_2$ did. If the cobweb is applied so that player 2 provides its marginal cost to player 1, it cycles indefinitely, whereas $IGCW_2$ was shown to converge.

6. Numerical Results

We now illustrate the $IGCW_2$ algorithm on realistic large scale examples involving one buyer and one seller. The seller is an integrated electricity producer who supplies the whole province of Quebec. It is represented by a linear program that minimizes total discounted cost by selecting supply options over the horizon. The buyer is the entire residential sector of the province. This sector is represented in much detail via a large linear program that minimizes discounted total cost by choosing among many technological options (some using electricity some using other fuels) to satisfy its useful demands. The other sectors consuming electricity are sketchily represented by fixed amounts of electricity demands. Technical and economic data are based on the database developed by the Energy group at GERAD.

The horizon studied extends from 1978 to 2022 and is divided in 9 periods of 5 years each. For each period, 6 time divisions are used: summer day, summer night, intermediate day, intermediate night, winter day and winter night. Hence, a total of 54 exchange variables ($q = 54$) are necessary. The sizes of the two linear programs are reported in Table 1.

	Buyer	Seller
# of rows	8055	2661
# of columns	8000	2687

Table 1. Sizes of the LP matrices

We ran two instances of the problem, with high and low demand scenarios, respectively.

Furthermore, we assumed two alternate pricing policies for electricity: marginal cost pricing, identified as COOP below, versus regulated (marginal cost plus taxes) pricing, identified as TAX below. We thus illustrate how the algorithm performs for these two cases. Recall from section 1 that when marginal cost pricing is used, the algorithm in fact computes a globally optimal solution for the two players acting cooperatively, whereas when other pricing schemes are used, there is no equivalent global linear program, although the equilibrium interpretation of the results still holds.

The cooperative equilibria were computed with algorithm $IGCW_2$, in which the residential sector (buyer) is the player chosen to compute the tentative equilibrium at each iteration of the algorithm. By way of verification, we also recomputed the final equilibrium by solving the global linear program (GLOBAL), which is still tractable with a good linear programming code (the exact size of the global LP is 10653 rows by 10624 columns).

Table 2 presents some key statistics on the performance of the algorithm when computing the cooperative equilibrium with the low scenario. The initial values of the exchange variables ($s^\ell(0)$) were all set at 0.0, a poor choice, but, as it turned out, without much bearing on the behaviour of the algorithm.

In Tables 2 to 5 we indicate, at each iteration, the number of pivots made by the simplex algorithm (phases 1 and 2) and the optimal values of the subproblems. The algorithm is stopped after two consecutive values of vector s verified

$$\sum_{\ell}(s^\ell(k+1) - s^\ell(k))^2 < eps$$

where eps is the zero of the simplex algorithm; the criterion value is printed for each iteration. The global value of the equilibrium is the sum of the optimal value of each player minus the amount

$$\sum_{\ell}\{p_+^{\ell}(k)u_1^{\ell} + p^{\ell}(k)u_2^{\ell} + p_-^{\ell}(k)u_3^{\ell}\}$$

computed by the buyer. The cpu time for the methods (in minutes on a Sunsparc workstation) compares favourably with the time obtained for the equivalent single linear program. The CPU time savings should be even greater when the sizes of the two linear programs are about the same. Of course, the main advantage of decomposition is not necessarily the reduction of CPU time, but rather the reduction of memory requirements and of various stability problems encountered when solving large LP's. The LP code used throughout is CPLEX.

		Iter. 1	Iter. 2	Iter. 3	Iter. 4	Total
Seller	# pivots ph.1	968	557	12	107	1644
	# pivots ph.2	757	1699	40	577	3073
	optimal value	33510.1	41361.9	40491.2	40983.8	–
Buyer	# pivots ph.1	3396	40	13	8	3457
	# pivots ph.2	4198	764	365	31	5358
	opt. value	19942.4	30256.1	24422.7	25135.6	–
	squared diff.	104103.0	948.5	653.9	0.0	–
	total value	–	55422.5	56338.8	55255.8	–
	cpu time	23.3	8.4	5.9	6.4	44.0
Global	# pivots ph.1					4282
	# pivots ph.2					14206
	optimal value					55255.8
	cpu time					55.6

Table 2. Performance of $IGCW_2$ on instance 1 (COOP–low demand)

Table 3 presents the results for the cooperative case with the high demand scenario. Again, algorithm $IGCW_2$ is used. In this case, the convergence was obtained after three iterations instead of four. The CPU savings are even more marked than in the first example.

The next two instances concern the same energy system, but a regulated equilibrium is sought, rather than a purely competitive one. We assume that electricity prices are obtained by applying an affine transformation to the marginal costs, meaning that the tariff has a proportional component of 15.56% and a fixed component of 1 ($/PJ), as per the following formula:

$$T(p) = 1.1556(p + 1.0)$$

As already discussed in section 4, step 4 of algorithm $IGCW_2$ is altered to correctly represent the true cost for the buyer. Of course for these instances, no equivalent global LP is available. Tables 4 and 5 summarize the results for the low and high scenarios respectively.

		Iter. 1	Iter. 2	Iter. 3	Total
Seller	# pivots ph. 1	1337	541	23	1901
	# pivots ph. 2	658	1440	41	2139
	optimal value	34293.7	43842.0	43408.4	–
Buyer	# pivots ph. 1	3459	26	20	3505
	# pivots ph. 2	3738	543	24	4305
	optimal value	20224.4	27923.6	27805.3	–
	squared diff.	104040.5	110.6	0.0	–
	total value	–	57903.2	57784.5	–
	cpu time	28.3	6.8	4.6	39.7
Global	# pivots ph. 1				4561
	# pivots ph. 2				15607
	optimal value				57784.5
	cpu time				63.3

Table 3. Performance of $IGCW_2$ on instance 2 (COOP-high demand)

		Iter. 1	Iter. 2	Iter. 3	Iter. 4	Total
Seller	# pivots ph. 1	968	573	14	1	1556
	# pivots ph. 2	757	1665	32	12	2466
	optimal value	33510.1	41098.3	40754.0	40775.6	–
Buyer	# pivots ph. 1	3396	24	6	4	3430
	# pivots ph. 2	4102	568	28	6	4704
	optimal value	24522.4	29546.4	29181.6	29675.8	–
	squared diff.	102062.3	209.3	3.2	0.0	–
	total value	–	55291.8	55306.1	55298.9	–
	total cpu	22.9	10.6	6.7	5.6	45.8

Table 4. Performance of $RIGCW_2$ on instance 3 (TAX-low demand)

		Iter. 1	Iter. 2	Iter. 3	Iter. 4	Iter. 5	Total
Seller	# pivots ph. 1	1337	679	39	67	7	2129
	# pivots ph. 2	658	1718	210	161	7	2754
	optimal value	34293.7	43627.6	42742.1	42882.8	42848.0	–
Buyer	# pivots ph. 1	3459	37	16	7	3	3522
	# pivots ph. 2	3923	666	82	31	0	4702
	optimal value	24846.2	33582.7	32084.8	31980.7	32194.6	–
	squared diff.	101467.7	694.2	36.9	1.9	0.0	–
	total value	–	57835.1	57925.3	57879.7	57889.9	–
	total cpu	29.1	8.2	4.4	4.1	4.5	50.3

Table 5. Performance of $RIGCW_2$ on instance 4 (TAX-high demand)

To give some idea of the behaviour of the exchanged capacities over the successive iterations, Table 6 presents a subset of these values (years 2000 and 2020) for the high demand scenario of the regulated equilibrium case. Note that the values quickly approach (at iteration 1) the equilibrium values, and that iterations 3 and 4 bring very minute changes in the exchange variables. This phenomenon seems to indicate that the initial values are of no great consequence on the behaviour of the algorithm.

Year	Division	Iter. 0	Iter. 1	Iter. 2	Iter. 3	Iter. 4
2000	Winter day	0.0	75.539	69.915	71.651	70.934
	Winter night	0.0	45.402	41.455	42.497	42.067
	Inter. day	0.0	49.149	43.523	45.259	44.542
	Inter. night	0.0	25.004	21.630	22.671	22.241
	Summer day	0.0	31.032	30.591	30.592	30.591
	Summer night	0.0	10.668	9.465	9.620	9.465
2020	Winter day	0.0	89.237	82.463	82.526	82.495
	Winter night	0.0	52.725	48.303	48.341	48.322
	Inter. day	0.0	59.213	53.125	53.156	53.125
	Inter. night	0.0	29.263	25.199	25.237	25.218
	Summer day	0.0	38.429	37.968	37.968	37.968
	Summer night	0.0	13.209	10.756	10.756	10.756

Table 6. Exchanged capacities for years 2000 and 2020 (PJ/yr) (TAX-high demand)

A few important remarks emerge from these experiments. First, the algorithms, although heuristic, have converged in all reported instances, and have required very few iterations, in spite of the fact that the initial exchange variable values were all set to zero. Secondly, the CPU times represent significant savings compared to solving directly the global LP, where applicable (i.e. instances 1 and 2 only). Saving CPU time is not the main reason for using decomposition, but is useful nevertheless. Finally, computing regulated equilibria appears to be just as easy as computing a global optimum, and does not require more effort. In fact, the algorithm is particularly useful for the regulated case, where very few other approaches seem available. It should however be remembered that the algorithms presented and illustrated in this section are not guaranteed to always converge. In section 7, we shall return to the question of convergence and discuss it further.

7. Locally Exact Approximations and Convergence of the *IGCW* Algorithms

Let us again consider a two-player situation and assume that player 1 computes the equilibrium, using some approximation $Z_2(\cdot)$ of the second player's value function. As seen in the previous section, the main cause of failure to converge is the fact that there are instances where the approximation Z_2 obtained via a separate parametric analysis of each right-hand-side gives erroneous information to the other player(s). In this section, we show that some classes of local approximations of the value function lead to a guarantee that if *IGCW* stops, then it stops at the true equilibrium (optimal solution). Furthermore, we describe a way to compute such a locally exact approximation.

Consider a generic algorithm similar to that of section 2, but adapted to the case of one seller (player 1) and one buyer (player 2).

Generic $IGCW_2$ algorithm
Step 1: $k \leftarrow 0$; choose some $s_i(0)$.
Step 2: Compute an approximation of player 2's optimal value function, $Z_2^k(\cdot)$.
Step 3: Compute a tentative equilibrium by solving the following LP

$$\begin{array}{lll} & \min & c_1 x - Z_2^k(b_2 - s) \\ (Q_k) & \text{s.t.} & A_1 x - E^{-1} s = b_1 \\ & & x \in L_1 \\ & & s \geq 0. \end{array}$$

Let $s(k+1)$ be the exchange vector found (i.e. the tentative equilibrium quantities).
Step 4: if $s(k+1) = s(k)$, STOP; else let $k \leftarrow k+1$ and GOTO *step 2.*

Theorem 2. *Let $Z_2^k(\cdot)$ be the approximation of the value function $Z_2(\cdot)$ of player 2 at the kth application of step 2 in the generic $IGCW_2$ algorithm, and assume that*

$$Z_2^k(\cdot) = Z_2(\cdot)$$

on a neighbourhood of current point $s(k)$. Then, if the algorithm stops (at step 4), $s(k)$ is an equilibrium quantity vector.

Proof. Let V be the neighbourhood of $s(k)$ over which the approximation is exact. The optimality of $s(k+1)$ for program Q_k means that $s(k+1)$ indeed minimizes the sum of the two players costs over V, i.e. $s(k+1)$ is a local optimum for the sum of the two players costs. But since the global program is a linear program, any local minimum is also global. □

The simple example of section 5 showed that there are instances where the approximation of the supply function obtained via a separate sensitivity analysis on the components of vector b_1 leads to a premature stopping of the algorithm, before the true equilibrium is obtained. This is precisely due to the fact that such player 2 approximation was not locally exact. More precisely, an approximation based on sensitivity analysis performed *separately* on the second player's demand constraints cannot be guaranteed to be locally exact, and this is due to the possible degeneracy of the second player's program at the current point $s(k)$. To remedy this, a much more elaborate scheme is needed for the computation of $Z_2(\cdot)$. We are able to describe such a scheme, but it requires performing a large number of sensitivity analyses (of the order of 2^q in the worst case), and is thus practical only for a small number of exchange variables.

8. Conclusion

In what precedes, we have presented a class of multi agent optimization problems that occurs in energy planning and other large scale economic trading problems. Each agent is represented by its own linear program, and may exchange commodities with the others. The cooperative solution to the trading problem is equivalent to the global minimization of joint costs, or, equivalently, to the computation of a competitive equilibrium in which the players exchange commodities at their marginal supply costs. Based on this equivalence, a number of variants of a heuristic algorithm were described, in which each agent's supply/demand function is approximated by a separable step function. A simple demonstration algorithm was used to show that the convergence properties of the suggested approaches rest on the non-degeneracy of some players' linear programs, and the 'heavy-duty' version of our algorithm was shown to palliate this difficulty. The same example showed that the simpler cob web may fail when our approach converges to the correct equilibrium. Finally, the same ideas were extended to the computation of regulated equilibria, thus extending the work of previous authors to the case of implicit rather than explicit demand functions.

Some large scale application to energy planning problems were presented. We are especially encouraged by the fact that the 'heavy-duty' version of the algorithm seems to avoid all the pitfalls due to degeneracy of the linear programs, both for computing competitive and regulated equilibria.

In spite of their practical success, our algorithms remain heuristic, although we have not been able to find counter examples where the 'heavy-duty' version of the algorithm failed. We also provided a sufficient condition for the algorithm to avoid being trapped at a point other than the true equilibrium. Although not presented in this article, there is a scheme which guarantees that the sufficient condition is always satisfied, but it may require an exponentially large number of computations in the worst case.

In conclusion, the algorithms presented in this article have proved to be rather robust when solving large scale realistic problems. They are being incorporated in the tool kit used at GERAD for the routine resolution of large scale energy planning problems. We plan next to use them to solve multi-regional optimization and equilibrium problems, cases that so far have eluded exact resolution by even the most powerful LP codes (especially because of memory and stability problems). We hope that this work will trigger renewed interest in an area of research that has been somewhat neglected in the last few years.

References

[AH82] Ahn, B.H. and W.W. Hogan, *On Convergence of the PIES Algorithm for Computing Equilibria*, Operations Research **30**(2), March-April 1982.

[G70] Geoffrion, A.M., *Elements of Large-Scale Mathematical Programming*, Management Science **16** (1970), 652–691.

[GM85] Greenberg, H.S. and F.H. Murphy, *Computing Market Equilibria with Price Regulations Using Mathematical Programming*, Operations Research **33** (1985), 935–954.

[HLS90] Haurie, A., R. Loulou et G. Savard, *A Two-level Systems Analysis Model of Power Cogeneration Under Asymmetric Pricing*, Proceedings of the 1990 American Control Conference, San Diego, 2095–2099, may 1990.

[HW83] Hogan, W.W. and J.P. Weyant, *Methods and Algorithms for Energy Model Composition: Optimization in a Network of Process Models*, in "Energy Models and Studies," (ed.) Lev, North-Holland, Amsterdam, 1983.

[LS91] Loulou, R. et G. Savard, *Computation of Cooperative and Stackelberg Solutions when Players are Described by Linear Programs*, Proceedings of the Fourth International Symposium on Differential Games and Applications (Helsinki, august 1990), Springer-Verlag's Lecture Notes in Control and Information Sciences, 285–292, 1991.

[M87] Murphy, F.H., *Equation Partitioning Techniques for Solving Partial Equilibrium Models*, European Journal of Operational Research **32** (1987), 380–392.

[TJ71] Takayama, T. and G.G. Judge, *Spatial and Temporal Price and Allocation Models*, North-Holland, Amsterdam, 1971.

[W81] Wagner, M.H., *Supply-Demand Decomposition of the National Coal Model*, Operations Research **29** (1981), 1137–1153.

Richard Loulou, Gilles Savard, and Denis Lavigne
GERAD
5255 Av. Decelles
Montréal, Québec
Canada, H3T 1V6

Convergent Stepsizes for Constrained Min-Max Algorithms

Berc Rustem

Abstract. We discuss two algorithms for the constrained discrete min-max problem. One is based on sequential quadratic min-max subproblem and the other on sequential quadratic programming. Both algorithms use an augmented Lagrangian formulation. One uses an adaptive strategy to determine the penalty parameter; the other assumes it to be constant. Two alternative stepsize strategies are studied to determine the progress along the direction computed by the quadratic subproblem. In this paper, we establish the conditions under which unit stepsizes can be achieved for both strategies. The achievement of unit stepsizes is of interest as this leads to the superlinear convergence rate analysis of the algorithms. We also demonstrate the superlinear convergence of the algorithms.

Key words. Constrained discrete min-max, nonlinear programming, augmented Lagrangian, adaptive penalty parameter, stepsize strategies.

1. Introduction

Let $x \in \mathbb{R}^n$ and let the functions $g: \mathbb{R}^n \to \mathbb{R}^e$, and $f^i: \mathbb{R}^n \to \mathbb{R}^1$ be twice continuously differentiable. Consider the discrete min-max problem

$$\min_{x} \max_{i \in \{1,2,\ldots,m\}} \{ f^i(x) \mid g(x) = 0 \} \tag{1.1}$$

where $f^i(x)$; $i = 1, \ldots, m$ are the objective functions and g is the set of equality constraints. Let $l \in \mathbb{R}^m$ denote the vector with every component unity and

$$\alpha \in \mathbb{E}_+^m = \{ \alpha \in \mathbb{R}^m \mid \alpha \geq 0;\ \langle l, \alpha \rangle = 1 \}.$$

It can be shown that (1.1) is equivalent to the min max problem

$$\min_{x} \max_{\alpha} \{ \langle f(x), \alpha \rangle \mid g(x) = 0,\ \alpha \in \mathbb{E}_+^m \} \tag{1.2}$$

(see e.g. Cohen, 1981). In (1.2), each element of the column vector $f: \mathbb{R}^n \to \mathbb{R}^m$, is one of the objective functions f^i in (1.1).

For $v \in \mathbb{R}^1$, problem (1.1) can be solved by the nonlinear programming problem

$$\min_{x,v}\{v \mid f(x) \leq l\, v;\ g(x) = 0\} \tag{1.3}$$

Algorithms for solving (1.2) or (1.3) have been considered by a number of authors, including Charalambous and Conn (1978), Coleman (1978), Conn (1979), Conn and Li (1988), Demyanov and Malomezov (1974), Demyanov and Pevnyi (1972), Dutta and Vidyasagar (1977), Hald and Madsen (1981), Han (1978, 1981), Murray and Overton (1980), Polak, Mayne and Higgins (1988), Womersley and Fletcher (1986). In the constrained case, discussed in some of these studies, convergence results to unit steplengths, global convergence and local convergence rates have not been established (e.g. Coleman, 1978 and Dutta and Vidyagasar, 1977). In this paper, the dual approach (1.2), formulated originally by Medanic and Andjelic (1971, 1972) and Cohen (1981), is initially adopted. Subsequently, both (1.2) and (1.3) are utilized to formulate the algorithm.

We discuss stepsize strategies for two quasi-Newton algorithms for the constrained min-max problem. The stepsizes are interchangeable between the algorithms. The quadratic subproblems solved sequentially by both algorithms are based on (1.2) and (1.3).

The global convergence properties of the algorithm based on (1.3) is discussed in Rustem (1992) and the one based on (1.2) is discussed in Rustem (1989). In this paper, we establish the convergence of the stepsizes to unity for both stepsize strategies and the superlinear convergence rates of the algorithms. The achievement of unit stepsizes is of interest as this leads to the the superlinear convergence rate analysis of the algorithms.

Consider the Lagrangian function associated with (1.2) with the multipliers $\lambda \in \mathbb{R}^e$, $\mu \in \mathbb{R}^m_+ = \{\mu \in \mathbb{R}^m \mid \mu \geq 0\}$ and $\eta \in \mathbb{R}^1$

$$L(x, \alpha, \lambda, \mu, \eta) = \langle f(x), , \alpha\rangle + \langle g(x), \lambda\rangle + \langle \alpha, \mu\rangle + (\langle l, \alpha\rangle - 1)\eta. \tag{1.4}$$

The characterization of the min-max solution of (1.2) as a saddle point requires the relaxation of convexity assumptions (see Demyanov and Malomezov, 1974 and Cohen, 1981). In order to achieve this characterization, we augment (1.4) with a penalty function. Hence, we define the augmented Lagrangian by

$$L^a(x, \alpha, \lambda\mu, \eta, c) = L(x, \alpha, \lambda, \mu, \eta) + \frac{c}{2}\langle g(x), g(x)\rangle \tag{1.5}$$

where the scalar $c \geq 0$ is the penalty parameter.

The penalty parameter c is either taken as a constant or adapted as the algorithm progresses. Examples of the latter strategy are Biggs (1974), Polak and Mayne (1981), Polak and Tits (1981), Rustem (1993). In one of the algorithms below, we take a constant penalty parameter and in the other we include an adaptive strategy. The departure of the adaptive strategy from other works is mainly in the relationship of c and the descent property of the direction of search, discussed in Lemmas (2.2) and (2.3) below. In particular, c is only adjusted to ensure that the direction of search is a descent direction for the penalty function that regulates the stepsize strategy. This is an extension of a nonlinear programming algorithm discussed in Rustem (1986, 1993).

We let $H(\cdot)$ denote the Hessian of L^a, with respect to x, evaluated at $(\cdot)$. We also denote by $N(\cdot)$ the matrix

$$N(x) = \left[\nabla_x g^1(x), \dots, \nabla_x g^e(x)\right].$$

Sometimes, $N(x)$ evaluated at x_k will be denoted by N_k and $g(x_k)$ will be denoted by g_k. Thus, a local linearization of $g(x)$ at x_k can be written as

$$g(x) = g_k + N_k^T\left[x - x_k\right].$$

Assumption (1.1) The columns of N_k are assumed to be linearly independent. □

This assumption is used to simplify the quadratic subproblem used in the algorithm in Section 2 for solving (1.2) and ensure that the system $g_k + N_k^T[x - x_k] = 0$ has a solution, $\forall\, x_k$. The assumption can be relaxed by increasing the complexity of the quadratic subproblem.

2. The Algorithms

Consider the objective function

$$\mathcal{F}(x,\alpha) = \langle \alpha, f(x)\rangle$$

and its linear approximations, with respect to x, at a point x_k

$$\mathcal{F}_k(x,\alpha) = \langle \alpha, f(x_k) + \nabla f(x_k)^T(x - x_k)\rangle. \tag{2.1}$$

where

$$\nabla f(x) = \left[\nabla f^1(x), \dots, \nabla f^m(x)\right].$$

We shall sometimes denote $f(x)$ and $\nabla f(x)$, evaluated at x_k, by f_k and ∇f_k respectively. Thus, for $d = x - x_k$ (2.1) can be written as

$$\mathcal{F}_k(x_k + d, \alpha) = \langle \alpha, f_k + \nabla f_k^T d\rangle.$$

The quadratic objective function used to compute the direction of progress is given by

$$\begin{aligned} q_k(x,\alpha,c) = {} & \mathcal{F}_k(x,\alpha) + c\langle N_k g_k, (x - x_k)\rangle \\ & + \frac{1}{2}\langle x - x_k, \hat{H}_k(x - x_k)\rangle \end{aligned}$$

or alternatively,

$$q_k(d,\alpha,c) = \mathcal{F}_k(x_k+d,\alpha) + c\langle N_k g_k, d\rangle + \frac{1}{2}\langle d, \hat{H}_k, d\rangle.$$

The matrix $\hat{H}_k$ is a positive semi-definite approximation to the Hessian of the augmented Lagrangian as required by the sequential quadratic programming subproblem[1]

$$H_k = \sum_{i=1}^{m} \alpha_k^i \nabla^2 f^i(x_k) + \sum_{j=1}^{e} \lambda_k^i \nabla^2 g^j(x_k) + cN_k N_k^T \tag{2.2}$$

The values α_k and λ_k are given by the solution of the quadratic subproblem in the previous iteration. The direction of progress at each iteration of the algorithm is determined by the quadratic subproblem

$$\min_d \max_\alpha \{q_k(d,\alpha,c_k) \mid N_k^T d + g_k = 0,\ \ \alpha \in \mathbb{E}_+^m\}. \tag{2.3}$$

The above quadratic min-max subproblem is the basis of one of the quadratic programming algorithms. Its solution is easy to compute. An equivalent subproblem is given by the quadratic program

$$\begin{aligned}\min_{d,v}\{v + c_k\langle N_k g_k, d\rangle + \frac{1}{2}\langle d, \hat{H}_k d\rangle \mid N_k^T d + g_k = 0,\\ \nabla f_k^T d + f_k \le l\,v\}.\end{aligned} \tag{2.4}$$

The multipliers associated with the inequalities in (2.4) are the values α and the solution of either subproblem satisfies common optimality properties.

Let the value of d, α, v solving (2.4) be denoted by d_k, α_{k+1}, v_{k+1}. The stepsize along d_k is defined using the equivalent min-max formulation (1.3). Thus, consider the function

$$\psi(x) = \max_{i\in\{1,2,\ldots,m\}} \{f^i(x)\} \tag{2.5}$$

and

$$\psi_k(x) = \max_{i\in\{1,2,\ldots,m\}} \{f^i(x_k) + \langle \nabla f^i(x_k),\ x - x_k\rangle\}.$$

Let $\psi_k(x_k+d_k)$ be given by

$$\psi_k(x_k+d_k) = \max_{i\in\{1,2,\ldots,m\}} \{f^i(x_k) + \langle \nabla f^i(x_k), d_k\rangle\}.$$

---

[1] The assumption of positive semi-definiteness will be tightened in the later stages of the discussion. The use of the Hessian of the augmented Lagrangian at this stage ensures unit stepsizes and superlinear rate of convergence of the algorithm.

The stepsize strategy determines τ_k as the largest value of $\tau = (\gamma)^j$, $j = 0, 1, 2, \dots$ such that x_{k+1} given by

$$x_{k+1} = x_k + \tau_k d_k$$

ensures that x_{k+1} simultaneously reduces the main objective and maintains or improves the feasibility with respect to the constraints. The penalty term used to improve feasibility is a quadratic consistent with the augmented Lagrangian (1.5). It is shown in Rustem (1992; Theorem 4.1) that the following stepsize strategy can always be fulfilled by the algorithm:

$$\begin{aligned}\psi(x_{k+1}) + \frac{c_{k+1}}{2}\langle g_{k+1}, g_{k+1}\rangle - \psi(x_k) - \frac{c_{k+1}}{2}\langle g_k, g_k\rangle \\ \leq \rho\tau_k \Phi(d_k, c_{k+1})\end{aligned} \tag{2.6a}$$

where $\rho \in (0,1)$ is a given scalar and $\Phi(d_k, c_{k+1})$

$$\begin{aligned}\Phi(d_k, c_{k+1}) = \psi_k(x_k + d_k) - \psi(x_k) + c_{k+1}\langle g_k, N_k^T d_k\rangle \\ + \frac{1}{2}\langle d_k, \hat{H}_k d_k\rangle.\end{aligned} \tag{2.6b}$$

The determination of the penalty parameter c is an important aspect of the algorithm. This is specified in the following description.

Algorithm 1

Step 0: Given x_0, $c_0 \in [0, \infty)$, and small[2] positive numbers δ, ρ, ϵ, γ such that $\delta \in (0, \infty)$, $\rho \in (0,1)$, $\epsilon \in \left(0, \frac{1}{2}\right]$, $\gamma \in (0,1)$, $\hat{H}_0$, set $k = 0$.

Step 1: Compute ∇f_k and N_k. Solve the quadratic subproblem (2.3) or (2.4) to obtain d_k, α_{k+1}, and the associated multiplier vector λ_{k+1}. In (2.3), we also compute μ_{k+1}, η_{k+1} and in (2.4) we also compute v_{k+1}.

Step 2: Test for optimality:[3] If optimality is achieved, stop. Else go to Step 3.

Step 3: If

$$\psi_k(x_k + d_k) - \psi(x_k) - c_k\langle g_k, g_k\rangle + \left(\epsilon + \frac{1}{2}\right)\langle d_k, \hat{H}_k d_k\rangle \leq 0,$$

then $c_{k+1} = c_k$. Else set

$$c_{k+1} = \max\left\{\frac{\psi_k(x_k + d_k) - \psi(x_k) + \left(\epsilon + \frac{1}{2}\right)\langle d_k, \hat{H}_k d_k\rangle}{\langle g_k, g_k\rangle}, c_k + \delta\right\}.$$

---

[2] It is desirable for these numbers to be small for numerical reasons. The convergence results for the algorithms are nevertheless ensured for the bounds specified in Step 0.

[3] By checking for the satisfaction of the first order optimality conditions of (1.3). Detailed termination criteria for constrained optimization algorithms are discussed in Gill, Murray and Wright (1981, Chapter 8).

Step 4: Find the smallest nonnegative integer j_k such that $\tau_k = \gamma^{j_k}$ with $x_{k+1} = x_k + \tau_k d_k$ such that the inequality (2.6) is satisfied.

Step 5: Update $\hat{H}_k$ using a quasi-Newton formula, such as Powell's (1978) modification to the BFGS formula, to compute $\hat{H}_{k+1}$, set $k = k+1$ and go to Step 1.

The penalty parameter c_{k+1} is adjusted to maintain progress towards feasibility. In particular, c_{k+1} is chosen to make the direction d_k computed by the quadratic subproblem is a descent direction for the penalty function $\psi(x_k) + \frac{c_{k+1}}{2}\langle g_k, g_k\rangle$.

Lemma (2.1) [Optimality Conditions of Quadratic Subproblems]. *The Lagrangian function for the quadratic subproblem* (2.3) *is given by*

$$\begin{aligned} L_k(x, \alpha, \lambda, \mu, \eta, c) = \mathcal{F}_k(x, \alpha) + c\langle g_k, N_k^T(x - x_k)\rangle \\ + \frac{1}{2}\langle x - x_k, \hat{H}_k(x - x_k)\rangle + \langle \alpha, \mu\rangle + (\langle l, \alpha\rangle - 1)\eta. \end{aligned} \tag{2.7a}$$

The first order necessary conditions of optimality for problem (2.3) *are*

$$\nabla f_k \alpha_{k+1} + N_k[c_k g_k + \lambda_{k+1}] + \hat{H}_k d_k = 0 \tag{2.7b}$$

$$f_k + \nabla f_k^T d_k + \mu_{k+1} + l\,\eta_{k+1} = 0 \tag{2.7c}$$

$$\alpha_{k+1} \in \mathbb{E}_+^m;\ \mu_{k+1} \geq 0;\ \alpha_{k+1}^i \mu_{k+1}^i = 0,\ \ \forall i \in \{1, \ldots, m\} \tag{2.7d}$$

$$N_k^T d_k + g_k = 0. \tag{2.7e}$$

and for the quadratic programming subproblem (2.4) *the necessary conditions are given by* (2.7,b), (2.7,e) *and*

$$\begin{aligned} f_k + \nabla f_k^T d_k \leq l\, v_{k+1};\ \alpha_{k+1}^i(f_k^i + \langle \nabla f_k^i, d_k\rangle - v_{k+1}) = 0; \\ i = 1, \ldots, m;\ \alpha_{k+1} \in \mathbb{E}_+^m. \end{aligned} \tag{2.7f}$$

Lemma (2.2) [The descent property of d_k for Algorithm 1]. *Let* $g_k \neq 0$ (*the case when* $g_k = 0$ *is discussed in Lemma 2.3*) *and let* $\hat{H}_k$ *be positive semi-definite. The direction* d_k *satisfying the linearized equality constraints* (2.7,e) *of the quadratic subproblem* (2.3) *or* (2.4) *and* c_{k+1} *chosen as in Step 3 of the above algorithm ensure the inequality*

$$\Phi(d_k, c_{k+1}) \leq -\epsilon\langle d_k, \hat{H}_k d_k\rangle. \tag{2.8}$$

The proof is given in Rustem (1992). □

Lemma (2.3) [The finiteness of the penalty parameter in Algorithm 1] *Let* f *and* g *be once differentiable,* $\hat{H}_k$ *be positive semi-definite and let*

$(d_k, \alpha_{k+1}, \lambda_{k+1}, \mu_{k+1}, \eta_{k+1})$ *generated by* (2.3) *and* $(d_k, v_{k+1}, \alpha_{k+1}, \lambda_{k+1})$ *computed by* (2.4) *be bounded.*[4] *Then*

(i) *If for some or all* k *the nonlinear constraints of* (1.2) *are satisfied, i.e.,* $g(x_k) = 0$, *then the descent property* (2.8) *of Lemma* (2.2) *are satisfied for any choice of* $c_k \in [0, \infty)$.

(ii) *Let* c_k *be bounded and* $g(x_k) = 0$, *for some* k. *Then, Step 3 of the algorithm chooses* $c_{k+1} = c_k$. *For this choice of* $c_k \in [0, \infty)$, *the direction* d_k *ensures the descent property* (2.8).[5]

(iii) $\forall\, k,\ \exists\, c_{k+1} \in [0, \infty)$ *satisfying Step 3 of the algorithm.*

(iv) *Let the sequence* $\{x_k\}$ *generated by the algorithm be bounded. Then,* c_k *is increased finitely often and* $\exists$ *an integer* $k_* \geq 0$ *and* $\exists\, c_* \in [0, \infty)$ *such that the algorithm chooses* $c_k = c_*$, $\forall\, k \geq k_*$.

The proof is given in Rustem (1992). □

The second algorithm is based on subproblem (2.3), although the equivalent formulation (2.4) may also be used. The former may be reformulated as

$$\min_{d}\max_{d^\alpha}\left\{\begin{bmatrix} d \\ d^\alpha \end{bmatrix}^T \left\{\frac{1}{2}\begin{bmatrix} \hat{H}_k & \nabla f_k \\ \nabla f_k^T & 0 \end{bmatrix}\begin{bmatrix} d \\ d^\alpha \end{bmatrix} + \begin{bmatrix} \nabla f_k \alpha_k + cN_k g_k \\ f_k \end{bmatrix}\right\}\middle|\, N_k^T d + h_k = 0;\ d^\alpha \geq 0,\ \langle l, d^\alpha \rangle = 0\right\}. \tag{2.9}$$

This can be rewritten as a positive quadratic programming problem by appropriate choice of Kuhn-Tucker multipliers. The notation to be used hereafter can be summarized as:

$$z = \begin{bmatrix} x \\ \alpha \\ \lambda \\ \mu \\ \eta \end{bmatrix};\ z_k = \begin{bmatrix} x_k \\ \alpha_k \\ \lambda_k \\ \mu_k \\ \eta_k \end{bmatrix};\ \overline{z} = \begin{bmatrix} \overline{x} \\ \overline{\alpha} \\ \overline{\lambda} \\ \overline{\mu} \\ \overline{\eta} \end{bmatrix};$$

$$d^z = \begin{bmatrix} d \\ d^\alpha \\ d^\lambda \\ d^\mu \\ d^\eta \end{bmatrix} = z - z_k;\ d_k^z = \overline{z} - z_k$$

[4] The boundedness invoked here is the property of the quadratic programming subproblem. It is ensured when the projected Hessian is positive definite and the active constraint normals are linearly independent at the solution of the quadratic program. Another implication is that the feasible region of the quadratic program is nonempty.

[5] c_k is the value used in the quadratic subproblems when computing d_k, α_{k+1}, λ_{k+1}, v_{k+1}.

with λ, μ, η denoting the associated Kuhn-Tucker multipliers. The vector $\overline{z}$ is the value of z that solves (2.9). Also, $\overline{\alpha} = \alpha_k + d^{\alpha}$ and the value of α_{k+1} is given in Algorithm 2 (Step 3) below.

Let the vector d_k^z denote the value of d^z that solves (2.9). The optimality conditions of (2.9) can be used to show that d_k^z satisfies

$$\nabla\hat{\Gamma}_k d_k^z + \Gamma_k = 0; \quad \overline{\alpha} \in \mathbb{E}_+^m, \ \overline{\mu} \geq 0 \tag{2.10}$$

where we also assume that strict complementarity holds for $\alpha \geq 0$. In (2.10) Γ_k and $\nabla\hat{\Gamma}_k$ are given by

$$\Gamma_k = \Gamma(z_k) = \begin{bmatrix} \Gamma^1(z_k) \\ \Gamma^2(z_k) \end{bmatrix} = \begin{bmatrix} \Gamma_k^1 \\ \Gamma_k^2 \end{bmatrix}$$

$$\Gamma_k^1 = \begin{bmatrix} \nabla f_k \alpha_k + N_k \lambda_k + c N_k g_k \\ f_k + \mu_k + l\,\eta_k \\ g_k \end{bmatrix} \tag{2.11a}$$

$$\Gamma_k^2 = \begin{bmatrix} \alpha_k \\ \langle l, \alpha_k \rangle - 1 \end{bmatrix} \tag{2.11b}$$

As the quadratic subproblem ensures that the constraints in (2.11,b) are always satisfied, we do not consider this term in the stepsize strategy. Furthermore, if we only consider the active nonnegativity constraints, $\Gamma_k^2 = 0$ is ensured. The rationale of the stepsize strategy used in Algorithm 2 below is to generate a monotonically decreasing sequence $\{\|\Gamma_k^1\|\}$ thereby ensuring the satisfaction of the first order optimality conditions of (1.2). We also have $\nabla\hat{\Gamma}_k$ given by

$$\nabla\hat{\Gamma}_k = \begin{bmatrix} \nabla\hat{\Gamma}_k^1 \\ \nabla\hat{\Gamma}_k^2 \end{bmatrix} = \begin{bmatrix} \nabla\hat{\Gamma}^1(z_k) \\ \nabla\hat{\Gamma}^2(z_k) \end{bmatrix}$$

where

$$\nabla\hat{\Gamma}_k^1 = \begin{bmatrix} \hat{H}_k & \nabla f_k & N_k & 0 & 0 \\ \nabla f_k^T & 0 & 0 & \tilde{I} & l \\ N_k^T & 0 & 0 & 0 & 0 \end{bmatrix} \tag{2.12a}$$

The matrix $\tilde{I}$ in (2.12a) consists of the columns of the identity matrix corresponding to the active nonnegativity constraints at $\overline{z}$. If the columns of (2.12,a) are linearly dependent, this can be overcome by eliminating the last column which corresponds to the equality $\langle l, \alpha \rangle = 1$. It can be shown that for k large, only the constraints active at the solution of (1.2) need to be considered (as strict equality constraints). Furthermore, for large k, the inequality constraints not active at the solution do not affect the computations in the algorithm any further (Rustem, 1989).

From (2.10), d_k^z satisfies

$$\nabla\hat{\Gamma}_k^1 d_k^z = -\Gamma_k^1 \tag{2.13a}$$

$$\begin{aligned}\langle \Gamma_k^1, \Gamma_k^1 \rangle &= \langle d_k^z, \nabla\hat{\Gamma}_k^{1^T} \nabla\hat{\Gamma}_k^1 d_k^z \rangle = \langle d_k^z, \nabla\hat{\Gamma}_k^T \nabla\hat{\Gamma}_k d_k^z \rangle \\ &= \langle \Gamma_k, \Gamma_k \rangle > 0.\end{aligned} \tag{2.13b}$$

We also define the exact $\nabla\Gamma_k$ as

$$\nabla\Gamma_k = \begin{bmatrix} \nabla\Gamma_k^1 \\ \nabla\Gamma_k^2 \end{bmatrix}$$

where $\nabla\Gamma_k^1$ is the same as $\nabla\hat{\Gamma}_k$ except that the approximation $\hat{H}$ in $\nabla\hat{\Gamma}^1$ is replaced by the exact value H_k in $\nabla\Gamma^1$. Clearly, the discussion related to (3.9) also applies to $\nabla\Gamma$.

Assumption (2.1). For a unique solution of (2.9) to exist, the columns of the matrix $\nabla\tilde{\Gamma}_k$ satisfying (2.13) must be linearly independent. Furthermore, we continue to assume that the system $N_k^T d + h_k = 0$ always has a feasible solution. □

The algorithm generates a sequence $\{z_k\}$ that ensures a sufficient decrease in $\|\Gamma_k^1\|$ at every iteration by means of a stepsize strategy.

Algorithm 2

Step 0: $\gamma \in (0,1)$, $\rho \in \left(0, \frac{1}{2}\right)$, $c \geq 0$, U_0, α_0, λ_0, μ_0, η_0, set $k = 0$.

Step 1: Check for optimality.

Step 2: Solve the quadratic subproblem (2.9) to obtain d_k^z.

Step 3: Determine $\tau_k = (\gamma)^{j_k}$ where $j_k \in \{0, 1, 2, \ldots\}$ is the smallest integer such that for $z_{k+1} = z_k + \tau_k d_k^z$ we have

$$\|\Gamma^1(z_{k+1})\|_2^2 - \|\Gamma^1(z_k)\|_2^2 \leq \tau\rho \langle \nabla\Gamma_k^{1^T} \Gamma_k^1, d_k^z \rangle. \tag{2.14}$$

Step 4: Set $k = k + 1$, go to Step 1.

Lemma (2.4) [The descent property of d_k^z for Algorithm 2]. *For d_k^z computed by* (2.9) *we have*

$$\begin{aligned}\langle \nabla\Gamma_k^{1^T} \Gamma_k^1, d_k^z \rangle &= \langle \nabla\hat{\Gamma}_k^{1^T} \Gamma_k^1, d_k^z \rangle = -\langle \Gamma_k^1, \Gamma_k^1 \rangle \\ &= -\langle d_k^2, \nabla\hat{\Gamma}_k^{1^T} \nabla\hat{\Gamma}_k^1 d_k^z \rangle\end{aligned} \tag{2.15}$$

Also, for

$$a_k \geq -\frac{1}{2}, \tag{2.16}$$

where

$$a_k = \frac{\langle d_k^z, \nabla\hat{\Gamma}_k^{1^T} \begin{Bmatrix} H_k - \hat{H}_k & 0 \\ 0 & 0 \end{Bmatrix} d_k^z \rangle}{\langle d_k^z, \nabla\hat{\Gamma}_k^{1^T} \nabla\hat{\Gamma}_k^1 d_k^z \rangle}, \tag{2.17}$$

we have

$$\langle \nabla \Gamma_k^{1^T} \Gamma_k^1, d_k^z \rangle \leq -\frac{1}{2} \langle d_k^z, \nabla \hat{\Gamma}_k^{1^T} \nabla \hat{\Gamma}_k^1 d_k^z \rangle. \tag{2.18}$$

Proof. To show (2.18) consider Premultiplying by Γ_k^1 we have

$$\langle \Gamma_k^1, \nabla \hat{\Gamma}_k^1 d_k^z \rangle = -\langle \Gamma_k^1, \Gamma_k^1 \rangle = -\langle \nabla \hat{\Gamma}_k^1 d_k^z, \nabla \hat{\Gamma}_k^1 d_k^z \rangle.$$

To show (2.18) consider

$$\begin{aligned} a_k &= \frac{\langle d_k^z, \nabla \hat{\Gamma}_k^{1^T} \{\nabla \Gamma_k^1 - \nabla \hat{\Gamma}_k^1\} d_k^z \rangle}{\langle d_k^z, \nabla \hat{\Gamma}_k^{1^T} \nabla \hat{\Gamma}_k^1 d_k^z \rangle} \\ &= \frac{-\langle \Gamma_k^1, \nabla \Gamma_k^1 d_k^z \rangle - \langle d_k^z, \nabla \hat{\Gamma}_k^{1^T} \nabla \hat{\Gamma}_k^1 d_k^z \rangle}{\langle d_k^z, \nabla \hat{\Gamma}_k^{1^T} \nabla \hat{\Gamma}_k^1 d_k^z \rangle} \end{aligned}$$

and for $a_k \geq -\frac{1}{2}$, (2.18) is established. □

Since $\frac{1}{2} \nabla_z \|\Gamma^1(z_k)\|_2^2 = \nabla \Gamma_k^{1^T} \Gamma_k^1$, it follows that d_k^z satisfying (2.16) is a descent direction for $\|\Gamma^1(z_k)\|$. Also, by (2.17), if $\hat{H}_k$ is exact, the descent condition (2.16) is immediately satisfied.

3. Convergent Stepsizes

The global convergence of algorithms 1 and 2 are established in Rustem (1989; 1992). In addition, it is shown in Rustem (1992) that c_k is not increased indefinitely and that $c_k = c_*$ after some stage, for some finite c_* (see Lemma, 2.3, iv). In this section, we consider the convergence of stepsizes to unity, assuming that the sequence $\{x_k\}$ converges to the solution x_* and that the inequality constraints, active at the solution, are also the constraints that are currently active (Rustem, 1992).

The conditions that ensure unit stepsizes turn out to be the closeness of the algorithm to the solution as well as the accuracy of the Hessian approximation used near this solution. For superlinear convergence rates, it is shown that increasing accuracy of the Hessian approximation is required.

For Algorithm 1, we first establish that $\langle g_k, \tilde{\lambda}_{k+1}(\tilde{c}_{k+1}) \rangle \geq 0$ for a set of specially constructed multipliers $\tilde{\lambda}_{k+1}(\tilde{c}_{k+1})$ and a specially constructed $\tilde{c}_{k+1}$. Also, it is shown that $\|\tilde{\lambda}_{k+1}(\tilde{c}_{k+1}) - \lambda_{k+1}\| \to 0$ and $(\tilde{c}_{k+1} - c_*) \to 0$ as the solution is approached. The values $\tilde{\lambda}$ and $\tilde{c}$ are constructed to show that unit stepsizes are attained such that $\tau_k = 1$, $\forall\, k \geq k_\tau \geq k_*$: they are *not* used in any computational procedure in the algorithm. The integer k_τ is the iteration at which $\tau_k = 1$ is achieved and maintained thereafter.

Assuming, as in Section 1, that the columns of N_k are linearly independent and that $\hat{H}_k$ is positive definite and using (2.7,b) and (2.7,e) we have

$$\tilde{\lambda}_{k+1}(\tilde{c}_{k+1}) = -\left[N_k^T\hat{H}_k^{-1}N_k\right]^{-1}\left[N_k^T\hat{H}_k^{-1}\nabla f_k\alpha_{k+1} - g_k\right] - \tilde{c}_{k+1}g_k \quad (3.1)$$
$$\lambda_{k+1} = -\left[N_k^T\hat{H}_k^{-1}N_k\right]^{-1}\left[N_k^T\hat{H}_k^{-1}\nabla f_k\alpha_{k+1} - g_k\right] - c_*g_k. \quad (3.2)$$

Lemma (3.1). *Let*

(i) $f, g \in \mathbb{C}^1(\mathbb{R}^n)$;
(ii) the approximate Hessian be such that, for each k,

$$m\|y\|_2^2 \leq y^T\hat{H}_k y \leq M\|y\|_2^2; \quad M \geq m > 0; \ 0 \neq y \in \mathbb{R}^n, \quad (3.3)$$

(iii) for each k, $\exists$ a bounded optimal point[6] $(d_k, \alpha_{k+1}, \lambda_{k+1}, \mu_{k+1}, \eta_{k+1}) \in \mathbb{R}^n\times\mathbb{R}^m\times\mathbb{R}^e\times\mathbb{R}^m\times\mathbb{R}^1$ of the quadratic subproblem (2.3) and similarly $\exists$ a bounded Kuhn-Tucker point $(d_k, v_{k+1}, \alpha_{k+1}, \lambda_{k+1}) \in \mathbb{R}^n \times \mathbb{R}^1 \times \mathbb{R}^m \times \mathbb{R}^e$ for subproblem (2.4).
(iv) $\{x_k\}$ be convergent and let the columns of N_k be linearly independent. Then
  (a) For λ_{k+1} given by (3.1), $\exists\tilde{c}_{k+1} \in (-\infty, +\infty)$ such that $\langle g_k, \tilde{\lambda}_{k+1}(\tilde{c}_{k+1})\rangle \geq 0$.
  (b) As $k \to \infty$, $k_* \leq k \in K$, and according to (iv), we have $\|d_k\| \to 0$ and

$$\left\|\lambda_{k+1} - \tilde{\lambda}_{k+1}(\tilde{c}_{k+1})\right\| \to 0.$$

Proof. To show (a), we premultiply the optimality condition (2.7,b) by d_k and use (2.7,e) to obtain

$$\langle d_k, \hat{H}_k d_k\rangle + \langle d_k, \nabla f_k\alpha_{k+1}\rangle - \tilde{c}_k\langle g_k, g_k\rangle = \langle g_k, \tilde{\lambda}_{k+1}(\tilde{c}_k)\rangle. \quad (3.4)$$

We note that the parameter $\tilde{c}$ does not affect the solution of the quadratic subproblem. Let $\tilde{c}_{k+1}$ be determined by

$$\tilde{c}_{k+1} = \begin{cases} c_* & \text{if } \langle g_k, \tilde{\lambda}_{k+1}(c_*)\rangle \geq 0 \text{ or } \|g_k\| = 0 \\ \frac{\langle d_k, \hat{H}_k d_k\rangle + \langle d_k, \nabla f_k\alpha_{k+1}\rangle}{\langle g_k, g_k\rangle} & \text{if } \langle g_k, \tilde{\lambda}_{k+1}(\tilde{c}_k)\rangle < 0 \end{cases} \quad (3.5)$$

Since $g_k = 0$, we have $\langle g_k, \tilde{\lambda}_{k+1}(c_*)\rangle = 0$, it can be verified that this choice of $\tilde{c}_{k+1}$ ensures (a). Furthermore, as $\{\|g_k\|\} \to 0$, it follows from (3.5) that $\{\tilde{c}_k\} \to c_*$.

[6] As mentioned above, the boundedness conditions can be equivalently associated with the properties of the quadratic subproblem.

To show (b), consider (3.1) and (3.2). We have

$$\left\|\lambda_{k+1} - \tilde{\lambda}_{k+1}(\tilde{c}_{k+1})\right\| = |c_* - \tilde{c}_{k+1}| \, \|g_k\| = 0$$

since $\|g_k\| \to 0$, c_* and $\tilde{c}_{k+1}$ are finite. □

Theorem (3.1) [Monotonic Decrease and Unit Stepsizes for Algorithm 1]. *Let assumptions* (i)–(iii) *of Lemma (3.1) be satisfied. Let $c_k = c_*$ for finite c_*. Then Stepsize τ_k computed in Step 3 of algorithm 1 is in the range $\tau_k \in (0, 1]$; hence the sequence $\{x_k\}$ computed by algorithm 1 generates a corresponding monotonically decreasing sequence $\{\psi(x_k) + \frac{c_* \langle g_k, g_k \rangle}{2}\}$.*

Let the assumptions of Lemma (3.1) be satisfied and let H_* denote the value of the Hessian (2.2) at the solution. There is a small number $\sigma > 0$ such that if $\phi_k \leq \sigma$, for all large k, with

$$\phi_k = \frac{\|P_k^T (H_* - \hat{H}_k) P_k d_k\|}{\|d_k\|} + \frac{\|g_k^T \tilde{N}_k^T (H_* - \hat{H}_k)\{\tilde{N}_k g_k - 2 P_k d_k\}\|}{\|d_k\|^2}$$

and P_k, $\tilde{N}_k$ given either by $P_k = I - \hat{H}_k^{-1} N_k (N_k^T \hat{H}_k^{-1} N_k)^{-1} N_k^T$, $\tilde{N}_k = \hat{H}_k^{-1} N_k (N_k^T \hat{H}_k^{-1} N_k)^{-1}$ or by $P_k = I - N_k (N_k^T N_k)^{-1} N_k^T$, $\tilde{N}_k = N_k (N_k^T N_k)^{-1}$, then stepsize strategy (2.6) is satisfied for $\tau_k = 1$, $\forall\, k \geq k_\tau \geq k_*$.

Proof. Consider the second order expansion of $\langle g(x), \lambda_{k+1} \rangle$

$$\begin{aligned}\langle g_{k+1}, \lambda_{k+1} \rangle &= \langle g_k, \lambda_{k+1} \rangle + \langle d_k, N_k \lambda_{k+1} \rangle \\ &\quad + \frac{1}{2} \langle d_k, \left[\sum_{i=1}^{e} \nabla^2 g_*^i \lambda_*^i\right] d_k \rangle + \nu_k^1\end{aligned} \tag{3.6a}$$

$$\begin{aligned}\nu_k^1 = \int_0^1 (1-t) \langle d_k, &\left[\left[\sum_{i=1}^{e} \nabla^2 g^i (x_k + t d_k) \lambda_{k+1}^i\right]\right. \\ &\left. - \left[\sum_{i=1}^{e} \nabla^2 g_*^i \lambda_*^i\right]\right] d_k \rangle dt\end{aligned} \tag{3.6b}$$

where $\left[\sum_{i=1}^{e} \nabla^2 g_*^i \lambda_*^i\right]$ is the constraint second derivatives at x_*, to be subsumed in H_*. Using the fact that d_k satisfies the linearized constraints, $N_k^T d_k + g_k = 0$, and using Lemma (3.1) we have

$$\begin{aligned}\langle g_{k+1}, \lambda_{k+1} \rangle &= \frac{1}{2} \langle d_k, \left[\sum_{i=1}^{e} \nabla^2 g_*^i \lambda_*^i\right] d_k \rangle + \nu_k^1 \\ &\leq \langle g_{k+1}, \tilde{\lambda}_{k+2}(\tilde{c}_{k+2}) \rangle + \frac{1}{2} \langle d_k, \left[\sum_{i=1}^{e} \nabla^2 g_*^i \lambda_*^i\right] d_k \rangle + \nu_k^1.\end{aligned} \tag{3.7}$$

Now consider the expression

$$\begin{aligned}\chi_k = \langle g_{k+1}, \tilde{\lambda}_{k+2}(\tilde{c}_{k+2})\rangle - \langle g_{k+1}, \lambda_{k+2}\rangle + \langle g_{k+1}, \lambda_{k+2}\rangle \\ - \langle g_{k+1}, \lambda_{k+1}\rangle.\end{aligned} \tag{3.8}$$

Using the second order expansion of $g(x)$ and (2.7,e) we can write

$$\|g_{k+1}\| \leq \|g_k + N_k^T d_k\| + e_1\|d_k\|_2^2 = e_1\|d_k\|_2^2$$

for some $e_1 \geq 0$. Furthermore, as $\|\lambda_k - \lambda_{k+1}\| \to 0$ and, by Lemma (3.1), $\|\lambda_{k+1} - \tilde{\lambda}_{k+1}(\tilde{c}_{k+1})\| \to 0$, $\|\chi_k\|$ can be expressed as

$$\begin{aligned}\|\chi_k\| \leq \beta_k\|d_k\|_2^2;\ \lim_{k\to\infty}\beta_k = \lim_{k\to\infty} e_1\big[\|\tilde{\lambda}_{k+2}(\tilde{c}_{k+2}) - \lambda_{k+2}\| \\ + \|\lambda_{k+2} - \lambda_{k+1}\|\big] = 0.\end{aligned} \tag{3.9}$$

Consider the expression

$$\begin{aligned}&\psi(x_{k+1}) + c_*\langle g(x_{k+1}), g(x_{k+1})\rangle \\ &\quad = \max_{\alpha\in\mathbb{E}_+^m}\{\langle\alpha, f(x_k + \tau_k d_k)\rangle\} + c_*\langle g_{k+1}, g_{k+1}\rangle.\end{aligned} \tag{3.10}$$

Let the value of corresponding to the solution of the right hand side be given by α_k^* thus, using the equivalence of (1.1) and (1.2)

$$\psi(x_{k+1}) + c_*\langle g(x_{k+1}), g(x_{k+1})\rangle = \langle\alpha_k^*, f(x_k + d_k)\rangle + c_*\langle g_{k+1}, g_{k+1}\rangle.$$

As $\{x_k\} \to x_*$, then clearly $\{\alpha_k^*\} \to \alpha_*$. Thus,

$$\begin{aligned}&\psi(x_{k+1}) + c_*\langle g(x_{k+1}, g(x_{k+1})\rangle \\ &\quad \leq \langle\alpha_k^*, f(x_k) + \nabla f_k^T d_k\rangle + \frac{1}{2}\langle d_k, [\hat{H}_k - \hat{H}_k + H_*^g]d_k\rangle \\ &\quad + \frac{c_*}{2}[\langle g_k, g_k\rangle + \langle g_k, N_k^T d_k\rangle] + \nu_k^2 \\ &\leq \max_{\alpha\in\mathbb{E}_+^m}\{\langle\alpha, f(x_k) + \nabla f_k^T d_k\rangle\} + \frac{1}{2}\langle d_k, [\hat{H}_k - \hat{H}_k + H_*^g]d_k\rangle \\ &\quad + \frac{c_*}{2}[\langle g_k, g_k\rangle + \langle g_k, N_k^T d_k\rangle] + \nu_k^2\end{aligned} \tag{3.11}$$

where, with $x(t) = x_k + td_k$,

$$\begin{aligned}\nu_k^2 = \int_0^1 (1-t)\langle d_k, \Big\{\Big\{\sum_{i=1}^m \tilde{\alpha}^i\nabla^2 f^i(x(t))\Big\} + c_* N(x(t))N(x(t))^T \\ + c_*\sum_{j=1}^e g^i(x(t))\nabla^2 g(x(t)) - H_*^g\Big\}d_k\rangle dt\end{aligned}$$

where H_*^g is the second derivative matrix evaluated at x_*, α_*

$$H_*^g = \sum_{i=1}^{m} \alpha_*^i \nabla^2 f^i(x_*) + c\, N_* N_*^T.$$

Using (2.1), we can write (3.10) as

$$\begin{aligned}
&\psi(x_{k+1}) + c_* \langle g(x_{k+1}), g(x_{k+1})\rangle \\
&\quad \le \max_{\alpha \in \mathbb{E}_+^m} \left\{\langle \alpha, f(x_k) + \tau_k \nabla f_k^T d_k\rangle\right\} + \frac{\tau_k^2}{2}\langle d_k, \hat{H}_k d_k\rangle \\
&\quad + \frac{c_*}{2}\left[\langle g_k, g_k\rangle + \tau_k \langle g_k, N_k^T d_k\rangle\right] + \xi_k \tau_k^2 \|d_k\|^2
\end{aligned}$$

where, with $x(t) = x_k + t\tau_k d_k$,

$$\begin{aligned}
\xi_k = \int_0^1 (1-t) \Bigg\| &\max_{\alpha \in \mathbb{E}_+^m} \left\{ \sum_{i=1}^{m} \alpha^i \nabla^2 f^i(x(t)) \right\} + c_* N(x(t)) N(x(t))^T \\
&+ c_* \sum_{j=1}^{e} g^i(x(t)) \nabla^2 g^i(x(t)) - \hat{H}_k \Bigg\| dt.
\end{aligned}$$

The matrix $\hat{H}_k$ is a symmetric positive definite approximation to (2.2).

$$\begin{aligned}
&\psi(x_{k+1}) + c_* \langle g(x_{k+1}), g(x_{k+1})\rangle \\
&\quad \le \max_{\alpha \in \mathbb{E}_+^m} \mathcal{F}_k(x_k + \tau_k d_k, \alpha) + \frac{\tau_k^2}{2}\langle d_k, \hat{H}_k d_k\rangle \\
&\quad + \frac{c_*}{2}\left[\langle g_k, g_k\rangle + \tau_k \langle g_k, N_k^T d_k\rangle\right] + \xi_k \tau_k^2 \|d_k\|^2.
\end{aligned} \tag{3.12}$$

Since the function $\max_{\alpha \in \mathbb{E}_+^m} \mathcal{F}_k(x, \alpha)$ is convex with respect to x, we have for $\tau_k \in [0, 1]$,

$$\begin{aligned}
\max_{\alpha \in \mathbb{E}_+^m} \mathcal{F}_k(x_k + \tau_k d_k, \alpha) &\le \max_{\alpha \in \mathbb{E}_+^m} \mathcal{F}_k(x_k, \alpha) \\
&+ \tau_k \left\{ \max_{\alpha \in \mathbb{E}_+^m} \mathcal{F}_k(x_k + d_k, \alpha) - \max_{\alpha \in \mathbb{E}_+^m} \mathcal{F}_k(x_k, \alpha) \right\}
\end{aligned} \tag{3.13}$$

furthermore, we have

$$\max_{\alpha \in \mathbb{E}_+^m} \mathcal{F}_k(x_k, \alpha) = \psi(x_k) \tag{3.14a}$$

$$\max_{\alpha \in \mathbb{E}_+^m} \mathcal{F}_k(x_k + d_k, \alpha) = \psi_k(x_k + d_k). \tag{3.14b}$$

Substituting (3.14) in (3.13), and (3.13) in turn, in (3.12) yields

$$\begin{aligned}
&\psi(x_{k+1}) + \frac{c_*}{2}\langle g(x_{k+1}), g(x_{k+1})\rangle \\
&\quad\le \psi(x_k) + \tau_k\{\psi_k(x_k+d_k) - \psi(x_k)\} + \frac{\tau_k^2}{2}\langle d_k, \hat{H}_k d_k\rangle \\
&\qquad + \frac{c_*}{2}\left[\langle g_k, g_k\rangle + \tau_k\langle g_k, N_k^T d_k\rangle\right] + \xi_k\tau_k^2\|d_k\|^2 \\
&\quad\le \psi(x_k) + \frac{c_*}{2}\langle g_k, g_k\rangle + \tau_k\Big\{\psi_k(x_k+d_k) - \psi(x_k) \\
&\qquad + c_*\langle g_k, N_k^T d_k\rangle + \frac{1}{2}\langle d_k, \hat{H}_k d_k\rangle + \xi_k\tau_k\|d_k\|^2\Big\}.
\end{aligned} \tag{3.15}$$

Using the descent property established in Lemma (2.2) and Lemma (2.3, i), we can express (3.15) as

$$\begin{aligned}
&\psi(x_{k+1}) + \frac{c_*}{2}\langle g(x_{k+1}), g(x_{k+1})\rangle \\
&\quad\le \psi(x_k) + \frac{c_*}{2}\langle g_k, g_k\rangle + \tau_k\Phi(d_k, c_*)\left\{1 - \frac{\tau_k}{\epsilon m}\xi_k\right\}.
\end{aligned} \tag{3.16}$$

The scalar $\rho \in (0,1)$ in the stepsize strategy (2.6) determines τ_k such that $\rho \le 1 - \frac{\tau_k}{\epsilon m}\xi_k \le 1$. Since by Lemma (2.2) and Lemma (2.3; i), the descent property $\Phi(d_k, c_{k+1}) \le 0$ holds, there exists a $\tau_k \in (0,1]$ to ensure that (3.16) and hence (2.6) holds. Suppose τ^0 is the largest $\tau \in [0,1]$ satisfying inequalities (3.16) and (2.6). It follows that all $\tau \le \tau^0$ also satisfy these conditions and that the strategy in Step 4 of the algorithm selects a $\tau_k \in [\gamma\tau^0, \tau^0]$. Since $\Phi(d_k, c_{k+1}) \le 0$, it follows that $\{\psi(x_k) + \frac{c_*\langle g_k, g_k\rangle}{2}\}$ is monotonically decreasing.

To establish convergence to unit stepsizes, we use (3.7), (3.8) and (3.11) to rewrite (3.15) as

$$\begin{aligned}
&\psi(x_{k+1}) + \frac{c_*}{2}\langle g(x_{k+1}), g(x_{k+1})\rangle \\
&\le \psi(x_k) + \frac{c_*}{2}\langle g_k, g_k\rangle + \Big\{\psi_k(x_k+d_k) - \psi(x_k) + c_*\langle g_k, N_k^T d_k\rangle \\
&\quad + \frac{1}{2}\langle d_k, [\hat{H}_k - \hat{H}_k + H_*] d_k\rangle + \chi_k + \nu_k^1 + \nu_k^2\Big\}.
\end{aligned} \tag{3.17}$$

We note that

$$\begin{aligned}
\|\nu_k^1 + \nu_k^2\| &= \|d_k\|_2^2\xi_k^* \\
\xi_k^* &= \Bigg\|\int_0^1 (1-t)\Bigg\{\Bigg\{\sum_{i=1}^m \overline{\alpha}^i\nabla^2 f^i(x(t))\Bigg\} \\
&\quad + c_* N(x(t))N(x(t))^T + \sum_{i=1}^e \nabla^2 g^i(x_k + td_k)\lambda_{k+1}^i \\
&\quad + c_*\sum_{j=1}^e g^i(x(t))\nabla^2 g(x(t)) - H_*^g - \sum_{i=1}^e \nabla^2 g_*^i\lambda_*^i\Bigg\}\Bigg\| dt
\end{aligned}$$

and

$$H_* = H_*^g + \sum_{i=1}^{e} \nabla^2 g_*^i \lambda_*^i.$$

Using (3.9), the descent property (2.8) in Lemma (2.2) and Lemma (2.3; i), we can write (3.17) as

$$\begin{aligned}
&\psi(x_{k+1}) + \frac{c_*}{2} \langle g(x_{k+1}), g(x_{k+1}) \rangle \\
&\quad \leq \psi(x_k) + \frac{c_*}{2} \langle g_k, g_k \rangle + \Big\{ \psi_k(x_k + d_k) - \psi(x_k) + c_* \langle g_k, N_k^T d_k \rangle \\
&\qquad + \frac{1}{2} \langle d_k, [\hat{H}_k - \hat{H}_k + H_*] d_k \rangle + \beta_k \|d_k\|_2^2 + \|d_k\|_2^2 \xi_k^* \\
&\quad \leq \psi(x_k) + \frac{c_*}{2} \langle g_k, g_k \rangle + \Phi(d_k, c_*) \left[1 - \frac{1}{m\epsilon} \left[\frac{1}{2} \phi_k + \beta_k + \xi_k^* \right] \right]
\end{aligned}$$

where the last expression is obtained by invoking the equality $d_k = P_k - \overline{N}_k g_k$. Since, $\{x_k\} \to x_*$, $\{\lambda_k\} \to \lambda_*$, $\{\mu_k\} \to \mu_*$, $\{\eta_k\} \to \eta_*$, $\{v_k\} \to v_*$, we have $\{\xi_k^*\} \to 0$ and $\{\beta_k\} \to 0$. The scalar $\rho \in (0,1)$ in (2.6) is bounded by

$$\rho \leq 1 - \frac{1}{m\epsilon} \left[\frac{1}{2} \phi_k + \beta_k + \xi_k^* \right]. \tag{3.18}$$

If $\sigma > 0$ is such that $\frac{1}{m\epsilon}\left[\frac{1}{2}\sigma + \beta_k + \xi_k^*\right] \leq 1 - \rho$ (in view of $\xi_k \to 0$, $\beta_k \to 0$, this defines the number ρ) then (3.18) holds, and therefore, because $\Phi(d_k, c_*) \leq 0$, (2.6) is satisfied with $\tau_k = 1$. □

We now consider Algorithm 2.

Theorem (3.2) [Monotonic Decrease of $\|\Gamma_k\|$ and Unit Stepsizes for Algorithm 2] *Let*

(i) $f, g \in \mathbb{C}^3(\mathbb{R}^n)$;
(ii) *the approximate Hessian* $\hat{H}_k$ *be uniformly positive definite;*
(iii) *there exists, for each* k, *a bounded vector* $(x, \alpha, \lambda, \mu, \eta)$ *solving* (1.2);
(iv) *let* $a_k \geq -\frac{1}{2}$ *as specified in Lemma* (2.4);
(v) *let the matrix* $\nabla \hat{\Gamma}_k^{1^T} \nabla \hat{\Gamma}_k^1$ *be uniformly bounded. Then*
  (a) *Stepsize* τ_k *computed in Step 3 of algorithm 2 is in the range* $\tau_k \in (0, 1]$; *hence the sequence* $\{z_k\}$ *computed by algorithm 2 generates a corresponding monotonically decreasing sequence* $\{\|\Gamma^1(z_k)\|\}$.
  (b) *Let the sequence* $\{x_k\} \to x_*$ *and hence* $\{z_k\} \to z_*$. *The sequence* $\{\tau_k\}$ *converges to unity if*

$$a_k - (b_k + r_k) \geq \rho - \frac{1}{2} \tag{3.19}$$

$\forall\, k \geq k_0$ *for some* $k_0 \geq 0$.

Proof. Using second order expansion $\|\Gamma^1(z)\|_2^2$ and (2.15)–(2.18) we can write

$$\|\Gamma^1(z_{k+1})\|_2^2 - \|\Gamma^1(z_k)\|_2^2 \\ \leq \tau_k \langle \nabla\hat{\Gamma}_k^{1^T}\Gamma_k^1, d_k^z\rangle \left\{1 + a_k - \frac{\tau_k}{2}[1 + 2(b_k + r_k)]\right\} \tag{3.20}$$

where

$$b_k = \frac{|\langle d_k^z, \{H_*^\Gamma - \nabla\hat{\Gamma}_k^{1^T}\nabla\hat{\Gamma}_k^1\}d_k^z\rangle|}{\langle d_k^z, \nabla\hat{\Gamma}_k^{1^T}\nabla\hat{\Gamma}_k^1 d_k^z\rangle} \tag{3.21}$$

$$r_k = \frac{\int_0^1(1-t)\left|\langle d_k^z, \{H^\Gamma(z_k + \tau_k t d_k^z) - H_*^\Gamma\}d_k^z\rangle\right|dt}{\langle d_k^z, \nabla\hat{\Gamma}_k^{1^T}\nabla\hat{\Gamma}_k^1 d_k^z\rangle} \tag{3.22}$$

and $H^\Gamma(\cdot)$ is the Hessian of $\|\Gamma^1(z)\|_2^2$ evaluated at $(\cdot)$ and H_*^Γ is its value at the solution of the problem. Furthermore,

$$H_*^\Gamma = \nabla\Gamma_*^{1^T}\nabla\Gamma_*^1 + \sum_i \nabla^2\Gamma_*^{1i}\Gamma_*^{1i} \tag{3.23}$$

where $\nabla^2\Gamma_*^{1i}$ is the Hessian of Γ_*^{1i}, the ith element of Γ_*^1. However, as $\Gamma_*^1 = 0$, by the optimality condition of (1.2), the second term in (3.23) vanishes and we have

$$H_*^\Gamma = \nabla\Gamma_*^{1^T}\nabla\Gamma_*^1. \tag{3.24}$$

For $\alpha_k \geq -\frac{1}{2}$, the scalar $\rho \in \left(0, \frac{1}{2}\right)$ in stepsize strategy (2.14) determines τ_k such that

$$\rho \leq 1 + \alpha_k - \frac{\tau_k}{2}\left[1 + 2(b_k + r_k)\right] \tag{3.25}$$

where

$$\tau_k \leq \frac{(1+a)}{\frac{1}{2}[1 + 2(b_k + r_k)]}$$

is satisfied for some $\tau_k \in [0,1]$. Since by (2.15)–(2.18) d_k^z is a descent direction and $\langle\nabla\hat{\Gamma}_k^{1^T}\Gamma_k^1, d_k^z\rangle \leq 0$, there exists a $\tau_k \in (0,1]$ to ensure that (3.20) and hence stepsize strategy (2.14) holds. Suppose τ^0 is largest $\tau \in [0,1]$ satisfying (3.20) and (2.14). It follows that all $\tau \leq \tau^0$ also satisfy these inequalities and that the strategy (2.14) selects a $\tau_k \in [\gamma\tau^0, \tau^0]$. Since by (2.18) we have a descent direction, it follows that $\{\|\Gamma_k^1\|\}$ is monotonically decreasing.

From (3.20) and (3.25) it can be verified that (3.19) ensures the satisfaction of stepsize strategy (2.14) with $\tau_k = 1$. It is clear from (3.22) that as $\{z_k\} \to z_*$ (i.e. $\{d_k^z\} \to 0$), $\{r_k\} \to 0$. Furthermore, considering (3.24), the quantities a_k and b_k are controlled (for $\{d_k^z\} \to 0$) by the accuracy of $\hat{H}_k$ since $\nabla\Gamma$ and $\nabla\hat{\Gamma}$ in (2.17) and (3.21) differ in this quantity. Hence, if $\nabla\hat{\Gamma}_k$ is approximated with increasing accuracy, as $\{z_k\} \to z_*$, we have

$$\{a_k - (b_k + r_k)\} \to 0$$

which also satisfies (3.19).

4. Superlinear Convergence

Having established that both algorithms attain unit stepsizes, we can unify the study of the convergence rates of both algorithms by invoking the equivalence of the quadratic subproblems used in either algorithm. We assume that Algorithm 1 has reached a stage where $c_k = c_*$.

The superlinear convergence of the algorithms is thus discussed only for the quadratic programming subproblem (2.4). An interesting aspect of the subproblem and the convergence rate is that the Hessian of the quadratic objective function is singular. The results below discuss the superlinear convergence of $\{x_k\}$, $\{x_k, v_k\}$, $\{x_k, \alpha_k, v_k\}$, $\{x_k, \alpha_k, \lambda_k, v_k\}$ and relate these to the accuracy of the projected Hessian. The reason for discussing the convergence rate of $\{x_k\}$, as well as the rates of $\{x_k, v_k\}$, $\{x_k, \alpha_k, v_k\}$, $\{x_k, \alpha_k, \lambda_k, v_k\}$ is that, as mentioned by Boggs, Tolle and Wang (1982), the Q-superlinear convergence of the latter three sequences only imply the R-superlinear convergence of the former. The situation is further complicated by the fact that, the superlinear convergence result for the latter two sequences are not necessarily the same as the one for the former two sequences when projected Hessians are being considered.

Lemma (4.1). *Let $\{x_k\} \to x_*$ and $\|d_k\| \le \kappa_k \|d_{k-1}\|$, $\lim_{k\to\infty} \kappa_k = 0$. Then the sequence $\{x_k\}$ is Q-superlinearly convergent, i.e.*

$$\lim_{k\to\infty} \frac{\|x_* - x_{k+1}\|}{\|x_* - x_k\|} = 0.$$

Proof. We have

$$\begin{aligned} \|x_* - x_k\| &\le \lim_{t\to\infty} \sum_{j=k}^{t-1} \|x_{j+1} - x_j\| \\ &\le \kappa_k \|d_{k-1}\| (1 + w + w^2 + w^3 + \cdots) \\ &\le \frac{\kappa_k}{1-w} \{ \|x_k - x_*\| + \|x_* - x_{k-1}\| \} \end{aligned}$$

for some $w \in [0, 1)$. As $\{\kappa_k\} \to 0$, is chosen such that $\kappa_k \le w < 1$, $\forall\, k \ge K_0$. K_0 is an integer and is such that $\kappa_k < 1$, $\forall\, k \ge K_0$. Rearranging the above expression yields

$$\frac{\|x_k - x_*\|}{\|x_k - x_*\| + \|x_{k-1} - x_*\|} \le \frac{\kappa_k}{1-w}$$

which yields the required result. □

Assumptions (4.1)

(i) $\exists\ x_*, \alpha_*, \lambda_*, \mu_*, \eta_*$ satisfying the first order conditions of (1.2) or, equivalently, $\exists\ x_*, \alpha_*, \lambda_*, v_*$ satisfying the first order conditions of (1.3);

(ii) Strict complementarity holds for the inequality constraints of the quadratic programming subproblem (2.4);

(iii) $v^T H_* v > 0;\ \forall\ v \in \mathbb{R}^n,\ N_*^T v = 0.$

(iv) The coefficient matrix

$$\nabla\hat{\Gamma}_k = \begin{bmatrix} \hat{H}_k & \nabla\overline{f}_k & N_k & 0 \\ \nabla\overline{f}_k^T & 0 & 0 & \overline{l} \\ N_k^T & 0 & 0 & 0 \\ 0 & \overline{l}^T & 0 & 0 \end{bmatrix} \tag{4.1}$$

arising from the first order necessary conditions of the quadratic subproblem (2.4), is nonsingular. The matrix $\nabla\overline{f}_k$ corresponds to the columns of ∇f_k of the inequality constraints in (2.4) satisfied as equalities at the solution x_* of the min-max problem. Similarly, the vector $\overline{l}$ corresponds to the right hands of the inequalities in (2.4) satisfied as equalities at x_*;

(v) The columns of N_k, $[\nabla f_k^T, \overline{l}]^T$ are linearly independent at the solution x_*.

As a result of the convergence of the algorithm, there is a stage of the algorithm, near the solution of the min-max problem, when the inequality constraints satisfied as equalities at the solution, are also satisfied as equalities for the quadratic subproblem (2.3) or (2.4) Rustem (1989; 1992). At this stage, with strict complementarity holding, the multipliers $\alpha_k^i > 0$ if the constraint i is satisfied as an equality at the solution x_*. Also, $\langle \nabla f_k^i, d_k\rangle + f_k^i < v_{k+1}$ and $\alpha_k^i = 0$ if the constraint is strictly satisfied at the solution. Thus, at this stage, the multipliers can predict the constraints active at the solution. These active constraints are written as $\langle \nabla\overline{f}_k^i, d_k\rangle + \overline{f}_k^i = v_{k+1}$. Since the other constraints do not affect the solution x_*, they can be ignored for the convergence rate analysis. Thus, the quadratic subproblem (2.4) can be rewritten as

$$\min_{d,v}\left\{ v + c_*\langle N_k g_k, d\rangle + \frac{1}{2}\langle d, \hat{H}_k d\rangle \mid N_k^T d + g_k = 0, \nabla\overline{f}_k^T d + \overline{f}_k = \overline{l}v \right\}. \tag{4.2}$$

The first order optimality conditions for (4.2) can be written as

$$\begin{bmatrix} \hat{H}_k & \nabla\overline{f}_k & N_k & 0 \\ \nabla\overline{f}_k^T & 0 & 0 & \overline{l} \\ N_k^T & 0 & 0 & 0 \\ 0 & \overline{l}^T & 0 & 0 \end{bmatrix} \begin{bmatrix} d_k \\ \overline{\alpha}_{k+1} - \overline{\alpha}_k \\ \lambda_{k+1} - \lambda_k \\ -(v_{k+1} - v_k) \end{bmatrix} = -\Gamma_k \tag{4.3}$$

where

$$\Gamma_k = \begin{bmatrix} \nabla \overline{f}_k \overline{\alpha}_k + N_k(\lambda_k + c_* g_k) \\ g_k \\ \overline{f}_k - \overline{l}\, v_k \\ 0 \end{bmatrix} \tag{4.4}$$

and $\overline{\alpha}_k$ relate to the strictly positive elements of α_k which correspond to the inequality constraints satisfied as equalities.

The quadratic subproblem computes d_k such that $N_{k-1}^T d_{k-1} + g_{k-1} = 0$. Thus, using the second order expansion of $g(x)$ at x_k, we have for some $e_1 \geq 0$,

$$\|g_k\| = e_1 \|d_{k-1}\|^2. \tag{4.5}$$

Since the inequality constraints are satisfied as equalities, we have

$$\overline{l}\, v_k = \nabla \overline{f}_{k-1}^T d_{k-1} + \overline{f}_{k-1}. \tag{4.6}$$

The second order expansion of $\overline{f}(x)$ about x_{k-1} yields, for some $e_2 \geq 0$,

$$\overline{f}_k - \overline{f}_{k-1} - \nabla \overline{f}_{k-1}^T d_{k-1} = e_2 \|d_{k-1}\|^2.$$

Thus, $\overline{f}_k - \overline{l}\, v_k$ in (4.4) can be written as

$$\overline{f}_k - \overline{l}\, v_k = e_2 \|d_{k-1}\|^2. \tag{4.7}$$

The first order expansion of the gradient, with respect to x, of the Lagrangian of either min-max formulation, $\nabla \overline{f}(x)\overline{\alpha} + N(x)(\lambda + c_* g(x))$, yields

$$\begin{aligned} \nabla \overline{f}_k \overline{\alpha}_k + N_k(\lambda_k + c_* g_k) = \nabla \overline{f}_{k-1} \overline{\alpha}_{k-1} + N_{k-1}(\lambda_{k-1} + c_* g_{k-1}) \\ + \hat{H}_{k-1} d_{k-1} + \Delta H(t) d_{k-1} \end{aligned} \tag{4.8a}$$

$$\Delta H(t) = \int_0^1 \{H(x_{k-1} + t d_{k-1}) - H_* + H_* - \hat{H}_{k-1}\} dt. \tag{4.8b}$$

Clearly, as $\{x_k\} \to x_*$,

$$\left\{ \int_0^1 \{H(x_{k-1} + t d_{k-1}) - H_*\} dt \right\} \to 0$$

and, as shown below, the condition for superlinear convergence ensures that the rest of (4.8,b) also vanishes at the solution.

The first order optimality condition (2.7,b), for either quadratic subproblem, at x_{k-1} is

$$\nabla \overline{f}_{k-1} \overline{\alpha}_{k-1} + N_{k-1}(\lambda_{k-1} + c_* g_{k-1}) = 0$$

and thus (4.8,a) can be written as

$$\nabla \overline{f}_k \overline{\alpha}_k + N_k(\lambda_k + c_* g_k) = \Delta H(t) d_{k-1}. \tag{4.9}$$

We now state the superlinear convergence theorems which are similar to the characterization of Boggs, Tolle and Wang (1982), Powell (1983) for nonlinear programming. The results below are established without invoking a prior linear convergence rate argument. As opposed to the necessary and sufficient result of Boggs, Tolle and Wang (1982), Theorem (4.1) only establishes the necessary part for min-max problems. The analogous necessary and sufficient condition is discussed in Theorem (4.2).

Lemma (4.2). *Let Assumptions* (2.1), (4.1) *and* $f, g \in \mathbb{C}^2(\mathbb{R}^n)$ *hold. Then, for some* $e_0 \in [0, \infty)$,

$$\|d_k\| \leq e_0 \, \|x_* - x_k\|.$$

Proof. The first order optimality condition of the quadratic programming subproblem is given by

$$[\nabla \hat{\Gamma}_k] \begin{bmatrix} d_k \\ \overline{\alpha}_{k+1} - \overline{\alpha}_* \\ \lambda_{k+1} - \lambda_* + c_* g_k \\ -(v_{k+1} - v_*) \end{bmatrix} = - \begin{bmatrix} \nabla \overline{f}_k \overline{\alpha}_* + N_k \lambda_* \\ g_k \\ \overline{f}_k - \overline{l}\, v_* \\ 0 \end{bmatrix}$$

Using the first order expansion of the right side of the above expression about $(x_*, \overline{\alpha}_*, \lambda_*, v_*)$; by the nonsingularity of the matrix on the left and by the first order optimality condition of (1.3), i.e.

$$\nabla \overline{f}(x_*) \overline{\alpha}_* + N_* \lambda_* = 0, \;\; g(x_*) = 0, \;\; \overline{f}(x_*) = \overline{l}\, v_*$$

we have

$$\|d_k\| \leq \left\| \begin{matrix} d_k \\ \overline{\alpha}_{k+1} - \overline{\alpha}_* \\ \lambda_{k+1} - \lambda_* + c_* g_k \\ -(v_{k+1} - v_*) \end{matrix} \right] \leq \hat{e} [\nabla \hat{\Gamma}_k]^{-1} \|x_* - x_k\|.$$

for some $\hat{e} \in [0, \infty)$. Hence, the result follows from the above expression.

□

Theorem (4.1). *Let Assumptions* (2.1), (4.1) *and* $f, g \in \mathbb{C}^2(\mathbb{R}^n)$ *be satisfied and let* $F_k = [N_k \;\; \nabla \overline{f}_k]$.

(i) *The sequence* $\{x_k\}$ *satisfies* $\|d_k\| \leq \kappa_k \|d_{k-1}\|$ *with* $\lim_{k \to \infty} \kappa_k = 0$ *and hence converges Q-superlinearly if, for* $P_k = I - N_k (N_k^T N_k)^{-1} N_k^T$,

$$\lim_{k \to \infty} \frac{\|P_k (H_* - \hat{H}_k) d_k\|}{\|d_k\|} = 0. \tag{4.10}$$

(ii) *Let* $\Delta\alpha_k = \alpha_{k+1} - \alpha_k$ *and* $\Delta v_k = v_{k+1} - v_k$. *The sequence* $\{x_k, \alpha_k, v_k\}$ *satisfies*

$$\left\| \begin{array}{c} d_k \\ \Delta\alpha_k \\ \Delta v_k \end{array} \right\| \leq \kappa_k \|d_{k-1}\| \leq \kappa_k \left\| \begin{array}{c} d_{k-1} \\ \Delta\alpha_{k-1} \\ \Delta v_{k-1} \end{array} \right\| \tag{4.11}$$

and hence converges Q-superlinearly if (4.10) *is satisfied.*

(iii) *The sequence* $\{x_k, \alpha_k, \lambda_k, v_k\}$ *converges at a two-step Q-superlinear rate if* (4.10) *holds.*

(iv) *The sequence* $\{x_k, \alpha_k, \lambda_k, v_k\}$ *converges at a Q-superlinear rate if*

$$\begin{aligned} lim_{k\to\infty} \frac{\|P_k(H_* - \hat{H}_k)d_k\|}{\|d_k\|} &= 0 \quad and \\ \lim_{k\to\infty} \frac{\|(F_k^T F_k)^{-1} F_k^T (H_* - \hat{H}_k)d_k\|}{\|d_k\|} &= 0. \end{aligned} \tag{4.12}$$

Proof. Using (4.9), expression (4.3) can be written as

$$\begin{aligned} [\nabla\hat{\Gamma}_k] & \left[\begin{array}{c} d_k \\ \Delta\alpha_k \\ \Delta\lambda_k + (N_{k-1}^T N_{k-1})^{-1} N_{k-1}^T \Delta H(t) d_{k-1} \\ -\Delta v_k \end{array} \right] \\ & = - \left[\begin{array}{c} P_{k-1}\Delta H(t) d_{k-1} \\ g_k \\ \overline{f}_k - \overline{l}\, v_k \\ 0 \end{array} \right]. \end{aligned} \tag{4.13}$$

Since the matrix $\nabla\hat{\Gamma}_k$ is nonsingular, (4.13), (4.10), (4.5), (4.7) and $\{\|d_k\|\} \to 0$ yield

$$\|d_k\| \leq \left\| \begin{array}{c} d_k \\ \Delta\alpha_k \\ \Delta v_k \end{array} \right\| \leq \kappa_k \|d_{k-1}\| \leq \kappa_k \left\| \begin{array}{c} d_{k-1} \\ \Delta\alpha_{k-1} \\ \Delta v_{k-1} \end{array} \right\|; \text{ with } \lim_{k\to\infty} \kappa_k = 0. \tag{4.14}$$

Lemma (4.1) establishes the Q-superlinear convergence of $\{x_k\}$, $\{x_k, \alpha_k, v_k\}$, and hence (i) and (ii).

In order to show (iii), we first show that

$$\left\| \begin{array}{c} \alpha_* - \alpha_k \\ \lambda_* - \lambda_k \\ v_* - v_k \end{array} \right\| \leq \kappa_k^1 \|x_* - x_{k-2}\|; \text{ with } \lim_{k\to\infty} \kappa_k^1 = 0. \tag{4.15}$$

From (4.3)–(4.5), (4.7) and (4.9) we have

$$\left\| \begin{array}{c} \Delta\alpha_k \\ \Delta\lambda_k \\ \Delta v_k \end{array} \right\| \leq e_3 \|d_k\| + e_4 \|d_{k-1}\| + e_5 \|d_{k-1}\|^2 \tag{4.16}$$

for some $e_3, e_4, e_5 \geq 0$. We can use $\|d_k\| \leq \kappa_k \|d_{k-1}\|$ in (4.16) to obtain

$$\left\| \begin{matrix} \Delta\alpha_k \\ \Delta\lambda_k \\ \Delta v_k \end{matrix} \right\| \leq \kappa_k^2 \|d_k\|; \quad \lim_{k\to\infty} \kappa_k^2 = 0 \tag{4.17}$$

This leads to

$$\begin{aligned} \left\| \begin{matrix} \alpha_* - \alpha_k \\ \lambda_* - \lambda_k \\ v_* - v_k \end{matrix} \right\| &\leq \lim_{t\to\infty} \sum_{j=k}^{t-1} \left\| \begin{matrix} \Delta\alpha_j \\ \Delta\lambda_j \\ \Delta v_j \end{matrix} \right\| \\ &\leq \lim_{t\to\infty} \sum_{j=k}^{t-1} \kappa_j^2 \|d_{j-2}\| \\ &\leq \kappa_k^2 \|d_{k-2}\| (1 + w + w^2 + w^3 \cdots) \end{aligned}$$

for some $w \in [0, 1)$. Combining this with Lemma (4.2) yields (4.15). Using (4.15) with the superlinear convergence of $\{x_k\}$, we have, for $\lim_{k\to\infty} \kappa_k^3 = 0$,

$$\left\| \begin{matrix} x_* - x_k \\ \alpha_* - \alpha_k \\ \lambda_* - \lambda_k \\ v_* - v_k \end{matrix} \right\| \leq \kappa_k \|x_* - x_{k-1}\| + \kappa_k^1 \|x_* - x_{k-2}\| \leq \kappa_k^3 \left\| \begin{matrix} x_* - x_{k-2} \\ \alpha_* - \alpha_{k-2} \\ \lambda_* - \lambda_{k-2} \\ v_* - v_{k-2} \end{matrix} \right\|. \tag{4.18}$$

To establish (iv), we consider the first equation of (4.3) with (4.5), (4.7) and (4.9). We have

$$\begin{aligned} F_k \begin{bmatrix} \Delta\alpha_k \\ \Delta\lambda_k \end{bmatrix} &= -\left[\hat{H}_k d_k + \Delta H(t) d_{k-1}\right] \\ \begin{bmatrix} \Delta\alpha_k \\ \Delta\lambda_k \end{bmatrix} &= -(F_k^T F_k)^{-1} F_k^T \left[\hat{H}_k d_k + \Delta H(t) d_{k-1}\right]. \end{aligned}$$

Using $\|d_k\| \leq \kappa_k \|d_{k-1}\|$ with $\lim_{k\to\infty} \kappa_k = 0$ established above, we have

$$\left\| \begin{matrix} \Delta\alpha_k \\ \Delta\lambda_k \end{matrix} \right\| \leq \left[\kappa_k \|(F_k^T F_k)^{-1} F_k^T \hat{H}_k\| + \frac{\|(F_k^T F_k)^{-1} F_k^T \Delta H(t) d_{k-1}\|}{\|d_{k-1}\|} \right] \|d_{k-1}\|.$$

which, with (4.11) and Lemma (4.1), yields the Q-superlinear convergence of $\{x_k, \alpha_k, \lambda_k, v_k\}$. □

We now consider an alternative Q-superlinear convergence theorem which establishes the necessary and sufficient condition for this rate. We define the following matrices which are used below:

$$\hat{\mathcal{H}}_k = \begin{bmatrix} \hat{H}_k & 0^{n\times 1} \\ 0^{1\times n} & 0^{1\times 1} \end{bmatrix}; \ \Delta\mathcal{H}(t) = \begin{bmatrix} \Delta H(t) & 0^{n\times 1} \\ 0^{1\times n} & 0^{1\times 1} \end{bmatrix}; \ \mathcal{F}_k = \begin{bmatrix} \nabla\overline{f}_k & N_k \\ \overline{l}^T & 0^{1\times e} \end{bmatrix}$$

and the corresponding projection operator given by $\mathcal{P}_k = I - \mathcal{F}_k(\mathcal{F}_k^T\mathcal{F}_k)^{-1}\mathcal{F}_k^T$. The superscript on the 0's indicate the dimensions.

Theorem (4.2). *Under the same assumptions of Theorem* (4.1), *the sequences* $\{x_k\}$ *and* $\{x_k, v_k\}$, *satisfy* $\|d_k\| \leq \kappa_k^4\|d_{k-1}\|$ *and* $\left\| \begin{matrix} d_k \\ \Delta v_k \end{matrix} \right\| \leq \kappa_k^5 \left\| \begin{matrix} d_{k-1} \\ \Delta v_{k-1} \end{matrix} \right\|$ *respectively and thereby converge Q-superlinearly iff*

$$\lim_{k\to\infty} \frac{\left\| \mathcal{P}_k \begin{bmatrix} H_* - \hat{H}_k \\ 0^{1\times n} \end{bmatrix} d_k \right\|}{\|d_k\|} = 0. \tag{4.19}$$

Proof. The optimality condition given by (4.3) can be rewritten using (4.5), (4.7) and (4.9) as

$$\begin{aligned} &\begin{bmatrix} \hat{\mathcal{H}}_k & \mathcal{F}_k \\ \mathcal{F}_k^T & 0 \end{bmatrix} \begin{bmatrix} d_k \\ v_k \\ \begin{bmatrix} \Delta\alpha_k \\ \Delta\lambda_k \end{bmatrix} + (\mathcal{F}_{k-1}^T\mathcal{F}_{k-1})^{-1}\mathcal{F}_{k-1}^T \begin{bmatrix} \Delta H(t) \\ 0^{1\times n} \end{bmatrix} d_{k-1} \end{bmatrix} \\ &\qquad = - \begin{bmatrix} \mathcal{P}_{k-1} \begin{bmatrix} \Delta H(t) \\ 0^{1\times n} \end{bmatrix} d_{k-1} \\ \bar{f}_k - \bar{l}\, v_k \\ g_k \end{bmatrix} \end{aligned} \tag{4.20}$$

Using the same arguments as in Theorem (4.1) for (4.13), the Q-superlinear convergence of $\{x_k\}$ and $\{x_k, v_k\}$ can be demonstrated if (4.19) is satisfied.

Suppose, conversely that $\|d_k\| \leq \kappa_k^4\|d_{k-1}\|$ and

$$\left\| \begin{matrix} d_k \\ \Delta v_k \end{matrix} \right\| \leq \kappa_k^5 \left\| \begin{matrix} d_{k-1} \\ \Delta v_{k-1} \end{matrix} \right\|$$

respectively for the two sequences. Premultiplying the first equation of (4.20) and using the identity

$$\mathcal{P}_k \begin{bmatrix} d_k \\ \Delta v_k \end{bmatrix} = \begin{bmatrix} d_k \\ \Delta v_k \end{bmatrix} - \mathcal{F}_k(\mathcal{F}_k^T\mathcal{F}_k)^{-1}\mathcal{F}_k^T \begin{bmatrix} d_k \\ \Delta v_k \end{bmatrix}$$

we have

$$\begin{aligned} &\mathcal{P}_k\hat{H}_k\mathcal{P}_k \begin{bmatrix} d_k \\ \Delta v_k \end{bmatrix} + \mathcal{F}_k(\mathcal{F}_k^T\mathcal{F}_k)^{-1}\mathcal{F}_k^T \begin{bmatrix} d_k \\ \Delta v_k \end{bmatrix} \\ &\qquad = -[\mathcal{P}_k - \mathcal{P}_{k-1}]\Delta\mathcal{H}(t) \begin{bmatrix} d_{k-1} \\ \Delta v_{k-1} \end{bmatrix} - \mathcal{P}_{k-1}\Delta\mathcal{H}(t) \begin{bmatrix} d_{k-1} \\ \Delta v_{k-1} \end{bmatrix}. \end{aligned}$$

Assumption (4.1; iv, v) ensures that $\|\mathcal{P}_k \hat{\mathcal{H}}_k \mathcal{P}_k\| \in (0, \infty)$. Also, as $\{\|d_k\|\} \to 0$, $\{\|\mathcal{P}_k - \mathcal{P}_{k-1}\|\} \to 0$. Using the facts that

$$\|d_k\| \leq \left\| \begin{matrix} d_k \\ \Delta v_k \end{matrix} \right\| \quad \text{and} \quad \Delta \mathcal{H}(t) \begin{bmatrix} d_k \\ \Delta v_k \end{bmatrix} = \begin{bmatrix} \Delta H(t) d_{k-1} \\ 0 \end{bmatrix}$$

and dividing the above expression by $\|d_{k-1}\|$ and $\left\| \begin{matrix} d_k \\ \Delta v_k \end{matrix} \right\|$ respectively yields (4.19) for both sequences as required.

Acknowledgments. The research reported in this paper was funded by ESRC. The author is grateful for the constructive comments of an anonymous referee.

References

R.G. Becker, B. Dwolatzky, E. Karakitsos, and B. Rustem, The simultaneous use of rival models in policy optimization, *The Economic Journal*, **96**, (1986), 425–448.

M.C.B. Biggs, The Development of a Class of Constrained Minimization Algorithms and their Application to the Problem of Power Scheduling, Ph.D. Thesis, University of London, (1974).

P.T. Boggs, J.W. Tolle, and P. Wang, On the local convergence of quasi-Newton methods for constrained optimization, *SIAM J. Control and Optimization*, **20**, (1982), 161–171.

C. Charalambous and J.W. Bandler, Nonlinear Minimax optimization as a sequence of least pth optimization with finite values of p, *International Journal of System Science*, **7**, (1976), 377–391.

C. Charalambous, and A.R. Conn, An efficient algorithm to solve the minmax problem directly, *SIAM J. Num. Anal.*, **15**, (1978), 162-187.

G. Cohen, An algorithm for convex constrained minmax optimization based on duality, *Appl. Math. Optim.*, **7**, (1981), 347–372.

T.F. Coleman, A note on 'New Algorithms for constrained minimax optimization', *Mathematical Programming*, **15**, (1978), 239–242.

A.R. Conn, An efficient second order method to solve the constrained minmax problem, Report. Department of Combinatorics and Optimization, University of Waterloo, January (1979).

A.R. Conn and Y. Li, The computational structure and characterization of nonlinear discrete chebychev problems, Technical Report CS-88-39, Dept. of Computer Science, University of Waterloo, Waterloo, Ontario, Canada N2L 3G1, (1988).

V.F. Demyanov and V.N. Malomezov, *Introduction to Minmax*, Wiley, New York (1974).

V.F. Demyanov and A.B. Pevnyi, Some estimates in minmax problems, *Kibernetika*, **1**, (1972), 107–112.

S.R.K. Dutta and M. Vidyasagar, New algorithms for constrained minmax optimization, *Mathematical Programming*, **13**, (1977), 140–155.

J. Fuhrer and J. Haltmaier, Minimum variance pooling of forecasts at different levels of aggregation, Special Studies Paper 208, Federal Reserve Board, Washington, D.C. (1986).

P.E. Gill, W. Murray, and M.H. Wright, *Practical Optimization*, Academic Press, New York (1981).

C. Granger and P. Newbold, *Forecasting Economic Time Series*, Academic Press New York (1977).

J.H. Hald and K. Madsen, Combined LP and quasi-Newton methods for minimax optimization, *Mathematical Programming*, **20**, (1981), 49–62.

S.P. Han, Superlinear convergence of a minimax method, Technical Report 78-336, Dept. of Computer Science, Cornell University, (1978).

S.P. Han, Variable metric methods for minimizing a class of nondifferentiable functions, *Math. Prog.*, **20**, (1981), 1–13.

M.J. Lawrence, R.H. Edmunson, and M.J. O'Connor, The accuracy of combining judgemental and statistical forecasts, *Management Science*, **32**, (1986), 1521–1532.

S. Makridakis and R. Winkler, Averages of Forecasts: some empirical results, *Management Science*, **29**, (1983), 987–996.

J. Medanic and M. Andjelic, On a class of differential games without saddle-point solutions, *J. Optim. Theory and Applic.*, **8**, (1971), 413–430.

J. Medanic and M. Andjelic, Minmax solution of the multiple target problem, *IEEE Trans. on Automatic Control*, **AC-17**, (1972), 597–604.

W. Murray and M.L. Overton, A projected Lagrangian algorithm for nonlinear minmax optimization, *SIAM J. Sci. Stat. Comput.*, **1**, (1980), 345–370.

E. Polak and D.Q. Mayne, A robust secant method for optimization poblems with inequality constraints, *J. Optim. Theory and Applic.*, **33**, (1981), 463–477.

E. Polak and A.L. Tits , A globally convergent, implementable multiplier method with automatic penalty limitation, *Appl. Math. and Optimization*, **6**, (1981), 335–360.

E. Polak, D.Q. Mayne, and J.E. Higgins, A superlinearly convergent minmax algorithm for min-max problems, Memorandum No: UCB/ERL M86/103, Berkeley, California (1988).

M.J.D. Powell, A fast algorithm for nonlinearly constrained optimization problems, in G.A. Watson (Ed.), *Numerical Analysis*, Lecture Notes in Mathematics, 630, Springer-Verlag, Berlin-Heidelberg (1978).

M.J.D. Powell, Variable Metric Methods for Constrained Optimization, in *Mathematical Programming: The State of The Art*, (Eds.) A. Bachem, M. Grotschel, B. Korte, Springer-Verlag, Berlin-Heidelberg (1983).

B. Rustem, Convergent Stepsizes for constrained optimization algorithms, *J. Optim. Theory and Applic.*, **49**, (1986), 136–160.

B. Rustem, Methods for the simultaneous use of multiple models in optimal policy design, *Developments in Control Theory for Economic Analysis*, (Eds.) C. Carraro and D. Sartore, Maritnus Nijhoff Kluwer Publishers, Dordrecht, (1987).

B. Rustem, A superlinearly convergent constrained min-max algorithm for rival models of the same system, *Computers Math. Applic.*, **17**, (1989), 1305–1316.

B. Rustem, A constrained min-max algorithm for rival models of the same economic system, *Mathematical Programming*, **53**, (1991), 279–295.

B. Rustem, Equality and inequality constrained optimization algorithms with convergent stepsizes, *J. Optim. Theory and Applic.*, **76**, (1993), 429–453.

R.S. Womersley and R. Fletcher, An algorithm for composite nonsmooth optimization problems, *J. Optim. Theory and Applic.*, **48**, (1986), 493–523.

Berc Rustem
Department of Computing
Imperial College of Science, Technology & Medicine
London SW7 2BZ, United Kingdom

Algorithms for the Solution of a Large-Scale Single-Controller Stochastic Game

Michèle Breton and Saeb El Hachem

Abstract. In this paper, we present three algorithms for the solution of a large-scale zero-sum two-player stochastic game in discrete time, with a finite state set and continuous action spaces, where one of the players controls the transition probabilities. This particular game has an application in multi-stage stochastic decision problems.

Résumé. Dans cet article, nous proposons trois algorithmes pour la solution d'un jeu stochastique à somme nulle en temps discret, comportant un ensemble fini d'états et un espace d'actions continu. Il s'agit d'un jeu dit à contrôleur unique, où l'un des joueurs décide des probabilités de transition entre états. On présente une application du jeu à des problèmes en environnement dynamique stochastique avec information incomplète sur les données.

1. Introduction

Non-cooperative stochastic games are mathematical models that can represent the behaviour of a group of competing decision makers, acting on a dynamic system at discrete points in time, under uncertainty. Discounted stochastic games were introduced by [Shapley (1953)]. His original model was a zero-sum, two-player game with finite state and action spaces, and his existence proof of saddle points in such a game provided a first iterative algorithm for the computation of the value in zero-sum stochastic games. However, in the general case, solutions to stochastic games lack the so-called *ordered field property* [Parthasarathy and Raghavan (1981)], which suggests that one cannot solve these games using finitely many arithmetic operations—hence the interest in finding algorithms for the solution of specially structured games for which the ordered field property holds. One of such class of games is the "single-controller" stochastic game, where one of the players controls the transition probabilities. In the case where the number of states and actions is finite, the single-controller stochastic game can be solved by linear programming [Vrieze (1981), Kallenberg (1983), Filar and Raghavan (1984)], where the size of the program is of the order of the number of states and the total number of actions available.

In this paper, we consider a finite-horizon single-controller stochastic game with a finite number of states, but continuous action spaces; the solution of this particular game can be transposed into a mathematical program, that is the maximization of a concave non-differentiable function.

One distinguishing feature of the game considered in this paper is the fact that admissible action spaces for one of the players at any stage of the game depend on "past commitments". Another distinguishing feature of the game is the fact that these past-commitment constraints linking decision variables in different stages make the underlying Markov decision program much too large to be handled by dynamic programming. We choose to constrain the strategy spaces instead of including past actions in the definition of the state in order to maintain the single-controller structure and the finiteness of the state set. The analysis of this particular game is motivated by an application to multi-period stochastic decision problems. In such applications, the scale of the mathematical program constructed can be very large. For that reason, the algorithms presented here rely on decomposition in *node deterministic sub-problems.*

The structure of the paper is as follows: in Section 2, we introduce the necessary notation and formulate the problem; Section 3 proposes a first algorithm based on a method of alternating directions of multipliers; Section 4 presents a second algorithm of a projected sub-gradient type, whereas Section 5 proposes a third algorithm using a bundle approach. Section 6 presents an application of this game to multi-stage decision problems and concludes the paper.

2. Problem and Notation

We consider a two-player finite-horizon sequential game played in discrete stages (decision times) t, $t = 1, \ldots, T$. We use an extensive form representation of the game and identify states with nodes in the game tree. Let S^t represent the finite set of possible states at stage t; the total number of states for the game is thus $S = \sum_{t=1}^{T} |S^t|$. At each stage t, both players observe the state of the system $s^t \in S^t$ and independently select their actions, denoted x_{s^t} and q_{s^t} for Player 1 and Player 2 respectively, $x_{s^t} \in \mathbb{R}^{n_t}$, $q_{s^t} \in \mathbb{R}^{|S^{t+1}|}$. Player 1 then incurs a cost $r_{s^t}(x_{s^t})$ which depends only on his own choice of action, and the system moves to state $j \in S^{t+1}$ according to the transition probability function $q(j|s^t, q_{s^t}) = [q_{s^t}]_j$, where $[q_{s^t}]_j$ denotes the jth element of the vector q_{s^t}, which depends only on Player 2's choice of actions. Player 2 does not play at the last stage T of the game, but chooses a probability vector for the initial states $s^1 \in S^1$, denoted q_{s^0}. Player 1 seeks to minimize his total expected cost over the finite horizon, whereas Player 2 seeks to maximize the same.

Given a Markov strategy for Player 2, whereby he takes the action q_{s^t} whenever, at stage t, the system in in state s^t, the joint probability distribution of all states can easily be deduced. A Markov strategy for Player 2 can thus be identified with a vector $p = (p_{s^t})_{s^t \in S^t, t=1,\ldots T} \in \mathbb{R}^S$, where p_{s^t} denotes the joint probability of state $s^t \in S^t$. Let P denote the set of admissible strategies for Player 2; the definition of P should include constraints ensuring that the vector p can indeed be identified with a joint probability distribution on the state set.

A Markov strategy for Player 1, whereby he takes the action x_{s^t} whenever, at stage t, the system is in state s^t, can be identified with a vector $x = (x_{s^t})_{s^t \in S^t, t=1,\ldots T}$. At each stage t of the game, the admissible domain for Player 1 is a compact convex set in $\mathbb{R}^{n_t}$ which depends on the current state of the system, but also on Player 1's past actions. In order to define precisely the set of admissible strategies for Player 1 and to obtain a formulation suitable for decomposition, we will need additional notation and modelling assumptions.

Given a fixed strategy for Player 2, Player 1 faces a finite-horizon Markov Decision Problem, which can be formulated as a stochastic dynamic program. We suppose that, as is often the case in multi-stage decision problems, the admissible domain for Player 1 is given by a set of constraints that can be partitioned according to the stage, where constraints in stage t pertain both to decision variables in stage t and to "linking" variables in stage $t-1$ (these linking variables often constitute only a small subset of the decision variables, but, for the sake of conciseness, we will not use this feature here). To fix ideas, in a deterministic setting, let $y_t \in \mathbb{R}^{m_t}$ represent the vector of decision variables of Player 1 in stage t; the set of admissible actions can thus be defined by constraints

$$\begin{aligned} & y_0 \in \mathbb{R}^{m_0} \quad \text{given}, \\ & (y_t, y_{t-1}) \in D_t \qquad \forall t = 1, \ldots, T, \end{aligned} \tag{1}$$

or, equivalently, for $z_t \in \mathbb{R}^{m_{t-1}}$,

$$\begin{aligned} & y_0 \ \text{given}, \\ & (y_t, z_t) \in D_t \quad \forall t = 1, \ldots, T \\ & y_{t-1} = z_t \quad \forall t = 1, \ldots, T. \end{aligned} \tag{2}$$

The artificial variables z_t are added in order to model inter-period dependencies indirectly, with equality constraints. At each stage $t = 1, \ldots, T-1$, denote $B^t_{s^t} \subset S^{t+1}$ the set of possible transition states from $s^t \in S^t$. At a given state s^t in stage t, the *node sub-problem* for Player 1 corresponding to a given stategy p of Player 2 is noted

$$\text{minimize}_{(y_{s^t}, z_{s^t}) \in D_{s^t}} \Big[p_{s^t} r_{s^t}(y_{s^t}, z_{s^t}) \Big], \tag{3}$$

with admissible domain $D_{s^t} \subset \mathbb{R}^{n_t}$, where $n_t = m_t + m_{t-1}$. The terms "node decomposition" and "node sub-problem" refer to the extensive form representation of the game. In that setting, we state the following set of additional *past-commitment constraints* to complete the definition of the domain of admissible strategies for Player 1:

$$\begin{aligned} z_{s^1} &= y_0 \\ z_{j^{t+1}} &= y_{s^t} \quad \forall j \in B^t_{s^t}, \;\; s^t \in S^t, \;\; t = 1, \ldots, T-1. \end{aligned} \tag{4}$$

Let $x_{s^t} \in \mathbb{R}^{n_t}$ denote the vector (y_{s^t}, z_{s^t}), let $n = n_1 + n_2 + \ldots + n_T$ and let N denote the set of vectors $x \in \mathbb{R}^n$ satisfying the past-commitment constraints (4). Denote $D = \prod_{t=1}^T \prod_{s^t \in S^t} D_{s^t} \subset \mathbb{R}^n$. The set of admissible strategies for Player 1 is thus $D \cap N \subset \mathbb{R}^n$.

We are now ready to state the problem formally: A strategy pair $(x^*, p^*) \in \mathbb{R}^{n+S}$ is a saddle point for the single-controller game defined above if it solves

$$\begin{aligned} \sum_{t=1}^{T} \sum_{s^t \in S^t} p^*_{s^t} r_{s^t}(x^*_{s^t}) &= \min_{x \in D \cap N} \max_{p \in P} \sum_{t=1}^{T} \sum_{s^t \in S^t} p_{s^t} r_{s^t}(x_{s^t}) \\ &= \max_{p \in P} \min_{x \in D \cap N} \sum_{t=1}^{T} \sum_{s^t \in S^t} p_{s^t} r_{s^t}(x_{s^t}). \end{aligned} \tag{5}$$

Define the inner product in $\mathbb{R}^{n+S}$ as follows: for $x, w \in \mathbb{R}^n$ and $p, v \in \mathbb{R}^S$

$$\begin{aligned} \langle (x, p) \cdot (w, v) \rangle &= \langle x \cdot w \rangle_n + \langle p \cdot v \rangle_S \\ &= \sum_{t=1}^{T} \sum_{s^t \in S^t} \frac{1}{S} \langle x_{s^t} \cdot w_{s^t} \rangle_t + \langle p \cdot v \rangle_S \end{aligned}$$

where $\langle \cdot \rangle_n$ defines the inner product in $\mathbb{R}^n$ and $\langle \cdot \rangle_t$ and $\langle \cdot \rangle_S$ are respectively the usual inner products in $\mathbb{R}^{n_t}$ and $\mathbb{R}^S$. Let $\| \cdot \|$, $\| \cdot \|_n$, $\| \cdot \|_t$ and $\| \cdot \|_S$ be the respective associated norms. In that setting, the objective function of the minmax problem (5) can be written $\langle p \cdot r(x) \rangle_S$, where the vector function $r(x) = (r_{s^t}(x_{s^t}))_{s^t \in S^t, t=1,\ldots T}$.

The orthogonal projection of a vector (x, p) on $N \times P$, denoted $(\hat{x}, \hat{p})$, is given by

$$\hat{x} = \text{Proj}_N(x) \quad \text{and} \quad \hat{p} = \text{Proj}_P(p)$$

where $\text{Proj}_N(\cdot)$ and $\text{Proj}_P(\cdot)$ are the orthogonal projections, in the spaces $\mathbb{R}^n$ and $\mathbb{R}^S$, on N and P respectively. The orthogonal projection $\hat{x} = (\hat{y}, \hat{z})$ on the particular sub-space N has a simple expression: the vector $(\hat{y}, \hat{z})$ is such that $\forall j \in B^t_{s^t}$, $s^t \in S^t$, $t = 1, \ldots, T$,

$$\begin{aligned} \hat{y}_{s^t} &= \hat{z}_{j^{t+1}} = \frac{y_{s^t} + \sum_{k \in B^t_{s^t}} z_{k^{t+1}}}{1 + |B^t_{s^t}|}, \\ \hat{y}_{s^T} &= y_{s^T} \\ \hat{z}_{s^1} &= y_0. \end{aligned}$$

The orthogonal projection of a vector $(x, p) \in \mathbb{R}^{n+S}$ on $N^{\perp} \times \mathbb{R}^S$, where $N^{\perp}$ is the complementary sub-space to N in $\mathbb{R}^n$, is given by $(x - \hat{x}, p)$.

Proposition 1. *Assume that*

A1 *P is a non-empty compact convex set, with $p \geq 0 \quad \forall p \in P$,*

A2 *$ri(D \cap N) \neq \emptyset$, or D is polyhedral and $D \cap N \neq \emptyset$,*

A3 *$\forall s^t \in S^t$, $t = 1, \dots, T$, $r_{s^t}(x_{s^t})$ is closed convex on the compact convex set D_{s^t}.*

Then,

i) *the objective function of problem (5) is closed convex with respect to x, and problem (5) has a saddle point,*

ii) *(5) has the same saddle value as*

$$\max_{p \in P, w \in N^{\perp}} \left[\min_{x \in D} \left[\langle p \cdot r(x) \rangle_S + \langle w \cdot x \rangle_n \right] \right] \tag{6}$$

iii) *problem (6) has a saddle point,*

iv) *for any saddle point (p^*, w^*, x^*) of problem (6), (p^*, x^*) is saddle point of problem (5).*

Proof. i) The verification of the closeness and convexity of $\langle p \cdot r(x) \rangle_S$ with respect to x is trivial. This, combined with Assumptions A1 and A2 and the fact that $\langle p \cdot r(x) \rangle_S$ is concave with respect to p yields the existence of a saddle-point of problem (5) [Rockafellar (1970)].

ii) Let $I_C(x)$ denote the indicator function of a vectorial sub-space C,

$$I_C(x) = \begin{cases} 0 & \text{if } x \in C \\ +\infty & \text{otherwise.} \end{cases}$$

The equivalence of problems (5) and problem (6) is obtained by the following argument:

$$\begin{aligned}
&\min_{x\in D\cap N}\ \max_{p\in P}\left[\langle p\cdot r(x)\rangle_S\right]\\
&\Leftrightarrow \max_{p\in P}\ \min_{x\in D\cap N}\left[\langle p\cdot r(x)\rangle_S\right]\\
&\Leftrightarrow \max_{p\in P}\left[\min_{x\in D}\left[\langle p\cdot r(x)\rangle_S+I_N(x)\right]\right]\\
&\Leftrightarrow \max_{p\in P}\left[\min_{x\in D,u}\left[\langle p\cdot r(x)\rangle_S+I_N(u):x-u=0\right]\right]\\
&\Leftrightarrow \max_{p\in P}\left[\max_{w}\ \min_{x\in D,u}\left[\langle p\cdot r(x)\rangle_S+I_N(u)+\langle w\cdot(x-u)\rangle_n\right]\right]\\
&\Leftrightarrow \max_{p\in P}\left[\max_{w}\left[\min_{x\in D}[\langle p\cdot r(x)\rangle_S+\langle w\cdot x\rangle_n]+\min_{u}[I_N(u)-\langle w\cdot u\rangle_n]\right]\right]\\
&\Leftrightarrow \max_{p\in P}\left[\max_{w\in N^\perp}\left[\min_{x\in D}[\langle p\cdot r(x)\rangle_S+\langle w\cdot x\rangle_n]\right]\right]\\
&\Leftrightarrow \max_{p\in P},w\in N^\perp\left[\min_{x\in D}\left[\langle p\cdot r(x)\rangle_S+\langle w\cdot x\rangle_n\right]\right],
\end{aligned}$$

where w is a vector of multipliers assigned to the constraints $x-u=0$. Thus the variables w have an interpretation in terms of dual prices of the past-commitment constraints.

iii) Let (p^*,x^*) be a saddle point of problem (5), whose existence is guaranteed according to i). From optimality conditions, (p^*,x^*) verifies

$$0\in\frac{\partial}{\partial p}\left(-\langle p^*\cdot r(x^*)\rangle_S+I_P(p^*)\right), \tag{7}$$

and, for $p=p^*$, x^* is an optimal solution of the inner (minimization) problem in (6), which is convex. From constraints qualification conditions assumed in A2, there exists [Minoux (1983)] a vector w^* of multipliers such that (w^*,x^*) is a saddle point on $N^\perp\times D$ of the Lagrangian

$$L_{p^*}(w,x)=\langle p^*\cdot r(x)\rangle_S+\langle w\cdot x\rangle_n.$$

From optimality conditions, (w^*,x^*) verifies

$$0\in\frac{\partial}{\partial x}\left(\langle p^*\cdot r(x^*)\rangle_S+I_D(x^*)+\langle w^*\cdot x\rangle_n\right) \tag{8}$$

$$0\in\frac{\partial}{\partial w}\left(\langle w^*\cdot x^*\rangle_n+I_{N^\perp}(w^*)\right). \tag{9}$$

The inclusions (7), (8) and (9) yield the optimality conditions of problem (6) which prove iii).

iv) Consider a saddle point (p^*, w^*, x^*) of (5), which then verifies the inclusions (7), (8) and (9). Since $\partial I_{N^\perp}(w^*) = N$, P is polyhedral and from Assumption A2, these conditions are equivalent to:

$$0 \in -r(x^*) + \mathcal{N}_P(p^*), \quad p^* \in P \tag{10}$$

$$0 \in p^* \frac{\partial}{\partial x} r(x^*) + \mathcal{N}_D(x^*) + \frac{1}{S} w^* \tag{11}$$

$$\frac{1}{S} x^* \in N, \quad w^* \in N^\perp \tag{12}$$

where $\mathcal{N}_D(x^*)$ and $\mathcal{N}_P(p^*)$ denote the appropriate normal cones to the convex sets D and P at points x^* and p^* respectively. The fraction $\frac{1}{S}$ appears in the inclusions here above because of the definition of the inner product. Since $x^* \in N$, a linear sub-space, inclusions (10)–(12) can be written

$$0 \in p^* \partial r(x^*) + \mathcal{N}_D(x^*) + N^\perp, \quad x^* \in D \cap N$$
$$0 \in -r(x^*) + \mathcal{N}_P(p^*), \quad p^* \in P$$

which prove part iv) of the proposition. □

The function $\langle p \cdot r(x) \rangle_S + \langle w \cdot x \rangle_n$ is a saddle function: separable convex with respect to x and concave (linear) with respect to (w, p). If $H(w, p)$ denotes the objective value of the inner optimization problem parametrized by (w, p)

$$H(w,p) = \min_{x \in D} \Big[\langle p \cdot r(x) \rangle_S + \langle w \cdot x \rangle_n \Big], \tag{13}$$

(6) amounts to the following maximization problem

$$\max_{(w,p) \in N^\perp \times P} H(w,p). \tag{14}$$

Under Assumptions A2 and A3, $H(w,p)$ is a finite non-differentiable concave function, and the maximization problem (14) can in principle be solved by numerical methods for non-differentiable convex problems.

3. An Alternating Direction of Multipliers Algorithm

In this section, we propose a first algorithm for the solution of problem (5). We will show that Algorithm 1 can be identified with the alternating direction of multipliers method (ADMM) applied to an augmented Lagrangian function derived from a particular dual formulation of problem (5). As a consequence, convergence results can be deduced from convergence properties of the ADMM.

Algorithm 1.

0. Initialization

Select $\hat{x}^0 \in R^n$, $w^0 \in N^\perp$, $\hat{p}^0 \in R^S$, $p^0 \in R^S$, $c > 0$, $\epsilon > 0$; let $k := 1$

1. **Sub-problem optimization**
For all $s^t \in S^t$, $t = 1, \ldots, T$, compute

$$x^k_{s^t} = \arg \min_{x_{s^t} \in D_{s^t}} \Big[\hat{p}^{k-1}_{s^t} r_{s^t}(x_{s^t}) + \frac{1}{S} \langle w^{k-1}_{s^t} \cdot x_{s^t} \rangle_t + \frac{1}{2cS} \| x_{s^t} - \hat{x}^{k-1}_{s^t} \|^2_t + \frac{1}{2c} \big(r_{s^t}(x_{s^t}) - c(p^{k-1}_{s^t} - \hat{p}^{k-1}_{s^t}) \big)^2 \Big]. \tag{15}$$

2. **Weighing and projecting**
Compute

$$\hat{x}^k = \mathrm{Proj}_N(x^k) \quad \text{and } \hat{p}^k = \mathrm{Proj}_P(p^k), \quad \text{where } p^k = \hat{p}^{k-1} + \frac{1}{c} r(x^k).$$

3. **Updating**
Set

$$w^k = w^{k-1} + \frac{1}{c}(x^k - \hat{x}^k).$$

4. **Stopping criterion**
If $\| r(x^k) - c(p^{k-1} - \hat{p}^{k-1}), x^k - \hat{x}^k \| < \epsilon$, stop;
otherwise, set $k := k + 1$ and return to step 1.

Numerically, the execution of Algorithm 1 involves two main tasks: the projection operation on P in step 2 and the optimization in step 1. Special methods are available for the projection operation. Step 1 amounts to the independent solution of all node sub-problems of type (3) for a given strategy of Player 2. These sub-problems are suitably modified by dual prices w related to the past-commitment constraints and by quadratic penalty terms at each iteration. Note that under Assumptions A1 and A3, every modified scenario sub-problem is convex and has a unique optimal solution.

Thus the execution of step 1 is open to the methods of convex programming; since the domain of sub-problems is unaffected by iterative modifications, former optimal solution may serve as starting points in order to avoid cold starts.

Convergence of Algorithm 1 to a solution of problem (5) will be proved by recognizing it as an instance of the alternating direction of multipliers method [Glowinski and Marroco (1975), Gabay and Mercier (1976)]. Convergence results for the ADMM can be found in [Eckstein (1989)].

We first write (14) in the form

$$\min_{u,v,w,\hat{p}} \Big[-H(u,v) + I_{N^\perp \times P}(w, \hat{p}) : u - w = 0,\ v - \hat{p} = 0 \Big], \tag{16}$$

where $I_{N^\perp \times P}(w, \hat{p})$ denotes the indicator function of the convex set $N^\perp \times P$, and assign multiplier vectors λ and μ to the constraints $u - w = 0$ and $v - \hat{p} = 0$ respectively. This yields the ordinary Lagrangian function

$$L(u, v, w, \hat{p}, \lambda, \mu) = -H(u,v) + I_{N^\perp \times P}(w, \hat{p}) + \langle \lambda \cdot (u - w) \rangle_n + \langle \mu \cdot (v - \hat{p}) \rangle_S$$

and one can form a dual problem by writing (16) in the form

$$\max_{(\lambda,\mu)\in R^{n+S}} \Big[\min_{u,v,w,\hat{p}} L(u,v,w,\hat{p},\lambda,\mu) \Big]. \tag{17}$$

For numerical stability, the ADMM uses the augmented Lagrangian to solve problem (17), that is

$$\begin{aligned} L_c(u,v,w,\hat{p},\lambda,\mu) = &-H(u,v) + I_{N^\perp \times P}(w,\hat{p}) + \langle \lambda \cdot (u-w)\rangle_n \\ &+ \langle \mu \cdot (v-\hat{p})\rangle_S + \frac{c}{2}\|u-w\|_n^2 + \frac{c}{2}\|v-\hat{p}\|_S^2. \end{aligned} \tag{18}$$

Alternating Direction of Multipliers Algorithm

0. **Initialization**
   Select $\lambda^0 \in \mathbb{R}^n$, $\mu^0 \in R^S$, $w^0 \in R^n$, $\hat{p}^0 \in R^S$, $c > 0$; let $k := 1$.
1. **Optimization**
   Compute

$$\begin{aligned} (u^k, v^k) = \arg\min_{u,v} \Big[&-H(u,v) + \langle \lambda^{k-1} \cdot u\rangle_n + \langle \mu^{k-1} \cdot v\rangle_S \\ &+ \frac{c}{2}\|u - w^{k-1}\|_n^2 + \frac{c}{2}\|v - \hat{p}^{k-1}\|_S^2 \Big] \end{aligned} \tag{19}$$

$$\begin{aligned} (w^k, \hat{p}^k) = \arg\min_{w,\hat{p}} \Big[&I_{N^\perp \times P}(w,\hat{p}) - \langle \lambda^{k-1} \cdot w\rangle_n - \langle \mu^{k-1} \cdot \hat{p}\rangle_S \\ &+ \frac{c}{2}\|u^k - w\|_n^2 + \frac{c}{2}\|v^k - \hat{p}\|_S^2 \Big]. \end{aligned} \tag{20}$$

2. **Updating**

$$(\lambda^k, \mu^k) = (\lambda^{k-1}, \mu^{k-1}) + c\left((u^k, v^k) - (w^k, \hat{p}^k)\right). \tag{21}$$

3.
   Set $k := k+1$ and return to step 1.

Proposition 2. *Algorithm* 1 *is exactly the alternating direction of multipliers method.*

Proof. Proof of this proposition proceeds in three steps: first, we write step (19) of the ADMM as optimization sub-problems in x_{s^t} for all $s^t \in S^t, t = 1, \ldots, T$; then we transform step (20) into projection operations; finally we adequately eliminate some variables.

i) With the definition of function H in (13), step (19) can be written

$$\begin{aligned} (u^k, v^k) = \arg\min_{u,v} \Big[&- \min_{x\in D} \Big[\langle v \cdot r(x)\rangle_S + \langle u \cdot x\rangle_n \Big] + \langle \lambda^{k-1} \cdot u\rangle_n \\ &+ \langle \mu^{k-1} \cdot v\rangle_S + \frac{c}{2}\|u - w^{k-1}\|_n^2 + \frac{c}{2}\|v - \hat{p}^{k-1}\|_S^2 \Big]. \end{aligned} \tag{22}$$

For a given (u^k, v^k), denote

$$x^k = \arg\min_{x\in D}\Big[\langle v^k \cdot r(x)\rangle_S + \langle u^k \cdot x\rangle_n\Big];$$

then (22) is equivalent to the following optimality conditions

$$\begin{aligned} 0 &= -x^k + \lambda^{k-1} + c(u^k - w^{k-1}) \\ 0 &= -f(x^k) + \mu^{k-1} + c(v^k - \hat{p}^{k-1}) \\ 0 &\in v^k_{s^t}\partial r_{s^t}(x^k_{s^t}) + \frac{1}{S}u^k_{s^t}, \quad x_{s^t} \in D_{s^t}\ \forall s^t \in S^t,\ t = 1, \dots T. \end{aligned} \tag{23}$$

From the first two optimality conditions in (23), one obtains the following expressions for u^k and v^k:

$$u^k = \frac{1}{c}(-\lambda^{k-1} + cw^{k-1} + x^k) \tag{24}$$

$$v^k = \frac{1}{c}(-\mu^{k-1} + c\hat{p}^{k-1} + r(x^k)). \tag{25}$$

Substituting (24) and (25) in the third condition in (23) and integrating yields, for all $s^t \in S^t$, $t = 1, \dots, T$,

$$\begin{aligned} x^k_{s^t} = \arg \min_{x_{s^t}\in D_{s^t}} \Big[&\hat{p}^{k-1}_{s^t} r_{s^t}(x_{s^t}) + \frac{1}{S}\langle w^{k-1}_{s^t} \cdot x_{s^t}\rangle_t \\ &+ \frac{1}{2c}(r_{s^t}(x_{s^t}) - \mu^{k-1}_{s^t})^2 + \frac{1}{2cS}\|x_{s^t} - \lambda^{k-1}_{s^t}\|^2\Big]. \end{aligned} \tag{26}$$

ii) Note that the optimization problem in step (20) of the ADMM is separable in w and $\hat{p}$. Hence, (20) amounts to finding

$$w^k = \arg \min_{w\in N^\perp} \Big[-\langle \lambda^{k-1} \cdot w\rangle_n + \frac{c}{2}\|u^k - w\|^2_n\Big]$$

and

$$\hat{p}^k = \arg\min_{p\in P}\Big[-\langle \mu^{k-1} \cdot \hat{p}\rangle_S + \frac{c}{2}\|v^k - \hat{p}\|^2_S\Big].$$

Now,

$$\begin{aligned} w^k &= \arg \min_{w\in N^\perp} \Big[-\langle \lambda^{k-1} \cdot w\rangle_n + \frac{c}{2}\|u^k - w\|^2_n\Big] \\ &= \arg \min_{w\in N^\perp} \Big[\frac{c}{2}\|w - \left(\frac{\lambda^{k-1}}{c} + u^k\right)\|^2_n\Big] \\ &= \text{Proj}_{N^\perp}\left(\frac{\lambda^{k-1}}{c} + u^k\right) \\ &= \text{Proj}_{N^\perp}\left(w^{k-1} + \frac{x^k}{c}\right) \qquad \text{from (24).} \end{aligned} \tag{27}$$

Since $N^{\perp}$ is a sub-space and the projection of the sum of two vectors is the sum of their projections, (27) yields

$$\begin{aligned} w^k &= \mathrm{Proj}_{N^{\perp}} w^{k-1} + \frac{1}{c}\mathrm{Proj}_{N^{\perp}}(x^k) \\ &= w^{k-1} + \frac{1}{c}(x^k - \hat{x}^k) \quad \text{when } w^0 \in N^{\perp}. \end{aligned} \tag{28}$$

Similarly, from the optimization problem in step (20) w.r.t. $\hat{p}$, using (25) one obtains

$$\hat{p}^k = \mathrm{Proj}_P(\hat{p}^{k-1} + \frac{1}{c}r(x^k)) \tag{29}$$

from which one uses the notation $p^k = \hat{p}^{k-1} + \frac{1}{c}r(x^k)$.

iii) Given the above developments, the updating formula in step 2 of the ADMM becomes:

$$\begin{aligned} \lambda^k &= \lambda^{k-1} + (-\lambda^{k-1} + cw^{k-1} + x^k) - (cw^{k-1} + x^k - \hat{x}^k) \\ &= \hat{x}^k \end{aligned}$$

and

$$\begin{aligned} \mu^k &= \mu^{k-1} + (-\mu^{k-1} + c\hat{p}^{k-1} + r(x^k)) - c\hat{p}^k \\ &= c\hat{p}^{k-1} + r(x^k) - c\hat{p}^k \\ &= c(p^k - \hat{p}^k). \end{aligned}$$

Now, replacing λ^k by $\hat{x}^k$ and μ^k by $c(p^k - \hat{p}^k)$ in the ADMM, together with (26), (28), (29) and the computation of the projection $\hat{x}^k$, yields Algorithm 1. □

One can now easily show infinite convergence of Algorithm 1 to a solution of problem (5), yielding a saddle point (x^*, p^*), by using the equivalence of Algorithm 1 and the ADMM. Moreover, it can be shown that Algorithm 1 converges to an ϵ-equilibrium in a finite number of steps, using results in [Breton and El-Hachem (1992)].

4. A Projected Sub-gradient Algorithm

In this section, we will show that the second algorithm we propose is an instance of the projected sub-gradient algorithm [Polyak (1966); Ermoliev (1966)] and will deduce some convergence properties from this result.

Algorithm 2

0. Initialization

Choose an appropriate sequence of parameters $\{\tau^k\}_{k=1,\ldots,\infty}$. Select $w^1 \in N^{\perp}$ and $p^1 \in P$; set $k := 1$.

1. **Sub-problem optimization**
   For all $s^t \in S^t$, $t = 1, \ldots, T$, compute

$$x_{s^t}^k \in \arg \min_{x_{s^t} \in D_{s^t}} \left[p_{s^t}^k r_{s^t}(x_{s^t}) + \langle w_{s^t}^k \cdot x_{s^t} \rangle_t \right].$$

2. **Weighing**
   Compute

$$\hat{x}^k = \text{Proj}_N x^k.$$

3. **Updating**
   Set

$$\begin{aligned} w^{k+1} &= w^k + \tau^k (x^k - \hat{x}^k) \\ p^{k+1} &= \text{Proj}_P(p^k + \tau^k r(x^k)) \end{aligned} \tag{30}$$

4. **Stopping criterion**
   If the stopping criterion fails, set $k := k + 1$ and return to step 1.

As in Algorithm 1, step 1 consists of solving modified node sub-problems. Assumption A2 ensures that the domain of each scenario sub-problem is not empty, and Assumption A3 ensures that each component of the vector $(x_{s^t}^k)_{s^t \in S^t, t=1,\ldots,T}$ takes finite values. Again, since each $p_{s^t}^k$ is positive, the computation of each $x_{s^t}^{k+1}$ requires the solution of a convex program, which domain remains unaffected from one iteration to the next. In the case where the r_{s^t} are linear functions of the decision variables x_{s^t}, step 1 amounts to the solution of one linear program for each possible state of the game. Given τ^k, step 3 updates the system of dual prices for the past-commitment constraints and the transition probabilities chosen by Player 2.

Proposition 3. *Algorithm 2 is an instance of the projected sub-gradient algorithm.*

Proof. The set of sub-gradients of $H(\cdot)$ at (w^k, p^k) is given by

$$\{(x, r(x)) : x_{s^t} \in \arg \min_{x_{s^t} \in D_{s^t}} \left[p_{s^t}^k r_{s^t}(x_{s^t}) + \langle w_{s^t}^k \cdot x_{s^t} \rangle_S \right] \quad \forall s \in S\}.$$

Step 1 in Algorithm 2 thus amounts to finding one sub-gradient of H at (w^k, p^k) which will be denoted γ^k.

Recall that $(x - \hat{x})$ is the projection of x on $N^\perp$. Since $w^1 \in N^\perp$, it is easy to show by induction that $w^k \in N^\perp$ for all $k \in \mathbb{N}$. Hence, since the projection on a sub-space of the sum of two vectors is equal to the sum of the projections, the updating step for w in (30) is

$$w^{k+1} = \text{Proj}_{N^\perp}(w^k + \tau^k x^k).$$

Given that $N^{\perp}$ and P are sub-sets of two complementary sub-spaces, (30) can be written

$$\begin{aligned}(w^{k+1}, p^{k+1}) &= \text{Proj}_{N^{\perp}\times P}(w^k + \tau^k x^k, p^k + \tau^k r(x^k)) \\ &= \text{Proj}_{N^{\perp}\times P}((w^k, p^k) + \tau^k (x^k, r(x^k)) \\ &= \text{Proj}_{N^{\perp}\times P}((w^k, p^k) + \tau^k \gamma^k)\end{aligned}$$

which is the updating formula of the sub-gradient algorithm. □

The convergence of Algorithm 1, as an instance of the projected sub-gradient, relies on an adequate selection of the sequence of step-sizes $\{\tau^k\}$ (see [Minoux (1983)] for a synthesis). [Kim et al. (1991)] suggest the following adaptative step-sizes selection

$$\tau^k = \frac{\rho(H^* - H(w^k, p^k))}{\|\gamma^k\|}$$

where $0 < \rho < 2$ is a relaxation coefficient and H^* is a variable target optimal value, corrected along the iterative process on the basis of information already collected about the optimization problem. The progressively improving estimation of H^* is achieved by adjusting iteratively the lower and upper bounds on the optimal value of H. Consequently, this method also provides a reliable optimality test, which assures finite termination: it consists of stopping the procedure as soon as the difference between the upper and lower bounds is sufficiently small.

Though this method provides a stopping criterion allowing the near identification of the value of the game, it can not guarantee convergence to near-equilibrium strategies. Indeed, even if a sub-sequence $\{(w^k, p^k)\}_{k=1}^{\infty}$ converges to a solution of problem (14), implying $\lim_{k\to\infty} \|\tau^k(x^k - \hat{x}^k)\|_n = 0$, this does not necessarily imply that x^k tends to verify the past-commitment constraints, since $\lim_{k\to\infty} \tau^k = 0$. But, if at the termination stage $\|x^k - \hat{x}^k\|_n \leq \epsilon$ and ϵ is relatively small, then (x^k, p^k) is a good approximation to a saddle point of the game.

5. A Bundle Algorithm

We now present a bundle-based decomposition algorithm for the solution of the single-controller game. This algorithm differs from Algorithm 2 in that it ensures convergence to a saddle point, at the cost of a more sophisticated computational effort. For the bundle algorithm presented here, we make the additional assumption that the set P is polyhedral; it is, however, possible to generalize the method to the compact convex case (see for instance [Kiwiel (1985)]).

We first introduce some additional notation. Let $\mathcal{M}$ be the affine variety specified by all the equality constraints defining P, denoted $Mp = d$, where M is a matrix of dimensions $\ell_1 \times S$ and $d \in \mathbb{R}^{\ell_1}$. Let $\mathcal{C}$ be the set of vectors p verifying the inequality constraints defining P, denoted $Cp \leq b$ where C is a matrix of dimensions $\ell_2 \times S$ and $b \in \mathbb{R}^{\ell_2}$.

For any vector $g \in \mathbb{R}^{n+S}$, denote $\bar{g}$ the projection of vector g on $N^{\perp} \times \mathcal{M}$. Note that the projection operation on the subspace $N^{\perp} \times \mathcal{M}$ amounts to left multiplication of g by a constant projection matrix denoted

$$Pr = \begin{bmatrix} Pr_1 & 0 \\ 0 & Pr_2 \end{bmatrix},$$

where Pr_1 is the projection matrix identified with the projection operator on $N^{\perp}$ and Pr_2 is the projection matrix identified with the projection operator on the unique sub-space parallel to the affine variety $\mathcal{M}$.

Denote $I_{\mathbb{R}^n \times \mathcal{C}}(\cdot)$ the indicator function of $\mathbb{R}^n \times \mathcal{C}$ and $Q(\cdot) \equiv -H(\cdot) + I_{\mathbb{R}^n \times \mathcal{C}}(\cdot)$. With these notations, the optimization problem (14) becomes

$$\min_{(w,p) \in N^{\perp} \times \mathcal{M}} Q(w,p) \tag{31}$$

and reduces to the minimization of a non-differentiable convex function on a sub-space, to which we can then apply bundle methods.

The rationale for bundle methods is to force a decrease in the objective function by using available information on sub-gradients at the current and neighbouring points. After a descent direction has been selected, the adaptative step-sizes are controlled by descent considerations. This procedure is repeated until a satisfactory approximate optimal solution, identifiable by sub-gradient norm consideration, is reached.

Selecting a descent d at (w^k, p^k) corresponds to the following minimax problem

$$\min_{\|d\| \leq 1} \max_{g \in \partial Q(w^k, p^k)} \langle g \cdot d \rangle. \tag{32}$$

Instead of selecting a direction among the vectors in $\partial Q(w^k, p^k)$, bundle methods use (an approximation of) what is called the ϵ-sub-differential of Q in order to avoid being infinitely trapped in the neighbourhood of a kink in the objective function. By definition, the ϵ-sub-differential of the function Q at (w^k, p^k), denoted by $\partial_\epsilon Q(w^k, p^k)$, is

$$\begin{aligned} \partial_\epsilon Q(w^k, p^k) = \{ g \in \mathbb{R}^{n+S} : \langle g \cdot ((w,p) - (w^k, p^k)) \rangle \\ \leq Q(w,p) - Q(w^k, p^k) + \epsilon \quad \forall (w,p) \in \mathbb{R}^{n+S} \}. \end{aligned} \tag{33}$$

The ϵ-sub-differential at (w^k, p^k) is a convex compact set and contains the sub-gradients for a whole neighborhood of (w^k, p^k); hence it provides a larger set than the usual sub-differential. For $\epsilon = 0$, definition (33) reduces to the usual sub-differential definition. The corresponding definition of the ϵ-directional derivative at (w^k, p^k) is then

$$Q'_\epsilon((w^k, p^k); d) = \inf_{t > 0} \frac{1}{t} [Q((w^k, p^k) + td) - Q(w^k, p^k) + \epsilon] \tag{34}$$

and it can be shown that

$$Q'_\epsilon((w^k,p^k);d) = \max_{g\in\partial_\epsilon Q(w^k,p^k)} \langle g \cdot d\rangle.$$

In our case, bundle algorithms have to be appropriately transposed to account for the fact that the domain is a sub-space. One way to deal with this is to force the direction d in (32) to be in the sub-space $N^\perp \times \mathcal{M}$. As shown in [Breton and El-Hachem (1991)], the identification at (w^k,p^k) of a descent direction d for Q in $N^\perp \times \mathcal{M}$ amounts to the following

$$g^* = \arg \min_{g\in\partial_\epsilon Q(w^k,p^k)} \|\bar{g}\|,$$

$$d = -\frac{\bar{g}^*}{\|\bar{g}^*\|}, \quad \|\bar{g}\| \neq 0.$$

Another distinguishing feature of problem (31) is that the sub-differential or ϵ-sub-differential of Q has to account for the fact that the objective function includes the indicator function of a polyhedral domain. A subgradient $g \in \partial Q(w^k,p^k)$ satisfies

$$g = -\gamma + \begin{bmatrix} 0 \\ C^T\mu \end{bmatrix}, \quad \mu_i \geq 0,\ i \in I, \quad \mu_i = 0,\ i \notin I,$$

where $\gamma \in \partial H(w^k,p^k)$, $\mu \in \mathbb{R}^{\ell_2}$ and I is the set of active inequality constraints at (w^k,p^k). The corresponding definition of the ϵ-sub-differential of Q at (w^k,p^k) is [Hiriart-Urruty (1980)]

$$\partial_\epsilon Q(w^k,p^k) \equiv \bigcup_{\substack{\epsilon_1+\epsilon_2=\epsilon,\\ \epsilon_1\geq 0,\epsilon_2\geq 0}} \{-\partial_{\epsilon_1} H(w^k,p^k) + \partial_{\epsilon_2} I_{\mathbb{R}^{n|S|}\times C}(w^k,p^k)\},$$

where, in our case where P is polyhedral,

$$\partial_{\epsilon_2} I_{\mathbb{R}^{n|S|}\times C}(w^k,p^k) = \left\{ \begin{bmatrix} 0 \\ C^T\mu \end{bmatrix} : \mu^T(b - Cp) \leq \epsilon_2, \quad \mu \geq 0 \right\}.$$

The following is a conceptual algorithm embodying the bundle method.

Algorithm 3. Conceptual Bundle Method

0. Select $(w^1,p^1) \in N^\perp \times P$; set $k := 1$.

1. Compute

$$g^k = \arg \min_{g\in\partial_\epsilon Q(w^k,p^k)} \|\bar{g}\|. \tag{35}$$

2. If $\bar{g}^k = 0$, stop; otherwise go on to step 3.

3. Set

$$d^k = -\frac{\bar{g}^k}{\|\bar{g}^k\|}, \tag{36}$$

$$(w^{k+1},p^{k+1}) = (w^k,p^k) + \tau^k d^k, \tag{37}$$

where

$$\tau^k \in \arg \min_{((w^k,p^k)+\tau d^k)\in\mathbb{R}^n\times\mathcal{C}} Q((w^k,p^k)+\tau d^k). \tag{38}$$

Set $k := k+1$ and return to step 2.

Algorithm 3 differs from the usual bundle methods in that the minimizing direction is selected in the projection of the set $\partial_\epsilon Q(w^k,p^k)$ on $N^\perp \times \mathcal{M}$ and that the step-size computation takes into account inequality constraints so that each p^k remains in the domain $\mathcal{C}$. The step-size computation amounts to a line search where $0<\tau\le\tau^{k*}$, with the maximum feasible step-size τ^{k*} defined by

$$\tau^{k*} = \begin{cases} \min\left\{\frac{b_j-(Cp^k)_j}{(Cd^k)_j} : (Cd^k)_j > 0\right\} & \text{if } Cd^k \not\leq 0, \\ +\infty & \text{if } Cd_k \leq 0. \end{cases}$$

This line search can be achieved with parametrization methods.

Proposition 4. *For any arbitrary starting point $(w^1,p^1)\in N^\perp\times P$, let the sequence $\{(w^k,p^k)\}$ be generated by* (35)–(38). *Then each (w^k,p^k) belongs to $N^\perp\times P$ and, under Assumptions* A1–A3, *for a given $\epsilon>0$, there exists a $k\in\mathbb{N}$ such that*

$$Q(w^k,p^k) \le Q^* + \epsilon, \tag{39}$$

where Q^* is the optimal value for problem (31).

Proof. It is easy to show by induction that, since $(w^1,p^1)\in N^\perp\times\mathcal{M}$ and since all $d^k\in N^\perp\times\mathcal{M}$, (w^k,p^k) also belongs to $N^\perp\times\mathcal{M}$ for all $k\in\mathbb{N}$. In order to prove the first claim, we still have to show that (w^k,p^k) also belongs to $\mathbb{R}^n\times\mathcal{C}$ for all k, since

$$N^\perp\times P = (N^\perp\times\mathcal{M})\cap(\mathbb{R}^n\times\mathcal{C}).$$

In other words, we need to show that d^k provides a feasible direction in $\mathbb{R}^n\times\mathcal{C}$, i.e. that there exists a $\tau>0$ such that $(w^k,p^k)+\tau d^k\in\mathbb{R}^n\times\mathcal{C}$.

For any vector $u=(u_1,u_2)\in\mathbb{R}^{n+S}$, denote u_1 and u_2 the subvectors of u relative to the sub-space $\mathbb{R}^n$ and $\mathbb{R}^S$ respectively. Since $g^k\in\partial_\epsilon Q(w^k,p^k)$, then there exist $\epsilon_1\ge0$, $\mu^k\ge0$ and $\gamma^k\in\partial_{\epsilon_1}H(w^k,p^k)$ such that

$$g^k = -\gamma^k + \begin{bmatrix} 0 \\ C^T\mu^k \end{bmatrix}$$

$$\mu^{kT}(b-Cp^k) \le \epsilon-\epsilon_1.$$

Thus, for a given value of ϵ_1, the solution to problem (35) satisfies

$$\bar{g}^k = -\bar{\gamma}^k + Pr\begin{bmatrix} 0 \\ C^T\mu^k \end{bmatrix}$$

where γ^k and μ^k solve the problem

$$\begin{aligned} &\text{minimize}_{\gamma,\mu}\frac{1}{2}\left\|-\bar{\gamma}+\begin{bmatrix}0\\ Pr_2C^T\mu\end{bmatrix}\right\|^2 \\ &\text{s.t.} \\ &\qquad \gamma \in \partial_{\epsilon_1} H(w^k,p^k) \\ &\qquad \mu^T(b-Cp^k)\le \epsilon-\epsilon_1 \\ &\qquad \mu\ge 0. \end{aligned} \tag{40}$$

Differentiating with respect to μ, we obtain from the first order optimality conditions

$$CPr_2(-\bar{\gamma}_2^k + Pr_2C^T\mu^k) + \xi(b - Cp^k) \geq 0,$$

where ξ is the vector of multipliers associated with the inequality constraints in (40). It follows that, whenever inequality constraint i is active at p^k, we have

$$[CPr_2(-\bar{\gamma}_2^k + Pr_2C^T\mu^k)]_i = [CPr_2\bar{g}_2^k]_i = [C\bar{g}_2^k]_i \geq 0.$$

Using (36), this in turn implies that $[Cd_2^k]_i \leq 0$ and d^k provides a feasible direction in $\mathbb{R}^n \times \mathcal{C}$.

If g^k solves (35), then $Q'_\epsilon((w^k,p^k);d^k) = -\|\bar{g}^k\|$. Thus, as long as $\bar{g}^k$ is not the zero-vector, $Q'_\epsilon((w^k,p^k);d^k) < 0$ and (34) implies that the move (37) provides a decrease of at least ϵ:

$$Q(w^{k+1},p^{k+1}) < Q(w^k,p^k) - \epsilon.$$

Thus, condition (39) is satisfied after a finite number of steps, provided that the sequence $\{\bar{g}^k\}$ does not contain the zero-vector.

Now, if $\bar{g}^k = 0$ at iteration k, then $g^k \in N \times \mathcal{M}^\perp$ and, since $(w^k,p^k) \in N^\perp \times \mathcal{M}$,

$$\langle g^k \cdot ((w,p) - (w^k,p^k))\rangle = 0 \quad \forall (w,p) \in N^\perp \times \mathcal{M}.$$

Since $g^k \in \partial Q_\epsilon(w^k,p^k)$, it follows from (33) that

$$\begin{aligned} Q(w^k,p^k) + \langle g^k \cdot ((w,p) - (w^k,p^k))\rangle &\leq Q(w,p) + \epsilon \;\forall (w,p) \in N^\perp \times P \\ \Leftrightarrow \qquad Q(w^k,p^k) &\leq Q(w,p) + \epsilon \;\forall (w,p) \in N^\perp \times P; \end{aligned}$$

hence (w^k,p^k) satisfies (39) and is thus ϵ-optimal. □

From a practical point of view, Algorithm 3 above cannot be implemented: indeed, the exact determination of the set $\partial_\epsilon Q(w^k,p^k)$ for a given ϵ is usually not possible. Instead, one can replace the set of ϵ-sub-gradients by an inner approximating polytope A constructed from a set of already computed sub-gradients. Given a collection of sub-gradients $g^j \in \partial Q(w^j, p^j)$, $j \in J$ obtained during the course of an algorithmic procedure, the *linearization error* $\alpha_{j,k}$ is defined as the difference, at (w^k, p^k), in the values of the objective function and its supporting hyperplane defined by g^j at (w^j, p^j). The approximating polytope A is formed as follows:

$$A = \left\{ \sum_{j \in J} -\lambda_j \gamma^j + \begin{bmatrix} 0 \\ C^T \mu \end{bmatrix} : \lambda_j \geq 0, \mu \geq 0, \right. \\ \left. \sum_{j \in J} \lambda_j = 1, \sum_{j \in J} \lambda_j \alpha_{j,k} + \mu^T (b - Cp) \leq \epsilon \right\}.$$

It is easy to see that A is contained in $\partial_\epsilon Q(w^k, p^k)$.

Replacing $\partial_\epsilon Q(w^k, p^k)$ by its approximation A in (35), direction d^k computed by (35)–(36) will not always imply a descent; in that event, $\tau^k = 0$, step (37) is designated "null-step" and the approximation of A is improved by an appropriate procedure. Otherwise, step (37) provides a significant descent and is designated "serious step".

This can be achieved if, in Algorithm 3, step 1 is replaced by the following quadratic program and step 3 is modified as follows:

1. **Direction finding**

   Find weights λ_j^k, $j \in J$ and vector μ^k that solve the problem

$$\min_{\lambda, \mu} \left\| \sum_{j \in J} -\lambda_j \bar{\gamma}^j + \begin{bmatrix} 0 \\ P_M C^T \mu \end{bmatrix} \right\|^2$$

s.t.

$$\sum_{j \in J} \lambda_j = 1 \\ \sum_{j \in J} \lambda_j \alpha_{j,k} + \mu^T (b - Cp^{k-1}) \leq \epsilon \\ \lambda_j \geq 0, \quad j \in J; \quad \mu \geq 0. \tag{41}$$

   Set

$$\bar{\gamma}^k = \sum_{j \in J} \lambda_j^k \bar{\gamma}^j \\ \bar{g}^k = -\bar{\gamma}^k + \begin{bmatrix} 0 \\ Pr_2 C^T \mu^k \end{bmatrix}. \tag{42}$$

3. Set

$$d^k = -\frac{\bar{g}^k}{\|\bar{g}^k\|}. \tag{43}$$

If d^k provides a significant descent, then set

$$(w^{k+1}, p^{k+1}) = (w^k, p^k) + \tau^k d^k, \tag{44}$$

where

$$\tau^k \in \arg \min_{((w^k,p^k)+\tau d^k) \in \mathbb{R}^n \times C} Q((w^k, p^k) + \tau d^k). \tag{45}$$

Otherwise, improve the approximating polytope A by adding an appropriate sub-gradient to the collection J.

Proposition 5. *The implementable version of Algorithm* 3, *where steps* 1 *and step* 3 *are modified as in* (41)–(45), *converges to an* ϵ*-optimal strategy pair in a finite number of steps.*

Proof. It suffices to show that the polytope improvement procedure in step 3 is finite (this is a well-known result in bundle methods; see for instance [Lemaréchal et al. (1981)]), and the result follows from Proposition 4. □

A more detailed version of an implementable algorithm, with the addition of a selection mechanism for successive sub-gradients, can be found in [Breton and El-Hachem (1992)].

6. Application to Multi-Stage Decision Problems

In many medium- and long-term multi-period planning situations, uncertainty regarding the realization of future events constitutes an important feature of the decision problem, often characterized by a stochastic process or a joint probability distribution on the problem data. In practice, the existence of a probability distribution for the uncertain parameters of an optimization problem may be a fairly strong hypothesis, owing to a lack of sufficient statistical information for a satisfactory estimation of this joint distribution.

A possible alternative consists of considering the problem in the context of stochastic programming with partial knowledge of information [Dupacova (1987), Ermoliev et al. (1985)], where the hypothesis of perfect knowledge of a probability distribution is relaxed. In the dynamic minimax approach in stochastic programming, the decision maker minimizes its expected cost for the most unfavourable probability distribution belonging to a class of possible joint probability distribution compatible with available information on random data; this class of joint probability distributions is defined by mathematical constraints modelling available information on mean, variance or other moments, correlation estimates, support, or any other information characterizing probability distributions.

The dynamic minimax approach in stochastic dynamic programming can be modelled as a game of the type presented in this paper, where Player

1 (the decision maker) minimizes his expected costs with respect to his dynamic decisions and Player 2 (nature) maximizes the same with respect to transition probabilities. In the model considered here, the discrete states correspond to different possible realizations of the random data, which are supposed to be known, whereas only partial information exists on their respective probabilities. The set P of possible strategies for Player 2 restricts the transition probabilities to be compatible with available partial information. The set $D \cap N$ of possible strategies for Player 1 includes past-commitment constraints, where the set of possible actions at stage t and for a realization s^t of the parameters is restricted by previous actions (e.g. investment - disinvestment decisions, stockpiling, etc.). For such particular applications, Assumptions A1–A3 are rather mild: for instance, cost functions $r_{s^t}(x_{s^t})$ are often linear or convex, where s^t is a possible set of parameters at time period t, and it is always possible to bound the decision variables for Player 1. It is worthwile to note that, to our knowledge, very few algorithms exist for the solution of dynamic minimax problems in stochastic programming, and that those which have been proposed are limited to two-period problems. Possible applications for the dynamic minimax approach include environmental modelling, where available statistical information is scarce and expert opinions differ (in that case, set P could represent the set of opinions on possible scenario probabilities).

We have presented here three algorithms for the solution of a particular single-controller stochastic game that has an application in multi-stage stochastic decision problems. All three algorithms are well adapted to large-scale problems, as they rely on decomposition in node sub-problems; since they involve the independent solution of sub-problems, these algorithms can be implemented in a parallel computing environment. Algorithm 1 requires the solution of convex sub-problems and projection operations. Algorithm 2 may be more appropriate in the case where the cost functions are linear, since it then amounts to the solution of independent linear programs; however, even if it converges to the value of the game, Algorithm 2 is not guaranteed to converge to the equilibrium strategy. Algorithm 3 implements a bundle method in order to guarantee descent and convergence to the equilibrium strategy. It involves the solution of a quadratic program at each iteration, but no longer requires a projection operation, which is replaced by a line search procedure achieved by parametric programming.

Let us point out finally that in the first two methods, parametric optimization may be used to speed up calculations by exploiting sub-problems similarities, either from one iteration to the next, or even within a given iteration. Indeed, passage from one iteration's objective to the next can be reduced to a single parameter change; this becomes especially interesting in the linear case, since commercial LP packages often include single-parametric procedures.

Acknowledgement. Research supported by NSERC-Canada grants no. OGP0004088 and no. GRS004-4901 and FCAR-Québec grant no. 99-NC-0252.

References

Breton, M. and S. El-Hachem (1991), Decomposition Algorithms for Stochastic Dynamic Programs, working paper, *Cahiers du GERAD G–91–15.*

Breton, M. and S. El-Hachem (1992), Algorithms for the Solution of Stochastic Dynamic Minimax Problems, working paper, *Cahiers du GERAD G–92–32.*

Dupacova , J. (1987), The Minimax Approach to Stochastic Programming and Illustrative Application, *Stochastics*, **28**, 73–88.

Eckstein, J. (1989), Splitting Methods for Monotone Operators with Applications to Parallel Optimization, PhD thesis, M.I.T.

Ermoliev, Y. (1966), Methods for Solving Non-Linear Extremal Problems, *Cybernetics*, **2**, 1–17.

Ermoliev, Y., A. Gaivoronsky and C. Nedeva (1985), Stochastic Optimization Problems with Partially Known Distribution Functions, *SIAM Journal on Control and Optimization*, **23**, 697–716.

Filar, J.A. and T.E.S. Raghavan (1984), A Matrix Game Solution of the Single-Controller Stochastic Game, *MOR*, **9**, 356–362.

Gabay, D. and B. Mercier (1976), A Dual Algorithm for the Solution of Nonlinear Variational Problems via Finite Element Approximations, *Computers and Mathematics with Applications*, **2**, 17–40.

Glowinski, R. and A. Marroco (1975), Sur l'approximation, par éléments finis d'ordre un, et la résolution, par pénalisation-dualité, d'une classe de problèmes de Dirichlet non linéaires, *Revue Française d'Automatique, Informatique et Recherche Opérationnelle*, **9**, 41–76.

Hiriart-Urruty, J.-B. (1980), Lipschitz r-continuity of the Approximate Subdifferential of a Convex Function, *Mathematica Scandinavica*, **47**, 123–134.

Kallenberg, L.C.M. (1983), Linear Programming and Finite Markovian Control Problems, Mathematical Center, Amsterdam, Holland (Chapter 6).

Kim, S., H. Ahn and S.-C. Cho (1991). Variable Target Value Subgradient Method, *Mathematical Programming*, **49**, 359–369.

Kiwiel, K.C. (1985), Methods of Descent for Nondifferentiable Optimization, *Lecture Notes in Mathematics*, **1133**, Springer-Verlag, Berlin.

Lemaréchal, C., J.-J. Strodiot and A. Bihain (1981), On a Bundle Algorithm for Non Smooth Optimization, in Nonlinear Programming 4, O.L. Mangasarian, R.R. Meyer and S.M. Robinson eds., *Academic Press*, New York, 245–282.

Minoux, M. (1983), Programmation mathématique, théorie et algorithmes, Dunod, Paris.

Parthasarathy, T. and T.E.S. Raghavan (1981), An Orderfield Property for Stochastic Games when One Player Controls Transition Probabilities, *Journal of Optimization Theory and Applications*, **33**, 375–392.

Polyak, B.T. (1966), A General Method for Solving Extremum Problems, *Soviet Mathematics*, **8**, 593–597.

Rockafellar, R.T. (1970), Convex Analysis, Princeton University Press, Princeton.

Shapley, L.S. (1953), Stochastic Games, *Proceedings of the National Academy of Sciences of USA*, **39**, 1095–1100.

Vrieze, O.J. (1981), Linear Programming and Undiscounted Games in Which One Player Controls Transitions, *OR Spectrum*, **3**, 29–35.

Michèle Breton
École des Hautes Études Commerciales
Affiliée à l'Université de Montréal
Montréal, Quebec Canada H3T 1V6

Saeb El Hachem
GERAD
5255 Av. Decelles
Montréal, Québec
Canada, H3T 1V6

PART IV

Stochastic Games: Differential, sequential and Markov Games

Stochastic Games with Average Cost Constraints

Nahum Shimkin

Abstract

The paper introduces the model of undiscounted stochastic games augmented by side constraints on the levels of a player's average costs. Several variants of this problem are suggested, and the special case of zero-sum games with constraints on one side is analyzed; under certain recurrence conditions, existence of the value and (non-stationary) optimal strategies are established.

1 Introduction

In undiscounted stochastic games the purpose of each player is to maximize his average reward, which is the long-term average of a scalar reward sequence. In this paper we consider the situation where, in addition, a player must obey some side constraints on average costs; that is, the player incurs certain costs on each stage, and must verify that the long-term average costs will not exceed pre-specified thresholds.

Dynamic optimization problems with average cost constraints have attracted considerable attention in the context of Markov decision processes (e.g. [7, 9, 2, 10, 14, 1, 5]). Indeed, such constraints follow naturally in various control problems, ranging from economics problems where resources are limited and supplied at a given rate, to communication networks with average delay constraints. The incorporation of such constraints in stochastic games is a natural extension of this line of work, and is similarly motivated.

From a strategic viewpoint, the average cost constraints effectively limit each player to a certain sub-class of his strategies, which generally depends on the strategies employed by other players. Constrained games of this nature have been first considered in [6]; and see also [12, 11]. In the context of dynamic games, a related model is considered in [13], where a two-person repeated matrix game with the average payoff criterion is augmented by the possibility of ruin which occurs if a player's initial wealth decreases to 0.

The general model of undiscounted stochastic games with average cost constraints is introduced formally in Section 2. Several variants are suggested and briefly discussed. For simplicity, only the case of two players and finite state and action spaces are considered. Section 3 presents the analysis of a particular case, namely the zero-sum game with constraints on

*Part of this work was performed when the author was with the department of Electrical Engineering at the Technion - Israel Institute of Technology

one side only. Under certain recurrence conditions, existence of a value and optimal strategies for both players are established. The optimal strategy for the constrained player turns out to be non-stationary in an essential way. The analysis relies on the ideas of Blackwell's approachability theory [4], extended to the stochastic game context. The paper is concluded with some remarks and problems for additional research.

2 The General Model

A standard two-person stochastic game is defined by the following elements. The player set $I = \{1,2\}$; a finite state-space S; a finite action space A^i for each player $i \in I$; transition law p, where $p(s'|s,a^1,a^2)$ is the probability of s' being the next state given the present state s and actions $a^1 \in A^1$, $a^2 \in A^2$; and one-step reward functions $r^i : S \times A^1 \times A^2 \to \mathbf{R}$, $i \in I$. The game is played in stages $n = 1,2,\cdots$. Let s_n and (a_n^1, a_n^2) denote the state and actions at stage n, and let $r_n^i \triangleq r^i(s_n, a_n^1, a_n^2)$ denote the reward of player i at that stage. $\bar{r}_n^i \triangleq n^{-1}\sum_{k=1}^n \bar{r}_k^i$ denotes the average reward up to stage n. Let Π^i be the set of behavioral strategies for player i; thus in each strategy $\pi^i \in \Pi^i$ the choice of action a_n^i at stage n may depend on the entire history $h_n = (s_1, a_1^1, a_2^1, \ldots\ s_n)$, and may include randomizations which are performed independently at each stage. Let F and G denote the set of stationary strategies for players 1 and 2, respectively. A strategy $f \in F$ may be identified with a function $f : S \to \mathcal{P}(A^1)$, where $\mathcal{P}(A^1)$ is the set of probability vectors over A^1, and similarly for $g \in G$. Let P_π^s and E_π^s be the probability measure and expectation associated with the strategy pair $\pi = (\pi^1, \pi^2) \in \Pi^1 \times \Pi^2$ and initial state $s_1 = s \in S$. The initial state will henceforth be assumed to be fixed, and will be omitted from the notation. The undiscounted (limit expected average) reward functional for each player $i \in I$ is now defined as

$$R^i(\pi) = \liminf_{n\to\infty} E_\pi(\bar{r}_n^i)\ . \tag{2.1}$$

We note that several other definitions for the undiscounted reward are possible and have been considered in the literature (see, e.g., [3]). For concreteness, only the limit expected average reward criterion will be used here.

We proceed to introduce the cost constraints into the model. Let $c^i : S \times A^1 \times A^2 \to \mathbf{R}^{k^i}$, $k^i \geq 1$, be a (vector valued) cost function for player $i \in I$. Let $\gamma^i \in \mathbf{R}^{k^i}$ be the allowed *cost level* vector for player i. Generally speaking, player i must keep his average costs below the maximal cost levels. The exact statement of this constraint may lead to several different definitions of the payoff, as discussed below.

a. Expected-Cost Constraints:

Let $\vec{c}^{\,i}_n = n^{-1} \sum_{k=1}^{n} c^i(s_k, a^1_k, a^2_k)$ denote the average cost for player i up to stage n, and let

$$C^i(\pi) \triangleq \limsup_{n\to\infty} E_\pi(\vec{c}^{\,i}_n) \,, \quad \pi \in \Pi^1 \times \Pi^2 \,, \tag{2.2}$$

(all limits of vector-valued sequences encountered in this paper are taken over each component separately). Define the following payoff function for player i:

$$\phi^i(\pi) = \begin{cases} R^i(\pi) & \text{if} \quad C^i(\pi) \leq \gamma^i \\ -\infty & \text{if} \quad C^i(\pi) \not\leq \gamma^i \quad . \end{cases} \tag{2.3}$$

The payoff functions ϕ^1 and ϕ^2 define a strategic-form game over the strategy sets Π^1, Π^2. The Nash equilibrium point for this game may now be defined in a standard manner. This definition of the payoff functions has the following interpretation: player i is primarily concerned with keeping his average expected costs (2.2) below the allowed level γ^i. A violation of these constraints is considered the worst possible outcome for him, as manifested by a payoff of $-\infty$. Otherwise, his payoff is simply given by his average expected reward.

b. Sample Path Constraints:

The expected-cost formulation has the appealing property and both the average reward and the average cost have a similar form. However, such constraints do not exclude the possibility that the actual (sample-path) average costs will exceed the allowed cost levels with positive probability. If this is not acceptable, one is led to the following sample-path formulation. Let

$$\vec{c}^{\,i}_\infty \triangleq \limsup_{n\to\infty} \vec{c}^{\,i}_n \tag{2.4}$$

denote the (upper) limit average cost vector. Define the payoff function for player i by

$$\phi^i(\pi) = \begin{cases} R^i(\pi) & \text{if} \quad P_\pi\{\vec{c}^{\,i}_\infty \leq \gamma^i\} = 1 \,. \\ -\infty & \text{if} \quad P_\pi\{\vec{c}^{\,i}_\infty \leq \gamma^i\} < 1 \,. \end{cases} \tag{2.5}$$

The payoff now reflects the fact that the allowed cost levels must be obeyed with probability 1.

We note that definitions (2.3) or (2.5) lead the standard definition of Nash equilibrium for constrained games [6, 12]. These definitions may be extended in a straightforward manner to n-person games.

c. Constraints with Ruin:

It has been assumed so far that the violation of a player's constraints affects only his own payoff, while the payoff for the other player is not directly affected. However, it is possible to interpret the violation of a player's constraints as leading to his ruin, or elimination from the game, thereby leading to an increased payoff for the other player. In the extreme case, the ruin of the other player may be a most desired result (cf. [13]). This obviously leads to the following payoff functional for player 1 (where expected-cost constraints are used):

$$\phi^1(\boldsymbol{\pi}) = \begin{cases} R^1(\boldsymbol{\pi}) & \text{if } C^1(\boldsymbol{\pi}) \leq \gamma^1,\, C^2(\boldsymbol{\pi}) \leq \gamma^2 \\ -\infty & \text{if } C^1(\boldsymbol{\pi}) \not\leq \gamma^1 \\ +\infty & \text{if } C^1(\boldsymbol{\pi}) \leq \gamma^1,\, C^2(\boldsymbol{\pi}) \not\leq \gamma^2 \end{cases} \tag{2.6}$$

The payoff for player 2 may be defined symmetrically. We note that the last definition of payoffs is not directly extendible to games of more than two players. However, it may capture the essence of some two-person conflicts, especially in cases of strict competition. We further note that this definition may be modified in an obvious manner to incorporate sample-path constraints (see the next section).

Our discussion has been limited to the case where the constraints are strict, in the sense that their violation is the worst possible outcome for the player irrespective of his reward. Obviously, one may think of different situations; for example, a player may pay a fixed penalty for violating his constraints, but may consider this worthwhile if appropriately compensated by an increased reward. More generally, the payoff (utility) for a player may be defined through some (partial) order on the multi-dimensional payoff/cost space. Such generalizations fall outside the scope of constrained games, and will not be considered in detail in this paper.

3 The Zero-Sum Game with Constraints on One Side

3.1 Model and Recurrence Conditions

This section focuses on the analysis of a special class of constrained stochastic games. Namely, we consider the case of zero-sum rewards ($r^1 = -r^2$), and constraints with ruin (in the sense of (2.6)) on player 1 only. For concreteness, we restrict attention to sample-path constraints (essentially the same analysis applies for expected cost constraints). Thus, the payoff ϕ^1 for player 1 is given by (2.5), with $i = 1$, and for player 2 by $\phi^2 = -\phi^1$. The game is said to have a value v if

$$\sup_{\pi^1} \inf_{\pi^2} \phi^1(\pi^1, \pi^2) = \inf_{\pi^2} \sup_{\pi^1} \phi^1(\pi^1, \pi^2) \triangleq v\,.$$

A strategy of player 1 is the *optimal* if it guarantees $\phi^1 \geq v$ (i.e., it achieves the supremum on the left). A strategy of player 2 is optimal if it guarantees $\phi^1 \leq v$, and is *ϵ-optimal* if it guarantees $\phi^1 \leq v + \epsilon$.

The avoid the case where $v = -\infty$ trivially, we assume the following.

Assumption A1: For every strategy of player 2, player 1 has a strategy which satisfies his constraints (w.p. 1).

The conditions for v and π^{1*}, π^{2*} to be the value and optimal strategies of player 1 and 2, respectively, are now equivalent to the following two requirements:

(*i*) $P_{(\pi^{1*},\pi^2)}\{\bar{c}^1_\infty \leq \gamma^1\} = 1$ for every $\pi^2 \in \Pi^2$.

(*ii*) $R^1(\pi^1, \pi^{2*}) \leq v \leq R^1(\pi^{1*}, \pi^2)$ for every $\pi^2 \in \Pi^2$, and for every $\pi^1 \in \Pi^1$ which satisfies $P_{(\pi^1,\pi^{2*})}\{\bar{c}^1_\infty \leq \gamma^1\} = 1$.

Existence of a value and optimal strategies will be established under the following recurrence assumption:

Assumption A2: There exists a fixed state, say state 0, which is accessible from any other state in the Markov chain induced by any pair of stationary non-randomized strategies. (In other words, each such Markov chain includes exactly one recurrent class, and state 0 is included in that class.)

3.2 Approachability for Stochastic Games

An approachability theory for repeated matrix games with vector payoffs was introduced in [4]. The basic results of this paper have been generalized to the stochastic games context in [15]. We review below some of these results, which will be applied in the next sub-section to solve the constrained game problem.

Consider the stochastic game model of Section 2, with a fixed initial state s, and the recurrence assumption **A2**. Introduce a vector-valued payoff function $m : S \times A^1 \times A^2 \rightarrow \mathbf{R}^b$, $b \geq 2$. Denote $m_n = m(s_n, a^1_n, a^2_n)$, $\bar{m}_n = n^{-1} \sum_1^n m_k$, and let $d(\cdot, \cdot)$ denote the Euclidean point-to-set distance in $\mathbf{R}^b$.

Definition: A set $B \in \mathbf{R}^b$ is *approachable* (by player 1 from initial state s) if there exists a *B-approaching strategy* $\pi^1 \in \Pi^1$ for which

$$\lim_{n\rightarrow\infty} d(\bar{m}_n, B) = 0 \text{ a.s.,} \quad \forall \pi^2 \in \Pi^2,$$

i.e., all limit points of $\{\bar{m}_n\}$ are in B.

A set B is *excludable* (by player 2 from s) if there exists a *B-excluding strategy* $\pi^2 \in \Pi^2$ and a constant $\delta > 0$ for which

$$\liminf_{n\rightarrow\infty} d(\bar{m}_n, B) \geq \delta \text{ a.s.,} \quad \forall \pi^1 \in \Pi^1,$$

i.e., all limit points of $\{\bar{m}_n\}$ are strictly outside B.

We now define an auxiliary family of stochastic games. Let U denote the set of unit vectors in $\mathbf{R}^b$, and let $\langle\cdot,\cdot\rangle$ denote the Euclidean inner product. For each $u \in U$, define $\Gamma(u)$ as the undiscounted (limit expected average) zero-sun stochastic game with the same dynamics as above, a *scalar* payoff function $m^u \triangleq \langle m, u\rangle$ (the projection of m in the direction of u), and player 1 the maximizer. Under assumption **A2**, each game $\Gamma(u)$ admits a value and stationary optimal strategies ([16, 8]); let $\mathrm{val}\,\Gamma(u)$ denote its value, and let $f^*(u)$ and $g^*(u)$ denote the stationary optimal strategy of player 1 and 2, respectively, in $\Gamma(u)$.

For every pair (f, g) of stationary strategies let

$$M(f,g) = \lim_{n\to\infty} E_{f,g}(\bar{m}_n)$$

(existence of the limit follows from **A2**), and define

$$M(F,g) = \{M(f,g) : f \in F\}\,,$$

which is the set of all average payoff vectors generated by stationary strategies of player 1 when player 1 employs g. It is well known ([7, p. 95]) that $M(F,g)$ is a closed convex set under **A2**.

The next theorem, which follows from Theorem 3.4 and Lemma 2.1 in [15], provides a complete characterization of approachability and excludability for convex sets.

Theorem 3.1 *Assume* **A2**. *Let B be a closed* convex *set in* $\mathbf{R}^b$. *Then*

(i) B is approachable if and only if

$$\mathrm{val}\,\Gamma(u) \geq \inf_{\beta\in B}\langle\beta, u\rangle\,, \quad \forall u \in U\,.$$

(ii) B is excludable if and only if $M(F,g)\cap B = \emptyset$ for some $g \in G$; in that case, the stationary strategy g is B-excluding.

(iii) B is either approachable or excludable (but obviously not both).

(iv) If approachable, then a B-approaching strategy for player 1 is given by the following 'piecewise stationary' strategy. At each time τ when state 0 is reached, a stationary strategy to be played until the next such time is determined as follows: compute the current average payoff $\bar{m}_\tau$; if $\bar{m}_\tau \notin B$, then play $f^*(u_\tau)$, where u_τ is a unit vector in the direction from $\bar{m}_\tau$ to the closest point in B; otherwise, play any strategy.

3.3 Main Results

Our main result for the game of Section 3.1 is the following.

Theorem 3.2 *Assume* **A1-A2**. *Define the vector-valued payoff function* $m \triangleq (r^1, c^1)$, *and let* ρ^* *be the maximal number* ρ *such that the convex set*

$$B(\rho) = \{(\bar{r}^1, \bar{c}^1) \in \mathbf{R}^{1+k^i} : \bar{r}^1 \geq \rho\,,\, \bar{c}^1 \leq \gamma^1\}$$

is approachable. Then

(i) The game has a finite value, which equals ρ^*.

(ii) An optimal strategy for player 1 is given by any $B(\rho^*)$*-approaching strategy.*

(iii) For player 2 (the unconstrained player), an ϵ*-optimal* stationary *strategy exists for any* $\epsilon > 0$.

Proof: We first establish that ρ^* is well defined and finite. Note first that the set $B(-\infty) = \{(\bar{r}^1, \bar{c}^1) : \bar{c}^1 \leq \gamma^1\}$ is approachable; this follows by assumption **A1**, which implies that $B(-\infty)$ is not excludable, and by Theorem 3.1(*ii*). Since the reward function r^1 is bounded, it follows immediately that $B(\rho)$ is approachable for ρ small enough, and not approachable for ρ large enough. Thus, the set $K = \{\rho : B(\rho) \text{ approachable}\}$ has a finite supremum ρ^*. Finally, it follows readily from Theorem 3.1(*i*) that K is closed (see also the proof of Corollary 3.1 below). Thus ρ^* is the maximal element and $B(\rho^*)$ is approachable.

To prove the theorem, it is now sufficient to establish that: (*i*) any $B(\rho^*)$-approaching strategy of player 1 guarantees a payoff $\phi^1 \geq \rho^*$, and (*ii*) for any $\epsilon > 0$ there exists a stationary strategy of player 2 which guarantees a payoff $\phi^1 \leq \rho^* + \epsilon$. Assume first that player 1 is using π^{1*}, a $B(\rho^*)$-approaching strategy. By definition of approachability and of the set $B(\rho^*)$, it follow that for every strategy of player 2:

$$\limsup_{n\to\infty} \bar{c}^1_n \leq \gamma^1 \;\text{ a.s.,} \quad \liminf_{n\to\infty} \bar{r}^1_n \geq \rho^* \;\text{ a.s.,}$$

and by Fatou's lemma the last inequality implies that $R^1 \geq \rho^*$. Thus π^{1*} guarantees for player 1 a payoff of ρ^* at least.

Next we identify an ϵ-optimal strategy for player 2. Fix $\epsilon > 0$. By our definition of ρ^*, the convex set $B(\rho^* + \epsilon)$ is not approachable, hence by Theorem 3.1 it is excludable, and there exists an excluding stationary strategy g^ϵ. (Note that this is not true for $\epsilon = 0$, since $B(\rho^*)$ is approachable.) Assume that player 2 employs the strategy g^ϵ, and player 1 any strategy π^1. Then by definition of excludability, all limit points of $\bar{m}_n = (\bar{r}^1_n, \bar{c}^1_n)$ are (almost surely) outside the set $B(\rho^* + \epsilon)$, hence contained in the set $\{\bar{r}^1 \leq \rho^* + \epsilon \text{ or } \bar{c}^1 \not\leq \gamma^1\}$. This implies that either (*i*) with positive probability, some limit point of $\bar{c}^1_n$ is in $\{\bar{c}^1 \not\leq \gamma^1\}$, or (*ii*) almost surely, all limit points of $\bar{r}^1_n$ are in $(-\infty, \rho^* + \epsilon]$. In case (*i*), we have $(\limsup_n \bar{c}^1_n \not\leq \gamma^1)$ with positive probability, hence the payoff ϕ^1 equals $-\infty$ by its definition. In case (*ii*) we have $(\limsup_n \bar{r}^1_n \leq \rho^* + \epsilon \text{ a.s.})$, which implies $R^1 \leq \limsup_n E(\bar{r}_n^1) \leq \rho^* + \epsilon$, and the payoff is again not larger than $\rho^* + \epsilon$. Thus, the stationary strategy g^ϵ is ϵ-optimal for player 2. □

Using the above characterization, a more explicit expression may be obtained for the value of the game.

Corollary 3.1 *The value of the game is given by*

$$\rho^* = \min_{x \leq 0} \{\mathrm{val}\,\Gamma((1,x)) - \langle \gamma^1, x \rangle\}.$$

Here x ranges over all vectors with non-positive entries in $\mathbf{R}^{k^1}$, and $(1,x)$ is the vector in $\mathbf{R} \times \mathbf{R}^{k^1}$ obtained from x by adding 1 as the first entry.

Proof: Partition each unit vector as $u = (u_1, u_2) \in \mathbf{R} \times \mathbf{R}^{k^i}$. For every real ρ, the set $B(\rho)$ is approachable if and only if

$$\mathrm{val}\,\Gamma(u) \geq \inf_{\beta \in B(\rho)} \langle \beta, u \rangle \tag{3.7}$$

is satisfied for every unit vector $u = (u_1, u_2)$ (Theorem 3.1(i)). Now, it may be verified that the right hand side of (3.7) is larger than $-\infty$ if and only if $u_1 \geq 0$ and $u_2 \leq 0$ (component-wise), and is then equal to $\langle (u_1, u_2), (\rho, \gamma^1) \rangle = u_1 \rho + \langle u_2, \gamma^1 \rangle$. Thus $B(\rho)$ is approachable if and only if

$$\mathrm{val}\,\Gamma((u_1, u_2)) \geq u_1 \rho + \langle u_2, \gamma^1 \rangle$$

for every such $u = (u_1, u_2)$. The case of $u_1 = 0$ need not be considered since ρ is canceled (the inequality is then satisfied as a consequence of **A1**). Thus, dividing by $u_1 > 0$ and noting that $x = u_2/u_1$ ranges over the entire set of vectors with non-positive entries, we get that $B(\rho)$ is approachable if and only if

$$\mathrm{val}\,\Gamma((1,x)) \geq \rho + \langle x, \gamma^1 \rangle \quad \forall\, 0 \geq x \in \mathbf{R}^{k^i}.$$

Since ρ^* is the largest ρ which satisfies this condition, the required expression is obtained. □

Remarks:

1. An optimal stationary strategy of player 2 can be given a nice geometric interpretation. Consider for simplicity the case of a single (scalar) cost constraint. An optimal stationary strategy g^* of player 2 is such that $M(F, g^*)$ just touches the boundary of $B(\rho^*)$. This is illustrated in Fig. 1. The set $M(F, g^*)$ represents all possible limit points of $(\bar{r}_n^1, \bar{c}_n^1)$ which player 1 can attain against g^* (indeed, interchanging the roles of the players in the definition of approachability, it easily follows from Theorem 3.1 that the set $M(F, g^*)$ is approachable by player 2). Thus, the figure clearly illustrates the interplay between reward and cost: although player 1 may be able to obtain a reward larger than ρ^*, then against g^* this can only be done at the expense of violating his cost constraint γ^1.

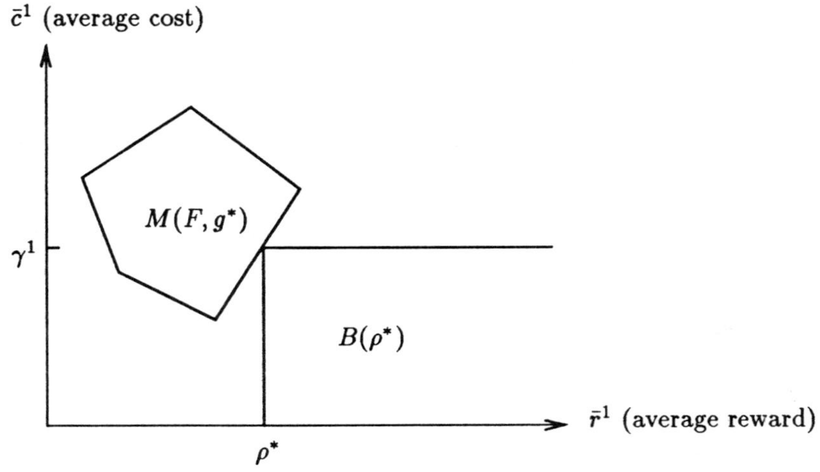

Figure 1: The geometric picture.

2. In this geometric picture, we may interpret the value ρ^* as follows. Considering the set $B(\rho)$, we translate this set to the right (by increasing ρ continuously), and stop just before it becomes disjoint with a set $M(F, g')$ for some g'. Then this point corresponds to ρ^*, and g' is an optimal strategy for player 2.

3. In general, an optimal or even ϵ-optimal strategy for player 1 does not exist within the class of stationary strategies. In fact, with two constraints or more player 1 may have to use a non-stationary (approachability) strategy even just to guarantee that his constraints are met.

4. Although player 2 does have ϵ-optimal stationary strategy, an optimal one may fail to exist (an example follows). We suspect however that generically it does. An obvious candidate for an optimal stationary strategy is the limit of any converging subsequence of the ϵ-optimal set $\{g^\epsilon\}$, as $\epsilon \to 0$. (Note that such a limit point exists since the strategy set G is compact. Also, it may be easily verified as in [7, Theorem 7.3] that under Assumption A2 the payoff vector $M(f, g)$ is continuous in g for each f.)

5. Here is a simple example where an optimal stationary strategy for player 2 does not exist. We assume only one state (a repeated matrix game), with two actions for each player which are also the pure stationary strategies, denoted f_0, f_1 (for player 1) and g_0, g_1. The corresponding payoff vectors in the combined reward-cost space are $M(f_0, g_0) = (-1, 0)$, $M(f_1, g_0) = (1, 0)$, $M(f_0, g_1) = (0, -1)$, $M(f_1, g_1) = (1, 1)$, and the (scalar) constraint level is $\gamma^1 = 0$, It

may be verified geometrically that the value is $v = 0$, and the strategy $g_\epsilon = (1-\epsilon)g_0 + \epsilon g_1$ is $(\epsilon/2)$-optimal (it guarantees $\phi^1 \leq \epsilon/2$). Obviously $g_\epsilon \to g_0$, but the latter guarantees only $\phi^1 \leq 1$.

6. The value and optimal strategies for both players as specified above do not depend on the initial state s.

7. The results above are not limited to the specific undiscounted reward functional (2.1), namely the limit expected average reward. Indeed, it was established in the proof that each player can guarantee an average reward of ρ^* in an almost-sure sense, namely the optimal strategy of player 1 guarantees $\liminf_n \bar{r}_n^1 \geq \rho^*$ (a.s.), while the optimal strategy of player 2 guarantees that $\limsup_n \bar{r}_n^1 \leq \rho^*$ (a.s.) for every strategy of player 1 which satisfies the constraints. Thus the main results remain valid under most common undiscounted reward criteria, such as the limit average criterion or the limit discount criterion (in the terminology of [3]).

8. It may also be shown that the main results hold if the sample-path constraints are replaced by (or combined with) expected cost constraints, as defined in Section 2.

9. The definition of approachability which was given here did not refer to the rate of convergence. In fact, it may be shown that the approaching strategy of Theorem 3.1 ensures the a.s. convergence of $\bar{m}_n$ to B *at a uniform rate over all strategies of the other player* [15]. That is, for any $\epsilon > 0$ there exists N such that

$$P\{d(\bar{m}_n, B) \leq \epsilon \ \forall n \geq N\} \geq 1 - \epsilon \, ,$$

for every strategy of player 2. This may be used in the present context to strengthen the results obtained above. Thus, if player 1 is using the strategy of Theorem 3.1 to approach $B(\rho^*)$, then for every $\epsilon > 0$ there exists N such that

$$P\{\bar{r}_n^1 \geq \rho^* - \epsilon, \, \bar{c}_n^1 \leq \gamma^1 + \epsilon e, \, \forall n \geq N\} \geq 1 - \epsilon$$

(where e is a vector with 1 at each entry) for every strategy of player 2. Similar comments apply with respect to the convergence of $(\bar{r}_n^1, \bar{c}_n^1)$ to $M(F, g^*)$ under g^*.

In conclusion, for the zero-sum constrained game of this section we have established the existence of the value and optimal strategies for both players. While the unconstrained player has an optimal stationary strategy, the constrained player benefits by using non-stationary strategies, and his optimal strategy is in fact an approaching strategy with respect to some set in the reward-cost space.

Obviously, in this paper we have just touched upon the analysis of the general classes of constrained stochastic games as defined in Section 2. We mention some open problems for research. In the zero-sum game of Section 3, it should be interesting to establish existence of the value after removing, or weakening, the recurrence assumption **A2**. We note that the game of Section 3 (constrains with ruin, on one player only) is the only case which has a zero-sum version; all others are essentially non zero-sum (even if the reward functions are zero-sum) due to the constraints. Thus, one may seek to characterize the set of Nash equilibria for these games (folk theorem). If constraints with ruin exist on both sides, then the above analysis implies that equilibrium strategies will be essentially non-stationary for both players, even under recurrence assumptions similar to **A2**. However, for constraints without ruin the existence of equilibrium points in stationary strategies is open for investigation.

REFERENCES

[1] E. Altman and A. Shwartz, "Markov decision problems and state-action frequencies," *SIAM J. Control and Optimization* **29** No. 4, July 1991.

[2] F. J. Beutler and K. W. Ross, "Optimal policies for controlled Markov chain with a constraint," *J. Math. Anal. Appl.* **12**, pp. 236-256, 1985.

[3] T. Bewly and E. Kohlberg, "On stochastic games with stationary optimal strategies," *Mathematics of Operations Research* **3**, pp. 104-125, 1978.

[4] D. Blackwell, "An analogue for the minimax theorem for vector payoffs," *Pacific J. Math.* **6**, pp. 1-8, 1956.

[5] V. S. Borkar, "Controlled Markov chains with constraints," *Proceedings of the Workshop on Recent Advances in Modeling and Control of Stochastic Systems*, Bangalore, Jan. 1991.

[6] G. Debreu, "A social equilibrium existence theorem," *Proceedings of the National Academy of Science* **38**, pp. 886-893, 1952.

[7] C. Derman, *Finite State Markovian Decision Processes*, Academic Press, New-York, 1970.

[8] A. Federgruen, "On n-person stochastic games with denumerable state space," *Adv. Appl. Prob.* **10**, pp. 452-471, 1978.

[9] A. Hordijk and L. C. M. Kallenberg, "Constrained undiscounted stochastic dynamic programming," *Mathematics of Operations Research* **9**, pp. 276-289, 1984.

[10] D.-J. Ma, A. M. Makowski and A. Shwartz, "Estimation and optimal control for constrained Markov chains," *Proc. 25th IEEE Conf. on Decision and Control,* pp. 994-999, 1986.

[11] H. Meister, *The Purification Problem for Constrained Games with Incomplete Information*, (Lecture notes in economics and mathematical systems; 295) Springer, Berlin, 1987.

[12] J. B. Rosen, "Existence and uniqueness of equilibrium points for concave n-person games," *Econometrica* **33**, pp. 520-534, 1965.

[13] R. W. Rosenthal and A. Rubinstein, "Repeated two-player games with ruin," *International Journal of Game Theory* **13**, pp. 155-177, 1984.

[14] K. W. Ross and R. Varadarajan, "Markov decision processes with sample path constraints: the communicating case," *Operations Research* **37**, pp. 780-790, 1989.

[15] N. Shimkin and A. Shwartz, "Guaranteed performance regions in Markovian systems with competing decision makers ," to appear in *IEEE Trans. Automat. Contr.*, Feb. 1993.

[16] M. A. Stern, *On Stochastic Games with Limiting Average Pay-Off.* Ph.D. dissertation, University of Illinois, Circle Campus, Chicago, 1975.

Nahum Shimkin
Institute for Mathematics and its Applications
University of Minnesota
Minneapolis, MN 55455, USA

Stationary Equilibria for Nonzero-Sum Average Payoff Ergodic Stochastic Games with General State Space

Andrzej S. Nowak

Abstract

This paper treats of nonzero-sum average payoff stochastic games with arbitrary state spaces. Such models of games very well fit in some studies in economic theory. A natural uniform geometric ergodicity condition, often used in control theory of Markov chains, is imposed on the transition probabilities of the games. A correlation of strategies of the players, involving "public signals", is allowed in this paper. The main result is an extension of the correlated equilibrium theorem proved recently by the author and Raghavan for dynamic games with discounting to the average payoff stochastic games. Also some special classes of games that possess Nash equilibria without public signals are discussed. This paper also provides a brief overview of the theory of nonzero-sum stochastic games which is very far from being complete.

AMS 1980 *Subject Classification.* Primary:90D15, Secondary: 93C30.
Key Words: Average payoff stochastic games, correlated stationary equilibria, nonzero-sum games.

1 Introduction and the stochastic game model

The theory of nonzero-sum stochastic games with the average payoffs per unit time for the players started with the papers by Rogers [1] and Sobel [2]. They considered finite state spaces only and assumed that the transition probability matrices induced by any stationary strategies of the players are unichain. Till now only special classes of nonzero-sum average payoff stochastic games are shown to possess Nash equilibria (or ϵ-equilibria). Parthasarathy and Raghavan [3] considered games in which one player is able to control transition probabilities and proved the existence of stationary equilibria in such a case. Non-stationary ϵ-equilibria were shown to exist in games with state independent transitions by Thuijsman [4] and in games with absorbing states by Vrieze and Thuijsman [5]. Parthasarathy [6] first considered nonzero-sum stochastic games with countable state spaces and proved that every discounted stochastic game always has a stationary Nash equilibrium solution. Federgruen [7] extended the works of Rogers and Sobel to average payoff nonzero-sum stochastic games with countably many states, satisfying a natural uniform geometric ergodicity condition. Federgruen's result [7] is strengthened in [8], where the overtaking optimality criterion in the class of stationary strategies of the players is adopted.

In many applications of stochastic games, especially in economic theory, it is desirable to assume that the state spaces are not discrete; see for example Duffie et al. [9], Dutta [10], Karatzas et al. [11], or Majumdar and Sundaram [12]. The mentioned papers deal with dynamic programming or discounted stochastic games only. There are also some papers devoted to nonzero-sum average payoff stochastic games with uncountable state spaces. Dutta and Sundaram [13] studied a class of dynamic economic games. They proved the existence of a stationary Nash equilibrium in a class of games satisfying a number of specific conditions (which are different from our assumptions in this paper) and a convergence condition imposed on discounted Nash equilibria as the discount factor tends to one. Lai and Tanaka [14, 15] considered two classes of nonzero-sum average payoff stochastic games with separable metric state spaces under a recurrence condition which is much stronger than uniform geometric ergodicity. It should be noted that there are some serious gaps in the proofs of the main results in [14, 15], mainly in the parts dealing with upper semi-continuity of the best response correspondences. With some minor changes Lai and Tanaka's proofs can be accepted for a class of nonzero-sum stochastic games with countable state spaces. Ghosh and Bagchi [16] studied games under some separability assumptions and a recurrence condition which is much stronger than uniform geometric ergodicity.

This paper treats of nonzero-sum average payoff stochastic games in arbitrary measurable state space and with compact metric action spaces for the players, satisfying some natural continuity conditions. A standard uniform geometric ergodicity condition is imposed on the transition probabilities induced by stationary strategies of the players. The main purpose of this paper is to extend a recent result of the author and Raghavan on the existence of stationary equilibria with public information in discounted stochastic games to the average payoff games. Nash equilibria are also shown to exist in some special cases. In the sequel, we give some further comments on our results and the existing literature on nonzero-sum stochastic games.

To describe our model, we need the following definition. Let X be a metric space, (S, Σ) a measurable space. A set-valued map or a correspondence F from S into a family of subsets of X is said to be *lower measurable* if for any open subset G of X the set $\{s \in S : F(s) \cap G \neq \emptyset\}$ belongs to Σ. For a broad discussion of lower measurable correspondences with some applications to control and optimization theory consult Castaing and Valadier [17]. In this paper we will also use some auxiliary facts concerning correspondences and their measurable selections stated in Nowak and Raghavan [18].

An *N-person nonzero-sum stochastic game* is defined by the following objects: $(S, \Sigma), X_k, A_k, r_k, q$ with the interpretation that

(i) (S, Σ) is a measurable space, where S is the *set of states* for the game, and Σ is a countably generated σ-algebra of subsets of S.

(ii) X_k is a non-empty compact metric *space of actions* for player k. We put $X = X_1 \times X_2 \times \cdots \times X_N$.

(iii) A_k's are lower measurable correspondences from S into non-empty compact subsets of X_k. For each $s \in S$, $A_k(s)$ represents the *set of actions available* to player k in state s. We put

$$A(s) = A_1(s) \times A_2(s) \times \cdots \times A_N(s), \quad s \in S.$$

(iv) $r_k : S \times X \to R$ is a bounded product measurable *payoff function* for player k. It is assumed that $r_k(s, \cdot)$ is continuous on X, for every $s \in S$.

(v) q is a product measurable transition probability from $S \times X$ to S, called the *law of motion* among states. If s is a state at some stage of the game and the players select an $x \in A(s)$, then $q(\cdot \mid s, x)$ is the probability distribution of the next state of the game. We assume that the transition probability q has a density function, say z, with respect to a fixed probability measure μ on (S, Σ), satisfying the following L_1 continuity condition:
For any sequence of joint action tuples $x^n \to x^0$,

$$\int_S \mid z(s,t,x^n) - z(s,t,x^0) \mid \mu(dt) \to 0 \quad \text{as } n \to \infty.$$

The L_1 continuity above is satisfied via Scheffe's theorem when $z(s,t,\cdot)$ is continuous on X. It implies the norm continuity of the transition probability $q(\cdot \mid s, x)$ with respect to $x \in X$.

The game is played in discrete time with past history as common knowledge for all the players. An *individual strategy* for a player is a map which associates with each given history a probability distribution on the set of available to him actions. A *stationary strategy* for player k is a map which associates with each state $s \in S$ a probability distribution on the set $A_k(s)$ of actions available to him at s, independent of the history that lead to the state s. A stationary strategy for player k can thus be identified with a measurable transition probability f from S to X_k such that $f(A_k(s) \mid s) = 1$, for every $s \in S$.

Let $H = S \times X \times S \times \cdots$ be the space of all infinite histories of the game, endowed with the product σ-algebra. For any profile of strategies $\pi = (\pi_1, \ldots, \pi_N)$ of the players and every initial state $s_1 = s \in S$, a probability measure P_s^π and a stochastic process $\{\sigma_n, \alpha_n\}$ are defined on H in a canonical way, where the random variables σ_n and α_n describe the state and the actions chosen by the players, respectively, on the n-th stage of the game (cf. Chapter 7 in Bertsekas and Shreve [19]). Thus, for each profile of strategies $\pi = (\pi_1, \ldots, \pi_N)$, any finite horizon T, and every initial state $s \in S$, the *expected T-stage payoff* to player k is

$$E_k^T(\pi)(s) = E_s^\pi(\sum_{n=1}^T r_k(\sigma_n, \alpha_n)).$$

Here E_s^π means the expectation operator with respect to the probability measure P_s^π. If β is a fixed real number in $(0,1)$, called the *discount factor*, then we can also consider the *β-discounted expected payoff* to player k defined as

$$E_k^\beta(\pi)(s) = E_s^\pi\left(\sum_{n=1}^\infty \beta^{n-1} r_k(\sigma_n, \alpha_n)\right).$$

The games with β-discounted payoffs will play a crucial role in this paper.

The *average payoff per unit time* for player k is defined as

$$E_k(\pi)(s) = \limsup_T \frac{1}{T} E_k^T(\pi)(s).$$

Let $\pi^* = (\pi_1^*, \ldots, \pi_N^*)$ be a fixed profile of strategies of the players. For any strategy π_k of player k, we write (π_{-k}^*, π_k) to denote the strategy profile obtained from π^* by replacing π_k^* with π_k.

A strategy profile $\pi^* = (\pi_1^*, \ldots, \pi_N^*)$ is called a *Nash equilibrium* for the average payoff stochastic game if no unilateral deviations from it are profitable, that is, for each $s \in S$,

$$E_k(\pi^*)(s) \geq E_k(\pi_{-k}^*, \pi_k)(s),$$

for every player k and any his strategy π_k. Of course, Nash equilibria are analogously defined for the β-discounted stochastic games.

It is still an *open problem* whether the β-discounted stochastic games with uncountable state space have stationary equilibrium solutions. A positive answer to this problem is known only for some special classes of games, where the transition probabilities satisfy certain additional separability assumptions (cf. Himmelberg et al., [20]), or some other specific conditions (cf. Majumdar and Sundaram [12], Dutta and Sundaram [13], Karatzas et al., [11]). Whitt [21] and Nowak [22] proved the existence of stationary ϵ-equilibrium strategies in discounted stochastic games using some (different) approximations by games with countably many states. The assumptions on the model in Nowak [22] are as in (i) - (v) above plus some extra integrability condition on the transition probability density. Whitt [21] assumed that the state spaces are separable metric and imposed some uniform continuity conditions on the payoffs and transition probabilities. Breton and L'Ecuyer [23] extended Whitt's result to games with a weaker form of discounting. Mertens and Parthasarathy [24] proved the existence of nonstationary Nash equilibria for discounted stochastic games with arbitrary state spaces. Finally, Nowak and Raghavan [18] obtained stationary equilibrium solutions in the class of correlated strategies of the players with symmetric information or "public signals" (see Theorem 1 in Section 2). A related result is reported in Duffie et al. [9]. They used some stronger assumptions about the primitive data of the game, but showed that there exists a stationary correlated equilibrium which induces an ergodic process. Nonstationary correlated equilibria in a class of dynamic games with weakly continuous transition probabilities were studied by Harris [25]. As already mentioned, Dutta and Sundaram [13] proved an existence theorem for stationary Nash equilibria in some average payoff dynamic economic games. We note that the assumptions and tools used in their proof are completely different from ours.

2 Correlated equilibria

In this section we extend the sets of strategies available to the players in the sense that we allow them to correlate their choices in a natural way described below. The resulting solution (which we obtain for the average payoff games) is a kind of extensive-form correlated equilibrium (cf. Forges [26]) and was already shown to exist for the β-discounted stochastic games by the author and Raghavan in [18].

Suppose that $\{\xi_n : n \geq 1\}$ is a sequence of so-called *signals*, drawn independently from $[0,1]$ according to the uniform distribution. Suppose that at the beginning of each period n of the game the players are informed not only of the outcome of the preceding period and the current state s_n, but also of ξ_n. Then the information available to them is a vector $h^n = (s_1, \xi_1, x_1, \ldots, s_{n-1}, \xi_{n-1}, x_{n-1}, s_n, \xi_n)$, where $s_i \in S$, $x_i \in A(s_i)$, $i = 1, \ldots, n-1$. We denote the set of such vectors by H^n.

An *extended strategy* for player k is a sequence $\pi_k = (\pi_k^1, \pi_k^2, \ldots)$, where every π_k^n is a (product) measurable transition probability from H^n to X_k such that $\pi_k^n(A_k(s_n) \mid h^n) = 1$ for any history $h^n \in H^n$. (Here s_n is the last state in h^n.) An *extended stationary strategy* for player k is a strategy $\pi_k = (\pi_k^1, \pi_k^2, \ldots)$ such that each π_k^n depends on the current state s_n and the last signal ξ_n only. In other words, a strategy π_k of player k is called stationary if there exists a transition probability f from $S \times [0,1]$ to X_k such that for every period n of the game and each history $h^n \in H^n$, we have $\pi_k^n(\cdot \mid h^n) = f(\cdot \mid s_n, \xi_n)$. Assuming that the players use extended strategies we actually assume that they play a stochastic game in the sense of Section 1, but with the extended state space $S \times [0,1]$. The law of motion, say $\bar{q}$, in the extended state space model is obviously the product of the original law of motion q and the uniform distribution η on $[0,1]$. More precisely, for any $s \in S$, $\xi \in [0,1]$, $a \in A(s)$, and any Borel sets $C \subseteq S$ and $D \subseteq [0,1]$, $\bar{q}(C \times D \mid s, \xi, a) = q(C \mid s, a)\eta(D)$.

For any profile of extended strategies $\pi = (\pi_1, \ldots, \pi_N)$ of the players, the average [β-discounted] payoff to player k is a function of the initial state s_1 and the first signal ξ_1 and is denoted by $E_k(\pi)(s_1,\xi_1)$ $[E_k^\beta(\pi)(s_1,\xi_1)]$.

We say that $f^* = (f_1^*,\ldots,f_N^*)$ is a *Nash equilibrium in the average payoff stochastic game in the class of extended strategies* if for each initial state $s_1 \in S$,

$$\int_0^1 E_k(f^*)(s_1,\xi_1)\eta(d\xi_1) \geq \int_0^1 E_k(f_{-k}^*,\pi_k)(s_1,\xi_1)\eta(d\xi_1), \tag{1}$$

for every player k and any his extended strategy π_k.

A Nash equilibrium in extended strategies will also be called a *correlated equilibrium with public signals.* The reason is that after the outcome of any period of the game, the players can coordinate their next choices by exploiting the next (known to all of them, i.e.,public) signal and using some coordination mechanism telling which (pure or mixed) action is to be played by each of them. In this paper, we are particularly interested in stationary equilibria. In such a case the coordination mechanism can be represented by a family of $N+1$ measurable functions $\lambda^1,\ldots,\lambda^{N+1} : S \to [0,1]$ such that $\sum_{i=1}^{N+1}\lambda^i(s) = 1$ for every $s \in S$. (We remind that N is the number of players. The number $N+1$ appears in our definition, because we apply Caratheodory's theorem in the proof of our main result.) A stationary Nash equilibrium in the class of extended strategies can be constructed then by using a family of $N+1$ stationary strategies $f_k^1,\ldots,f_k^{N+1}$, given for each player k, and the following coordination rule. If the game is at a state s on the n-th stage and a random number ξ_n is selected, then each player k is suggested to use $f_k^m(\cdot \mid s)$, where m is the least index for which $\sum_{i=1}^m \lambda^i(s) \geq \xi_n$. The λ^i's and f_k^i's fixed above induce an extended stationary strategy f_k^* for each player k as follows

$$f_k^*(\cdot \mid s,\xi) = f_k^1(\cdot \mid s) \qquad \text{if} \quad \xi \leq \lambda^1(s),\ s \in S,$$

and (2)

$$f_k^*(\cdot \mid s,\xi) = f_k^m(\cdot \mid s) \qquad \text{if} \quad \textstyle\sum_{i=1}^{m-1}\lambda^i(s) < \xi \leq \sum_{i=1}^m \lambda^i(s),$$

for $s \in S$, $2 \leq m \leq N+1$. Because the signals are independent and uniformly distributed in $[0,1]$, it follows that at any period of the game and for any current state s, the number $\lambda^i(s)$ can be interpreted as the probability that player k is suggested to use $f_k^i(\cdot \mid s)$ as his mixed action. Now it is quite obvious that a strategy profile $(f_1^*,\ldots,f_N^*)$ obtained by the above construction is a stationary Nash equilibrium in the class of extended strategies of the players in a game iff no player k can unilaterally improve upon his expected payoff by changing any of his strategies f_k^i, $i = 1,\ldots,N+1$.

The following result was proved by Nowak and Raghavan [18].

Theorem 1 *Every nonzero-sum discounted stochastic game satisfying (i) - (v) has a stationary correlated equilibrium with public signals.*

To state our main result, we need some additional assumptions on the transition probability q. For any stationary strategy profile f and $n \geq 1$, let $q^n(\cdot \mid s, f)$ denote the n-step transition probability determined by q and f. The following condition is used in the theory of Markov decision processes (cf. Tweedie [27], Hernández-Lerma et al. [28, 29] and their references):

C1 (*Uniform Geometric Ergodicity*): There exist scalars $\alpha \in (0,1)$ and $\gamma > 0$ for which the following holds: For any profile f of stationary strategies of the players, there exists a probability measure p_f on S such that

$$||q^n(\cdot \mid s, f) - p_f(\cdot)||_\nu \leq \gamma\alpha^n \qquad \text{for each } n \geq 1.$$

Here $|| \cdot ||_\nu$ denotes the total variation norm in the space of probability measures on S.

Note that **C1** implies that for every two states $s, t \in S$, we have

$$||q^n(\cdot \mid s, f) - q^n(\cdot \mid t, f)||_\nu \leq 2\gamma\alpha^n \qquad \text{for each } n \geq 1.$$

Using this, we easily show that, for each discount factor $\beta \in (0,1)$, every player k, any stationary strategy profile f, and for all $s, t \in S$,

$$\begin{aligned} \mid E_k^\beta(f)(s) - E_k^\beta(f)(t) \mid &\leq 2||r_k||[1 + \gamma(\alpha\beta + \alpha^2\beta^2 + \cdots)] \\ &\leq L := \max_k 2||r_k|| \frac{(1-\alpha+\gamma)}{(1-\alpha)}, \end{aligned} \tag{3}$$

where $||r_k||$ is the supremum norm of the bounded payoff function r_k of player k.

It is well-known that **C1** follows from the following assumption (cf. Theorem 6.15 and Remark 6.1 in Nummelin [30] or page 185 in Neveu [31]):

M (*Minorization Property*): There exists a positive integer p, a constant $\vartheta > 0$, and a probability measure δ on S, such that

$$q^p(D \mid s, f) \geq \vartheta\delta(D),$$

for every stationary strategy profile $f, s \in S$, and for each Borel set $D \subseteq S$.

Condition **M** was used in stochastic dynamic programming (one person stochastic game) by Kurano [32] who proved only the existence of "p-periodic" optimal strategies in his model. It is satisfied and easy to verify in some inventory models (cf. Yamada [33]) and some control of water resources problems (cf. Yakovitz [34]). In Section 4, we give some additional comments on **C1**.

We now arrived at our main result.

Theorem 2 *Every nonzero-sum average payoff stochastic game satisfying (i) – (v) and* **C1** *has a stationary correlated equilibrium with public signals.*

Proof. For each bounded R^N-valued function v on S and any state $s \in S$, we define the game $\Gamma_\beta(v)(s)$, in which the payoff for player k is given by

$$r_k(s,x) + \beta \int_S v_k(s')q(ds' \mid s,x), \tag{4}$$

where $x = (x_1, \ldots, x_N) \in A(s)$ and v_k is the k-th component of v. By $coP_\beta(v)(s)$, we denote the convex-hull of all Nash equilibrium payoff vectors in the game $\Gamma_\beta(v)(s)$. From Nowak and Raghavan [18], it follows that for any $\beta \in (0,1)$, there exists a measurable $v_\beta : S \to R^N$ such that

$$v_\beta(s) \in coP_\beta(v_\beta)(s) \qquad \text{for all } s \in S. \tag{5}$$

We fix some state $t \in S$, and for each β, we fix a v_β satisfying (5). Now we define

$$w_\beta(s) = v_\beta(s) - v_\beta(t), \qquad s \in S.$$

We now define W as the space of all μ-equivalence classes of measurable functions $v : S \to R$ such that $\mid v(s) \mid \leq L$ μ-a.e., where L is the constant fixed in (3). It is obvious that W is a compact and metrizable subset of $L_\infty = L_\infty(S, \Sigma, \mu)$, when endowed with the weak star-topology $\sigma(L_\infty, L_1)$. (We remind that Σ is assumed to be countably generated, which implies that $L_1 = L_1(S, \Sigma, \mu)$ is separable and, consequently, that W is metrizable.) Let $W^N = W \times W \times \cdots \times W$ (N times). We endow W^N with the product topology, so W^N is a compact metrizable space too. Assuming **C1**, we can use (3), and conclude that $w_\beta \in W^N$ for every β. By compactness of W^N, there exists a sequence $\{\beta_n\}$ converging to one such that w_{β_n} converges to some $w^* \in W^N$ as $n \to \infty$.

Because the payoffs r_k are bounded, $\{(1-\beta_n)v_{\beta_n}(t)\}$ is a bounded subset of R^N and thus has a subsequence, converging to some vector $J^* \in R^N$. Without loss of generality, we assume that $(1-\beta_n)v_{\beta_n}(t) \to J^*$ as $n \to \infty$. From (5), it follows that

$$(w_{\beta_n}(s) + (1-\beta_n)v_{\beta_n}(t)) \in coP_{\beta_n}(w_{\beta_n})(s), \qquad \text{for all } s \in S. \tag{6}$$

Because w_{β_n} converges to w^* in W^N, some sequence of convex combinations of the functions w_{β_n} converges to w^* μ-a.e.. Using the arguments employed in the proof of Lemma 7 in Nowak and Raghavan [18], we conclude from (6) that

$$w^*(s) + J^* \in coP(w^*)(s) \quad \mu\text{-a.e.}, \tag{7}$$

where $coP(w^*)(s)$ is the convex-hull of Nash equilibrium payoff vectors in the game (4) with $\beta = 1$ and $v = w^*$. (Here we use the assumptions formulated in (v) and saying that the transition probability q satisfies the L_1 continuity condition.) From (7) and a random version of

Caratheodory's theorem (cf. Castaing and Valadier [17]), there exist $N+1$ measurable functions $\lambda^1, \ldots, \lambda^{N+1} : S \to [0,1]$ such that $\sum_{i=1}^{N+1} \lambda^i(s) = 1$, $s \in S$, and there exist measurable maps $u^1, \ldots, u^{N+1} : S \to R^N$ such that $u^i(s) \in P(w^*)(s)$, the set of all Nash equilibrium payoff vectors in the game $\Gamma(w^*)(s) = \Gamma_\beta(w^*)(s)$ with $\beta = 1$, and $w^*(s) + J^* = \sum_{i=1}^{N+1} \lambda^i(s) u^i(s)$ μ-a.e.. By a measurable implicit function theorem (cf. Himmelberg [35], or Lemma 4 in Nowak and Raghavan [18]), for every i, there exist stationary strategies $f_1^i, \ldots, f_N^i$ for the players such that $f^i(\cdot \mid s)$ defined as the product measure $f_1^i(\cdot \mid s) \times \cdots \times f_N^i(\cdot \mid s)$ is a Nash equilibrium in the game $\Gamma(w^*)(s)$, for each $s \in S$, and there exists a measurable subset B of S with $\mu(B) = 1$ such that for all $s \in B$,

$$u_k^i(s) = r_k(s, f^i) + \int_S w_k^*(s') q(ds' \mid s, f^i), \qquad i = 1, \ldots, N+1. \tag{8}$$

Here u_k^i, w_k^* are the k-th components of u^i and w^*, respectively, and $r_k(s, f^i)$ is the expectation of $r_k(s, \cdot)$ with respect to $f^i(\cdot \mid s)$. Obviously, B can be taken in such a way that $w^*(s) + J^* = \sum_{i=1}^{N+1} \lambda^i(s) u^i(s)$ for all $s \in B$. If we put $f^\lambda(\cdot \mid s) = \sum_{i=1}^{N+1} \lambda^i(s) f^i(\cdot \mid s)$, we get from (8) that

$$w_k^*(s) + J_k^* = r_k(s, f^\lambda) + \int_S w_k^*(s') q(ds' \mid s, f^\lambda), \quad s \in B. \tag{9}$$

Here J_k^* is the k-th component of J^*. Now using the f_k^i's introduced above, we define the extended stationary strategies f_k^* $(k = 1, \ldots, N)$ according to (2) and set $f^* = (f_1^*, \ldots, f_N^*)$. For every state $s \in S$ and for each signal ξ, we write $r_k(s, \xi, f^*)$ and $q(\cdot \mid s, \xi, f^*)$ for the one-stage payoff to player k and the transition probability, respectively, which correspond with the extended stationary strategy profile f^*. Similarly, we define $r_k(s, \xi, f_{-k}^*, \varphi_k)$ and $q(\cdot \mid s, \xi, f_{-k}^*, \varphi_k)$ for any extended stationary strategy φ_k of player k. Now (9) can be rewritten as

$$\begin{aligned} & w_k^*(s_1) + J_k^* \\ = & \int_0^1 \left[r_k(s_1, \xi_1, f^*) + \int_S w_k^*(s_2) q(ds_2 \mid s_1, \xi_1, f^*) \right] \eta(d\xi_1), \end{aligned} \tag{10}$$

for every player k and each $s_1 \in B$. For each player k and any his extended stationary strategy φ_k, we can conclude from the fact that every $f^i(\cdot \mid s)$ is a Nash equilibrium in the game $\Gamma(w^*)(s)$ that

$$\begin{aligned} & r_k(s_1, \xi_1, f^*) + \int_S w_k^*(s_2) q(ds_2 \mid s_1, \xi_1, f^*) \\ \geq \; & r_k(s_1, \xi_1, f_{-k}^*, \varphi_k) + \int_S w_k^*(s_2) q(ds_2 \mid s_1, \xi_1, f_{-k}^*, \varphi_k), \end{aligned}$$

for all $s_1 \in S$ and $\xi_1 \in [0,1]$. This and (10) imply

$$
\begin{aligned}
& w_k^*(s_1) + J_k^* \\
\geq & \int_0^1 \left[r_k(s_1, \xi_1, f_{-k}^*, \varphi_k) + \int_S w_k^*(s_2) q(ds_2 \mid s_1, \xi_1, f_{-k}^*, \varphi_k) \right] \eta(d\xi_1),
\end{aligned} \tag{11}
$$

for each extended stationary strategy φ_k of player k and $s_1 \in B$. Because the transition probability q is absolutely continuous with respect to μ, we can iterate (10) and, in such a way, we obtain for each finite horizon T that

$$
\begin{aligned}
& w_k^*(s_1) + TJ_k^* \\
= & \int_0^1 \left[E_k^T(f^*)(s_1, \xi_1) + \int_S w_k^*(s_{T+1}) \bar{q}^T(ds_{T+1} \mid s_1, \xi_1, f^*) \right] \eta(d\xi_1),
\end{aligned} \tag{12}
$$

for all $s_1 \in B$, where $\bar{q}^T(\cdot \mid s_1, \xi_1, f^*)$ is the T-step transition probability in the game with the extended state space $S \times [0,1]$, mentioned earlier, determined by q, η, and f^*. $E_k^T(f^*)(s_1, \xi_1)$ is the T-stage payoff of player k corresponding to f^*, s_1 and ξ_1. Because w_k^* is bounded, from (12), we infer that

$$
J_k^* = \int_0^1 E_k(f^*)(s_1, \xi_1) \eta(d\xi_1), \quad s_1 \in B, \tag{13}
$$

for every player k. Using (11) and a similar iteration argument, we get

$$
J_k^* \geq \int_0^1 E_k(f_{-k}^*, \varphi_k)(s_1, \xi_1) \eta(d\xi_1), \quad s_1 \in B, \tag{14}
$$

for every player k and any his extended stationary strategy φ_k. We remind that $q \ll \mu$. Thus, if the game starts at some state in $S \backslash B$, then the chance that it will remain in $S \backslash B$ on the next stage is zero no matter what the players do. Because we deal with the (limiting) average payoffs, it follows that (13) and (14) hold for every $s \in S$, which in turn implies that f^* is a Nash equilibrium in the class of all extended stationary strategies of the players. It is quite obvious that (10) and (11) correspond to the optimality equation in the average payoff dynamic programming. Thus, using standard arguments from dynamic programming (cf. Ross [36], Hernández-Lerma et al. [28] or Kurano [32]) one can show that f^* is a Nash equilibrium in the class of all extended strategies of the players, that is, (1) holds.

□

3 Stationary Nash equilibria

In this section we study some special classes of nonzero-sum average payoff stochastic games, for which there exist Nash equilibria without public signals. First, we consider games satisfying the following separability conditions:

SC1: For each player k and any $s \in S$, $x = (x_1, \ldots, x_N) \in A(s)$,

$$r_k(s,x) = \sum_{j=1}^{N} r_{kj}(s, x_j),$$

where each r_{kj} is bounded and $r_{kj}(s, \cdot)$ is continuous on X_j.

SC2: For any $s \in S$, $x = (x_1, \ldots, x_N) \in A(s)$,

$$q(\cdot \mid s, x) = \sum_{j=1}^{N} q_j(\cdot \mid s, x_j)/N,$$

where $q(\cdot \mid s, x_j)$ is a transition probability from $S \times X_j$ to S, norm continuous with respect to $x_j \in X_j$.

Himmelberg et al. [20] and Parthasarathy [37] already showed that nonzero-sum β-discounted stochastic games satisfying **SC1** and **SC2** possess stationary Nash equilibria. Our next result extends their theorem to the average payoff stochastic games.

Theorem 3 *Every nonzero-sum average payoff stochastic game satisfying (i) – (v),* **C1** *and separability conditions* **SC1** *and* **SC2** *has a stationary Nash equilibrium without public signals.*

Proof. As in the proof of Theorem 2, we can find some w^* satisfying (7), and a transition probability f^λ satisfying (9). Next, for each player k and any $s \in S$, we may define $f_k^*(\cdot \mid s)$ to be the marginal of $f^\lambda(\cdot \mid s)$ on X_k and put $f^* = (f_1^*, \ldots, f_N^*)$. Observe that, under our separability assumptions, (9) implies that

$$w_k^*(s) + J_k^* = r_k(s, f^*) + \int_S w_k^*(t) q(dt \mid s, f^*), \quad \mu \text{ - a.e.},$$

and $f_1^*(\cdot \mid s) \times \cdots \times f_N^*(\cdot \mid s)$ is a Nash equilibrium in the game $\Gamma(w^*)(s)$, for any $s \in S$. The rest of the proof uses standard iteration arguments. □

A version of Theorem 3 with a recurrence assumption which is much stronger than the uniform geometric ergodicity was independently proved (by a different method) in Ghosh and Bagchi [16].

Parthasarathy and Sinha [38] showed that β-discounted stochastic games with state independent transitions and finite action spaces have stationary Nash equilibria. We extend their result in this paper to the average payoff stochastic games, assuming the uniform geometric ergodicity condition.

Theorem 4 *Assume that the action spaces X_k are finite sets and $A_k(s) = X_k$ for each $s \in S$. Assume also that the transition probability $q(\cdot \mid s, x)$ depends on x only and is non-atomic for each $x \in X$. If (i), (iv), (v), and* **C1** *are also satisfied, then the nonzero-sum average payoff stochastic game has a stationary Nash equilibrium without public signals.*

Proof. From the proof of Theorem 2, we can carry over (7) and (9) to the present situation. The basic idea then is to replace f^λ in (9) by some stationary Nash equilibrium for the games $\Gamma(w^*)(s)$ for μ-a.e. $s \in S$. This can be done, similarly as in Parthasarathy and Sinha [38], by applying Lyapunov's theorem. Here we use the assumption that X is finite and q is independent of s and non-atomic. The concluding part of the proof consists of standard iteration arguments.

□

We do not know if condition **C1** can be dropped from Theorem 4. When we deal with zero-sum average payoff stochastic games with state independent transitions, then no ergodicity properties of the transition probability q are relevant (cf. Thuijsman [4] for the finite state space case or Theorem 2 in Nowak [39] for general state space games).

4 Comments

The last section of this paper is devoted to a discussion of **C1** - the uniform geometric ergodicity assumption. It has often been used (even in some stronger versions) in control theory of Markov chains (cf. Georgin [40], Hernández-Lerma et al. [28, 29], and the references therein). We now present some conditions which are known to be equivalent to **C1**. By F we denote the set of all stationary strategy N-tuples of the players.

C2 (*Uniform Ergodicity*): For each $f \in F$, there exists a probability measure p_f on S such that, as $n \to \infty$,

$$||q^n(\cdot \mid s, f) - p_f(\cdot)||_\nu \to 0, \quad \text{uniformly in } s \in S \text{ and } f \in F.$$

C3: There exist a positive integer r and a positive number $\delta < 1$ such that

$$||q^r(\cdot \mid s, f) - q^r(\cdot \mid t, f)||_\nu \leq 2\delta, \quad \text{for all } s, t \in S \text{ and } f \in F.$$

Obviously **C1** implies **C2** and **C3** follows immediadely from **C2** and the triangle inequality for the norm $|| \cdot ||_\nu$. Finally, **C3** implies **C1** by Ueno's lemma [41]. For details consult pages 275-276 in [40].

Another equivalent version of **C1**, called the *simultaneous Doeblin condition*, was used by Hordijk [42] in control theory and Federgruen [7] in stochastic games with countably many states. It can also be formulated for general state space stochastic games following pages 192 and 221 in Doob [43].

C4: There is a probability measure ψ on S, a positive integer r, and a positive ϵ, such that

$$q^r(B \mid s, f) \leq 1 - \epsilon \quad \text{for each } s \in S \text{ and } f \in F \text{ if } \psi(B) \leq \epsilon.$$

Moreover, for each $f \in F$, the Markov chain induced by q and f has a single ergodic set and this set contains no cyclically moving subsets.

It turns out that **C1** is equivalent to **C4**; see Chapter V in Doob [43] for details. For a further discussion of several recurrence and ergodicity conditions which have beed used in the theory of Markov decision processes in a general state space consult Hernández-Lerma et al. [29].

Acknowledgement. I would like to thank T.E.S. Raghavan for several stimulating discussions on undiscounted stochastic games and the referees for some helpful comments.

Andrzej S. Nowak
Institute of Mathematics
Technical University of Wrocław
Wybrzeże Wyspiańskiego 27
PL-50-370 Wrocław, Poland

REFERENCES

[1] P. D. Rogers, *Nonzero-Sum Stochastic Games.* PhD thesis, University of California, Berkeley, 1969. Report ORC 69-8.

[2] M. Sobel, "Noncooperative stochastic games," *Ann. Math. Statist.*, vol. 42, pp. 1930 – 1935, 1971.

[3] T. Parthasarathy and T. E. S. Raghavan, "An orderfield property for stochastic games when one player controls transition probabilities," *J. Optim. Theory Appl.*, vol. 33, pp. 375 – 392, 1981.

[4] F. Thuijsman, *Optimality and Equilibria in Stochastic Games.* PhD thesis, University of Limburg, Maastricht, The Netherlands, 1989.

[5] O. J. Vrieze and F. Thuijsman, "On equilibria in repeated games with absorbing states," *Internat. J. Game Theory*, vol. 18, pp. 293 – 310, 1989.

[6] T. Parthasarathy, "Discounted, positive, and non-cooperative stochastic games," *Internat. J. Game Theory*, vol. 2, pp. 25 – 37, 1973.

[7] A. Federgruen, "On n-person stochastic games with denumerable state space," *Adv. Appl. Probab.*, vol. 10, pp. 452 – 471, 1978.

[8] A. S. Nowak, "Stationary overtaking equilibria for non-zero-sum stochastic games with countable state spaces," mimeo, 1992.

[9] D. Duffie, J. Geanakoplos, A. Mas-Colell, and A. McLennan, "Stationary Markov equilibria," Technical Report, Dept. of Economics, Harvard University, 1988.

[10] P. K. Dutta, "What do discounted optima converge to? A theory of discount rate asymptotics in economics models," *J. Economic Theory*, vol. 55, pp. 64 – 94, 1991.

[11] I. Karatzas, M. Shubik, and W. D. Sudderth, "Construction of stationary Markov equilibria in a strategic market game," Technical Report 92-05-022, Santa Fe Institute Working Paper, Santa Fe, New Mexico, 1992.

[12] M. Majumdar and R. Sundaram, "Symmetric stochastic games of resource extraction: The existence of non-randomized stationary equilibrium," in *Stochastic Games and Related Topics*, pp. 175 – 190, Dordrecht, The Netherlands: Kluwer Academic Publishers, 1991.

[13] P. K. Dutta and R. Sundaram, "Markovian equilibrium in a class of stochastic games: Existence theorems for discounted and undiscounted models," *Economic Theory*, vol. 2, 1992.

[14] H. C. Lai and K. Tanaka, "A noncooperative n-person semi-Markov game with a separable metric state space," *Appl. Math. Optim.*, vol. 11, pp. 23 – 42, 1984.

[15] H. C. Lai and K. Tanaka, "On an n-person noncooperative Markov game with a metric state space," *J. Math. Anal. Appl.*, vol. 101, pp. 78 – 96, 1984.

[16] M. K. Ghosh and A. Bagchi, "Stochastic games with average payoff criterion," Technical Report 985, Faculty of Applied Mathematics, University of Twente, Enschede, The Netherlands, 1991.

[17] C. Castaing and M. Valadier, *Convex Analysis and Measurable Multifunctions*, vol. 580 of *Lecture Notes in Mathematics*. New York: Springer-Verlag, 1977.

[18] A. S. Nowak and T. E. S. Raghavan, "Existence of stationary correlated equilibria with symmetric information for discounted stochastic games," *Math. Oper. Res.*, vol. 17, pp. 519 – 526, 1992.

[19] D. P. Bertsekas and S. E. Shreve, *Stochastic Optimal Control: The Discrete Time Case.* New York: Academic Press, 1978.

[20] C. J. Himmelberg, T. Parthasarathy, T. E. S. Raghavan, and F. S. van Vleck, "Existence of p-equilibrium and optimal stationary strategies in stochastic games," *Proc. Amer. Math. Soc.*, vol. 60, pp. 245 – 251, 1976.

[21] W. Whitt, "Representation and approximation of noncooperative sequential games," *SIAM J. Control Optim.*, vol. 18, pp. 33 – 48, 1980.

[22] A. S. Nowak, "Existence of equilibrium stationary strategies in discounted noncooperative stochastic games with uncountable state space," *J. Optim. Theory Appl.*, vol. 45, pp. 591 – 602, 1985.

[23] M. Breton and P. L'Ecuyer, "Noncooperative stochastic games under a n-stage local contraction assumption," *Stochastics and Stochastic Reports*, vol. 26, pp. 227 – 245, 1989.

[24] J.-F. Mertens and T. Parthasarathy, "Equilibria for discounted stochastic games," Technical Report 8750, CORE Discussion Paper, Université Catholique de Louvain, 1987.

[25] C. Harris, "The existence of subgame-perfect equilibrium in games with simultaneous moves: a case for extensive-form correlation," mimeo, Nuffield College, Oxford, U.K., 1990.

[26] F. Forges, "An approach to communication equilibria," *Econometrica*, vol. 54, pp. 1375 – 1385, 1986.

[27] R. L. Tweedie, "Criteria for rates of convergence of Markov chains, with an application to queueing and storage theory," in *Papers in Probability Statistics and Analysis* (J. F. C. Kingman and G. E. H. Reuter, eds.), pp. 260 – 276, Cambridge, U. K.: Cambridge University Press, 1983.

[28] O. Hernández-Lerma, J. C. Hennet, and J. B. Lasserre, "Average cost Markov decision processes: Optimality conditions," *J. Math. Anal. Appl.*, vol. 158, pp. 396 – 406, 1991.

[29] O. Hernández-Lerma, R. Montes-de-Oca, and R. Cavazos-Cadena, "Recurrence conditions for Markov decision processes with Borel state space," *Ann. Oper. Res.*, vol. 28, pp. 29 – 46, 1991.

[30] E. Nummelin, *General Irreducible Markov Chains and Non-Negative Operators.* London: Cambridge Univ. Press, 1984.

[31] J. Neveu, *Mathematical Foundations of the Calculus of Probability.* San Francisco: Holden-Day, 1965.

[32] M. Kurano, "Markov decision processes with a Borel measurable cost function - the average case," *Math. Oper. Res.*, vol. 11, pp. 309 – 320, 1986.

[33] K. Yamada, "Duality theorem in Markovian decision problems," *J. Math. Anal. Appl.*, vol. 50, pp. 579 – 595, 1975.

[34] S. Yakovitz, "Dynamic programming applications in water resources," *Water Resources Res.*, vol. 18, pp. 673 – 696, 1982.

[35] C. J. Himmelberg, "Measurable relations," *Fund. Math*, vol. 87, pp. 53 – 72, 1975.

[36] S. M. Ross, "Arbitrary state Markovian decision processes," *Ann. Math. Statist.*, vol. 39, pp. 2118 – 2122, 1968.

[37] T. Parthasarathy, "Existence of equilibrium stationary strategies in discounted stochastic games," *Sankhya Series A*, vol. 44, pp. 114 – 127, 1982.

[38] T. Parthasarathy and S. Sinha, "Existence of stationary equilibrium strategies in non-zero-sum discounted stochastic games with uncountable state space and state independent transitions," *Internat. J. Game Theory*, vol. 18, pp. 189 – 194, 1989.

[39] A. S. Nowak, "Zero-sum average payoff stochastic games with general state space," *Games and Economic Behavior*, (to appear) 1992.

[40] J. Georgin, "Controle de chaines de Markov sur des espaces arbitraires," *Ann. Inst. H. Poincare*, vol. 14, pp. 255 – 277, 1978.

[41] T. Ueno, "Some limit theorems for temporally discrete Markov processes," *J. Fac. Science Univ. Tokyo*, vol. 7, pp. 449 – 462, 1957.

[42] A. Hordijk, *Dynamic Programming and Markov Potential Theory.* Amsterdam: Math. Centrum, 1977.

[43] J. L. Doob, *Stochastic Processes.* New York: Wiley, 1953.

Overtaking Equilibria for Switching Regulator and Tracking Games

D. Carlson, A. Haurie, and A. Leizarowitz

Abstract. We consider a nonzero-sum stochastic differential game with linear dynamics and quadratic costs. The random disturbances are defined as a Markov chain which switches the dynamics' modes and the targets which are tracked by the different players. Due to these random switches the overtaking equilibrium concept has to be used when the time horizon is infinite. We characterize an affine feedback overtaking equilibrium for the p-player case, prove existence for the completely symmetric game case and establish a comparison theorem for the solution of the coupled Riccati equations in the two-player deterministic case.

1. Introduction

This paper deals with a class of infinite horizon differential games where p players control a linear dynamical system, subject to jump random disturbances and strive to minimize the expectation of quadratic cost functions. These quadratic costs correspond to the tracking by the different players of different stationary nominal states called *targets*. We assume that the jump disturbances affect the system's dynamics and also the nominal state values entering in the definition of the quadratic cost functions. Due to these switches between the different nominal stationary states the expected quadratic cost tends to infinity with the time horizon for every player. This motivates the use of the *overtaking optimality* concept which is introduced in Section 2.

Although the linear-quadratic differential game has been studied extensively (see [4]) the introduction of the overtaking optimality concept is quite recent, even for single-player control systems [9]. Our motivation for studying this class of games stems from the problems of policy coordination in international economics or international relations. One can easily imagine a set of p nations aiming at stabilizing their respective economies toward different target levels of, say *unemployment* and *inflation.* The piecewise deterministic structure permits one to consider random variations in the economies' dynamics as well as changes in the targets of the different countries. For example these random switches may correspond to sudden political changes.

The overtaking optimality concept has been introduced in the economic growth literature by Von Weizäcker [16], then further studied by Gale [7], Brock [5] and many others (for a recent survey of the theory of infinite horizon optimal control we refer to the monograph [6]). The particular case of the regulation and tracking of stochastic systems has been fully developed in [9].

The extension of the overtaking optimality concept to noncooperative dynamic games has been first proposed by Rubinstein [14]. It has been used in [8] for the study of a class of open-loop deterministic differential games and in [10] in the context of dynamic games with memory strategy equilibria. More recently Ichikawa [11] extended the theory of tracking a periodic signal developed by Artstein and Leizarowitz [1], with the help of the overtaking criterion concept, to the case of a zero-sum linear quadratic differential game.

The type of stochastic disturbances introduced in this paper, described as a jump Markov process, has been considered for a long time in the optimal control literature (see [15], [17], [13]). This stochastic structure has been introduced in a differential game context by Başar and Haurie [3] who characterized *feedback Stackelberg equilibria.*

The paper is organized as follows: in Section 2 we define the class of games under consideration, we introduce the concept of overtaking optimality and we study the basic dynamic programming equation which characterizes a *feedback equilibrium* in the particular case where the target values are all equal to 0. In Section 3 we investigate the *completely symmetric game case* for which we are able to prove the existence of a solution described by a system of matrix Riccati equations. In Section 4 we characterize the *overtaking equilibrium solution* for the case of a p-player tracking game with random switching targets. Finally In Section 5 we investigate, via a comparison theorem, the dependence of outcomes on the initial cost data in the two player deterministic case.

2. Linear Quadratic Piecewise Deterministic Games

2.1. Overtaking Equilibria

We consider a game with p players. The dynamics of the game is governed by the following linear system of equations

$$\begin{aligned} \dot{x}(t) &= A_{j_t}x(t) + \sum_{k=1}^{p} B_{j_t}^k u_k(t) \text{ a.e. } t \geq 0 \\ x(0) &= x_0,\ j_0 = i,\ x(t) \in \mathbb{R}^n,\ u_k(t) \in \mathbb{R}^{m_k}. \end{aligned} \tag{2.1}$$

Here $\{j_t\}_{t\geq 0}$ is a Markov process on the finite state space $E = \{1, 2, \dots N\}$ with generator $G = (g_{ij})_{i,j=1}^N$. This process is such that any two states communicate on some finite time interval. The matrices A_j, $1 \leq j \leq N$, are $n \times n$ and the matrices B_j^k; $1 \leq j \leq N$, are $n \times m_k$ for $k = 1, 2, \ldots p$. With this notation we define an admissible pair as follows.

Definition 2.1. A trajectory-control pair

$$\{x(\cdot), \mathbf{u}(\cdot)\} \doteq \{x(\cdot), (u_1(\cdot), u_2(\cdot), \ldots u_p(\cdot))\}$$

is called admissible if $x(\cdot) \in AC_{\text{loc}}([0, \infty); \mathbb{R}^n)$ and $u_k(\cdot)$, $1 \leq k \leq p$, is Lebesgue measurable such that (2.1) holds and $t \to E\|x(t)\|^2$ is bounded on $[0, \infty)$. We say that $u(\cdot)$ is an admissible control if $\{x(\cdot), \mathbf{u}(\cdot)\}$ is admissible for some $x(\cdot)$.

The cost over $[0, T]$ for the kth player associated with the controls $u_l(\cdot) : [0, \infty) \to \mathbb{R}^{m_k}$, $l = 1, 2, \ldots, p$ is given by

$$C_T^k(i, x_0; \mathbf{u}(\cdot)) = E_{i,x_0}\left[\int_0^T \|x(t) - \Gamma_{j_t}^k\|_{Q_{j_t}^k}^2 + \|u_k(t)\|_{R_{j_t}^k}^2 \, dt\right] \tag{2.2}$$

where for $k = 1, 2, \ldots p$ and $j = 1, 2, \ldots N$ the matrices Q_j^k and R_j^k are positive definite and symmetric with dimensions $n \times n$ and $m_k \times m_k$ respectively, Γ_j^k are targets associated with different modes and players, and for any $z \in \mathbb{R}^d$ and $d \times d$ matrix K, $\|z\|_K^2 \doteq z' K z$.

Definition 2.2.

(i) A stationary feedback strategy for the k-th player is defined by a continuous feedback function $\gamma_k(\cdot, \cdot) : E \times \mathbb{R}^n \to \mathbb{R}^{m_k}$, such that

$$u_k(t) = \gamma_k(j_t, x(t)) \text{ a.e. } t \geq 0; \tag{2.3}$$

(ii) We say that the strategy $\gamma(\cdot, \cdot) = \{\gamma_k(\cdot, \cdot)\}_{k=1}^p$ is admissible if the resulting $u_k(\cdot)$ in (2.3) form an admissible p-tuple $\{u_k(\cdot)\}_{k=1}^p$. If all players $l = 1, 2, \ldots, p$ use stationary feedback strategies $\gamma_l(\cdot, \cdot)$ then for the initial condition (i, x_0) the cost functionals (2.2) are rewritten in the *normal form*

$$V_T^k(i, x_0; \gamma_1, \gamma_2, \ldots, \gamma_p) = C_T^k(i, x_0; \mathbf{u}(\cdot)). \tag{2.4}$$

Our goal is to study the existence of stationary feedback equilibria for these games in normal form when the time horizon T tends to infinity. Let us introduce the notation $V_T^k(i, x_0; \gamma^{(u_k)})$ to represent the cost for player k when all other players use the feedback $\gamma_l(\cdot, \cdot)$ and player k uses the control $u_k(\cdot)$. Due to the switching targets entering into the cost functional we can not assume *a priori* that the map $T \to C_T^k(\mathbf{u}(\cdot))$ is bounded on $[0, \infty)$.

This leads us to investigate the existence of *overtaking optimal* feedback equilibria as defined below.

Definition 2.3. An admissible strategy $\gamma_*(\cdot,\cdot)$ is called an overtaking feedback equilibrium if for any initial condition (i, x_0) we have

$$\liminf_{T\to\infty}\left[V_T^k(i,x_0;\gamma_*^{(u_k)}) - V_T^k(i,x_0;\gamma_*)\right] \geq 0. \tag{2.5}$$

for every $k = 1, 2, \ldots p$ and for any measurable control function $u_k(\cdot)$: $[0,\infty) \to \mathbb{R}^{m_k}$ which, for which $\gamma_*^{(u_k)}$ is admissible.

Remark 2.1. Since we are dealing with a linear-quadratic game it is natural to look for linear feedback equilibria. However, in Başar [2] an example (on the finite horizon) of a linear-quadratic game in which a non-linear Nash equilibrium exists is given. Additionally, one could also have equilibria based on memory strategies. For such equilibria a threat would be incorporated in the strategy. This threat should be the feedback equilibrium characterized as in this paper. Thus while we focus our attention on linear feedbacks here, we remind the reader of the possibility of the existence of equilibria involving more general strategies. We also direct the reader to Remark A-2 of Başar and Olsder [4].

2.2. The Switching Regulator Game

If we assume that the targets are all zero (i.e., $\Gamma_j^k \equiv 0$) we have a special case of the above in which the overtaking criterion is not necessary, provided sufficient controllability and stabilizability properties hold. The theory developed for this particular case will be utilized in subsequent sections. Motivated by the classical regulator problem and the work of Haurie and Leizarowitz [9] we assume there exists positive definite symmetric matrices M_j^l, $l = 1, 2, \ldots, p$ and $j = 1, 2, \ldots N$ so that the players follow the feedback responses

$$u_l(t) = -\left(R_{j_t}^l\right)^{-1}\left(B_{j_t}^l\right)' M_{j_t}^l x(t) \tag{2.6}$$

$l = 1, 2, \ldots p$. From the kth player's point of view the dynamics of the system becomes

$$\dot{x}(t) = \mathcal{A}_{j_t}^k x(t) + B_{j_t}^k v_k(t), \tag{2.7}$$

where $v_k(\cdot)$ is the control she uses and

$$\mathcal{A}_{j_t}^k \doteq \left[A_{j_t} - \sum_{\substack{l=1\\ l\neq k}}^{p} B_{j_t}^l \left(R_{j_t}^l\right)^{-1}\left(B_{j_t}^l\right)' M_{j_t}^l\right].$$

In this way she can view (2.7) and her cost function (2.2) as a piecewise deterministic linear-quadratic regulator problem. Thus, from Proposition

2.5 of [9], if for some j the pair $(\mathcal{A}_j^k, Q_j^k)$ is observable then the stabilization of (2.7) is a necessary and sufficient condition for the existence of positive definite matrices $\left\{\tilde{M}_j^k\right\}_{j=1}^N$ satisfying the system of Riccati equations

$$\begin{aligned}
0 = &\left[A_j - \sum_{\substack{l=1 \\ l\neq k}}^{p} B_j^l \left(R_j^l\right)^{-1} \left(B_j^l\right)' M_j^l\right]' \tilde{M}_j^k \\
&+ \tilde{M}_j^k \left[A_j - \sum_{\substack{l=1 \\ l\neq k}}^{p} B_j^l \left(R_j^l\right)^{-1} \left(B_j^l\right)' M_j^l\right] \\
&- \tilde{M}_j^k B_j^k \left(R_j^k\right)^{-1} \left(B_j^k\right)' \tilde{M}_j^k + Q_j^k + \sum_{i=1}^{N} g_{ji} \tilde{M}_i^k, \\
&j = 1, 2, \ldots, N.
\end{aligned}$$

Introducing the notation

$$S_j^l = B_j^l (R_j^l)^{-1} (B_j^l)', \quad l = 1, 2, \ldots, p \quad j = 1, 2, \ldots, N, \tag{2.8}$$

we can rewrite the above as

$$\begin{aligned}
0 = &A_j' \tilde{M}_j^k + \tilde{M}_j^k A_j - \sum_{l=1}^{p} M_j^l S_j^l \tilde{M}_j^k - \tilde{M}_j^k \sum_{l=1}^{p} S_j^l M_j^l + \tilde{M}_j^k S_j^k \tilde{M}_j^k \\
&+ [M_j^k S_j^k \tilde{M}_j^k + \tilde{M}_j^k S_j^k M_j^k - 2\tilde{M}_j^k S_j^k \tilde{M}_j^k] \\
&+ Q_j^k + \sum_{i=1}^{N} g_{ji} \tilde{M}_i^k
\end{aligned} \tag{2.9}$$

for $j = 1, 2, \ldots N$ (observe that the term in the brackets in (2.9) vanishes if $M_j^k = \tilde{M}_j^k$). Further, the feedback control defined by

$$u_k^*(t) = -\left(R_{j_t}^k\right)^{-1} \left(B_{j_t}^k\right)' \tilde{M}_{j_t}^k x^*(t) \tag{2.10}$$

is stabilizing with

$$E_{i,x_0} \|x^*(t)\|^2 \to 0 \text{ as } t \to \infty, \tag{2.11}$$

and in fact is optimal, by a direct application of Proposition 2.5 in [9], for the piecewise deterministic linear-quadratic regulator problem described by (2.7) and (2.2) since

$$C_\infty(i, x_0; \mathbf{u}^*(\cdot)) = x_0 \tilde{M}_i^k x_0. \tag{2.12}$$

Thus we are led to consider the system of Riccati equations

$$
\begin{aligned}
0 = A_j' M_j^k + M_j^k A_j - \sum_{l=1}^{p} M_j^l S_j^l M_j^k - M_j^k \sum_{l=1}^{p} S_j^l M_j^l + M_j^k S_j^k M_j^k \\
+ Q_j^k + \sum_{i=1}^{N} g_{ji} M_i^k \qquad\qquad (2.13)\\
j = 1, 2, \dots, N \text{ and } k = 1, 2, \dots, p,
\end{aligned}
$$

for the unknown matrices $\{M_j^k\}_{k,j=1}^{p,N}$. This is a system of pN matrix Riccati equations. Our first result provides sufficient conditions for the existence of a linear feedback equilibrium. The optimality of (2.10) follows from the fact that the minimal cost of the kth player for the system (2.7) over the interval $[0, \infty)$ is given by $x_0' M_i^k x_0$, if the process starts at x_0 and mode i. Moreover this minimal cost is realized by the control (2.10). (For details see [9], the proof of Proposition 2.5.)

Theorem 2.1. *If there exists a set of pN positive definite $n \times n$ matrices $\{M_j^k\}_{k,j=1}^{p,N}$ solving the system* (2.13), *then the feedback responses $u_k^*(\cdot) : [0, \infty) \to \mathbb{R}^{m_k}$, $k = 1, 2, \dots p$, defined by*

$$
u_k^*(t) = -\left(R_{j_t}^k\right)^{-1} \left(B_{j_t}^k\right)' M_{j_t}^k x^*(t) \qquad (2.14)
$$

are stabilizing and define a feedback equilibrium.

Remark 2.2. In the above theorem, we do not require the use of overtaking feedback equilibria since in the case considered here the stabilizing properties provide us with finite costs for each player.

Proof. We fix our attention to the kth player and recall that we must show that $u_k^*(\cdot)$ as defined by (2.14) is the optimal solution of the piecewise deterministic regulator problem with cost (2.2) and state equation (2.7). The system of matrix Riccati equations associated with this control problem is

$$
\begin{aligned}
0 = \left[A_j - \sum_{\substack{l=1 \\ l \neq k}}^{p} B_j^l \left(R_j^l\right)^{-1} \left(B_j^l\right)' M_j^l \right]' M_j^k \\
+ M_j^k \left[A_j - \sum_{\substack{l=1 \\ l \neq k}}^{p} B_j^l \left(R_j^l\right)^{-1} \left(B_j^l\right)' M_j^l \right]
\end{aligned}
$$

$$
\begin{aligned}
&- M_j^k B_j^k \left(R_j^k\right)^{-1} \left(B_j^k\right)' M_j^k + Q_j^k + \sum_{i=1}^{N} g_{ji} M_i^k \\
&= \left[A_j - \sum_{l=1}^{p} B_j^l \left(R_j^l\right)^{-1} \left(B_j^l\right)' M_j^l\right]' M_j^k \\
&\quad + M_j^k \left[A_j - \sum_{l=1}^{p} B_j^l \left(R_j^l\right)^{-1} \left(B_j^l\right)' M_j^l\right] \\
&\quad + M_j^k B_j^k \left(R_j^k\right)^{-1} \left(B_j^k\right)' M_j^k + Q_j^k + \sum_{i=1}^{N} g_{ji} M_i^k.
\end{aligned}
$$

which, since the matrices $\{M_j^l\}_{l,j=1}^{p,N}$ solve (2.13), implies that the matrices $\{M_j^k\}_{j=1}^{N}$ are solutions of the Riccati system for the piecewise deterministic regulator problem, and this for every $1 \leq k \leq p$. Thus it follows, from Proposition 2.5 of [9], that if $x(0) = x_0$ and $j_0 = i$, then the feedback control $u_k^*(\cdot)$ defined by (2.14) is stabilizing and the response $x^*(\cdot)$ satisfies

$$E_{i,x_0} \|x^*(t)\|^2 \to 0 \text{ as } t \to \infty.$$

Additionally we have that

$$
\begin{aligned}
C_\infty^k(i, x_0; \mathbf{u}^*(\cdot)) &= \lim_{T \to \infty} C_T^k(i, x_0; \mathbf{u}^*(\cdot)) \\
&= \lim_{T \to \infty} x_0' M_i^k x_0 - E_{i,x_0} x^*(T)' M_{j_T}^k x^*(T) \\
&= x_0' M_i^k x_0
\end{aligned}
$$

giving us that $u_k^*(\cdot)$ is optimal. □

The existence of solutions to coupled systems of algebraic Riccati equations has been addressed in the past, (e.g. Papavassilopoulos, et. al. [12]). Here we adapt the methods of [12] to give analogous results. For simplicity and ease of notation we restrict our attention to the case when $p = 2$ and $N = 2$. From our discussion it will be clear how to generalize this result to an arbitrary number of players and jump states. We begin by defining the following block diagonal matrices

$$M = \operatorname{diag}\left[M_1^1, M_1^2, M_2^1, M_2^2\right], \; F = \operatorname{diag}\left[A_1, A_1, A_2, A_2\right],$$

$$Q = \operatorname{diag}\left[Q_1^1, Q_1^2, Q_2^1, Q_2^2\right], \text{ and } S = \operatorname{diag}\left[S_1^1, S_1^2, S_2^1, S_2^2\right]$$

where S_j^i is defined as in (2.8) for $i, j = 1, 2$. Additionally define the following block matrices

$$
G = \begin{bmatrix} g_{11}I & 0 & g_{12}I & 0 \\ 0 & g_{11}I & 0 & g_{12}I \\ g_{21}I & 0 & g_{22}I & 0 \\ 0 & g_{21}I & 0 & g_{22}I \end{bmatrix}, \; P = \begin{bmatrix} I \\ I \\ I \\ I \end{bmatrix}
$$

and

$$J = \begin{bmatrix} 0 & I & 0 & 0 \\ I & 0 & 0 & 0 \\ 0 & 0 & 0 & I \\ 0 & 0 & I & 0 \end{bmatrix}.$$

With this notation, it is easy to see that the system of matrix Riccati equations (2.13) may be compactly written as

$$\begin{aligned} 0 = F'M + MF + \frac{1}{2}\left(\mathrm{Diag}[GMP] + (\mathrm{Diag}[GMP])'\right) + Q \\ - MSM - JMSJM - MJSMJ. \end{aligned} \tag{2.15}$$

Here the notation Diag$[GMP]$ means that if

$$GMP = \begin{bmatrix} A \\ B \\ C \\ D \end{bmatrix}$$

where A, B, C, and D are $n \times n$ matrices, then

$$\mathrm{Diag}\,[GMP] = \begin{bmatrix} A & 0 & 0 & 0 \\ 0 & B & 0 & 0 \\ 0 & 0 & C & 0 \\ 0 & 0 & 0 & D \end{bmatrix}.$$

We now let $\mathcal{M}$ denote the subspace of all $npN \times npN$ block diagonal matrices having $n \times n$ blocks with the usual operator norm and define $\mathcal{R}(\cdot) : \mathcal{M} \to \mathcal{M}$ by the right hand side of (2.15). In this way a solution of (2.15) can be viewed as a fixed point of the continuous map $\Phi(\cdot) : \mathcal{M} \to \mathcal{M}$ defined simply as

$$\Phi(M) = \mathcal{R}(M) + M.$$

Using the Brouwer fixed point theorem one can establish the following result by proceeding as exactly as in [12]. For brevity we omit the proof of this result.

Theorem 2.2. *Assume that $\|S\| > 0$ and suppose that there exists $\alpha \neq 0$ such that*

$$1 \geq \|I + 2\alpha F\| + |\alpha| \left(\|G\| + 2\,(3\|Q\|\,\|S\|)^{1/2} \right).$$

Then for every R, $R_1 \leq R \leq R_2$, there exists M_j^i (depending on R) satisfying (2.13) *(for $i, j = 1, 2$) such that*

$$\|M_j^i\| \leq |\alpha| R \leq |\alpha| R_2 \leq \frac{2\|F\| + \|G\|}{3\|S\|}$$

where R_1 *and* R_2 *are given as*

$$R_1 = \frac{1}{6\alpha^2\|S\|}\{1 - \|I + 2\alpha F\| - |\alpha|\,\|G\| \\ -\sqrt{\left[(1 - \|I + 2\alpha F\| - |\alpha|\,\|G\|)^2 - 12\alpha^2\|S\|\,\|Q\|\right]}\right\}$$

and

$$R_2 = \frac{1}{6\alpha^2\|S\|}\{1 - \|I + 2\alpha F\| - |\alpha|\,\|G\| \\ +\sqrt{\left[(1 - \|I + 2\alpha F\| - |\alpha|\,\|G\|)^2 - 12\alpha^2\|S\|\,\|Q\|\right]}\right\}.$$

Remark 2.3. The above result provides sufficient conditions for the system of Riccati equations (2.13) to have a solution. If one desires symmetric solutions it is enough to further require the matrix G to be symmetric. This is an incomplete result since it does not provide for the existence of positive definite solutions. In the next section we show that this existence can be established in the so called completely symmetric game case. For the nonsymmetric case we can only provide (see Section 5) a comparison result which may be useful in establishing the positiveness. We further remark that the above result gives a uniform bound on the solutions that is independent of α.

3. The Completely Symmetric Game Case

In Theorem 2.1 we have already established that the existence of positive definite solutions to the coupled Riccati system implies stabilizability of the differential game. The conditions insuring the existence of positive definite solutions to the coupled Riccati equations for the general case is apparently still an open problem, even for the deterministic p-player regulator game. In the completely symmetric game case considered below we have a class of games where indeed stabilizability of an associated linear system implies the existence of positive definite solutions.

Definition 3.1. We call the game completely symmetric if the following holds for all k

$$Q_j^k \equiv Q_j,\ R_j^k \equiv R_j,\ \text{and}\ B_j^k \equiv B_j.$$

In the completely symmetric game case we look for solutions of the coupled Riccati system (2.9) which do not depend on k, and we denote them by $\{M_j\}$. Denote $S_j = B_j(R_j)^{-1}(B_j)'$. The system (2.9) then takes the simplified form

$$M_jA_j + A_j'M_j - (2p-1)M_jS_jM_j + Q_j + \sum_{i=1}^{N} g_{ji}M_i = 0,\ 1 \le j \le N. \quad (3.1)$$

Proposition 3.1. *If for some $1 \leq j \leq N$ the pair $\{A_j, Q_j\}$ is observable, then stabilizability of the system*

$$\dot{x} = A_{j_t}x(t) + B_{j_t}u(t) \tag{3.2}$$

implies the existence of positive definite solutions to (3.1).

Proof. It suffices to notice that the system (3.1) has the form of the Riccati equation of the associated single-player control system with the cost expression

$$C_T(i, x_0; u(\cdot)) = E_{i,x_0} \int_0^T [||x(t)||^2_{Q_{j_t}} + \frac{1}{2p-1}||u(t)||^2_{R_{j_t}}]dt.$$

and state equation given by (3.2). One now applies Proposition 2.5 from [9]. □

4. Overtaking Equilibrium with Switching Targets

In this section we use the results obtained in Section 2 to solve the tracking problem when the target states are changing with the dynamics, at random times, according to a Markov chain. In the first subsection we shall consider a set of coupled dynamic programming equations and establish the existence of a solution. Then, in a second subsection we shall prove that this solution to the coupled dynamic programming equations indeed defines an overtaking feedback equilibrium.

4.1. Infinite Horizon Bellman Equations

Inspired by the case of a single-player control system, treated in Ref. [9] we consider the set of generic infinite horizon coupled dynamic programming equations defined below

$$\begin{aligned} &\min_{u_k} \left\{ \frac{1}{2}||x - \Gamma_i^k||^2_{Q_i^k} + \frac{1}{2}||u_k||_{R_i^k} + \frac{\partial \Phi_k}{\partial x}(i,x)(A_i x + \sum_{\ell=1}^{p} B_i^\ell u_\ell) \right\} \\ &\quad + \sum_{j=1}^{N} g_{ij}\Phi_k(j,x) = \mu_k \\ &\qquad i = 1, \ldots, N; k = 1, \ldots, p. \end{aligned} \tag{4.1}$$

In these equations the functionals $\Phi_k(\cdot,\cdot)$ can be interpreted as the *incremental* equilibrium cost-to-go functionals while the scalars μ_k are the average equilibrium cost rates for each player $k = 1, 2, \ldots, p$. The minimum in the LHS of (4.1) is reached for

$$\begin{aligned} u_k = -(R_i^k)^{-1} \left\{ \frac{\partial \Phi_k}{\partial x}(i,x) B_i^k \right\}' \\ i = 1, \ldots, N; k = 1, \ldots, p. \end{aligned} \tag{4.2}$$

If we assume the following form for the value functions $\Phi_k(\cdot,\cdot)$

$$\begin{aligned}\Phi_k(i,x) = \frac{1}{2}x'M_i^k x + (q_i^k)'x + \nu_i^k \\ i = 1,\ldots,N; k = 1,\ldots,p\end{aligned} \tag{4.3}$$

then the control of Player k is determined as the affine feedback law

$$\begin{aligned}u_k = -(R_i^k)^{-1}(B_i^k)'[M_i^k x + q_i^k] \\ i = 1,\ldots,N; k = 1,\ldots,p,\end{aligned} \tag{4.4}$$

so that (4.1) becomes

$$\begin{aligned}&\frac{1}{2}\|x - \Gamma_i^k\|^2_{Q_i^k} + \frac{1}{2}[M_i^k x + q_i^k]'B_i^k(R_i^k)^{-1}(B_i^k)'[M_i^k x + q_i^k] \\ &+ [M_i^k x + q_i^k]'(A_i x - \sum_{\ell=1}^{p} B_i^\ell(R_i^k)^{-1}(B_i^\ell)'[M_i^\ell x + q_i^\ell]) \\ &+ \sum_{j=1}^{N} g_{ij}[\frac{1}{2}x'M_j^k x + (q_j^k)'x + \nu_j^k] = \mu_k \\ &i = 1,\ldots,N; k = 1,\ldots,p.\end{aligned} \tag{4.5}$$

The quadratic terms in (4.5) vanish if the matrices M_j^k satisfy the set of coupled Riccati equations (2.13). Collecting the linear terms and equating to 0 yields

$$\begin{aligned}Q_i^k\Gamma_i^k = [M_i^k S_i^k + A_i' - \sum_{\ell=1}^{p} M_i^\ell S_i^\ell]q_i^k - M_i^k \sum_{\ell=1}^{p} S_i^\ell q_i^\ell + \sum_{j=1}^{N} g_{ij}q_j^k \\ i = 1,\ldots,N; k = 1,\ldots,p,\end{aligned} \tag{4.6}$$

where we have used the notation $S_i^k = B_i^k(R_i^k)^{-1}(B_i^k)'$.

A control system is determined by a collection of matrices

$$\{A_j, B_j, Q_j^k, R_j^k, G\}$$

which we consider as elements in some high-dimensional parameter space $\mathbb{R}^N$. Let $U \subset \mathbb{R}^N$ be a domain in which there exists a unique positive definite solution to the Riccati equation (2.10) for every parameter in U. Then the determinant of the linear system (4.6) is an analytic function of the parameters in U. Thus if (4.6) has a unique solution for *some parameter*, then it has a unique solution for almost every parameter in U.

Finally the constant terms yield

$$\begin{aligned}\frac{1}{2}\|\Gamma_i^k\|^2_{Q_i^k} + \frac{1}{2}(q_i^k)'S_i^k q_i^k - (q_i^k)'\sum_{\ell=1}^{p} S_i^\ell q_i^\ell + \sum_{j=1}^{N} g_{ij}(\nu_j^k)' = \mu_k \\ i = 1,\ldots,N; k = 1,\ldots,p.\end{aligned} \tag{4.7}$$

Once the vectors $\{q_i^k\}$ are computed from (4.6), then (4.7) has the form

$$\sum_{j=1}^{N} g_{ij}\nu_j^k = \mu_k - \alpha_i^k, \quad i = 1, \ldots, N; k = 1, \ldots, p. \tag{4.8}$$

We can consider (4.8) for every k separately since the equations are uncoupled with respect to k. Then if $\bar{\nu}^k$ and $\bar{\alpha}^k \in \mathbf{R}^N$ are such that $(\bar{\nu}^k)_i = \nu_i^k$ and $(\bar{\alpha}^k)_i = \alpha_i^k$ then we want to solve

$$G\bar{\nu}^k = \mu_k e - \bar{\alpha}^k, \tag{4.9}$$

where $e_i = 1$ for $1 \leq i \leq N$. This equation has a unique solution pair $\bar{\nu}^k \in \mathbf{R}^N$ and $\mu_k \in \mathbf{R}$ if the following conditions hold:

(i) $\text{rank}(G) = N - 1$,
(ii) $e \notin G(\mathbf{R}^N)$.

Furthermore, as shown in the next result, the condition (i) implies (ii) allowing us to provide sufficient conditions for the existence of $\bar{\nu}_k$ and $\bar{\alpha}_k$.

Proposition 4.1. *If* $\text{rank}(G) = N - 1$, *then* $e = (1, 1, \ldots, 1)' \in \mathbf{R}^N$ *is not in the range of* G.

Proof. Suppose that $e \in G(\mathbf{R}^N)$. Then for some $v \in \mathbf{R}^N$ we have $e = Gv$. As G is a generator for a Markov process, we have $G^k v = 0$ for $k \geq 2$. Thus for every $T \in \mathbf{R}$

$$e^{GT} v = v + TGv = v + Te. \tag{4.10}$$

This however leads to a contradiction since the left hand side of equation (4.10) remains bounded as $T \to +\infty$ (since e^{GT} is a transition probability) while the right hand side clearly does not. □

Using the above equations we are able to find a solution to the set of coupled dynamic programming equations (4.1). In the next subsection we check that the affine feedback strategies given in (4.4) define an overtaking equilibrium.

4.2. Tracking Switching Targets

In this section we construct an overtaking feedback equilibrium for the p-player switching target tracking game. Again we consider the game from the k-th player point of view, assuming that the other players $l = 1, \ldots, p$, $l \neq k$, use a feedback strategy defined by

$$\gamma_l(j_t, x(t)) = -(R_{j_t}^l)^{-1}(B_{j_t}^l)'[M_{j_t}^l x(t) + q_{j_t}^l]$$

We want to establish that the affine feedback strategy

$$\gamma_k(j_t, x(t)) = -(R^k_{j_t})^{-1}(B^k_{j_t})'[M^k_{j_t}x(t) + q^k_{j_t}] \tag{4.11}$$

is an overtaking optimal response for player k. For doing this we integrate over the time interval $[0, T]$, for arbitrary $T \in [0, \infty)$, the derivative of the quadratic function $\Phi_k(i, x)$, defined in Eqn. (4.3), along a trajectory generated by a control $u_k(\cdot)$. We then get

$$\begin{aligned} &\frac{1}{2}x(T)'M^k_{j_T}x(T) + (q^k_{j_T})'x(T) + \nu^k_{j_T} \\ &\quad = \frac{1}{2}x(0)'M^k_i x(0) + (q^k_i)'x(0) + \nu^k_i \\ &\quad + \int_0^T \{[A_{j_t}x(t) - \sum_{l \neq k} S^l_{j_t}x(t) + B^k_{j_t}u^k(t)]'(M^k_{j_t}x(t) + q^k_{j_t}) \\ &\quad + \sum_{j=1}^{N} g_{j_t j}\Phi_k(j, x(t))\}dt. \end{aligned} \tag{4.12}$$

Using the equations (4.4), (4.5) and (4.6) we see that the integrand in Eqn. (4.12) can be written as

$$\begin{aligned} &\frac{1}{2}\|u_k(t) + (R^k_{j_t})^{-1}(B^k_{j_t})'[M^k_{j_t}x(t) + q^k_{j_t}]\|^2_{R^k_{j_t}} - \frac{1}{2}(u_k(t))'R^k_{j_t}u_k(t) \\ &\quad - \frac{1}{2}\|x(t) - \Gamma^k_{j_t}\|^2_{Q_{j_t}} + \mu_k. \end{aligned}$$

Therefore the cost accumulation process for the k-th player,

$$\{C^k_T(i, x_0; u_k(\cdot))\}_{T \geq 0},$$

is given by

$$\begin{aligned} C_T(i, x_0; u_k(\cdot)) = &\int_0^T \mu_k dt \\ &+ \frac{1}{2}\int_0^T \|u_k(t) + (R^k_{j_t})^{-1}(B^k_{j_t})'[M^k_{j_t}x(t) + q^k_{j_t}]\|^2_{R^k_{j_t}} dt \\ &+ \frac{1}{2}x_0'M^k_i x_0 + (q^k_i)'x_0 - \frac{1}{2}x(T)'M^k_{j_T}x(T) \\ &- (q^k_{j_T})'x(T) + \nu^k_i - \nu^k_{j_T}. \end{aligned}$$

Now it should be clear that for all trajectories $x(\cdot)$, for which $\|x(t)\|^2$ remains bounded over $[0, \infty)$, the optimal control, in the overtaking sense, is given by Eqn. (4.11). This follows from the results in [9] when considering the control problem for player k while fixing the feedback controls of all other players. In particular we have the result [9], Lemma 3.2) that all the controls $u_k(\cdot)$ for which $\mathrm{E}[\int_0^\infty \|u_k + (R^k_{j_t})^{-1}(B^k_{j_t})'(M^k_{j_t}x(t) + q^k_{j_t})\|^2 \, dt < \infty$ have the same value for $\lim_{T\to\infty} x(T)'M^k_{j_T}x(T)$ and $\lim_{T\to\infty} q^k_{j_T}x(T)$. Thus it only remains to show that we have

$$\lim_{t\to\infty} \|x(t)\|^2 < \infty$$

for the strategy defined by this choice of feedback. This is a direct consequence of the Riccati equations (2.13) when we assume the existence of positive definite solutions (see Lemma 5.2 in [9]). We may summarize the above discussion in the following result.

Theorem 4.1. *Assume that the set of coupled Riccati equations* (2.13) *admits as a solution a set of positive definite matrices. Then the affine strategies defined by* (4.11), (4.4), (4.5) *and* (4.6), *if these equations admit a solution, constitute an overtaking feedback equilibrium for the switching tracking game.*

Corollary 4.1. *If the switching tracking game is completely symmetric and if one pair* (A_j, Q_j) *is observable then there exists an overtaking feedback equilibrium.*

Remark 4.1. Notice that the symmetry condition does not imply that all the players track the same targets.

5. A Comparison Theorem for the Two-player Case

In the linear-quadratic control problem the positive definite matrix M which is a solution of the Riccati equation is such that $x_0'Mx_0$ is the minimal cost corresponding to the initial state x_0 at time $t = 0$. This immediately implies monotonicity (in the positive definite sense) of M as a function of the weight matrices Q and R. This argument is not applicable while considering equilibrium points of differential games. Due to the coupling between the equilibrium controls corresponding to various players, we do not expect such monotonicity to hold in general. In fact, Example 5.1 below describes a situation where this monotonicity fails. However, under certain conditions we can establish a comparison theorem which yields a monotonicity property of the kind described above.

We remark that the comparison theorems given below may be utilized to obtain *a priori* estimates for solutions of the Riccati system, e.g. by comparing with a completely symmetric system which bounds the system of interest above. We will not elaborate on this issue in this paper.

The comparison theorems of this section are that under certain conditions, if the weight matrices in the cost expression corresponding to player k_1 are larger (in the positive definite sense) than those corresponding to player k_2, then $M_j^{k_1} \geq M_j^{k_2}$ for every $1 \leq j \leq N$. This assertion is concerned only with two players (once we have fixed the linear feedback controls of all the other players). This leads us to consider differential games with only two players. Moreover, in order to simplify the presentation we consider deterministic differential games, so that $N = 1$.

We denote $S^1 = B^1(R^1)^{-1}(B^1)'$, $S^2 = B^2(R^2)^{-1}(B^2)'$ and consider

$$\begin{cases} M^1A + A'M^1 - M^1S^2M^2 - M^2S^2M^1 - M^1S^1M^1 + Q^1 = 0 \\ M^2A + A'M^2 - M^2S^1M^1 - M^1S^1M^2 - M^2S^2M^2 + Q^2 = 0. \end{cases} \tag{5.1}$$

We look for conditions under which the inequalities $Q^2 > Q^1$ and $S^2 < S^1$ imply $M^2 > M^1$. That some condition is needed for this implication to hold is evident from the following example.

Example 5.1. We consider the scalar case $n = 1$

$$\begin{cases} 2am^1 - 2s^2m^1m^2 - s^1(m^1)^2 + q^1 = 0 \\ 2am^2 - 2s^1m^1m^2 - s^2(m^2)^2 + q^2 = 0. \end{cases} \tag{5.2}$$

We choose numbers $\tilde{m}^1, \tilde{m}^2 > 0$ such that $1 < \frac{\tilde{m}^1}{\tilde{m}^2} < 2$, and choose $s^2 < s^1$ which satisfy

$$2(s^1 - s^2) > \left(\frac{\tilde{m}^1}{\tilde{m}^2}\right) s^1 - \left(\frac{\tilde{m}^2}{\tilde{m}^1}\right) s^2.$$

Let k be a large positive number and define

$$m^1 = k\tilde{m}^1,\ m^2 = k\tilde{m}^2,\ q^1 = k^2\tilde{q}^1,\ q^2 = k^2\tilde{q}^2.$$

Substituting this into (3.4) yields

$$\begin{cases} \tilde{q}^1 = s^1(\tilde{m}^1)^2 + 2s^2\tilde{m}^1\tilde{m}^2 - \dfrac{2a}{k}\tilde{m}^1 \\ \tilde{q}^2 = s^2(\tilde{m}^2)^2 + 2s^1\tilde{m}^1\tilde{m}^2 - \dfrac{2a}{k}\tilde{m}^2. \end{cases}$$

If we now *define* $\tilde{q}^1$ and $\tilde{q}^2$ by the last two equations, then for large enough k we have $\tilde{q}^1 < \tilde{q}^2$, hence $q^1 < q^2$. Therefore for large k the Riccati system (5.2) is such that $q^1 < q^2$ and $s^1 > s^2$, while the solutions m^1 and m^2 are positive and satisfy $m^1 > m^2$.

For the proof of the comparison theorem we need an estimate on $||M^1||$ and $||M^2||$, the norms of the solutions of (5.1). We employ the Euclidean norm for vectors, and the corresponding operator norm for matrices.

Lemma 5.1. *Assume that A is stable so that $||e^{At}|| \leq \gamma e^{-\alpha t}$ for some $\alpha, \gamma > 0$ and every $t \geq 0$. If*

$$||S^1||, ||S^2|| \leq \frac{\alpha^2}{3\gamma^4 \max\{||Q^1||, ||Q^2||\}} \tag{5.3}$$

then

$$||M^1||, ||M^2|| \leq \frac{\gamma^2}{\alpha} \max\{||Q^1||, ||Q^2||\}.$$

Proof. We define an operator $T : R^{n\times n} \times R^{n\times n} \to R^{n\times n} \times R^{n\times n}$ by $T(M^1, M^2) = (\tilde{M}^1, \tilde{M}^2)$ where $\tilde{M}^1$ and $\tilde{M}^2$ are solutions of the (uncoupled) linear equations

$$\begin{cases} \tilde{M}^1 A + A'\tilde{M}^1 - M^1 S^2 M^2 - M^2 S^2 M^1 - M^1 S^1 M^1 + Q^1 = 0 \\ \tilde{M}^2 A + A'\tilde{M}^2 - M^2 S^1 M^1 - M^1 S^1 M^2 - M^2 S^2 M^2 + Q^2 = 0 \end{cases} \tag{5.4}$$

Then a fixed-point of T is a solution of (5.1) and vice versa.

Denote $q = \max\{||Q^1||, ||Q^2||\}$. If $||M^1||, ||M^2|| \leq \frac{\gamma^2}{\alpha} q$, then it follows from Eqn. (5.3) that

$$||M^1 S^2 M^2 + M^2 S^2 M^1 + M^1 S^1 M^1|| \leq \frac{\alpha^2}{\gamma^4 q}\left(\frac{\gamma^2 q}{\alpha}\right)^2 = q.$$

and

$$||M^2 S^1 M^1 + M^1 S^1 M^2 + M^2 S^2 M^2|| \leq \frac{\alpha^2}{\gamma^4 q}\left(\frac{\gamma^2 q}{\alpha}\right)^2 = q.$$

It follows from (5.4) that

$$\tilde{M}^1 = \int_0^\infty e^{A't}[M^1 S^2 M^2 + M^2 S^2 M^1 + M^1 S^1 M^1 - Q^1]e^{At}\, dt$$

and

$$\tilde{M}^2 = \int_0^\infty e^{A't}[M^2 S^1 M^1 + M^1 S^1 M^2 + M^2 S^2 M^2 - Q^2]e^{At}\, dt$$

which, in view of $\|e^{At}\| \leq \gamma e^{at}$ and the estimates for $M^1S^2M^2 + M^2S^2M^1 + M^1S^1M^1$ and $M^2S^1M^1 + M^1S^1M^2 + M^2S^2M^2$ implies that the solutions $\tilde{M}^1$ and $\tilde{M}^2$ of Eqn. (5.4) satisfy

$$||\tilde{M}^1|| \leq \frac{\gamma^2}{2\alpha}[q + ||Q^1||], \; ||\tilde{M}^2|| \leq \frac{\gamma^2}{2\alpha}[q + ||Q^2||],$$

which implies that

$$||\tilde{M}^1||, ||\tilde{M}^2|| \leq \frac{\gamma^2}{\alpha} q.$$

Thus T maps the convex and compact set

$$\{(M^1, M^2) : ||M^1|| \leq \frac{\gamma^2}{\alpha} q, ||M^2|| \leq \frac{\gamma^2}{\alpha} q\}$$

into itself, hence by the Brouwer fixed-point Theorem, Eqn. (5.1) has a solution (M^1, M^2) such that $||M^1||, ||M^2|| \leq \frac{\gamma^2}{\alpha} \max\{||Q^1||, ||Q^2||\}$, concluding the proof of the lemma. □

Remark 5.1.

(i) The more stable the matrix A is the less restrictive is condition (5.3).

(ii) The condition (5.3) is invariant under multiplying all the matrices Q^1, Q^2, R^1 and R^2 by a positive constant.

We denote by $\rho(M)$ the spectral norm of a matrix M, and let $k > 1$ be such that

$$\frac{1}{k} ||M|| \leq \rho(M) \leq k||M||$$

for every symmetric matrix M.

Theorem 5.1. *Let A be a stable matrix such that $||e^{At}|| \leq \gamma e^{-\alpha t}$ for some $\alpha, \gamma > 0$ and every $t \geq 0$. Assume that*

$$||S^1|| \max\{||Q^1||, ||Q^2||\} < \frac{\alpha^2}{3(2k^2+1)\gamma^4} [||Q^2 - Q^1|| \cdot ||(Q^2 - Q^1)^{-1}||]^{-1}. \tag{5.5}$$

Then the relations $Q^2 > Q^1$ and $0 \leq S^2 \leq S^1$ imply $M^2 > M^1$.

Remark 5.2. Notice that we do not assume that the matrices Q^1 and Q^2 are positive semidefinite, but only assume that $Q^2 - Q^1$ is positive definite. This fact is significant while considering, e.g., two-players, zero-sum games.

Moreover, the condition that A is stable can be replaced by the condition that the pair (A, B) is controllable, where B is the $n \times (m_1 + m_2)$ matrix obtained by putting B^1 and B^2 side by side so that the first m_1 columns of B are the columns of B_1 and the last m_2 columns of B are the columns of B_2, respectively.

Proof. For every $0 \leq t \leq 1$ we denote

$$Q^{(t)} = tQ^2 + (1-t)Q^1, \; S^{(t)} = tS^2 + (1-t)S^1$$

and consider the coupled Riccati equations

$$\begin{cases} F^{(t)}A + A'F^{(t)} - F^{(t)}S^1G^{(t)} - G^{(t)}S^1F^{(t)} - F^{(t)}S^{(t)}F^{(t)} + Q^{(t)} = 0 \\ G^{(t)}A + A'G^{(t)} - G^{(t)}S^{(t)}F^{(t)} - F^{(t)}S^{(t)}G^{(t)} - G^{(t)}S^1G^{(t)} + Q^1 = 0 \end{cases} \tag{5.6}$$

for the matrices $F^{(t)}$ and $G^{(t)}$. Let H be the unique symmetric positive definite solution of $HA + A'H - 3HS^1H + Q^1 = 0$ and choose $F^{(0)} = GF^{(0)} = H$, which indeed satisfy (5.6) for $t = 0$. We consider a continuous path $(F^{(t)}, G^{(t)})_{0\le t\le 1}$ of solutions of (5.6). The theorem will be proved once we have established

$$F^{(1)} > G^{(1)}. \tag{5.7}$$

For $0 \le t < 1$ we consider $\delta > 0$ such that $t + \delta < 1$, and let the matrices $f^{(t)}$ and $g^{(t)}$ be such that

$$\begin{cases} F^{(t+\delta)} - F^{(t)} = \delta f^{(t)} + o(\delta) \\ G^{(t+\delta)} - G^{(t)} = \delta g^{(t)} + o(\delta) \end{cases} \tag{5.8}$$

as $\delta \to 0$. Since $F^0 = G^0$ it is sufficient, in order to establish Eqn. (5.7), to prove that

$$f^{(t)} - g^{(t)} > 0 \text{ for every } 0 < t < 1. \tag{5.9}$$

Substituting Eqn. (5.8) in Eqn. (5.6), dividing by δ and letting $\delta \to 0$ yields

$$\begin{cases} f^{(t)}A + A'f^{(t)} - [f^{(t)}S^1G^{(t)} + F^{(t)}S^1g^{(t)} + G^{(t)}S^1f^{(t)} \\ \quad + g^{(t)}S^1F^{(t)} + f^{(t)}S^{(t)}F^{(t)} + F^{(t)}S^{(t)}f^{(t)}] + (Q^2 - Q^1) = 0 \\ g^{(t)}A + A'g^{(t)} - [g^{(t)}S^{(t)}F^{(t)} + G^{(t)}S^{(t)}f^{(t)} + f^{(t)}S^{(t)}G^{(t)} \\ \quad + F^{(t)}S^{(t)}g^{(t)} + g^{(t)}S^1G^{(t)} + G^{(t)}S^1g^{(t)}] = 0. \end{cases} \tag{5.10}$$

We first establish, as in the proof of Lemma 5.1, bounds for $||f^{(t)}||$ and $||g^{(t)}||$.

If the following holds

$$||f^{(t)}||, ||g^{(t)}|| \le \frac{(2k^2+1)\gamma^2}{4\alpha k^2}||Q^2 - Q^1||$$

then in view of $||S^2|| < ||S^1||$ (which follows by positive definiteness and symmetry of these matrices), we obtain from Eqn. (5.5) and Lemma 5.1 that

$$\begin{aligned} &||f^{(t)}S^1G^{(t)} + G^{(t)}S^1f^{(t)} + g^{(t)}S^1F^{(t)} + F^{(t)}S^1g^{(t)} \\ &+ f^{(t)}S^{(t)}F^{(t)} + F^{(t)}S^{(t)}f^{(t)}|| \\ &\le \frac{6\alpha^2\gamma^4 q(2k^2+1)||Q^2 - Q^1||}{12\alpha 2k^2(2k^2+1)q||Q^2 - Q^1||||(Q^2 - Q^1)^{-1}||} \\ &= \frac{1}{2k^2||(Q^2 - Q^1)^{-1}||} \\ &\le \frac{1}{2k^2}||Q^2 - Q^1|| \end{aligned} \tag{5.11}$$

and similarly

$$
\begin{aligned}
&||g^{(t)}S^{(t)}F^{(t)} + F^{(t)}S^{(t)}g^{(t)} + f^{(t)}S^{(t)}G^{(t)} \\
&+ G^{(t)}S^{(t)}f^{(t)} + g^{(t)}S^1G^{(t)} + G^{(t)}S^1g^{(t)}|| \\
&\le \frac{1}{2k^2||(Q^2 - Q^1)^{-1}||} \\
&\le \frac{1}{2k^2}||Q^2 - Q^1||.
\end{aligned} \tag{5.12}
$$

In the above computation we applied Lemma 5.1 to the matrices S^1, $S^{(t)}$, Q^1, $Q^{(t)}$ which clearly satisfy the conditions of this lemma. We obtain from Eqn. (5.11) and Eqn. (5.12) that the solutions $f^{(t)}$ and $g^{(t)}$ of Eqn. (5.10) satisfy

$$
||f^{(t)}||, ||g^{(t)}|| \le \frac{(2k^2+1)\gamma^2}{4\alpha k^2}||Q^2 - Q^1||.
$$

We conclude this by viewing the solution of Eqn. (5.10) as a fixed-point of an operator from the set

$$
\{(f, g) : ||f||, ||g|| \le \frac{(2k^2+1)\gamma^2}{4k^2\alpha}||Q^2 - Q^1||\}
$$

into itself, as in the proof of Lemma 5.1. We subtract the second equation in Eqn. (5.10) from the first one obtaining

$$
(f^{(t)} - g^{(t)})A + A'(f^{(t)} - g^{(t)}) + E^{(t)} + (Q^2 - Q^1) = 0 \tag{5.13}
$$

where, by the computation of Eqn. (5.11) and Eqn. (5.12) we have

$$
||E^{(t)}|| \le [k^2||(Q^2 - Q^1)^{-1}||]^{-1}.
$$

This implies that

$$
\rho(E^{(t)}) \le k||E^{(t)}|| < \frac{1}{k||(Q^2 - Q^1)^{-1}||} \le \frac{1}{\rho([Q^2 - Q^1]^{-1})},
$$

and we conclude that $E^{(t)} + (Q^2 - Q^1)$ is positive definite. Now (5.13) yields $f^{(t)} - g^{(t)}) = \int_0^\infty e^{A's}[E^{(t)} + Q^2 - Q^1]e^{As}\,ds$ implying that $f^{(t)} - g^{(t)}$ is positive definite.

Once a continuous path $f^{(t)} - g^{(t)}$ of solutions to (5.10) is established, it justifies (5.8). We choose a large integer m and consider $\{F_{i/m}, G_{i/m}\}_{i=1}^m$ and it follows that $F^2 - G^2 = \frac{1}{m}\sum_{i=1}^m (f_{i/m}, g_{i/m}) + m \times o(\frac{1}{m})$ which establishes the theorem. □

The following is an obvious consequence of Theorem 3.5, which we present in order to emphasize the homogeneity invariance of the condition on the weight matrices. Let $||B|| = b$ (so that also $||B'|| = b$) and denote

$$
\beta = \frac{\alpha^2}{3(2k^2+1)\gamma^4 b^2}. \tag{5.14}
$$

Corollary 5.1. *Let A be a stable matrix such that $||e^{At}|| \leq \gamma e^{-\alpha t}$ for some $\alpha, \gamma > 0$ and all $t \geq 0$, and let β be defined by Eqn.* (5.14). *Assume that*

$$||(R^1)^{-1}|| \cdot ||(Q^2 - Q^1)^{-1}|| \cdot ||Q^2 - Q^1|| \cdot \max\{||Q^1||, ||Q^2||\} < \beta.$$

Then the relations $Q^2 > Q^1$ and $R^2 > R^1 > 0$ imply $M^2 > M^1$.

Example 5.2.

(i) The scalar case $n = 1$. Then $\dot{x} = ax + b^1u^1 + b^2u^2$,

$$C_T^1(u^1(\cdot), u^2(\cdot)) = \int_0^t (q^1(x(t))^2 + r^1(u^1(t))^2)dt$$
$$C_T^2(u^2(\cdot), u^2(\cdot)) = \int_0^t (q^2(x(t))^2 + r^2(u^2(t))^2)dt$$

and we assume that $q^1 < q^2$ and $(b^1)^2/r^1 > (b^2)^2/r^2$. We then have $\gamma = 1$, $\alpha = -a$, $k = 1$ and $||(Q^2 - Q^1)^{-1}|| \cdot ||Q^2 - Q^1|| = 1$. The condition in Eqn. (5.5) is thus

$$\frac{\max\{|q^1|, |q^2|\}}{r^1} < \left(\frac{a}{3b^1}\right)^2.$$

(ii) If Q^1 and Q^2 are positive semidefinite then the condition Eqn. (5.5) may be replaced by

$$||(R^1)^{-1}|| \cdot ||Q^2||^2 \cdot ||(Q^2 - Q^1)^{-1}|| < \frac{\alpha^2}{3(2k^2 + 1)\gamma^4 b^2}$$

where $b = ||B|| = ||B'||$.

6. Conclusion

In this paper we have considered a class of linear quadratic stochastic differential games defined over an infinite time horizon. These games would occur naturally if one models a problem of economic or environmental policy coordination among a set of countries. We have shown that the concept of overtaking equilibrium has to be introduced if the players want to track randomly switching targets. In the case of completely symmetric games we have shown that an overtaking equilibrium exists, defined by affine strategies. For the more general case the existence of such an equilibrium is linked to the possibility to obtain positive definite solutions to a set of coupled algebraic Riccati equations. The comparison theorem established in the last section of this paper may be helpful to check this condition. A natural extension of this research would be to consider a diffusion noise in addition to the jump disturbances.

Acknowledgment. This research was supported by FNRS and a visiting professor grant from the University of Toledo.

References

[1] Artstein, Z. and Leizarowitz, A., Tracking Periodic Signals with the Overtaking Criterion, *IEEE Transactions Automatic Control*, **AC-30**, pp. 1122–1126, 1985.

[2] Başar, T., A Counterexample in Linear-Quadratic Games: Existence of Non-Linear Nash Solutions, *Journal of Optimization Theory and Applications* , **18**, pp. 425–430, 1974.

[3] Başar, T. and Haurie, A., Feedback Equilibria in Differential Games with Structural and Modal Uncertainties, in Jose B. Cruz Jr. (ed.) *Advances in Large Scale Systems* , **160**, JIT Press, New York, 1984.

[4] Başar, T. and Olsder, G. J., *Dynamic Noncooperative Games*, **160**, Academic Press, New York, 1982.

[5] Brock W., On Existence of Weakly Maximal Programmes in a Multisector Economy, *Review of Economic Studies* , **37**, pp. 275–280, 1970.

[6] Carlson, D., Haurie, A. and Leizarowitz. A., *Infinite Horizon Control Theory: Deterministic and Stochastic Systems*, Springer Verlag, New York, 1992.

[7] Gale, D., On Optimal Development in a Multisector Economy, *Review of Economic Studies*, **34**, pp. 1–19, 1967.

[8] Haurie, A. and Leitmann, G., On the Global Stability of Equilibrium Solutions for Open-loop Differential games, *Large Scale systems*, **6**, pp. 107–122, 1984.

[9] Haurie, A. and Leizarowitz, A., Overtaking Regulation and Tracking of Piecewise Diffusion Linear Systems, *S.I.A.M. J. Control and Optimization*, **30**, May 1992.

[10] Haurie, A. and Tolwinski, B., Definition and Properties of Cooperative Equilibria in a Two-Player Game of Infinite Duration, *Journal of Optimization Theory and Applications* , **46**, pp. 525–534, 1985.

[11] Ichikawa, A., Quadratic Differential Games with Overtaking Optimality, in H. Ethamo and R. Hämäläinen (eds.) *Differential Games-Development in Modeling and Computation* , Springer Verlag, New York, pp. 207–215 1990.

[12] Papavassilopoulos, G. P., Medanic, J. V., and Cruz, J. B., On the Existence of Nash Strategies and Solutions to Coupled Riccati Equations in Linear-Quadratic Games, *Journal of Optimization Theory and Applications* , **28**, pp. 49–76, 1979.

[13] Rishel, R., Control of Systems with Jump Disturbances, *IEEE Transactions Automatic Control*, **AC-20**, pp. 241–244, 1975.

[14] Rubinstein, A., Equilibrium in Supergames with the Overtaking Criterion, *Journal of Economic Theory*, **21**, pp. 1–9, 1979.

[15] Sworder, D. D., Feedback Control of a Class of Linear Systems with Jump Parameters, *IEEE Transactions Automatic Control*, **AC-14**, pp. 9–14, 1969.

[16] Von Weizäcker, C.C., On Existence of Optimal Programs of Accumulation for an Infinite Time Horizon, *Review of Economic Studies*, **32**, pp. 85–104, 1965.

[17] Wonham, W.M., Random Differential Equations in Control Theory, in A.T. Bharucha-Reid (ed.) *Probabilistic Methods in Applied Mathematics* , Academic Press, New York, pp. 131–217, 1970.

D. Carlson
Department of Mathematics
University of Toledo
Toledo, Ohio 43606, USA

A. Haurie
Department of Management Studies (COMIN)
University of Geneva
102 Carl Vogt
CH-1211 Geneva, Switzerland

A. Leizarowitz
Department of Mathematics
Technion
32000 Haifa, Israel

Monotonicity of Optimal Policies in a Zero Sum Game: A Flow Control Model

Eitan Altman

Abstract. The purpose of this paper is to illustrate how value iteration can be used in a zero-sum game to obtain structural results on the optimal (equilibrium) value and policy. This is done through the following example. We consider the problem of dynamic flow control of arriving customers into a finite buffer. The service rate may depend on the state of the system, may change in time and is unknown to the controller. The goal of the controller is to design a policy that guarantees the best performance under the worst case service conditions. The cost is composed of a holding cost, a cost for rejecting customers and a cost that depends on the quality of the service. We consider both discounted and expected average cost. The problem is studied in the framework of zero-sum Markov games where the server, called player 1, is assumed to play against the flow controller, called player 2. Each player is assumed to have the information of all previous actions of both players as well as the current and past states of the system. We show that there exists an optimal policy for both players which is stationary (that does not depend on the time). A value iteration algorithm is used to obtain monotonicity properties of the optimal policies. For the case that only two actions are available to one of the players, we show that his optimal policy is of a threshold type, and optimal policies exist for both players that may need randomization in at most one state.

1. Introduction

There are many known computational methods for solving stochastic games, see e.g. the survey paper of Raghavan and Filar [10] and references therein. An alternative method that could reduce computations would be to prove in some way that the optimal strategies for all players are restricted to a small class of policies that have some structural properties. If we are lucky, this class of policies may depend only on one parameter (e.g. some real number), and the calculation of the performance under these policies might be easily done. In that case the original dynamic game may be reduced to a simpler optimization problem over the parameter space.

Several methods have been used in different stochastic games to obtain structural results. Altman and Shimkin considered in [3] a non zero-sum game with an infinite number of players, to solve the problem of choosing between the use of an individual personal computer, and a central computer whose capacity is simultaneously shared between different users. Using coupling and sample-path methods, all Nash optimal policies were shown to be of threshold type. This then enabled the computation of an optimal threshold. McNamara et al. [8] obtained a threshold type equilibrium policy for a dynamic version of the Hawk-Dove game using dynamic programming arguments. Hsiao and Lazar [4] obtained threshold equilibrium policies for a decentralized flow control into a network using the product form of the network as well as Northon's equivalent. The threshold policy is then obtained through a Linear Program. Küenle [7] used dynamic programming and especially the value iteration to solve an inventory control problem under worst case demand conditions. He modeled the problem as a stochastic zero-sum game with full information and identified the structure of an optimal policy of the controller, known as the (s,S) policy. Other results on worst case control in queueing models with routing and scheduling have recently been obtained by Altman and Koole [2], using again tools from stochastic games. Their results include as special case the optimality of "join the shortest queue" for the routing problem, and the well known μc-type policy for the problem of scheduling of the server.

In the present paper we consider a system where one user controls dynamically the flow of arriving customers into a finite buffer. The presence of other users as well as congestion phenomena is modeled by allowing the service rate to depend on the state of the system, and to change in time in a way that is unknown to the controller. Our goal is to design a control strategy that guarantees the best performance under the worst case service conditions. We formulate this problem as a zero-sum stochastic game, where the server is player one and the flow controller is player two. Using the value iteration technique, we are able to establish monotonicity properties of the policies that are optimal for both players, i.e. we provide not only the structure of an optimal policy for the flow controller, but identify also the structure of a worst case service conditions. We show that both policies are monotone decreasing in the state; the flow controller decreases the input flow as the number of customers increases, and under the worst service conditions, the service quality decreases with the number of customers in the queue. If one of the players has only two available actions, then this monotonicity is shown to imply that his optimal policy is of a threshold type, with randomization for both players in at most one state. Related results for the case of an infinite buffer with only two actions to each player were obtained by Altman in [1]. The case of finite buffer seems more delicate since additional boundary conditions appear in the value iteration.

The structure of the paper is as follows: in Section 2 we describe the model. Then basic tools for solving the problem with discounted cost are described in Section 3, and the properties of the optimal policies and value are derived in Section 4. The results are extended to the expected average cost in Section 5. In Section 6 we restrict to the case where two actions are available to each player; we use the structural results from Section 5 to reduce the original dynamic game into a simpler static one.

2. The Model

Considered is a discrete-time single-server queue with a buffer of size L. We assume that at most one customer may join the system in a time slot. This arrival (if any) is assumed to occur at the beginning of the time slot.

Let X_t denote the number of customers in the system at time t, $t = 0, 1, ...$; the state space is denoted by $\mathbf{X} = \{0, 1, ..., L\}$.

Let g_{max} be a real number satisfying $0 < g_{max} < 1$. At the beginning of each time slot, if the state is x then the flow control mechanism, called player 2, chooses in a finite set $\mathbf{G}_x \subset [0, g_{max}]$ an action, which is interpreted as the probability of having one arrival in this time slot. Therefore, if action g is chosen at time t then a customer will enter the system in $[t, t+1)$ with the probability g. We assume that when the buffer is full, no arrivals are possible, and thus $\mathbf{G}_L = \{0\}$. In all states other than L we assume that the available actions are the same, and we denote them by $\mathbf{G}_x = \mathbf{G}$. We further assume that $0 \in \mathbf{G}$.

Let b_{min} and b_{max} be two real numbers satisfying $0 < b_{min} \leq b_{max} < 1$. At the end of each slot, if the state at the beginning of the slot is x, a successful service of a customer occurs, with some probability $b \in \mathbf{B}$ where $\mathbf{B}$ is a finite subset of $[b_{min}, b_{max}]$. If the service fails the customer remains in the queue, and if it succeeds then the customer leaves the system. The value of b, which may represent the quality of service, may change in each time slot, and is not known to player 2. The objective of player 2, to be described below, is to find a best strategy under the worst case service conditions. We model the system as a zero-sum Markov game, where player 1 controls the service quality.

Actions b and g are assumed to be taken independently, based on the information on the current state as well as the information of all past states and actions of both players.

We assume that a customer that enters an empty system may leave the system (with the probability b) at the end of this same time slot.

The state X_t denotes the number of customers in the system at time t, $t \in \mathbb{N}$, and B_t and G_t denote the actions of players 1 and 2 respectively. Let $M(Y)$ be the set of probability measures on a set Y.

The transition law q is:

$$q(y \mid x; b; g) := \begin{cases} \bar{g}b, & \text{if } L \geq x \geq 1,\ y = x - 1; \\ gb + \bar{g}\bar{b}, & \text{if } L \geq x \geq 1,\ y = x; \\ g\bar{b}, & \text{if } L > x \geq 0,\ y = x + 1; \\ 1 - g\bar{b}, & \text{if } y = x = 0\ ; \end{cases}$$

(for any number $\xi \in [0,1]$, $\bar{\xi} := 1 - \xi$).

We define an immediate payoff

$$C(x, b, g) := c(x) + \theta(b) + \rho(g) \tag{2.1}$$

for all $x \in \mathbf{X}$, $b \in \mathbf{B}$ and $g \in \mathbf{G}$. $C(x,b,g)$ is the cost that player 2 pays to player 1 when the state is x, and the actions of the players are b and g. C generalizes a cost frequently encountered in the literature on flow control models. In (2.1) $c(x)$ is any real-valued *increasing convex* function on $\mathbf{X}$, θ is a real function on $\mathbf{B}$ and ρ is a real function on $\mathbf{G}$. It is natural to assume that θ is increasing in b and $\theta \geq 0$ whereas ρ is decreasing in g and $\rho \leq 0$. $c(x)$ can be interpreted as a holding cost per unit time, ρ as a reward related to the acceptance of incoming customer, and θ as a cost per quality of service.

Let U (V) be the class of policies of player 1 (player 2 resp.). A policy $u \in U$ ($v \in V$) is a sequence $u = (u_0, u_1, ...)$ ($v = (v_0, v_1, ...)$ resp.) where u_n (resp. v_n) is a conditional probability on $\mathbf{B}$ (resp. $\mathbf{G}$) given the history of all states and actions of both players as well as the current state. Thus each player is assumed to have the information of all last actions of both players as well as the current and past states of the system. Both players know the action sets, the immediate cost C, the initial state and the transition probabilities q.

Let u be a policy of player 1 and v a policy of player 2. Let ξ be a fixed number in $[0,1)$. Define the discounted cost:

$$V_\xi(x, u, v) := E^{u,v}\left[\sum_{t=0}^{\infty} \xi^t\, C(X_t, B_t, G_t) \mid X_0 = x\right], \tag{2.2}$$

Define the following problem ($\mathbf{Q}_\xi$): Find u, v that achieve

$$V_\xi(x) := \sup_{u \in U} \inf_{v \in V} V_\xi(x, u, v), \qquad \forall x \in \mathbf{X}. \tag{2.3}$$

We know ([10] Section 2) that there exists a pair of stationary policies (u^*, v^*) that achieves (2.3), and

$$\begin{aligned} \sup_{u \in U} \inf_{v \in V} V_\xi(x, u, v) &= \inf_{v \in V} \sup_{u \in U} V_\xi(x, u, v) = \sup_{u \in U} V_\xi(x, u, v^*) \\ &= \inf_{v \in V} V_\xi(x, u^*, v) = V_\xi(x, u^*, v^*). \end{aligned}$$

$V_\xi(x)$ is called the ξ-discounted *value* of the game, and the policies (u^*, v^*) are called optimal policies.

A pair of policies (u, v) are said to be stationary if they depend only on the current state. In that case, we use the notation $u = \{u_x, x \in \mathbf{X}\}$, $u_x \in M(\mathbf{B})$, where $u_x(b)$ is the probability of choosing b when in state x; similarly we use the notation $v = \{v_x, x \in \mathbf{X}\}$, $v_x \in M(\mathbf{G})$, where $v_x(g)$ is the probability of choosing g in state x.

For any $\beta \in M(\mathbf{B})$ we denote $b(\beta) := E_\beta[b] = \sum_{b\in\mathbf{B}} b \cdot \beta(b)$. Similarly, for any $\gamma \in M(\mathbf{G})$ we denote $g(\gamma) := E_\gamma[g] = \sum_{g\in\mathbf{G}} g \cdot \gamma(g)$.

3. Preliminary Results

Let $\mathcal{K}$ be the set of all real-valued functions on $\mathbf{X}$. Define the operator $R : \mathbf{X} \times \mathbf{B} \times \mathbf{G} \times \mathcal{K} \to \mathbb{R}$ as

$$R(x, b, g, f) := E\left[f(X_{t+1}) | X_t = x, B_t = b, G_t = g\right], \qquad g \in \mathbf{G}_x$$

we get:

$$R(x, b, g, f) = \begin{cases} (1 - g\bar{b})\, f(x) + g\bar{b} f(x + 1) & x = 0 \\ \bar{g} b f(x - 1) + (gb + \bar{g}\bar{b}) f(x) + g\bar{b} f(x + 1) & 1 \leq x \leq L \end{cases} \tag{3.1}$$

(in the above equation we shall understand $0 \cdot f(L + 1) := 0$). Let $R(x, f)$ denote the matrix whose entries are $R(x, b, g, f)$.

Define the operator $S : \mathbf{X} \times \mathbf{B} \times \mathbf{G} \times \mathcal{K} \to \mathbb{R}$ as

$$S(x, b, g, f) := C(x, b, g) + \xi R(x, b, g, f), \qquad g \in \mathbf{G}_x$$

and let $S(x, f)$ denote the matrix whose entries are $S(x, b, g, f)$.

For any x, and functions $D : \mathbf{B} \times \mathbf{G}_x \to \mathbb{R}$, $\beta \in M(\mathbf{B})$ and $\gamma \in M(\mathbf{G}_x)$ define

$$\beta D \gamma := \sum_{b\in\mathbf{B}} \sum_{g\in\mathbf{G}_x} \beta(b)\gamma(g) D(b, g)$$

The value of the "matrix game D" is defined as

$$val(D) := \sup_{\beta\in M(\mathbf{B})} \inf_{\gamma\in M(\mathbf{G}_x)} \beta D \gamma.$$

It is known to satisfy $val(D) = \inf_{\gamma\in M(\mathbf{G}_x)} \sup_{\beta\in M(\mathbf{B})} \beta D \gamma$, and there are measures $\beta^* \in M(\mathbf{B})$, $\gamma^* \in M(\mathbf{G}_x)$ such that

$$val(D) = \inf_{\gamma\in M(\mathbf{G}_x)} \beta^* D \gamma = \sup_{\beta\in M(\mathbf{B})} \beta D \gamma^* = \beta^* D \gamma^*$$

β^* and γ^* are said to be optimal for the matrix game D. We shall use the following properties of matrix games. Given some $\gamma \in M(\mathbf{G}_x)$, let *supp*$(\gamma)$

be the set of actions in the support of γ, i.e. actions that are chosen with positive probability by γ. Define similarly $supp(\beta)$.

Lemma 3.1. (i) *Let (γ^*, β^*) be optimal for a matrix game D. Then for any $g \in supp(\gamma^*)$ and any $b \in supp(\beta^*)$,*

$$\sum_{g \in \mathbf{G}} \gamma^*(g) D(b, g) = val(D) = \sum_{b \in \mathbf{B}} \beta^*(b) D(b, g).$$

(ii) *Let (γ^*, β^*) and $(\hat{\gamma}^*, \hat{\beta}^*)$ be optimal solutions for a matrix game D. Then $(\hat{\gamma}^*, \beta^*)$ is also optimal.*

Proof. (i) follows from [12], p. 36. (ii) is straightforward. □

For any $\beta \in M(\mathbf{B}), \gamma \in M(\mathbf{G}_x)$ we shall understand with some abuse of notation

$$R(x, \beta, \gamma, f) := \beta R(x, f) \gamma, \qquad S(x, \beta, \gamma, f) := \beta S(x, f) \gamma.$$

Recall the definitions of $b(\beta)$ and $g(\gamma)$. We extend $f : \mathbf{X} \to \mathbb{R}$ to $\mathbf{X} \cup \{-1\} \to \mathbb{R}$, and set $f(-1) = f(0)$. With these definitions we have for any $0 \leq x \leq L$ and stationary u and v,

$$\begin{aligned} R(x, u_x, v_x, f) &= u_x R(x, f) v_x \\ &= \bar{g}(v_x) b(u_x) f(x-1) + (g(v_x) b(u_x) + \bar{g}(v_x) \bar{b}(u_x)) f(x) \\ &\quad + g(v_x) \bar{b}(u_x) f(x+1). \end{aligned} \tag{3.2}$$

Let $T_\xi : \mathcal{K} \to \mathcal{K}$ be the DP (Dynamic Programming) operator associated with $\mathbf{Q}_\xi$:

$$T_\xi f(x) := val\, S(x, f), \qquad x \in \mathbf{X}. \tag{3.3}$$

Let $(u(f), v(f))$ be stationary policies such that the probability measures $(u_x(f), v_x(f))$ are optimal for the matrix game $S(x, f)$ for all $x \in \mathbf{X}$. We shall use the following tools for solving $\mathbf{Q}_\xi$:

Proposition 3.1. (i) *V_ξ satisfies $V_\xi(x) = T_\xi V_\xi(x)$.*

(ii) *Let (u^*, v^*) be stationary policies of player 1 and 2 respectively such that for each $x \in \mathbf{X}$, the probability measures u_x and v_x are optimal for the matrix game $S(x, V_\xi)$. Then (u^*, v^*) are optimal for $\mathbf{Q}_\xi$.*

(iii) *For every $f \in \mathcal{K}$, $\lim_{n \to \infty} T_\xi^n f = V_\xi$.*

Proof. See Shapely [11]. □

Remark 3.1. If either u_x^* or v_x^* do not randomize in some state x, then neither of them needs randomization in that state. This follows from the fact that u_x^* and v_x^* are solutions for the matrix game $S(x, V_\xi)$ (we make use of [12], Theorem 1.16.3 p. 26).

4. Monotonicity of the Optimal Policies

We begin by defining the monotonicity of policies. Let u, v be stationary. Denote $b_x^{sup}(u) :=$ the greatest b in the support of u_x, i.e. the greatest $b \in \mathbf{B}$ that is chosen by u with positive probability when in state x. Denote $b_x^{inf}(u) :=$ the smallest b in the support of u_x, and define similarly $g_x^{sup}(v)$ and $g_x^{inf}(v)$.

We say that a stationary policy $u \in U$ is **strongly monotone** if for any $x \leq L$ and y with $y < x$, $b_y^{inf}(u) \geq b_x^{sup}(u)$. We say that a stationary policy $v \in V$ is **strongly monotone** if for any $x \leq L$ and y with $y < x$, $g_y^{inf}(v) \geq g_x^{sup}(v)$.

The monotonicity of a policy u means that the service quality is nonincreasing (in a probabilistic sense) as the number of customers in the buffer becomes larger. The monotonicity of a policy v means that the input flow is nonincreasing as the number of customers in the buffer becomes larger.

The following is a straight forward consequence from the definition of strongly monotone policies.

Lemma 4.1. *If u is strongly monotone then it randomizes in at most $|\mathbf{B}| - 1$ states. If v is strongly monotone then it randomizes in at most $|\mathbf{G}| - 1$ states.*

We shall say that $f \in \mathcal{K}$ satisfies assumption:
WC (weakly convex) if for all $0 \leq x < L - 1$,

$$f(x+2) - f(x+1) \geq f(x+1) - f(x). \tag{4.1}$$

SC(x) (strongly convex) if for x given,

$$f(x+2) - f(x+1) > f(x+1) - f(x). \tag{4.2}$$

MI if $f(x)$ is monotone increasing in x, i.e. for any $0 \leq x < L$,

$$f(x+1) \geq f(x) \tag{4.3}$$

Let U^* be the set of stationary policies for the service controller such that $u \in U^*$ if and only if for any $x \in \mathbf{X}$ u_x is optimal for player 1 in the matrix game $S(x, V_\xi)$. Let V^* be the set of stationary policies for the flow controller such that $v \in V^*$ if and only if for any $x \in \mathbf{X}$ v_x is optimal for player 2 in the matrix game $S(x, V_\xi)$. It follows from Proposition 3.1 (ii) and Lemma 3.1 (ii) that any pair (u, v) such that $u \in U^*$ and $v \in V^*$ is optimal for problem $\mathbf{Q}_\xi$.

We are ready to present the main result.

Theorem 4.1. *If the holding cost c satisfies* **MI**, **WC** *and either $c(1) > c(0)$ or* **SC(0)** *then any of the optimal policies $u \in U^*$ and $v \in V^*$ are strongly monotone.*

In order to prove Theorem 4.1 we need first to introduce the following two technical lemmas.

Lemma 4.2. *Let $h : \mathbf{X} \cup \{-1\} \to \mathbb{R}$ be a nondecreasing function. Let $\zeta_1, \zeta_2 \in [0,1]$. Then, for all $0 \le x < L$,*

$$F(x) := \zeta_2 h(x+1) + \bar{\zeta}_2 h(x) - \zeta_1 h(x) - \bar{\zeta}_1 h(x-1) \ge 0 \tag{4.4}$$

Moreover, if (i) $h(x+1) > h(x)$ and $\zeta_2 \neq 0$, or (ii) if $h(x) > h(x-1)$ and $\zeta_1 \neq 1$, then $F(x) > 0$.

Proof.

$$F(x) \ge h(x) - \zeta_1 h(x) - \bar{\zeta}_1 h(x-1) = \bar{\zeta}_1 [h(x) - h(x-1)] \ge 0,$$

and the second claim follows similarly. □

Lemma 4.3. *Assume that the holding cost c satisfies* **WC** *and* **MI**.

(i) *Assume that f satisfies* **WC** *and* **MI**. *Then $T_\xi f$ satisfies* **WC** *and* **MI**.

(ii) *The value function V_ξ satisfies* **WC** *and* **MI**.

(iii) *If V_ξ satsifies* **SC(x)** *in one state $x < L-1$, then it satisfies* **SC(y)** *for all $y \ge x$. If $V_\xi(1) - V_\xi(0) > 0$ then V_ξ satisfies* **SC(y)** *for all states y, $0 \le y < L-1$.*

Finally, assume that the holding cost c satisfies **WC**, **MI** *and either $c(1) > c(0)$ or* **SC(0)**. *Then*

(iv) *V_ξ satisfies* **SC(y)** *for all states y, $0 \le y < L-1$.*

Proof. (i) Choose arbitrary $u \in U^*$ and $v \in V^*$. We begin by establishing **MI**. Recall the definitions of $u_x(f)$ and $v_x(f)$ before Proposition 3.1. Choose any $0 \le x \le L-1$; let $b := b(u_x(f))$ and $g := g(v_{x+1}(f))$ ($g(\gamma)$ and $b(\beta)$ were defined in the end of Section 2).

$$\begin{aligned} T_\xi f(x+1) - T_\xi f(x) &= valS(x+1, f) - valS(x, f) \qquad\qquad (4.5)\\ &\ge S(x+1, u_x(f), v_{x+1}(f), f) - S(x, u_x(f), v_{x+1}(f), f) \\ &= c(x+1) - c(x) + \xi \Big\{ \bar{g}b[f(x) - f(x-1)] \\ &\quad + (gb + \bar{g}\bar{b})[f(x+1) - f(x)] \\ &\quad + g\bar{b}[f(x+2) - f(x+1)] \Big\} \\ &\ge c(1) - c(0) \ge 0. \qquad\qquad (4.6) \end{aligned}$$

(The equation above holds indeed for $x = L-1$ too since in that case $g = 0$; in that case, we shall understand $gf(x+2) := 0$).

Next we check **WC**. Choose any $0 \le x \le L-2$; let $b := b(u_{x+1}(f))$, $g_1 := g(v_x(f))$ and $g_2 := g(v_{x+2}(f))$. Denote

$$F(x) = valS(x+2, f) - valS(x+1, f) - [valS(x+1, f) - valS(x, f)],$$

We have

$$\begin{aligned} F(x) \ge\ & S(x+2, u_{x+1}(f), v_{x+2}(f), f) - S(x+1, u_{x+1}(f), v_x(f), f) \qquad (4.7)\\ & - [S(x+1, u_{x+1}(f), v_{x+2}(f), f) - S(x, u_{x+1}(f), v_x(f), f)]\\ =\ & c(x+2) - c(x+1) - c(x+1) + c(x) +\\ & \xi \{ b\,[g_2 f(x+2) + \bar{g}_2 f(x+1) - g_1 f(x+1) - \bar{g}_1 f(x)]\\ & + \bar{b}\,[g_2 f(x+3) + \bar{g}_2 f(x+2) - g_1 f(x+2) - \bar{g}_1 f(x+1)]\\ & - b\,[g_2 f(x+1) + \bar{g}_2 f(x) - g_1 f(x) - \bar{g}_1 f(x-1)]\\ & -\bar{b}\,[g_2 f(x+2) + \bar{g}_2 f(x+1) - g_1 f(x+1) - \bar{g}_1 f(x)]\}\\ \ge\ & \xi \{ b\,[g_2(f(x+2) - f(x+1)) + \bar{g}_2(f(x+1) - f(x)) \qquad (4.8)\\ & - \ g_1(f(x+1) - f(x)) - \bar{g}_1(f(x) - f(x-1))]\}\\ & + \xi \{ \bar{b}\,[g_2(f(x+3) - f(x+2)) + \bar{g}_2(f(x+2) - f(x+1))\\ & - g_1(f(x+2) - f(x+1)) - \bar{g}_1(f(x+1) - f(x))]\}\\ \ge\ & 0 \qquad (4.9) \end{aligned}$$

which follows by applying twice Lemma 4.2 with $\zeta_i = g_i$, once with $h(x) = f(x+1) - f(x)$ for the term in the first curly brackets, and once with $h(x) = f(x+2) - f(x-1)$ for the term in the second curly brackets. (The equation above holds indeed for $x = L-2$ too since in that case $g_2 = 0$; in that case, we shall understand $g_2 f(x+3) := 0$).

(ii) Choose $f(x) = 0, \forall x \in \mathbf{X}$. By repeated application of Lemma 4.3 (i), it follows that $T_\xi^n f$ satisfies **MI** and **WC** for $n = 1, 2, \ldots$; moreover, $\lim_{n\to\infty} T_\xi^n f$ satisfies **MI** and **WC**. Hence by Proposition 3.1 (iii), V_ξ satisfies **MI** and **WC**.

(iii) Suppose that V_ξ satsifies **SC(x-1)** for some fixed $0 < x < L-1$. By substituting V_ξ instead of f in (4.7) and applying again Lemma 4.2 (this time we apply the second part of the Lemma; indeed condition (ii) there holds since g_1 cannot be equal to one, and $h(x) = V_\xi(x+1) - V_\xi(x)$ satisfies $h(x) > h(x-1)$ by the assumption). We thus get strict inequality in (4.9). Hence V_ξ satisfies **SC(x)** as well, and similarly we conclude that it satisfies **SC(y)** for any $y \ge x$.

To prove the second claim, we substitute again V_ξ instead of f in (4.7) and consider $x = 0$. Again we have the case of the strict inequality in Lemma 4.2 since $h(x) := V_\xi(x+1) - V_\xi(x)$ satisfies indeed $h(x) - h(x-1) = V_\xi(1) - V_\xi(0) > 0$ (recall that $h(0) := 0$ since $V_\xi(-1) := V_\xi(0)$). We thus get again strict inequality in (4.9). It follows that $V_\xi(0)$ satisfies **SC(0)**, and hence by the first claim, it satisfies **SC(y)** for all $0 \le y < L-1$.

(iv) Fix $x = 0$. Assume $c(1) > c(0)$. It follows that (4.6) holds with strict inequality for any f satisfying **MI** and in particular for $f = V_\xi$. Hence V_ξ which, by Proposition 3.1 (i), is equal to $valS(x, V_\xi)$, satisfies

$$V_\xi(1) - V_\xi(0) \geq c(1) - c(0) > 0.$$

The proof is then established by applying the second part of (iii).

Next assume that c satisfies **SC(0)**. Substituting V_ξ into (4.7) and considering $x = 0$ we get $F(x) > 0$ since we have a strict inequality in (4.8). Hence V_ξ satisfies **SC(0)**. The proof is then established by applying the first part of (iii). □

Proof of Theorem 4.1. (i) Choose some $v \in V^*$. In order to establish the monotonicity of v, it suffices to show that for any $x < L$, $g_1 < g_2$ and any $u \in U^*$,

$$\begin{aligned}\Delta(u, x) := &S(x, u_x, g_2, V_\xi) - S(x, u_x, g_1, V_\xi)\\ &- [S(x-1, u_{x-1}, g_2, V_\xi) - S(x-1, u_{x-1}, g_1, V_\xi)]\end{aligned} \tag{4.10}$$

is positive. Indeed, we show that this implies that $g_x^{sup}(v) \leq g_{x-1}^{inf}(v)$. Suppose $\Delta(x)$ is positive but the latter does not hold, i.e. $g_x^{sup}(v) > g_{x-1}^{inf}(v)$. Set $g_1 = g_{x-1}^{inf}(v)$ and $g_2 = g_x^{sup}(v)$. It follows from Lemma 3.1 (i) that for any $u \in U^*$

$$valS(x, V_\xi) = S(x, u_x, g_2, V_\xi) \leq S(x, u_x, g_1, V_\xi).$$

Hence since $\Delta(x, u)$ is positive, we have by (4.10)

$$S(x-1, u_{x-1}, g_2, V_\xi) < S(x-1, u_{x-1}, g_1, V_\xi) = valS(x-1, V_\xi)$$

where the last equality follows from Lemma 3.1 (i). This however contradicts the definition of the value of the matrix game $S(x-1, V_\xi)$. Hence it is indeed sufficient to show that $\Delta(u, x) > 0$, $\forall u \in U^*$ in order to prove the monotonicity of v. Fix some $u \in U^*$.

$$\begin{aligned}\Delta(x, u) = &\xi(g_2 - g_1)\big(b(u_x)[V_\xi(x) - V_\xi(x-1)] + \bar{b}(u_x)[V_\xi(x+1) - V_\xi(x)]\\ &- \big(b(u_{x-1})[V_\xi(x-1) - V_\xi(x-2)] + \bar{b}(u_{x-1})[V_\xi(x) - V_\xi(x-1)]\big)\big)\\ > &\,0\end{aligned} \tag{4.11}$$

where the last inequality follows from Lemma 4.2 with $\zeta_2 = \bar{b}(u_x)$, $\zeta_1 = \bar{b}(u_{x-1})$ and $h(x) = V_\xi(x) - V_\xi(x-1)$, and since, by Lemma 4.3 (iv), V_ξ satisfies **SC(x)** for all x. This establishes the monotonicity of v.

Choose some $u \in U^*$. From similar arguments as in the first part of the proof, it suffices in order to establish the monotonicity of u, to show that for any $x \leq L$, $b_2 > b_1$, $v \in V^*$,

$$\begin{aligned}\hat{\Delta}(x, v) := &S(x, b_2, v_x, V_\xi) - S(x, b_1, v_x, V_\xi) - [S(x-1, b_2, v_{x-1}, V_\xi)\\ &- S(x-1, b_1, v_{x-1}, V_\xi)] < 0\end{aligned} \tag{4.12}$$

Indeed, we show that this implies that $b_x^{sup}(u) \le b_{x-1}^{inf}(u)$. Suppose $\hat{\Delta}(x,v)$ is negative but the latter does not hold, i.e. $b_x^{sup}(u) > b_{x-1}^{inf}(u)$. Set $b_1 = b_{x-1}^{inf}(u)$ and $b_2 = b_x^{sup}(u)$.

It follows from Lemma 3.1 (i) that for any $v \in V^*$ we have

$$valS(x, V_\xi) = S(x, b_2, v_x, V_\xi) \ge S(x, b_1, v_x, V_\xi).$$

Hence since $\hat{\Delta}(x)$ is negative, we have by (4.12)

$$S(x-1, b_2, v_{x-1}, V_\xi) > S(x-1, b_1, v_{x-1}, V_\xi) = valS(x-1, V_\xi)$$

where the last equality follows from Lemma 3.1 (i). This however contradicts the definition of the value of the matrix game $S(x-1, V_\xi)$. Hence it is indeed sufficient to show that $\hat{\Delta}(x) < 0$ in order to prove the monotonicity of u. Fix some $v \in V^*$.

$$\begin{aligned}\hat{\Delta}(x) = &-\xi(b_2 - b_1)\,(\bar{g}(v_x)[V_\xi(x) - V_\xi(x-1)] + g(v_x)[V_\xi(x+1) - V_\xi(x)] \\ &- (\bar{g}(v_{x-1})[V_\xi(x-1) - V_\xi(x-2)] + g(v_{x-1})[V_\xi(x) - V_\xi(x-1)])) \\ < &\,0 \end{aligned} \tag{4.13}$$

where the last inequality follows from Lemma 4.2 with $\zeta_2 = g(v_x)$, $\zeta_1 = g(v_{x-1})$ and $h(x) = V_\xi(x) - V_\xi(x-1)$, and since V_ξ satisfies **SC(x)** for all x. (Recall that $\mathbf{G}_L = \{0\}$. Hence in the above equation, when $x = L$, $g(v_x)[V_\xi(x+1) - V_\xi(x)]$ is understood to be zero). This establishes the monotonicity of u. □

Lemma 4.1, Theorem 4.1 and Remark 3.1 imply the following

Corollary 4.1. *If the holding cost c satisfies* **MI, WC** *and either* $c(1) > c(0)$ *or* **SC(0)** *then there exist optimal stationary* u^* *and* v^* *that require randomization in not more than* $\min(|\mathbf{B}|, |\mathbf{G}|) - 1$ *states.*

In the next Corollary, we specify two cases where one of the players (or both) have in fact a threshold type optimal policy. The proof is a direct application of Theorem the previous Corollary as well as Remark 3.1.

We consider the case where $\mathbf{B} = \{b_1, b_2\}$ (i.e. the server has only two possible actions); we then use the notation $u_x = (u_x(1), u_x(2))$. We show for this case that the server has an optimal stationary policy of a threshold type.

Similarly, we may consider the case where $\mathbf{G} = \{g_1, g_2\}$, i.e. the flow controller has two actions in all states excluding state L (where the only available action is 0), and $g_1 = 0$. We show that v^* is of threshold type. We use the notation $v_x = (v_x(1), v_x(2))$ for this case.

Corollary 4.2. *Assume that the holding cost c satisfies* **MI, WC** *and either* $c(1) > c(0)$ *or* **SC(0)**. (i) *Assume that* $\mathbf{B} = \{b_1, b_2\}$ *where* $b_1 < b_2$. *Then there exists* $m_u \in \mathbf{X}$ *such that*

$$u_x^* = \begin{cases} (1,0) & \text{if } x > m_u \\ (q_u, \bar{q}_u) & \text{if } x = m_u \\ (0,1) & \text{if } x < m_u \end{cases} \tag{4.14}$$

where $q_u \in [0,1]$ *is some constant. Moreover,* v_x^* *needs no randomizations in any state except for (perhaps)* $x = m_u$.

(ii) *Assume that* $\mathbf{G} = \{0, g\}$ *where* $g > 0$. *Then there exists* $m_v \in \mathbf{X}$ *such that*

$$v_x^* = \begin{cases} (1,0) & \text{if } x > m_v \\ (q_v, \bar{q}_v) & \text{if } x = m_v \\ (0,1) & \text{if } x < m_v \end{cases} \tag{4.15}$$

where $q_v \in [0,1]$ *is some constant. Moreover,* u_x^* *needs no randomizations in any state except for (perhaps)* $x = m_v$.

Remark 4.1. It follows from Corollary 4.2 that if $\mathbf{B} = \{b_1, b_2\}$ and $\mathbf{G} = \{g_1, g_2\}$ and if $m_u \neq m_v$ then no randomization is needed at any state by both u^* and v^*. If $m_u = m_v$ then randomization may be needed only at $x = m_u = m_v$.

Corollary 4.3. *Assume that the holding cost c satisfies* **MI, WC** *and either* $c(1) > c(0)$ *or* **SC(0)**. *Assume that* $\mathbf{G} = \{0, g\}$. *Let* u^*, v^* *be optimal policies where* v^* *is as in Corollary 4.2. Let* m_v *be the threshold used by* v^* *and assume that* $m_v < L - 2$. *Consider the problem with all parameters the same, except that the buffer size* $\hat{L}$ *satisfies* $m_v < \hat{L} < L$. *Then* $\hat{u}^*, \hat{v}^*$ *are optimal policies for the new problem, where* $\hat{u}_x^* := u_x^*$, $\hat{v}_x^* := v_x^*$ *for all* $0 \le x \le \hat{L}$.

Proof. We shall use "hat" to denote quantities that correspond to the buffer $\hat{L}$. Choose any x, $0 \le x \le \hat{L}$. Since $m_v < \hat{L}$, it follows that for any stationary u, the policies $\hat{u}$ and $\hat{v}^*$ (that are the restriction of u and v^* to the states $\{0, ..., \hat{L}\}$) satisfy $V_\xi(x, u, v^*) = \hat{V}_\xi(x, \hat{u}, \hat{v}^*)$. Hence

$$V_\xi(x, u^*, v^*) = \sup_{u \in U} V_\xi(x, u, v^*) = \sup_{\hat{u} \in \hat{U}} \hat{V}_\xi(x, \hat{u}, \hat{v}^*) = \hat{V}_\xi(x, \hat{u}^*, \hat{v}^*).$$

Consider the class of policies (for the system L) denoted by V' where player two always chooses 0 at any $x \ge \hat{L}$. Then for any $v \in V'$ and $x \le \hat{L}$, $V_\xi(x, u^*, v) = \hat{V}_\xi(x, \hat{u}^*, \hat{v})$. Since $v^* \in V'$, we have for $x \le \hat{L}$

$$\begin{aligned} \hat{V}_\xi(x, \hat{u}^*, \hat{v}^*) &= V_\xi(x, u^*, v^*) = \inf_{v \in V} V_\xi(x, u^*, v) \\ &= \inf_{v \in V'} V_\xi(x, u^*, v) = \inf_{\hat{v} \in \hat{V}} \hat{V}_\xi(x, \hat{u}^*, \hat{v}). \end{aligned}$$

Hence $\hat{u}^*, \hat{v}^*$ are optimal when the buffer size is $\hat{L}$. □

5. The Average Cost

Define the expected average cost

$$W(x,u,v) := \overline{\lim_{s\to\infty}} \frac{1}{s} E^{u,v} \left[\sum_{t=0}^{s-1} C(X_t, B_t, G_t) \,|\, X_0 = x \right], \tag{5.1}$$

Define the problem $\overline{\mathbf{Q}}$: Find u, v that achieve

$$W(x) := \sup_{u\in U} \inf_{v\in V} W(x,u,v), \qquad \forall x \in \mathbf{X}. \tag{5.2}$$

Theorem 5.1. (i) *There exists a pair of stationary policies* (u^*, v^*) *that achieves* (5.2) *such that* $W = W(x)$ *does not depend on* x, *and*

$$\begin{aligned} \sup_{u\in U} \inf_{v\in V} W(x,u,v) &= \inf_{v\in V} \sup_{u\in U} W(x,u,v) = \sup_{u\in U} W(x,u,v^*) \\ &= \inf_{v\in V} W(x,u^*,v) = W(x,u^*,v^*). \end{aligned} \tag{5.3}$$

(ii) *If the holding cost* c *satisfies* **MI**, **WC** *and either* $c(1) > c(0)$ *or* **SC(0)** *then there exist stationary* u^* *and* v^* *satisfying (5.3) which are strongly monotone policies.*

In the above theorem, $W(x)$ is called the expected average *value* of the game and the policies (u^*, v^*) are said to be optimal policies.

Proof. (i) follows from [9], Theorem 2.2. The only condition that should be verified is that there exists some state x_0 such that under any policies u and v, the state process reaches eventually x_0. It is easily seen that this condition is indeed verified, with $x_0 = 0$.

(ii) Due to (i), we may restrict the search of u^* and v^* to stationary policies (where the limsup in (5.1) is achieved as a limit). The claim then follows from Theorem 4.1 and [5], Corollary 3.2. that shows that any limit of the discounted optimal policies converges to a policy which is expected average optimal, as the discount factor goes to one. It remains to check Assumption A1 or A2 of that Corollary. However, it is shown in [5], Theorem 3 that the following weaker conditions imply A1:

A5(i) there exists a set $K \subset \mathbf{X}$ and a finite number $\mathcal{B}$ such that under any pair of stationary policies u and v and any initial state x, the mean first passage time to K is at most $\mathcal{B}$.

A5(ii) For each stationary u and v, the Markov chain has no two disjoint closed sets.

In our model, A5(ii) clearly holds, since from every state we can reach 0 under any stationary u and v. A5(i) is satisfied too by choosing $K = \{0\}$. To see that, define $T(x)$ to be the supremum over all policies in (U, V) of the expected hitting time of state zero starting from state x. Consider the transition probabilities

$$\tilde{q}(y \mid x; g; b) := \begin{cases} q(y \mid x; g; b) & \text{if } x \neq 0 \\ 1\{y = 0\} & \text{if } x = 0 \end{cases}$$

Then $T(x) = \sup_{u,v} \sum_{s=1}^{\infty} P_x^{u,v}(X_s \neq 0)$. It follows by Theorem 3.2.1 in Kallenberg [6] that there exist stationary (u,v) that achieve that sup. However, for any stationary (u,v), 0 is a recurrent state that is reachable from any other state, the expected hitting time of 0 under (u,v) is finite, and hence $T(x) < \infty$. □

If one of the players has only two actions, we get in particular the following.

Corollary 5.1. Assume that the holding cost c satisfies **MI**, **WC** and either $c(1) > c(0)$ or **SC(0)**. Assume that $\mathbf{B} = \{b_1, b_2\}$ (or $\mathbf{G} = \{g_1, g_2\}$). Then a stationary optimal policy u^* (or v^* respectively) exists which has the threshold structure described in Corollary 4.2.

It is easily seen that Corollary 4.1 and 4.2 extend also to the expected average case.

Remark 5.1. All the results of this Section hold also for the case that the expected average cost (5.1) is defined through a liminf instead of limsup.

6. Calculating the Optimal Policies

In this Section we assume that both players have only two actions. In particular, the flow controller may either use action 0 or action $g = g_{max}$. The server may choose between b_1 and b_2, where $b_1 = b_{min}$ and $b_2 = b_{max}$. We show how to use the previous structural results in order to compute the optimal policies. We restrict to the expected average cost criterion and assume that the conditions of Corollary 5.1 hold. We know from Corollary 5.1 that there exist optimal threshold policies for both players (given in Corollary 4.2). Hence we may restrict our problem to searching for optimal stationary policies among all threshold ones. We thus calculate the cost under a pair u, v of threshold policies that are characterized by the parameters m_u, q_u and m_v, q_v respectively (see definitions in Corollary 4.2).

We first calculate the costs for the case that $q_u = q_v = 1$, and then obtain the general case. Clearly, we need only to consider the case that $m_v \leq L$. Note that the policy where the flow controller always chooses g

can be identified with $m_v = L$. We note that for a given policy m_v of the flow controller, all the policies such that $m_u > m_v$ have the same cost. In particular, for $q_u = q_v = 1$, all the policies such that $m_u \geq m_v$ have the same cost.

The steady state probabilities. From standard balance arguments we have the following relations between the steady state probabilities $\pi(m_u, 1, m_v, 1)$ (i.e., the steady state probabilities when using m_u, m_v, $q_u = q_v = 1$).

$$\begin{aligned}
\pi_x(m_u, 1, m_v, 1) g\bar{b}_2 &= \pi_{x+1}(m_u, 1, m_v, 1)\bar{g}b_2 \\
&\qquad 0 \leq x < \min(m_u, m_v) - 1 \\
\pi_x(m_u, 1, m_v, 1) g\bar{b}_2 &= \pi_{x+1}(m_u, 1, m_v, 1)\bar{g}b_1 \\
&\qquad x = m_u - 1, m_u < m_v \\
\pi_x(m_u, 1, m_v, 1) g\bar{b}_1 &= \pi_{x+1}(m_u, 1, m_v, 1)\bar{g}b_1 \\
&\qquad m_u \leq x < m_v - 1 \\
\pi_x(m_u, 1, m_v, 1) g\bar{b}_1 &= \pi_{x+1}(m_u, 1, m_v, 1)b_1 \\
&\qquad x = m_v - 1, m_u < m_v \\
\pi_x(m_u, 1, m_v, 1) g\bar{b}_2 &= \pi_{x+1}(m_u, 1, m_v, 1)b_1 \\
&\qquad x = m_v - 1, m_u = m_v \\
\pi_x(m_u, 1, m_v, 1) g\bar{b}_2 &= \pi_{x+1}(m_u, 1, m_v, 1)b_2 \\
&\qquad x = m_v - 1, m_u > m_v
\end{aligned}$$

Denote

$$\begin{aligned}
\zeta_1 &:= \frac{g\bar{b}_2}{\bar{g}b_2}, \\
\zeta_2 &:= \frac{g\bar{b}_2}{\bar{g}b_1} \text{ if } m_u < m_v, \quad 1 \text{ otherwise;} \\
\zeta_3 &:= \frac{g\bar{b}_1}{\bar{g}b_1} \text{ if } m_u < m_v - 1, \quad 1 \text{ otherwise;} \\
\zeta_4 &:= \frac{g\bar{b}_1}{b_1} \text{ if } m_u < m_v, \ \frac{g\bar{b}_2}{b_1} \text{ if } m_u = m_v, \ \frac{g\bar{b}_2}{b_2} \text{ if } m_u > m_v.
\end{aligned}$$

We get

$$\pi_x(m_u, 1, m_v, 1) = \begin{cases} \pi_0(m_u, 1, m_v, 1)\zeta_1^x & 0 \leq x < \min(m_u, m_v) \\ \pi_0(m_u, 1, m_v, 1)\zeta_1^{\min(m_u,m_v)-1}\zeta_2\zeta_3^{x-m_u} & m_u \leq x < m_v \\ \pi_0(m_u, 1, m_v, 1)\zeta_1^{\min(m_u,m_v)-1}\zeta_2\zeta_3^{m_v-m_u-1}\zeta_4 & x = m_v \\ 0 & \text{otherwise} \end{cases} \tag{6.1}$$

Since the steady state probabilities sum to one, (6.2) yields

$$\pi_0(m_u,1,m_v,1) = \left[\sum_{x=0}^{min(m_u,m_v)-1} \zeta_1^x + \zeta_1^{min(m_u,m_v)-1}\zeta_2\left(\sum_{x=0}^{m_v-m_u-1}\zeta_3^x + \zeta_3^{m_v-m_u-1}\zeta_4\right)\right]^{-1} = \left[\frac{1-\zeta_1^{min(m_u,m_v)}}{1-\zeta_1} + \zeta_1^{min(m_u,m_v)-1}\zeta_2\left(\frac{1-\zeta_3^{m_v-m_u}}{1-\zeta_3} + \zeta_3^{m_v-m_u-1}\zeta_4\right)\right]^{-1} \tag{6.2}$$

The steady state probabilities are now obtained by substituting (6.2) into (6.1).

The expected average costs. When both u and v are of threshold type, we denote (with some abuse of notation) $W(m_u, q_u, m_v, q_v) := W(x, u, v)$. The expected average cost is given by

$$\begin{aligned} W(m_u,1,m_v,1) &= \sum_{x=0}^{m_v} c(x)\pi_x(m_u,1,m_v,1) \\ &\quad + \theta(b_2)\sum_{x=0}^{m_u-1}\pi_x(m_u,1,m_v,1) + \theta(b_1)\sum_{x=m_u}^{m_v}\pi_x(m_u,1,m_v,1) \\ &\quad + \rho(g)(1-\pi_{m_v}(m_u,1,m_v,1)) + \rho(0)\pi_{m_v}(m_u,1,m_v,1) \end{aligned}$$

Next we calculate for any q_u, and $q_v = 1$. To do that, we consider the regeneration points as the times that $X_t = m_u$. We call a "cycle" the duration between two consecutive visits to that state. The expected cost is then given by the expected cost per cycle divided by the expected cycle duration. We observe that with probability q_u, (respectively $\bar{q}_u$) the expected cost per cycle is equal to the one obtained if the server uses the policy $(m_u, 1)$, (respectively, $(m_u, 0) = (m_u + 1, 1)$); moreover, with probability q_u, (respectively $\bar{q}_u$) the expected cycle duration is equal to the one obtained if the server uses the policy $(m_u, 1)$, (respectively, $(m_u, 0) = (m_u + 1, 1)$). Finally, we note that the expected cycle durations are just the inverse of the steady state probabilities of visiting state m_u. This yields

$$\begin{aligned} &W(m_u,q_u,m_v,1) = \\ &\frac{q_u W(m_u,1,m_v,1)\pi_{m_u}^{-1}(m_u,1,m_v,1)}{\pi_{m_u}^{-1}(m_u,q_u,m_v,1)} \\ &+ \frac{\bar{q}_u W(m_u+1,1,m_v,1)\pi_{m_u}^{-1}(m_u+1,1,m_v,1)}{\pi_{m_u}^{-1}(m_u,q_u,m_v,1)} \end{aligned}$$

where

$$\pi_{m_u}^{-1}(m_u, q_u, m_v, 1) = q_u \pi_{m_u}^{-1}(m_u, 1, m_v, 1) + \bar{q}_u \pi_{m_u}^{-1}(m_u + 1, 1, m_v, 1)$$

Finally, we get by similar arguments the cost for any q_u and q_v. We consider the regeneration points as the times that $X_t = m_v$. We thus get

$$\begin{aligned} &W(m_u, q_u, m_v, q_v) = \\ &\frac{q_v W(m_u, q_u, m_v, 1)\pi_{m_v}^{-1}(m_u, q_u, m_v, 1)}{q_v \pi_{m_v}^{-1}(m_u, q_u, m_v, 1) + \bar{q}_v \pi_{m_v}^{-1}(m_u, q_u, m_v + 1, 1)} \\ &+ \frac{\bar{q}_v W(m_u, q_u, m_v + 1, 1)\pi_{m_v}^{-1}(m_u + 1, q_u, m_v, 1)}{q_v \pi_{m_v}^{-1}(m_u, q_u, m_v, 1) + \bar{q}_v \pi_{m_v}^{-1}(m_u, q_u, m_v + 1, 1)} \end{aligned} \tag{6.3}$$

Thus the original dynamic game reduces to the problem of

$$\max_{m_u, q_u} \min_{m_v, q_v} W(m_u, q_u, m_v, q_v) \tag{6.4}$$

where $W(m_u, q_u, m_v, q_v)$ is given in (6.3). It follows from Corollary 5.1 that in order to solve (6.4) it suffices to restrict to the cases where either $m_u = m_v$ or $q_u = q_v = 1$.

Acknowledgment. The work of this author was supported by the Chateaubriand fellowship from the French embassy in Israel.

References

[1] E. Altman, Flow control using the theory of zero-sum Markov games, *Proceedings of the 31st IEEE Conference on Decision and Control*, Tucson, Arizona, pp. 1632–1637, December 1992.

[2] E. Altman and G. Koole, Stochastic Scheduling Games with Markov Decision Arrival Processes, *Journal Computers and Mathematics with Appl.*, 3rd special issue on Differential Games, pp. 141–148, 1993.

[3] E. Altman and N. Shimkin, Individually Optimal Dynamic Routing in a Processor Sharing System: Stochastic Game Analysis, EE Pub No. 849, August 1992. Submitted.

[4] M. T. Hsiao and A. A. Lazar, Optimal Decentralized Flow Control of Markovian queueing Networks with Multiple Controller, *CTR Technical Report*, CUCTR-TR-19, Columbia University, 1986.

[5] A. Federgruen, On N-person stochastic Games with denumerable state space, *Adv. Appl Prob.* **10**, pp. 452–471, 1978.

[6] L. C. M. Kallenberg, *Linear Programming and Finite Markovian Control Problems*, Math. Centre Tracts 148, Amsterdam, 1983.

[7] H.-U. Küenle, On the optimality of (s,S)-strategies in a minimax inventory model with average cost criterion, *Optimization* **22** No. 1, pp. 123–138, 1991.

[8] J. M. McNamara, S. Merad and E. J. Collins, The Hawk-Dove game as an average-cost problem, *Adv. Appl. Prob.* **23**, pp. 667–682, 1991.

[9] T. Parthasarathy and M. Stern, Markov games – a survey, *Differential Games and Control Theory II*, Roxin, Liu and Sternberg, 1977.

[10] T.E.S. Raghavan and J.A. Filar, Algorithms for Stochastic Games - A survey, *Zeitschrift für OR*, vol 35, pp. 437–472, 1991.

[11] L. S. Shapely, Stochastic games, *Proceeding of the National Academy of Sciences USA* **39**, pp. 1095–1100, 1953.

[12] N. N. Vorob'ev, *Game Theory*, Lectures for Economists and Systems Scientists, Springer-Verlag, 1977.

Eitan Altman
INRIA
Centre Sophia Antipolis
06565 Valbonne Cedex, France

PART V

Applications

Capital Accumulation Subject to Pollution Control: A Differential Game with a Feedback Nash Equilibrium

David W. K. Yeung and Michael Tow Cheung

Abstract. This paper studies a differential game of capital accumulation subject to pollution control. In the industrial sector, capital is used to produce output, which also creates pollution. Investment is chosen to maximise the present value of net profit. The policy maker (government) values output positively but pollution negatively. Its objective is to tax industrial output and spend the proceeds on pollution abatement, to maximise a social welfare function containing consumption and pollution as arguments. We solve the game for its feedback Nash equilibrium, and the resulting paths of capital and pollution accumulation are obtained explicitly. Finally, the feedback Nash and open-loop Nash solutions of the game are compared.

1. Introduction

In the past two decades, increasing attention has been paid to the management of the physical waste which accompanies production and consumption. A number of considerations combine to make theoretical analysis in this area difficult and complicated. One is the dynamic (intertemporal) nature of pollution and the economic activities associated with it, especially production for the purpose of capital accumulation. Environmental studies have generally been couched in the framework of optimal control, with pollution being the operational state variable [1–5]. In Lin [6] (see also Hartl [14]), pollution management is analysed together with optimal investment and employment over time. Another consideration is the game theoretic nature of the relationship between the policy maker(s) and the individual(s) responsible for pollution. The application of game theory to pollution is a recent development. Misiolek [7], Yao [8], Millerman and Price [9] approached the problem in terms of (static) principal-agent theory, while in Yeung [10] industrial pollution management is studied in a differential game without capital accumulation.

In the present paper we construct a differential game with capital accumulation, to incorporate the dynamic and game theoretic aspects of the problem of pollution. Investment is chosen to maximise the present value of net profit in the industrial sector. Capital is used in production, which yields output and generates pollution. The policy maker (or government) values output and consumption positively but pollution negatively. Its objective is to tax industrial output and spend the proceeds on pollution abatement, to maximise a social welfare function containing consumption and pollution as arguments.

Two dynamic processes—one for capital accumulation and the other for pollution accumulation—are modelled in the resulting differential game. Capital accumulation depends on investment in the industrial sector and on the rate of depreciation of capital. Pollution dynamics is determined by industrial output, the level of pollution abatement, and the pollutant's natural rate of decay. We solve the game for its feedback Nash equilibrium, and derive a set of state-dependent feedback Nash equilibrium strategies. The resulting time paths of capital and pollution accumulation are obtained explicitly. This solution is then compared with the open-loop Nash solution to the game.

The paper is organized as follows. Section 2 describes the structure of the model. The feedback Nash equilibrium solution of the game is obtained in Section 3. Properties of the time paths of capital and pollution are studied in Section 4. Section 5 compares the feedback Nash and open-loop Nash solutions of the game. Section 6 concludes.

2. Model Structure

Since solving for a feedback Nash equilibrium solution to a differential game requires the explicit solution of a system of partial differential equations, for analytical and tractability reasons we explicitly specify the structure of our model. Consider, therefore, an economy with two decision makers—a government and an entrepreneur. In the industrial sector, the entrepreneur uses capital to produce output, which also creates pollution. Part of output is taxed by the government, to finance pollution abatement. The entrepreneur's decision problem in the industrial sector is to choose an investment strategy to maximize the present value of net profit, which is equal to:

$$\int_0^\infty e^{-rs} R\left[gK(s)^{\frac{1}{2}} - T(s) - u(s)^{\frac{3}{2}}\right] ds, \tag{1}$$

subject to the capital accumulation dynamics:

$$\dot{K}(s) = u(s)^{\frac{1}{2}} K(s)^{\frac{1}{2}} - \delta K(s), \tag{2}$$

where:

$K(s)$ is the capital stock at time s,

$u(s)$ is investment input at time s,

$T(s)$ is the industrial output taxed by the government,

δ is the rate of depreciation of capital,

R is net revenue per unit of output,

r is the discount rate,

$gK(s)^{\frac{1}{2}}$ is the production function for output,

g is a productivity parameter,

$u(s)^{\frac{1}{2}}K(s)^{\frac{1}{2}}$ is the production function for capital, and

$u(s)^{\frac{3}{2}}$ gives the user cost of investment inputs in terms of output.

The government chooses a taxation strategy to maximize the social welfare function:

$$\int_0^{\infty} e^{-rs}\left[\beta\left(gK(s)^{\frac{1}{2}} - T(s) - u(s)^{\frac{3}{2}}\right) - \omega P(s)\right] ds, \tag{3}$$

subject to the pollution accumulation dynamics:

$$\dot{P}(s) = agK(s)^{\frac{1}{2}} - \left[T(s)P(s)\right]^{\frac{1}{2}} - \Omega P(s), \tag{4}$$

where:

$P(s)$ is the stock of pollution at time s,

β is the welfare weight of output,

$\omega < \beta$ is the welfare weight of pollution,

$a < 1$ is the pollution created when one unit of output is produced,

Ω is the natural rate of decay of the pollutant, and

$T(s)^{\frac{1}{2}}P(s)^{\frac{1}{2}}$ is the pollution clean-up function.

The initial values of the stocks of capital and pollution $K(0) = K_0$ and $P(0) = P_0$ are assumed to be given.

3. Feedback Nash Equilibrium Solution

To avoid the problem of time inconsistency, we consider the feedback Nash equilibrium (subgame perfect) solution of the game described in the above section.

Definition 3.1. A pair of strategies:

$$\{u^*(t) = \phi^*(K,P,t),\ T^*(t) = \gamma^*(K,P,t)\}$$

constitutes a feedback Nash equilibrium solution for the game (1)–(4), if there exist functionals $V(K,P,t)$ and $U(K,P,t)$ which satisfy the following conditions:

$$\begin{aligned} V(K,P,t) &= \int_t^\infty e^{-rs} R\left[K^*(s)^{\frac{1}{2}} - \gamma^*(K,P,s) - [\phi^*(K,P,s)]^{\frac{3}{2}}\right] ds \\ &\geq \int_t^\infty e^{-rs} R\left[K^*(s)^{\frac{1}{2}} - \gamma^*(K,P,s) - [\phi(K,P,s)]^{\frac{3}{2}}\right] ds, \end{aligned}$$

$\forall\ \phi(K,P,s)$ in the feasible set of choices of $u(t)$, which satisfies the accumulation dynamics (2) and (4);

$$\begin{aligned} U(K,P,t) &= \int_t^\infty e^{-rs}\left[\beta\left(gK^*(s)^{\frac{1}{2}} - \gamma^*(K,P,s) - [\phi^*(K,P,s)]^{\frac{3}{2}}\right)\right. \\ &\qquad \left. - \omega P^*(s)\right] ds \\ &\geq \int_t^\infty e^{-rs}\left[\beta\left(gK^*(s)^{\frac{1}{2}} - \gamma(K,P,t) - [\phi^*(K,P,s)]^{\frac{3}{2}}\right)\right. \\ &\qquad \left. - \omega P^*(s)\right] ds, \end{aligned}$$

$\forall\ \gamma(K,P,s)$ in the feasible set of choices of $T(t)$, which satisfies the accumulation equations (2) and (4); and where:

$$\begin{aligned} \dot{K}^*(s) &= \phi^*(K,P,s)^{\frac{1}{2}} K^*(s)^{\frac{1}{2}} - \delta K^*(s), \\ \dot{P}^*(s) &= agK^*(s)^{\frac{1}{2}} - [\gamma^*(K,P,s)P^*(s)]^{\frac{1}{2}} - \Omega P^*(s). \end{aligned}$$

□

Definition 3.1 implies that the following Hamilton-Jacobi-Bellman equations must be satisfied (Basar and Olsder [11]):

$$
\begin{aligned}
-V_t = \max_u \Big\{ & e^{-rt} R\Big[gK(t)^{\frac{1}{2}} - T(t) - u(t)^{\frac{3}{2}}\Big] \\
& + V_K\Big[u(t)^{\frac{1}{2}} K(t)^{\frac{1}{2}} - \delta K(t)\Big] \\
& + V_P\Big[agK(t)^{\frac{1}{2}} - \big[T(t)P(t)\big]^{\frac{1}{2}} - \Omega P(t)\Big]\Big\}, \\
-U_t = \max_T \Big\{ & e^{-rt}\Big[\beta\Big(gK(t)^{\frac{1}{2}} - T(t) - u(t)^{\frac{3}{2}}\Big) - \omega P(t)\Big] \\
& + U_K\Big[u(t)^{\frac{1}{2}} K(t)^{\frac{1}{2}} - \delta K(t)\Big] \\
& + U_P\Big[agK(t)^{\frac{1}{2}} - \big[T(t)P(t)\big]^{\frac{1}{2}} - \Omega P(t)\Big]\Big\}.
\end{aligned}
\tag{5}
$$

Maximizing the terms inside the curly brackets in (5) yields:

$$
u(t) = \frac{e^{rt}}{3R} V_K K(t)^{\frac{1}{2}}, \tag{6}
$$

$$
T(t) = \frac{e^{2rt}}{4\beta^2} U_P^2 P(t). \tag{7}
$$

Substituting (6) and (7) into (5) we obtain a pair of simultaneous non-linear partial differential equations, which yields the solution:

$$
\begin{aligned}
V(K,P,t) &= e^{-rt}\left[\mathcal{A}_1 K(t)^{\frac{1}{2}} + \mathcal{C}_1 P(t) + \mathcal{D}_1\right], \\
U(K,P,t) &= e^{-rt}\left[\mathcal{A}_2 K(t)^{\frac{1}{2}} + \mathcal{C}_2 P(t) + \mathcal{D}_2\right],
\end{aligned}
\tag{8}
$$

where:

$$
\begin{aligned}
\mathcal{C}_2 &\equiv 2\beta\left\{(r+\Omega) - \sqrt{(r+\Omega)^2 + \omega/\beta}\right\}, \\
\mathcal{A}_2 &\equiv \frac{g(\beta + a\mathcal{C}_2)}{r+\delta/2}, \\
\mathcal{C}_1 &\equiv \frac{-R\mathcal{C}_2^2}{4\beta^2(r+\Omega-\mathcal{C}_2/2\beta)}, \\
\mathcal{A}_1 &\equiv \frac{g(R + a\mathcal{C}_1)}{r+\delta/2}, \\
\mathcal{D}_2 &\equiv \frac{1}{r}\left[\frac{1}{2}\frac{\mathcal{A}_1^{\frac{1}{2}}\mathcal{A}_2}{(6R)^{\frac{1}{2}}} - \frac{\beta\mathcal{A}_1^{\frac{3}{2}}}{(6R)^{\frac{3}{2}}}\right], \\
\mathcal{D}_1 &\equiv \frac{1}{r}\left[\left(\frac{1}{2} - \frac{1}{6R}\right)\frac{\mathcal{A}_1^{\frac{3}{2}}}{(6R)^{\frac{1}{2}}}\right].
\end{aligned}
$$

Upon evaluating V_K and U_P in (8) and substituting into (6) and (7), we obtain the feedback Nash equilibrium strategies:

$$\begin{aligned} u^*(t) = \phi^*(K,P,t) &= \frac{A_1}{6R} \\ &= \frac{g\left[1 - \frac{a\{r+\Omega-\sqrt{(r+\Omega)^2+\omega/\beta}\}^2}{\sqrt{(r+\Omega)^2+\omega/\beta}}\right]}{6(r+\delta/2)} \end{aligned} \tag{9}$$

$$\begin{aligned} T^*(t) = \gamma^*(K,P,t) &= \frac{C_2^2}{4\beta^2}P(t) \\ &= \{r+\Omega-\sqrt{(r+\Omega)^2+\omega/\beta}\}^2 P(t). \end{aligned} \tag{10}$$

Remark 3.1. The feedback Nash equilibrium is subgame perfect (Selton [12]), in which the pair of solution strategies (9) and (10) constitutes an equilibrium for every possible subgame starting at $t \in [0,\infty)$.

Remark 3.2. Although strictly speaking $T(t)$ is not a Pigouvian tax, the government devises a levy clean-up scheme to divert output (in the form of tax) into clean-up activities, so that pollution, which is a by-product of output external to the firm's profit maximizing decision, is reduced.

4. Capital and Pollution Accumulation in Game Equilibrium

In this section, we characterise capital accumulation and the build-up of pollution in the game equilibrium. Substituting (9) and (10) into (2) and (4):

$$\dot{K}(s) = \left(\frac{A_1}{6R}\right)^{\frac{1}{2}} K(s)^{\frac{1}{2}} - \delta K(s) \tag{11}$$

$$\dot{P}(s) = agK(s)^{\frac{1}{2}} - \left(\Omega + \frac{C_2}{2\beta}\right) P(s). \tag{12}$$

To solve this system of non-linear differential equations, let us introduce the transformation $k(s) = \sqrt{K(s)}$. (11) becomes:

$$\dot{k}(s) = \frac{1}{2}\left[\left(\frac{A_1}{6R}\right)^{\frac{1}{2}} - \delta k(s)\right]. \tag{11'}$$

In addition, let:

$$\hat{k}(s) = k(s) - \frac{\left(\frac{A_1}{6R}\right)^{\frac{1}{2}}}{\delta},$$

$$\hat{P}(s) = P(s) - \frac{ag\left(\frac{A_1}{6R}\right)^{\frac{1}{2}}}{(\Omega + C_2/2\beta)\delta}.$$

We can then re-write (11), (11') and (12) as:

$$\dot{\hat{k}}(s) = -\frac{1}{2}\delta\hat{k}(s), \tag{13}$$

$$\dot{\hat{P}}(s) = ag\hat{k}(s) - \left(\Omega + C_2/2\beta\right)\hat{P}(s). \tag{14}$$

(13) and (14) yield a pair of homogeneous linear differential equations, with characteristic roots $-\delta/2$ and $-\left(\Omega + C_2/2\beta\right)$. The solution can then be obtained as:

$$\begin{aligned}
\hat{k}(s) &= \exp\left[-\left(\frac{\delta}{2}\right)s\right]\left[K_0^{\frac{1}{2}} - \frac{(A_1/6R)^{\frac{1}{2}}}{\delta}\right],\\
\hat{P}(s) &= \exp\left[-\left(\frac{\delta}{2}\right)s\right]\left[\frac{-ag}{1+\delta/2}\left(K_0^{\frac{1}{2}} - \frac{(A_1/6R)^{\frac{1}{2}}}{\delta}\right)\right]\\
&\quad + \exp\left[-\left(\Omega + \frac{C_2}{2\beta}\right)s\right]\left[P_0 - \frac{ag(A_1/6R)^{\frac{1}{2}}}{(\Omega + C_2/\beta)\delta}\right.\\
&\quad \left. + \frac{ag}{1+\delta/2}\left(K_0^{\frac{1}{2}} - \frac{(A_1/6R)^{\frac{1}{2}}}{\delta}\right)\right]
\end{aligned} \tag{15}$$

(15) gives the time paths of $\hat{k}$ and $\hat{P}$ in feedback Nash equilibrium.

Since both characteristic roots are real and negative, $\hat{k} \to 0$ and $\hat{P} \to 0$ as $s \to \infty$, so that system (15) has the stable node $(0,0)$. Transforming back to the original variables, we obtain the steady state values of the stocks of capital and pollution as:

$$K^{ss} = \lim_{s\to\infty} K(s) = \frac{g\left[1 - \frac{a\left\{r+\Omega-\sqrt{(r+\Omega)^2+\omega/\beta}\right\}^2}{\sqrt{(r+\Omega)^2+\omega/\beta}}\right]}{6(r+\delta/2)\delta^2}, \tag{16}$$

$$P^{ss} = \lim_{s\to\infty} P(s) = \frac{ag^{\frac{3}{2}}\left[1 - \frac{a\left\{r+\Omega-\sqrt{(r+\Omega)^2+\omega/\beta}\right\}^2}{\sqrt{(r+\Omega)^2+\omega/\beta}}\right]^{\frac{1}{2}}}{\delta\sqrt{6(r+\delta/2)}\left\{-r + \sqrt{(r+\Omega)^2 + \omega/\beta}\right\}}. \tag{17}$$

Since $r + \Omega - \sqrt{(r+\Omega)^2 + \omega/\beta} \leq \sqrt{\omega/\beta}$, $\omega < \beta$ and $a < 1$, in (16) and (17), we have $K^{ss} > 0$, $P^{ss} > 0$.

Properties of the game equilibrium in our model can be obtained from the closed form solution (15), (16) and (17). In particular, we note the following propositions:

Proposition 4.1. *An increase in productivity in the industrial sector, represented by an increase in the parameter g, increases the equilibrium rate of investment and the stationary equilibrium levels of capital and pollution.*

Proof. From (1), (16) and (17):

$$\frac{\partial u^*(t)}{\partial g} = \frac{1 - \frac{a\left\{r+\Omega-\sqrt{(r+\Omega)^2+\omega/\beta}\right\}}{\sqrt{(r+\Omega)^2+\omega/\beta}}}{6(r+\delta/2)} > 0,$$

$$\frac{\partial K^{ss}}{\partial g} = \frac{1 - \frac{a\left\{r+\Omega-\sqrt{(r+\Omega)^2+\omega/\beta}\right\}}{\sqrt{(r+\Omega)^2+\omega/\beta}}}{6(r+\delta/2)\delta^2} > 0,$$

$$\frac{\partial P^{ss}}{\partial g} = \frac{\frac{3a\sqrt{g}}{2}\left[1 - \frac{a\left\{r+\Omega-\sqrt{(r+\Omega)^2+\omega/\beta}\right\}^2}{\sqrt{(r+\Omega)^2+\omega/\beta}}\right]^{\frac{1}{2}}}{\delta\sqrt{6(r+\delta/2)}\left\{-r+\sqrt{(r+\Omega)^2+\omega/\beta}\right\}} > 0. \qquad \square$$

Proposition 4.2. *A decrease in the pollution generated by a unit of production, represented by a decrease in the parameter a, increases the steady state capital stock, while the steady state level of pollution decreases (increases) if:*

$$\frac{a\left\{r+\Omega-\sqrt{(r+\Omega)^2+\omega/\beta}\right\}^2}{\sqrt{(r+\Omega)^2+\omega/\beta}} < (>)\frac{2}{3}.$$

Proof. From (16) and (17), we have:

$$\frac{\partial K^{ss}}{\partial a} = \frac{\left\{r+\Omega-\sqrt{(r+\Omega)^2+\omega/\beta}\right\}^2}{6\delta^2(r+\delta/2)\sqrt{(r+\Omega)^2+\omega/\beta}} < 0,$$

$$\frac{\partial P^{ss}}{\partial a} = \left[1 - \frac{3a\left\{r+\Omega-\sqrt{(r+\Omega)^2+\omega/\beta}\right\}^2}{2\sqrt{(r+\Omega)^2+\omega/\beta}}\right]\left(\frac{g^{\frac{3}{2}}}{\mathcal{M}}\right)$$

$$\begin{matrix} > \\ < \end{matrix} 0 \text{ if } \frac{a\left\{r+\Omega-\sqrt{(r+\Omega)^2+\omega/\beta}\right\}^2}{\sqrt{(r+\Omega)^2+\omega/\beta}} \begin{matrix} < \\ > \end{matrix} \left(\frac{2}{3}\right),$$

where

$$\mathcal{M} \equiv \delta\sqrt{6(r+\delta/2)}\left\{-r+\sqrt{(r+\Omega)^2+\omega/\beta}\right\} \left[1 - \frac{a\left\{r+\Omega-\sqrt{(r+\Omega)^2+\omega/\beta}\right\}^2}{\sqrt{(r+\Omega)^2+\omega/\beta}}\right]^{-\frac{1}{2}}.$$

$\partial P^{ss}/\partial a$ can clearly be of either sign. $\square$

Proposition 4.3. *A reduction in the rate of depreciation of capital, represented by a fall in the parameter δ, increases the steady state stocks of capital and pollution.*

Proof. We have from (16) and (17):

$$\frac{\partial K^{ss}}{\partial \delta} = -g\left[1 - \frac{a\{r+\Omega-\sqrt{(r+\Omega)^2+\omega/\beta}\}^2}{\sqrt{(r+\Omega)^2+\omega/\beta}}\right]\frac{6[2\delta(r+\delta/2)+\delta^2/2]}{[6(r+\delta/2)\delta^2]^2}$$
$$< 0,$$

$$\frac{\partial P^{ss}}{\partial \delta} = \left\{-ag^{\frac{3}{2}}\left[1 - \frac{a\{r+\Omega-\sqrt{(r+\Omega)^2+\omega/\beta}\}^2}{\sqrt{(r+\Omega)^2+\omega/\beta}}\right]^{\frac{1}{2}}\right.$$
$$\left.\{-r+\sqrt{(r+\Omega)^2+\omega/\beta}\}\left[\sqrt{6(r+\delta)}+\frac{\delta}{4\sqrt{6(r+\delta/2)}}\right]\right\}$$
$$\div\ [\delta\sqrt{6(r+\delta/2)}\{-r+\sqrt{(r+\Omega)^2+\omega/\beta}\}] < 0. \quad \|$$

Proposition 4.4. *An increase in the welfare weight the government attaches to pollution, represented by an increase in the parameter ω, reduces the steady state stocks of capital and pollution.*

Proof. From (16), we have:

$$\frac{\partial K^{ss}}{\partial \omega} = \frac{-ag(\omega/\beta)}{12\beta\delta^2(r+\delta/2)[(r+\Omega)^2+\omega/\beta]^{\frac{3}{2}}} < 0.$$

Since in (17) can be written

$$P^{ss} = \frac{ag(K^{ss})^{\frac{1}{2}}}{\delta\sqrt{6(r+\delta/2)}\{-r+\sqrt{(r+\Omega)^2+\omega/\beta}\}},$$

we have:

$$\frac{\partial P^{ss}}{\partial \omega} = \frac{1}{\delta\sqrt{6(r+\delta/2)}\{-r+\sqrt{(r+\Omega)^2+\omega/\beta}\}}$$
$$\left[ag\delta\sqrt{(r+\delta/2)}\{-r+\sqrt{(r+\Omega)^2+\omega/\beta}\}\frac{1}{2}(K^{ss})^{-\frac{1}{2}}\frac{\partial K^{ss}}{\partial \omega}\right.$$
$$\left.-\frac{ag\delta\sqrt{6(r+\delta/2)}}{2\beta\sqrt{(r+\Omega)^2+\omega/\beta}}(K^{ss})^{\frac{1}{2}}\right] < 0.$$

□

5. Comparing the Feedback Nash and Open-Loop Nash Solutions

We can compare the solution obtained above with the open-loop Nash equilibrium of the game (1)–(4). To derive the latter, we make use of the transformation $k(s) = \sqrt{K(s)}$ to write (2) as:

$$\dot{k}(s) = \frac{1}{2}\left[u(s)^{\frac{1}{2}} - \delta k(s)\right]. \tag{18}$$

Using $k(s)$ as the state variable, we obtain the current value Hamiltonians as:

$$\begin{aligned}\mathcal{H}_F &= R\big[gk(t) - T(t) - u(t)^{\frac{3}{2}}\big] + \lambda_F(t)\left[\frac{1}{2}\big(u(t)^{\frac{1}{2}} - \delta k(t)\big)\right] \\ &\quad + \theta_F(t)\big[agk(t) - T(t)^{\frac{1}{2}}P(t)^{\frac{1}{2}} - \Omega P(t)\big], \\ \mathcal{H}_G &= \beta\big[gk(t) - T(t) - u(t)^{\frac{3}{2}}\big] - \omega P(t) + \lambda_G(t)\left[\frac{1}{2}\big(u(t)^{\frac{1}{2}} - \delta k(t)\big)\right] \\ &\quad + \theta_G(t)\big[agk(t) - T(t)^{\frac{1}{2}}P(t)^{\frac{1}{2}} - \Omega P(t)\big],\end{aligned} \tag{19}$$

where $\lambda_F(t)$, $\theta_F(t)$, $\lambda_G(t)$ and $\theta_G(t)$ are co-state variables.

First order conditions require:

$$\begin{aligned}u(t) &= \frac{\lambda_F(t)}{6R}, \\ T(t) &= \frac{\theta_G(t)^2 P(t)}{4\beta^2}.\end{aligned} \tag{20}$$

The adjoint equations are:

$$\begin{aligned}\dot{\lambda}_F(t) &= \left(r + \frac{\delta}{2}\right)\lambda_F(t) - gR - ag\theta_F(t), \\ \dot{\theta}_F(t) &= \left[r + \Omega + \frac{1}{2}T(t)^{\frac{1}{2}}P(t)^{-\frac{1}{2}}\right]\theta_F(t), \\ \dot{\lambda}_G(t) &= \left(r + \frac{\delta}{2}\right)\lambda_G(t) - \beta g - ag\theta_G(t), \\ \dot{\theta}_G(t) &= \left[r + \Omega + \frac{1}{2}T(t)^{\frac{1}{2}}P(t)^{-\frac{1}{2}}\right]\theta_G(t) + \omega.\end{aligned} \tag{21}$$

The steady state values of the co-state variables are then:

$$\begin{aligned}\lambda_F^{S*} &= \frac{gR}{r + \delta/2}, \\ \theta_F^{S*} &= 0, \\ \lambda_G^{S*} &= \frac{\beta g + ag\{r + \Omega - \sqrt{(r+\Omega)^2 + \omega/\beta}\}}{r + \delta/2}, \\ \theta_G^{S*} &= 2\beta\left[r + \Omega - \sqrt{(r+\Omega)^2 + \omega/\beta}\right].\end{aligned} \tag{22}$$

The steady state values of k, K and P are obtained as:

$$k^{S*} = \frac{g^{\frac{1}{2}}}{\delta\sqrt{6(r+\delta/2)}}, \tag{23}$$

$$K^{S*} = [= (k^{S*})^2] = \frac{g}{\delta^2 6(r+\delta/2)}, \tag{24}$$

$$P^{S*} = \frac{ag^{\frac{3}{2}}}{\delta\sqrt{(6(r+\delta/2)}\{-r+\sqrt{(r+\Omega)^2+\omega/\beta}\}}. \tag{25}$$

Proposition 5.1. *The rate of investment and stationary levels of capital and pollution in the open-loop Nash solution are higher than the corresponding values in the feedback Nash equilibrium solution.*

Proof. Since $1 - \frac{a\left\{r+\Omega-\sqrt{(r+\Omega)^2+\omega/\beta}\right\}^2}{\sqrt{(r+\Omega)^2+\omega/\beta}} \leq 1$, comparing $u(t)$ in (20) with $u(t)$ in (6), K^{S*} in (24) with K^{ss} in (16), and P^{S*} in (25) with P^{ss} in (17), Proposition 5.1 obtains. □

6. Conclusions

This paper develops a differential game of capital accumulation and pollution control in a model with a government and an industrial sector. We solve for the (subgame perfect) feedback Nash equilibrium solution of the game, and obtain explicitly the game equilibrium accumulation dynamics of capital and pollution. Various properties of the equilibrium follow from this closed form solution. An increase in productivity would lead to both higher output and higher pollution. An increase in the government's welfare weight on pollution or an increase in the rate of depreciation of capital decreases the steady state levels of capital and pollution. A decrease in the amount of pollution generated per unit of output (e.g. brought about by a cleaner technology) would definitely increase steady state capital stock, but the steady state level of pollution may increase or decrease. Finally, the feedback Nash equilibrium is compared to the open-loop Nash solution of the game, and it is shown that stationary levels of output and pollution are higher in the latter case.

References

[1] d'Arge, R.C. and K.C. Kogiku (1972), Economic growth and environment. *Review of Economic Studies*, **40**, 61–78.

[2] Plourde, C.G. (1972), A model of waste accumulation and disposal. *Canadian Journal of Economics*, **5**, 119–125.

[3] Luski, R. (1976), A model of recycling and pollution control. *Canadian Journal of Economics*, **9**, 91–101.

[4] Plourde, C.G. and D. Yeung (1989), A model of industrial pollution in a stochastic environment. *Journal of Environmental Economics and Management*, **16**, 91–105.

[5] Forster, B.A. (1977), On a one state variable pptimal control problem — consumption-pollution trade-offs. In S.J. Turnovsky ed., *Applications of Control Theory to Economic Analysis*, New York: North Holland.

[6] Lin, W.T. (1987), The control of environmental pollution and optimal investment and employment decisions. *Optimal Control Applications and Methods*, **8**, 21–36.

[7] Misiolek, W.S. (1988), Pollution control through price incentives: the role of rent seeking costs in monopoly markets. *Journal of Environmental Economics and Management*, **15**, 1–8.

[8] Yao, D.A. (1988), Strategic responses to automobile emissions control: A game theoretic analysis. *Journal of Environmental Economics and Management*, **15**, 419–438.

[9] Millerman, S.R. and P. Prince (1989), Firm incentives to promote technological change in pollution control. *Journal of Environmental Economics and Management*, **17**, 247–265.

[10] Yeung, D. (1992), A differential game of industrial pollution management. *Annals of Operations Research*, **37**, 297–311.

[11] Basar, T. and G.J. Olsder (1982), *Dynamic Noncooperative Game Theory*. New York: Academic Press.

[12] Selton, R. (1975), Reexamination of the perfectness concept for equilibrium points in extensive games. *International Journal of Game Theory*, **4**, 25–55.

[13] Mehlman, A. (1988), *Applied Differential Games*, New York: Plenum Press.

[14] Hartl, R.F. (1988), The control of environmental pollution and optimal investment and employment decisions: a comment. *Optimal Control Applications and Methods*, **9**, 337–339.

David. W.K. Yeung and Michael Tow Cheung
School of Economics and Finance
Centre of Urban Planning & Environmental Management
University of Hong Kong,
Hong Kong

Coastal States and Distant Water Fleets Under Extended Jurisdiction: The Search for Optimal Incentive Schemes

Gordon R. Munro

Abstract. This paper addresses one of the major fishery issues arising from the New Law of the Sea, namely the establishment of optimal access arrangements by individual coastal states for distant water fleets operating in coastal state Exclusive Economic Zones. The analysis takes the form of principal-agent analysis in recognition of the coastal state's inability to exercise full control over the distant-water fleets. Depending upon circumstances, it may be optimal for the coastal state to encourage through time apparent "overexploitation" or "underexploitation" of the fishery resources within the EEZ.

1. Introduction

There have for centuries been fishing nations whose fleets have sailed far beyond home waters in search of harvests. The position of these fishing nations, now referred to as distant water fishing nations, vis a vis the coastal states off whose shores they harvested, was radically altered by the U.N. Third Conference on the Law of the Sea and the resultant Law of the Sea Convention. Under the New Law of the Sea, coastal states were able to extend their property rights over fishery resources from 3 to 12 miles to 200 miles off shore. Vast amounts of renewable resource wealth, which hitherto had been international common property, became coastal state property. Distant water fishing nations found that their free access to major fishery resources had vanished. Access was either eliminated entirely or was dependent upon the good graces of the coastal states.

At the inception of Extended Fisheries Jurisdiction in the late 1970's, the prevailing view appeared to be that coastal states would gradually eliminate distant water fishing nations from their 200 mile zones and replace the activities of these nations' fleets with those of the domestic coastal state fishing industry. This view has now been revised.

First it has been proven uneconomic for some coastal states, particularly developing ones, to attempt to replace distant water fishing nation activity entirely. Secondly, under the New Law of the Sea, the coastal state clearly has the right, upon granting a distant water fishing nation access to the 200 mile zone, to demand remuneration from the aforementioned distant water fishing nation. Consequently, it became clear that even developed coastal states could find that it would make more economic sense to have some of the harvesting and or processing activities in their 200 mile zones undertaken by distant water fishing nations. This author has argued on many occasions (e.g., Munro, 1989) that it is appropriate to think of a coastal state which grants such access as "importing" harvesting/processing services, or "hiring" these services, from distant water fishing nations.

If a coastal state chooses to grant a distant water fishing nation access to its 200 mile zone, it faces a problem in designing appropriate terms and conditions of access. The problem arises from the fact that it is quite unrealistic for the coastal state to assume that it can exercise complete control over the distant water fishing nation fleet, when the fleet is in the coastal state's 200 mile zone.

This fact has led the writer, in collaboration with F.H. Clarke, to attempt to bring principal–agent analysis to bear on the problem. The coastal state, as owner of the fishing resource, is seen as the principal. The distant water fishing nation granted access to the resource is viewed as the agent.

Principal–agent analysis is not game theory per se. It is, however, a close relative. Roy Radner (1981) has described the typical principal–agent relationship as being similar to a two stage non-cooperative game with the solution to the game having the characteristics of a Nash equilibrium. Moreover, the particular principal–agent relationship being examined here is inherently dynamic in nature. Incentive schemes imposed by the principal on the agent today have resource management implications for the future. Hence, one can argue that the analysis to follow comes under the heading of the theory and application of dynamic games, broadly defined.

This author and his aforementioned colleague have now published two articles on the topic (Clarke and Munro, 1987; 1991). This paper draws heavily upon the two articles and is, in a sense, a progress report on the application of principal–agent analysis to this major resource management issue arising under Extended Fisheries Jurisdiction.

This author would readily concede that only the first step has been achieved in the applications of the analysis to this issue. Many simplifications are introduced. It is assumed that the fishery resource in question is wholly confined to the waters of the coastal state and that the coastal state is negotiating with but one distant water fishing nation. The presumed nature of the distant water fishing operations in the coastal state zone involves "fee fishing" in which the distant water fishing nation both harvests the fish and processes the catch. The coastal state is assumed to obtain remuneration from the distant water fishing nation through the imposition of fees or taxes. The most commonly used taxes in practice are taxes on catch and taxes on fishing effort, e.g., a tax per vessel boat day.[1] It is assumed that the heart of the access agreement consists of one or both of these taxes. The access agreements are assumed to be long term. In a principal–agent context, the access agreement can be seen as the incentive scheme.

Perhaps the strongest assumption employed is that the world in which the coastal state and distant water fishing nation operate is deterministic. This appears to run counter to the standard application of principal–agent analysis in Economics in which uncertainty is seen to play a major role. Our response would be that what is critical to the relevance of principal–agent analysis is the power and ability of the principal to control the action of the agent. Where the principal's power is imperfect, principal–agent analysis is relevant, with or without the presence of uncertainty.

The primary motivation for the applications by this author and his colleague of principal–agent analysis to the aforementioned issue has been this author's involvement with the fishery management problems of the Pacific Islands Region of the western and central Pacific. The region, containing 23 island states and territories, has within it the richest tropical tuna fisheries in the world. For many of the Island states, the tuna fisheries are of fundamental economic importance (Munro, 1990).

The Pacific Island countries are currently capable of taking no more than 20 per cent of the total tuna harvest in their region. The rest is taken by distant water fishing nations under negotiated access arrangements. Most of the distant water fishing nations' activity in the zone is "fee fishing" in nature as described earlier. The Pacific Island countries make it clear that they regard the taxes imposed by them upon the distant water fishing nations, not just as a device to extract revenue or resource rent, but also as means for exercising resource management control through time. The tax system is thus very much an incentive scheme (Clark, 1985; Doulman, 1987). Finally the Pacific Islands look forward to establishing long term arrangements with distant water fishing nations.

This paper does not, let it be noted in passing, address itself to compliance schemes. It can in fact be said that the Pacific Islands Region does

[1] The seemingly most obvious tax to use in a "fee fishing" arrangement is a profits tax. The tax is seldom used, however, because it is far too easy for the distant water fishing nation fleet to disguise profits. Another form of tax which is used, from time to time and to which we make no other reference, is the lump sum tax. At a later point in the paper, we shall talk about cases in which the social rates of discount of principal and agent are, or are not, equal. In the case in which the discount rates are equal, the lump sum tax will in fact work effectively. In the case in which the discount rates are not equal, the lump sum tax will prove to be entirely inappropriate.

collectively have a very effective compliance scheme, which rests heavily upon incentives (Doulman and Terawasi, 1990).

2. The Principal–Agent Paradigm

Prior to establishing the basic model, a comment is in order on the principal–agent paradigm as viewed by this author and his colleague and co-author, F.H. Clarke (see in particular Clarke and Munro, 1987). The principal, in our case a country, selects an incentive scheme π which determines, together with the action γ chosen by the agent, the remuneration received by the agent, a quantity we shall label $R_A(\pi, \gamma)$. There is a corresponding return to the principal $R_P(\pi, \gamma)$. In our model, both R_A and R_P take a monetary form.

The ideal situation from the perspective of the principal is one in which it can both specify and control the action of the agent. In this so called First Best Situation, the principal's problem is to select both π and γ so as to maximize R_P. It is assumed, however, that the agent has a reservation price, a certain minimum amount which it must be assured of receiving if it is to serve as the principal's agent. Denote the reservation price as L. The principal thus faces the constraint that:

$$R_A(\pi, \gamma) \geq L. \tag{1}$$

together with $\pi \in \Pi$ and $\gamma \in G$, where Π and G are the sets of feasible incentive schemes and actions respectively. The optimal revenue or return for the principal is given by:

$$B_1 = \max\{R_P(\pi, \gamma) : \pi \in \Pi, \gamma \in G, R_A(\pi, \gamma) \geq L\} \tag{2}$$

Now suppose that the principal is faced with a so called Second Best Situation in which it cannot control the agent's action, γ, but rather can influence γ only indirectly through the selection of the incentive scheme, π. Once the incentive scheme is in place, the agent determines γ. It is assumed that the agent will select γ so as to maximize R_A over G. The optimal revenue for the principal in this situation is given by:

$$\begin{aligned} B_2 &= \max\{R_P(\pi, \gamma) : \pi \in \Pi, \gamma \in G, R_A(\pi, \gamma) \\ &= \max_{\tilde{\gamma} \in G} R_A(\pi, \tilde{\gamma}) \geq L\} \end{aligned} \tag{3}$$

It follows that B_2 cannot exceed B_1. The difference $B_1 - B_2$ is termed the incentive gap and is a measure of the loss of economic benefits experienced by the principal as a consequence of the insufficiency of the incentive scheme in compensating for the inability of the principal to exercise direct and full control over the agent's action.

We turn now to the basic model.

3. The Basic Model

It is assumed that the relevant fishery is based upon a single and independent fishery resource wholly confined to the coastal state's 200 mile Exclusive Economic Zone (EEZ). It is assumed that there is but one feasible distant water fishing nation partner.

It is assumed further that the fishery can be modelled by the standard "general production" fishery model (see for example Schaefer, 1957). We have:

$$\dot{x} = F(x) - h(t), \quad x(0) = x_0 \tag{4}$$

and

$$h(t) = qE^{\alpha}(t)x^{\beta}(t) \tag{5}$$

where

$$\begin{aligned} x(t) &= \text{fish biomass at time } t \\ x_0 &= \text{fish biomass at time } t = 0 \\ F(x) &= \text{natural growth rate} \\ h(t) &= \text{harvest rate at time } t \\ E(t) &= \text{rate of fishing effort at time } t. \end{aligned}$$

and where q, α, β are constants. For the sake of simplicity, and with no loss of generality, it will be assumed that $q = \alpha = \beta = 1$.

It is assumed, as usual, that $F(x) > 0$, for $0 < x < \bar{x}$, $F(0) = F(\bar{x}) = 0$, and $F''(x) < 0$ for all $x > 0$, where $\bar{x}$ denotes the carrying capacity of the resource (Clark and Munro, 1975). It shall, as well, be assumed that:

$$0 \leq E(t) \leq E_{\max}$$

where $E_{\max}$ is a prescribed upper bound on $E(t)$ (Clark and Munro, 1975).

It is now assumed that the processing of harvested fish is negligible and that the demand for harvested fish and the supply of fishing effort to the agent are infinitely elastic. Consequently, both the price of harvested fish and the unit cost of fishing effort are constants, which we denote as p_0 and c_0 respectively.

Next denote the social rates of discount of the coastal state as principal and the distant water fishing nation as agent as δ_P and δ_A respectively. We allow for the possibility that δ_P is less than, equal to, or greater than δ_A. If it should be the case that $\delta_P > \delta_A$, then we would have to concede that the easiest solution for the coastal state would be to sell off the resource to the distant water fishing nation. We shall suppose, however, that political, social or other factors make this simple solution infeasible.

We now suppose that the coastal state is prepared to enter into an access arrangement with the distant water fishing nation of indefinite term. As part of the terms and conditions of access, the coastal state contemplates imposing a tax on catch – a royalty – or a tax on fishing effort, or both.

Denote the unit taxes on catch and effort as r and m respectively. The taxes, r and m, constitute the incentive scheme, π, while the agent's action, γ, consists of its fishing effort profile through time, $E(t)$.

We let $p = p_0 - r$ and $c = c_0 + m$, where p and c can be viewed as the realized price of harvested fish and the effective unit cost of fishing effort to the agent respectively.

In light of our assumption that the access agreement is of indefinite term, the return function for the agent can be expressed as:

$$R_A(p, c, E) = \int_0^\infty e^{-\delta_A t}\{px(t) - c\}E(t)\,dt \tag{6}$$

while the return function for the principal can be expressed as:

$$\begin{aligned} R_P(r, m, E) &= \int_0^\infty e^{-\delta_P t}\{rx(t) + m\}E(t)\,dt \\ &= \int_0^\infty e^{-\delta_P t}\{p_0 x(t) - c_0\}E(t)\,dt \\ &\quad - \int_0^\infty e^{-\delta_P t}\{px(t) - c\}E(t)\,dt \end{aligned} \tag{7}$$

The return to the principal can be viewed as the present value of the flow of tax revenue or as the present value of the net global economic returns from the fishery, minus the present value of the flow of payments to the agent for its services (discounted at the rate δ_P).

4. First Best Situations

In a First Best Situation, the principal will determine and dictate the fishing effort profile through time $E(t)$. The agent will slavishly follow the principal's dictates. The tax scheme will be used for the sole purpose of extracting a share of the net economic benefit, or resource rent, arising from the fishery for the principal.

The principal's policy in this case is straightforward. The principal must set a fishing effort profile through time that will maximize the present value of the net global economic return from the fishery through time (as perceived by the principal). The taxes must be manipulated such that $R_A = L$. The employment of one of the two taxes alone should be entirely adequate for this purpose.

The appropriate fishing effort profile through time, $E(t)$, to be determined by the principal will be one which will move the biomass toward a steady state which we shall denote by X_0^p, and which is given as the unique solution of the following equation (see Clark and Munro, 1975)):

$$\frac{1}{\delta_P}\frac{d}{dx}\{(p_0 - \frac{c_0}{X_0^p})F(X_0^p)\} = p_0 - c_0/X_0^p \tag{8}$$

or alternatively:

$$F'(X_0^p) + \frac{c_0/(X_0^p)^2 F(X_0^p)}{p_0 - c_0/X_0^p} = \delta_P \tag{9}$$

The sense of equation (8) is that the optimal biomass will have been achieved when the present value of the sustainable net economic return arising from the incremental addition to, or investment in, the resource stock is equal to the forgone returns from current harvesting at the margin.

For future reference let $Q(\delta_P, c_0, p_0, x)$ denote:

$$\frac{1}{\delta_P}\frac{d}{dx}\left\{\left(p_0 - \frac{c_0}{x}\right)F(x)\right\} - (p_0 - c_0/x)$$

We can define $Q(\delta_P, c_0, p_0, x)$ as the "net" global economic return arising from an incremental addition to or investment in the resource stock, as perceived by the principal. When the optimal biomass level, X_0^p, is achieved, we have $Q(\delta_P, p_0, c_0, X_0^p) = 0$

The optimal approach to X_0^p is the most rapid, or "bang-bang" one. That is to say, if we denote the optimal fishing rate at time t, as perceived by the coastal state (principal), as $E^*(t)$, then:

$$E^*(t) = \begin{cases} E_{\max} & \text{whenever } x(t) > X_0^p \\ \frac{F(X_0^p)}{X_0^p} & \text{whenever } x(t) = X_0^p \\ 0 & \text{whenever } x(t) < X_0^p \end{cases} \tag{10}$$

Two comments are now in order. If $\delta_P < \delta_A$, the agent must receive L, in its entirety, at the first instant of the resource management program. Any return received by the agent after the first instant will be worth more to the principal than to the agent. If $\delta_P > \delta_A$, the principal will wish to postpone the payment to the agent for as long as possible. Either situation raises obvious moral hazard problems, which can be suppressed only if the principal can maintain absolute control over the agent.

5. Second Best Situations

Under Second Best, and far more realistic, situations, the principal cannot dictate the agent's fishing effort profile, i.e. actions, through time. Rather the principal must now influence the agent's fishing effort profile by the imposition of taxes under the access agreement. The set of taxes is thus called upon to play a dual role, to extract a share of the resource rent for the principal, and to serve as the principal's instrument for managing the fishery resource through time.

Once the tax system is in place, the agent will, it is assumed, attempt to maximize R_A. This will be achieved by moving the biomass towards a steady state, X^A, given by the following equation:

$$\frac{1}{\delta_A}\frac{d}{dx}\{(p - c/X^A)F(X^A)\} = p - c/X^A \tag{11}$$

or alternatively:

$$F'(X^A) + \frac{c/(X^A)^2 F(X^A)}{p - c/X^A} = \delta_A \tag{12}$$

The optimal approach to X^A is the most rapid or "bang-bang" one.

For the sake of future reference, let X_0^A denote the agent's perception of the optimal biomass level when it is subject to no taxes, i.e. when $p = p_0$, $c = c_0$. Let us also note for future reference that

$$X_0^A \begin{cases} < X_0^P & \text{if } \delta_P < \delta_A \\ = X_0^P & \text{if } \delta_P = \delta_A \\ > X_0^P & \text{if } \delta_P > \delta_A \end{cases} \tag{13}$$

Two remarks are in order. Return to equation (12). The second term on the L.H.S. is referred to as the "marginal stock effect" and is a measure of the impact of stock density upon the unit cost of harvesting. The magnitude of the marginal stock effect is determined by the ratio of prices and costs, i.e., p/c. By varying this ratio through its taxation policy, the principal can influence, indeed determine, the agent's fishing effort profile, $E(t)$ over time. Thus the principal appears to have, at first glance, perfect control.

Secondly, let $Q(\delta_A, p, c, x)$ denote:

$$\frac{1}{\delta_A}\frac{d}{dx}\{(p - c/x)F(x)\} - (p - c/x),$$

which we shall refer to as the "net" economic return to the agent arising from an incremental investment in the resource stock. When X^A is achieved we have:

$$Q(\delta_A, p, c, X^A) = 0$$

Let us also introduce the expression

$$Q(\delta_P, p, c, x) = \frac{1}{\delta_P}\frac{d}{dx}\{(p - c/x)F(x)\} - (p - c/x),$$

which we shall refer to as the principal's "net" hiring cost burden arising from an incremental investment in the resource stock or biomass. When a positive marginal addition is made to the resource stock, the principal is relieved of a current payment at the margin to the agent, as current harvests are reduced, but must look forward to a marginal stream of additional future payments to the agent.

In this Second Best Situation, the principal faces the problem, as it does in a First Best Situation, of maximizing R_P, subject to the constraint, $R_A \geq L$. Now, however, the principal has the further problem in that it can influence, but not determine, $E(t)$. The power to determine $E(t)$ lies with the agent.

The analysis which follows could be applied both to situations in which the initial biomass $x(0)$ is greater than X_A and in which the reverse holds true. It proves to be the case that the results are particularly transparent in situations in which $x(0) < X^A$. We shall, therefore, for ease of exposition, assume throughout that $x(0) < X^A$.[2]

There is one aspect of the analysis to which we shall give only modest attention. This aspect concerns the case in which the coastal state, as principal, restricts itself to the use of only one of the two taxes. The results are highly unsatisfactory from the principal's perspective, basically because the single tax is overwhelmed by the dual tasks imposed upon it, namely that of extracting resource rent for the principal and that of influencing the agent's fishing effort profile through time. Indeed it is quite possible that an optimal policy (from the principal's perspective) in the single tax case will be one resulting in $R_A > L$, i.e., in the agent receiving in excess of its reservation price (Clarke and Munro, 1991).

The case of the single tax is useful, however, in illustrating some of the essential distinctions between First Best and Second Best situations. Consider the following in which $\delta_P = \delta_A$, and in which the principal restricts itself to the use of a tax on catch, r. Under a First Best Situation, the principal would simply determine the fishing effort profile through time which would maximize the present value of the stream of net economic benefits from the fishery, and then order the agent to follow that effort profile. The principal would then set the level of r (lump sum payments are obviously unnecessary) such that $R_A = L$. There is, in effect, a decoupling of the harvesting program through time and the extraction of resource rent by the principal.

Under Second Best conditions, the principal can influence the agent's fishing effort profile, and hence harvesting program, only through the tax system, which serves as the incentive scheme. Given that $\delta_A = \delta_P$, we can determine without difficulty the level of tax on catch which will lead to the agent maximizing the net economic return from the fishery through time. The tax level is simply $r = 0$ (Clarke and Munro, 1987). The principal would then, of course, receive precisely nothing.

If the principal is to receive a share of the resource rent, i.e., if $r > 0$, the agent will be driven to "underexploiting" the resource (Clarke and Munro, 1987). The greater is r, the greater will be the degree of "underexploitation." Thus the principal is faced with a clear tradeoff between the

---

[2] In the Appendix it is shown how an a priori upper bound on $x(0)$ will ensure that we shall have $x(0) < X^A$.

extraction of resource rent at each moment in time and the desired nature of the harvest program over time. It might be added in passing, that these conclusions require only minor modification when $\delta_A \neq \delta_P$ (Clarke and Munro, 1991).

We turn now to the more fruitful case in which the principal is unconstrained with respect to the use of the two taxes under Second Best situations. Let us make one observation before examining the detailed analysis. Under Second Best situations the principal must always be concerned in keeping the agent committed to the harvest program. Thus, for example, in the case in which $\delta_P < \delta_A$, offering the agent a lump sum payment at the commencement of the harvest program is wholly unacceptable. The agent must be offered a stream of payments through time. We might refer to this as a "continued motivation" constraint facing the principal (Clarke and Munro, 1991).

With this observation being kept in mind, it is found that, in addressing the two tax incentive scheme problem, the Lagrangian technique proves to be helpful.

The Lagrangian for the problem of maximizing R_P, subject to the constraint $R_A \geq L$, is given by:

$$\mathcal{L} = R_P + \lambda R_A \tag{14}$$

Here, however, R_A must be interpreted as the agent's revenue following its own maximization.

We then have the Lagrange multiplier rule:

$$\partial \mathcal{L}/\partial r = 0 \tag{15}$$

$$\partial \mathcal{L}/\partial m = 0 \tag{16}$$

$$R_A \geq L \text{ with } \begin{cases} \lambda = 0 & \text{if } R_A > L \\ \lambda \geq 0 & \text{otherwise} \end{cases} \tag{17}$$

We thus have three unknowns, r, m, and λ, and the requisite system of three equations.

Upon solving the system of equations, a set of optimality conditions will be forthcoming, yielding an optimal set of taxes, which we shall denote as r^* and m^*.

We present the optimality conditions in the following theorem. The proof of the theorem is to be found in an appendix. The proof is drawn in its entirety from Clarke and Munro, 1991.

Theorem. *When the principal employs both a tax on catch and a tax on fishing effort, the optimality conditions, prove to be as follows:*

$$R_A(p, c) = L \tag{18}$$

$$Q(\delta_A, p, c, X^A) = 0 \tag{19}$$

$$Q(\delta_P, r^*, -m^*, X^A) = 0 \tag{20}$$

where

$$Q(\delta_P, r^*, -m^*, X^A) = \frac{1}{\delta_P}\frac{d}{dx}\{(r^* + m^*/X^A)F(X^A)\} - (r^* + m^*/X^A)$$

The first optimality condition states that the agent is always to receive no more than its reservation price. The second optimality condition simply states that the agent must be found to maximize its return from the fishery over time, subject to the incentive scheme imposed upon it. We might refer to Equation (19) as the agent's resource investment rule.

The third optimality condition can be referred to as the principal's resource investment rule. It states that investment in the resource should proceed up to the point that the net economic benefit to the principal, in terms of tax revenue arising from the incremental resource investment, is zero. The nature of the principal's resource investment rule is made more apparent when it is observed that Eq. (20) can be re-expressed as:

$$Q(\delta_P, p_0, c_0, X^A) - Q(\delta_P, p, c, X^A) = 0 \tag{21}$$

Seen in this light, the principal's resource investment rule requires the principal to balance off the net global economic return from an incremental investment in the resource stock against the net "hiring" cost accompanying the resource investment.

Now let us examine the implications of Eq. (21), considering first the case when $\delta_P = \delta_A$. It will then be true that, for a given r, m and x, we shall have:

$$Q(\delta_P, p, c, x) = Q(\delta_A, p, c, x)$$

From this it follows that, if the agent maximizes its return from the resource, subject to the tax system, i.e. if the second optimality condition is met, we shall have $Q(\delta_P, p, c, X^A) = 0$. At X^A, the net "hiring" cost to the principal from an incremental investment in the resource stock is zero. This in turn, implies that, for the third optimality condition to be satisfied, we must have: $Q(\delta_P, p_0, c_0, X^A) = 0$. The principal should introduce an incentive scheme (a set of taxes) which will drive the agent to maximize the present value of the global net economic returns from the fishery through time.

The first optimality condition states that $R_A = L$. We thus have the equivalent of a First Best Situation. We conclude, therefore, that the principal can, with the tax incentive scheme, achieve the equivalence of full control if: 1. the principal can and does employ both taxes;[3] 2. the social rates of discount of the principal and agent are equal.

---

[3] It is worth noting in passing that, while r^* will always be positive, it is quite possible that m^* will be negative, i.e. that it may be optimal for the principal to subsidize the agent's fishing effort. See Clarke and Munro, 1991.

Next suppose that δ_P and δ_A are not equal, but rather that $\delta_P < \delta_A$. It will now be found that optimality requires a set of taxes, $r^*$and m^*, which will cause the agent to "overexploit" the resource in the sense that $X^A < X_0^P$. The reason lies in the net "hiring" cost to the principal of an incremental investment in the resource stock. For any given r, m and x, it will be found that $Q(\delta_P, p, c, x) > Q(\delta_A, p, c, x)$. Thus, if the second optimality condition is satisfied, we shall have $Q(\delta_P, p, c, X^A) > 0$. The principal's resource investment rule, the third optimality condition, can then be satisfied only if $Q(\delta_P, p_0, c_0, X^A) > 0$, which implies that $X^A < X_0^P$. Current payments to the agent are obviously of equal value to both agent and principal. Future payments to the agent are, however, of greater value to the principal than to the agent. Consequently, it is to the principal's advantage to induce the agent to "overexploit" the resource.

An incentive gap thus exists. The difference between X^A and X_0^P, the extent to which "overexploitation" of the resource is encouraged by the principal's incentive scheme, is a measure of the incentive gap. As a general rule it can be said that the greater is the extent to which the agent discounts the future in comparison with the principal (i.e. the greater is the extent to which δ_A/δ_p exceeds 1), and the larger is L, the greater will be the degree of induced "overexploitation" (Clarke and Munro, 1991).

From a slightly different perspective, we can observe that incentives do, in a sense, flow in both directions. The wishes and desires of the principal are imposed upon the agent through the incentive scheme. Yet the wishes and the desires of the agent do, in this instance, clearly influence the principal.

Next consider the case in which $\delta_P > \delta_A$. If the coastal state is a capital poor developing country, while the distant water fishing country is a capital rich developed country, then it is indeed possible that $\delta_P > \delta_A$.

If $\delta_P > \delta_A$, the outcome will be symmetrical to that of the previous case. It will now be found that, for any given r, m and x, $Q(\delta_P, p, c, x) < Q(\delta_A, p, c, x)$. If the second optimality condition is satisfied, then we shall have $Q(\delta_P, p, c, X^A) < 0$. Hence, if the third optimality condition is to be satisfied, then it must be true that $Q(\delta_P, p_0, c_0, X^A) < 0$, which implies in turn that $X^A > X_0^p$. The principal places less value on future payments to the agent than does the agent itself. Consequently the principal has an incentive to induce the agent to "over-invest" in the resource. Thus an incentive gap will emerge, one which we might refer to as a "virtuous" incentive gap.

As an aside, we can remark upon the fact that there are often expressions of concern in developed countries about the lack of conservationist incentives in developing countries. Translated into terms used in this paper, we would say that the social rate of discount applied to natural resources in developing countries is higher than the comparable rate in developed countries. Our analysis suggests that in such circumstances, co-exploitation arrangements between low discount rate developed country interests and a high discount developing country could result in the developing country adopting a more conservationist stance.

6. Conclusions and Observations

This paper draws heavily upon two articles which the writer has co-authored (Clarke and Munro, 1987; 1991). The paper, and the articles which provide its foundation, represent no more than a first attempt to address one of the more difficult resource management issues to have arisen under the New Law of the Sea. The assumption used in this paper of a deterministic world is, for example, a particularly heroic and restrictive one.

Nonetheless, it has been demonstrated that, even under seemingly ideal circumstances, the coastal state must expect a considerable amount of "slippage," i.e. the emergence of incentive gaps, in its dealing with distant water fishing nations. This will be particularly true if the coastal state relies upon the use of single tax schemes. Among developing coastal states, such as the Pacific Islands, this approach appears to be standard, in considerable part because of the difficulty in constructing multitax schemes.

The analysis also has implications for the term of the coastal state – distant water fishing nation arrangement and the security of the arrangements. The Pacific Islands, for example, have prided themselves on the length of term and stability of the arrangements with distant water fishing nations (Clark, 1985). The shorter the term and the more unstable the arrangements, the more heavily will the distant water fishing nation partner discount the future, with all that that implies.

Finally, the analysis has implications for management and sovereignty within the Exclusive Economic Zone. One of the most contentious issues to have arisen from the U.N. Third Conference on the Law of the Sea is the extent of the property rights which coastal states have over fishery resources within their Exclusive Economic Zones (EEZs). The resolution of the issue appears to be that the coastal state has sovereign rights over all fishery resources within the EEZ and, as a consequence, has full management powers over these resources. Many coastal states, developing ones in particular, guard these management powers jealously. Once again, the Pacific Islands provide an example.

Yet, what our analysis suggests is that coastal states, in establishing arrangements with distant water fishing nations, may find that these management powers have been eroded. The implication of the cases, considered in the paper, in which $\delta_P \neq \delta_A$ is that the distant water fishing nation will, for good or for ill, exercise an influence, perhaps a considerable one, over resource management programs within the exclusive Economic Zone.

Acknowledgement. Funding for this paper was provided in part by the Donner Canadian Foundation through a grant to the Institute of International Relations research project on "Canada and International Regulatory Regimes" at the University of British Columbia.

Parts of this paper have drawn heavily upon two articles published in the journal *Natural Resource Modeling*, (Clarke and Munro, 1987; 1991). This has been done with the express permission of the Resource Modeling Association.

Appendix

In this appendix we verify that one can ensure that $x(0) < X^A$ by the introduction of an a priori upper bound on $x(0)$. We then turn and offer a proof of the single theorem contained in the text. The proofs are drawn in their entirety from Clarke and Munro, 1991.

For simplicity we shall assume that $F(x)$ is of the logistic form: $\omega x(1-x/\bar{x})$, where ω, a constant, is the "intrinsic" proportional growth rate of the biomass and where, as before, $\bar{x}$ denotes the carrying capacity of the biomass. We should add that the nature of the conclusions is robust for growth functions satisfying the usual concavity assumptions (see Part 3). For future reference, we assume that $\delta_A < \omega$. A *feasible* tax scheme is a choice of taxes r, m such that the unit price p and the unit cost c are positive, and having the property that the agent, acting optimally in the presence of p and c, acquires a revenue R_A at least equal to L (which we assume to be positive), while the resulting revenue R_P to the principal is positive.

Proposition. *Let r, m be a feasible tax scheme. The agent's optimal steady biomass level X^A satisfies $X^A > \frac{\bar{x}}{2}[1 - \frac{\delta_A}{w}]$.*

Proof. The solution X^A to equation (11) can be written explicitly by the quadratic formula:

$$X^A = \frac{1}{4}\Big\{\bar{x}(1 - \frac{\delta_A}{w}) + \frac{c}{p} + [(\bar{x}(1 - \frac{\delta_A}{w}) + \frac{c}{p})^2 + 8\frac{c\bar{x}}{pw}\delta_A]^{1/2}\Big\} \tag{22}$$

Since c/p is positive, the required inequality is immediate.

A corollary of this result is that when δ_A is less than w, which we have assumed, then taking x_0 to be less than $\frac{\bar{x}}{2}[1 - \frac{\delta_A}{w}]$ will guarantee that we have $x(0) < X^A$ as supposed in the text.

The principal's problem is to find the feasible tax scheme which maximizes its own return R_P. We shall not dwell on the issue of the existence of an optimal tax scheme r^*, m^*, other than to remark that by continuity considerations based on the relatively simple analysis of the case $\delta_A = \delta_P$ (Clarke and Munro, 1987), it follows that an optimal feasible tax scheme certainly exists when δ_A and δ_P are relatively close.

Theorem. *The revenue to the agent, R_A, resulting from the implementation of the optimal tax scheme, r^* and m^*, is always equal to L. Moreover, in addition to the agent's optimality condition, Equation (19), Equation (20) holds.*

Proof. Let us denote X^A simply as X. Then from (22) we find

$$\frac{dX}{dr} > 0, \quad \frac{dX}{dm} > 0, \quad p\frac{dX}{dr} = c\frac{dX}{dm} \tag{23}$$

These relations hold for r and m in a neighborhood of the optimal tax scheme.

Given our assumption that $x(0) < X$[4], we can restate the principal's return function as follows. Let $\tau(X)$ denote the time required for the biomass to climb to the level X from its initial level x_0, in the absence of any harvesting. Then we have:

$$R_P = \frac{e^{-\delta_P \tau(X)}}{\delta_P} \{rx + m\} \frac{F(X)}{X} \tag{24}$$

or

$$R_P = \frac{e^{-\delta_P \tau(X)}}{\delta_P} \{(p_0 - p)X - (c_0 - c)\} \frac{F(X)}{X} \tag{25}$$

subject to $R_A \geq L$.

The return function for the agent, given our assumptions, can be expressed as:

$$R_A = \frac{e^{-\delta_A \tau(X)}}{\delta_A} \{pX - c\} \frac{F(X)}{X}. \tag{26}$$

Now let $x(t)$ be the solution to

$$\dot{x}(t) = F(x(t)), \quad x(0) = x_0.$$

We have by definition $x(\tau(X)) = X$, so that differentiating through leads to

$$\frac{d\tau(X)}{dr} = \frac{dX}{dr}/F(X), \quad \frac{d\tau(X)}{dm} = \frac{dX}{dm}/F(X). \tag{27}$$

As in the text, let us define $Q(\delta, p, c, x)$ to be the expression

$$\frac{1}{\delta}\frac{d}{dx}\Big\{(p - \frac{c}{x})F(x)\Big\} - (p - \frac{c}{x}).$$

[4] The reader should recall that $x(0) = x_0$.

Armed with the above, it is routine to calculate formulas for the derivatives of R_P and R_A, as given above, with respect to r and m; The result:

$$\begin{aligned}
\frac{d}{dr}R_P &= e^{-\delta_P T}\Big\{\frac{dX}{dr}Q(\delta_P,r,-m,X)+F(X)/\delta_P\Big\}\\
\frac{d}{dm}R_P &= e^{-\delta_P T}\Big\{\frac{dX}{dm}Q(\delta_P,r,-m,X)+F(X)/(X\delta_P)\Big\}\\
\frac{d}{dr}R_A &= e^{-\delta_A T}\Big\{\frac{dX}{dr}Q(\delta_A,p,c,X)+F(X)/\delta_A\Big\}\\
\frac{d}{dm}R_A &= e^{-\delta_A T}\Big\{\frac{dX}{dm}Q(\delta_A,p,c,X)+F(X)/(X\delta_A)\Big\}
\end{aligned}$$

Observe that at the optimal tax scheme, i.e. for r^*, m^*, the term $Q(\delta_A,p,c,X)$ vanishes: this is just the content of the agent's optimality rule (19). It follows that $\frac{d}{dr}R_A$ and $\frac{d}{dm}R_A$ do not vanish at the optimal tax values, which permits application of the normal (or Kuhn-Tucker) form of the Lagrange multiplier rule as given in (15)–(17).

Suppose now that λ were zero. Then the derivatives of R_P with respect to r and m must vanish, whence

$$\begin{aligned}
\frac{dX}{dr}Q(\delta_P,r,-m,X)+F(X)/\delta_P &= 0\\
\frac{dX}{dm}Q(\delta_P,r,-m,X)+F(X)/(X\delta_P) &= 0.
\end{aligned}$$

It follows that Q is nonzero. Substituting from (23), we arrive at

$$\frac{dX}{dr}Q(\delta_P,r,-m,X)(pX-c)=0,$$

which yields $pX-c=0$. But then $R_A=0$, which is a contradiction. This establishes that λ is strictly positive, and hence that $R_A=L$ at optimality.

Upon eliminating λ from the equations (15) (16) and invoking (23), regrouping leads to the equation

$$\frac{dX}{dm}(pX-c)\Big\{\delta_A Q(\delta_A,p,c,X)+\delta_p Q(\delta_P,r,-m,X)\Big\}=0,$$

whence we conclude that the expression in braces vanishes at r^*,m^*. Recalling that $Q(\delta_A,p,c,X)$ is zero leads to:

$$Q(\delta_P,r^*,-m^*,X)=0.$$

This is precisely Equation (20). Thus the proof is complete.

References

[1] Colin W. Clark and Gordon R. Munro, The Economics of Fishing and Modern Capital Theory: A Simplified Approach, *Journal of Environmental Economics and Management*, vol. 2, (1975), 92–106.

[2] L. Clark, Fisheries Issues in the Pacific Islands," in Ian Clark (ed.), *Proceedings of the Second Conference of the Institute of Fisheries Economics and Trade*, vol. 1,2 IIFET, Corvalis, (1985), 19–26.

[3] F.H. Clarke and G.R. Munro, Coastal States, Distant Water Fishing Nations and Extended Jurisdiction: A Principal-Agent Analysis, *Natural Resource Modeling*, vol. 2, (1987), 81–107.

[4] F.H. Clarke and G.R. Munro, Coastal States and Distant Water Fishing Nations: Conflicting Views of the Future, *Natural Resource Modeling*, vol. 5, (1991), 345–370.

[5] David Doulman, Licensing Distant-Water Tuna Fleets in Papua New Guinea, *Marine Policy*, vol. 11, (1987), 16–28.

[6] David Doulman and Peter Terawasi, The South Pacific Regional Register of Foreign Fishing Vessels, *Marine Policy*, vol. 14, (1990), 324–332.

[7] Gordon R. Munro, Coastal State–Distant Water Fishing Nation Relations, *Marine Fisheries Review*, vol. 51, (1989), 3–10.

[8] Gordon R. Munro, Extended Jurisdiction and the Management of Pacific Highly Migratory Species, *Ocean Development and International Law*, vol. 21, (1990), 289–307.

[9] Roy Radner, Monitoring Cooperative Agreements in a Repeated Principal–Agent Relationship, *Econometrica*, vol. 49, (1981), 1127–1148.

[10] M.B. Schaefer, Some Considerations of Population Dynamics and Economics in Relation to the Management of Marine Fisheries, *Journal of the Fisheries Research Board of Canada*, vol. 14, (1957), 669–681.

Gordon R. Munro
Department of Economics
University of British Columbia
Vancouver, B.C. Canada

Stabilizing Management and Structural Development of Open-Access Fisheries

Mikael Hildén, Veijo Kaitala and George Leitmann

Abstract. In this paper we model and analyse a robust control approach to sustainable resource use. In particular, we study the entry and exit of enterprises in a regulated open-access common-property fishery. The economics are exposed to uncertain and unpredictable disturbances, and there is a possibility that the resource will be depleted as a result of overexploitation. Fishing is a combination of labor and capital inputs. We derive a unified approach to model labor-dominated and capital-dominated fishing. We show that applying a price subsidizing system in a fishery where the harvesting activity is composed of both labor and capital inputs may lead to structural changes in the fishery. A price subsidizing policy planned to stabilize fishermen's income and employment in a traditional labor-dominated fishery will ultimately result in capital accumulation in the other fishing sector. Uncontrolled entry of capital will eventually lead to considerable increase in catch, and the resource will decline. We propose that the regulation of labor- and capital-dominated fishing be differentiated from each other when these two forms of harvesting coexist. We also show that there may exist a need for active regulation policy even though the current numbers of the enterprises match the target values. A simulation example dealing with stabilizing regulation and structural development of the Baltic herring fishery of Finland is studied. We conclude that identifying production functions and entry/exit dynamics in resource economics are important for regulation purposes. These topics have received, however, only minor attention in the fisheries literature.

1. Introduction

Stable conditions are generally perceived to be beneficial. Fisheries is no exception and the demand for stability has been voiced by the fishing industry and by fisheries managers (e.g., [1], [2]). Stabilization of all variables of interest to the fishing industry is, however, impossible as long as there are no mechanisms for controlling the fluctuations in the resource base (e.g., [3]). Therefore it is interesting to explore the consequences and limitations of management measures which can be used for achieving stability in in the European Community, the possibilities for stabilization or an active social policy are limited [5]. Attempts to overcome these problems have been made by introducing economic considerations in the setting of quotas [6], [7]. There is, however, also a genuine concern not only for economic efficiency but for the wellbeing of particular types of fisheries [5]. O'Boyle et al. [8] have demonstrated that optimal economic management may be incompatible with preserving different types of fisheries. The challenge to the management scheme is that the productivities and the profits in the different types of fishing differ from each other. For example the increased efficiency of capital-dominated fishing may threaten the existence of a small scale activity which is considered important from a regional welfare point of view, before it threatens the sustainability of the resource use. Therefore, it is worthwhile to use the stabilization of particular structure in a fishery as an explicit objective in order to examine what type of management intervention is required in order to achieve the objective.

In this paper we model and analyse a robust control approach to sustainable resource use. In particular, we study the entry and exit of fishing enterprises in a regulated open-access common-property fishery. We make two important assumptions in our model. First, we assume that the economics are exposed to uncertain and unpredictable disturbances. Second, fishing is a combination of labor and capital inputs. In many forms of traditional fishing the catch level is largely determined by the amount of work done in harvesting. In modern fishing activities, e.g. trawl fishing, the catch level is determined primarily by the amount of capital invested.

The approach used in the regulation scheme is based on the theory of robust control of uncertain dynamic systems. Kaitala and Leitmann [9-11] developed a control theory for the stabilization problem in resource economics. The theory deals with uncertain discrete time systems dynamics. It is assumed the realization of the uncertainties is not known at the time of decision making. The theory differs from the common modeling practice in the resource literature by assuming that no statistical description of the fluctuations is available or used in decision making. Instead, it is assumed that bounds of the uncertainties are known. The main result is that it is possible to stabilize the responses of a resource economics system around the nominal solution obtained in the absence of uncertainties. The theory is flexible in that it allows the management agency to choose, for example, between stabilizing the resource level, the income level of fishermen, or the number of enterprises in the fishery. Kaitala [12] showed that the theory can also be applied for "maximizing" the yield in an uncertain fishery.

Kaitala et al. [13] undertook a preliminary theoretical study of the problems of stabilizing a structurally heterogeneous fishery. They showed that an attempt to stabilize a fishery by controlling prices may create problems not only because of economic uncertainties but also because it may have effects which are in conflict with overall stabilization of the fishery. They did not consider, however, the effects of resource depletion on the have effects which are in conflict with overall stabilization of the fishery. They did not consider, however, the effects of resource depletion on the possibility to stabilize the fishery. Furthermore, they used different types of production models in which the catch was determined only by the labor input or by the capital input but a combination of the inputs was not possible. In this study we analyze the importance these additional complications.

The paper is organized as follows. In Section 2 we derive a unified approach to model labor-dominated and capital-dominated fishing. We also derive models for the entry and exit behavior of the fishing enterprises. In Section 3 we present a control-theoretic definition for a stabilizing regulation problem. In Section 4 we derive and discuss schemes for regulating the entry and exit of fishing enterprises. In Section 5 we present a detailed example based on a Baltic herring fishery.

2. Modeling Harvesting

We begin the section by modeling the harvest production function. We then develop a model for entry/exit dynamics of labor and capital in open access fisheries under uncertainty.

2.1. Fishing effort and production function

In modeling the dependence of the catch on fishing effort and stock level the traditional fisheries literature relies usually on the Schaefer [14] production function of the form

$$H = qeS, \tag{1}$$

where H is the catch, q is a catchability coefficient, e is the fishing effort, and S denotes the resource level. In particular, the harvest is linear in both the effort, e, and the stock level, S.

The conditions under which the Schaefer model holds are stringent [15]. The main assumption underlying the model is that the distribution of fish is uniform, which clearly fails in fisheries for schooling species such as herring. Application of (1) in fisheries for schooling species has been criticized [15], and consequently attempts to develop more general production functions have been made [15], [16].

Fishing effort, e, is usually a combination of labor and capital inputs. Bjørndal [15] and Hannesson [16] propose that the production functions in fisheries for schooling species be modeled by a Cobb–Douglas type production function of the form

$$H(k) = q_K N(k) D L^{\alpha} K^{\beta} S^{\gamma}(k), \tag{2}$$

where $H(k)$ is the catch in year k, q_K is a catchability coefficient as above, $N(k)$ is the number of vessels participating in the fishery in year k, D is the number of fishing days, L is the daily per vessel labor input, K is the daily capital input per vessel and $S(k)$ is the stock level in the fishery in year k. The constants α, β and γ are the elasticity coefficients. Capital input K represents the investments in a vessel and the fishing gear and reflects the size and the quality of the vessel. Labor input L is the number of fishermen on a vessel. Both K and L are assumed to be constants. The catchability coefficient q_K is a measure of the productivity of a unit effort and it is related to the fishing techniques used. We assume further that the total catch depends linearly on the number of the vessels.

It is of special interest to study the elasticities of the daily per vessel effort production function

$$E(L, K) = q_K L^{\alpha} K^{\beta}. \tag{3}$$

In the effort production function (3) the vessel-level returns to scale can be decreasing, constant, or increasing ($\alpha + \beta < 1$, $= 1$ or > 1).

Fishing enterprises are heterogeneous and the relative share of capital and labor may vary among fishing techniques. Below we consider a fishery subject to two different types of fishing in which the role of labor input and capital input differ; this difference will be seen in the estimates of α, β, and γ in different types of fishing. The crucial difference is not, however, in the elasticity parameters but in the share of the total labor costs and total capital costs. If the capital costs dominate, then we are dealing with a fishing activity in which the capital dynamics determine the development of the fishing activity. If the labor costs dominate, then the development of the sector is determined by the labor markets in the society. We next turn to model entry/exit dynamics in fisheries.

2.2. Opportunity costs and profit functions

In this subsection we first model in detail the entry/exit dynamics of capital-dominated fishing and then briefly that of labor-dominated fishing. The harvest by the fishing vessels is sold to fish merchants (first hand buyers) at a per unit average dock-side price $\pi(k)$. Thus, the dock-side value of the catch per vessel is $\pi(k)H(k)$, where $H(k)$ is given by (2) Let c_L denote the daily labor costs and c_K the operating costs per vessel. Then, the profit per vessel in year k is

$$D\big[\pi(k) q_K L^{\alpha} K^{\beta} S^{\gamma}(k) - c_L L - c_K K\big]. \tag{4}$$

Labor cost c_L is assumed equal to the opportunity cost of labor. However, in order to deal with the entry/exit dynamics we also need to include the opportunity costs of capital in the profit. Thus, the pure profit is defined as

$$\Pi_K(k) = D\left\{\pi(k) q_K L^{\alpha} K^{\beta} S^{\gamma}(k) - c_L L - [c_K + a] K\right\}, \tag{5}$$

where a denotes the opportunity cost of capital expressed as a daily rate.

The pure profit of a vessel depends crucially on the labor and capital costs. Equation (5) can, however, be used to determine the optimal combination of labor and capital as follows [16]. Assume that a vessel owner plans to utilise daily a fixed amount of input combination $g(L,K) = L^\alpha K^\beta$. Setting the marginal products of labor and capital equal to their price ($\partial\Pi_K/\partial L = 0$, $\partial\Pi_K/\partial K = 0$) we obtain, respectively,

$$\alpha\pi(k)q_K L^{\alpha-1}K^\beta S^\gamma(k) = c_L, \tag{6}$$

and

$$\beta\pi(k)q_K L^\alpha K^{\beta-1}S^\gamma(k) = c_K + a, \tag{7}$$

from which we get the optimal per vessel capital/labor ratio

$$\frac{K}{L} = \frac{c_L}{c_K + a}\frac{\beta}{\alpha}. \tag{8}$$

From (8) we see that the ratio β/α is equal to the the ratio of the total capital costs to the total labor costs. Utilizing (8) in (5) we get

$$\Pi_K(k) = D\left[\pi(k)\hat{q}_K K^{\alpha+\beta}S^\gamma(k) - [c_K + a]\left(1 + \frac{\alpha}{\beta}\right)K\right], \tag{9}$$

where

$$\hat{q}_K = q_K\left(\frac{c_K + a}{c_L}\frac{\alpha}{\beta}\right)^\alpha. \tag{10}$$

Since we are interested in this paper in the relationship of two completely different fishing activities (i.e., trawl fishery and trapnet fishery) we proceed by assuming that in each of the fishery there are vessels of only one size. However, the vessel sizes can differ between the fisheries.

Occasional economic studies on entry/exit dynamics in resource management deal with homogeneous capital inputs [17]. There are reasons to expect, however, that the entry/exit dynamics may differ crucially between different types of fisheries. We propose here that cost ratio may be an important factor in determining the entry and exit of capital and labor inputs in fisheries. In particular, we assume that when the capital costs clearly dominate the labor costs, i.e.,

$$(c_K + a)K >> c_L L, \tag{11}$$

the entry and exit of fishing firms are largely determined by capital dynamics (7).

Noting (8), condition (11) can be rewritten as

$$\beta >> \alpha. \tag{12}$$

Thus, in a capital-dominated fishery in which the labor/capital ratio is optimized the capital elasticity is higher than the labor elasticity.

When the labor costs dominate the capital costs, then we assume that the entry and exit of fishing firms are largely determined by labor dynamics. A similar derivation as above gives an equation for the labor dynamics

$$\Pi_L(k) = D'\left[\pi(k)\hat{q}'_L L^{\alpha'+\beta'} S^{\gamma'}(k) - c'_L\left(1+\frac{\beta'}{\alpha'}\right)L\right], \tag{13}$$

where

$$\hat{q}_L = q'\left(\frac{c'_L}{c'_K + a}\frac{\beta'}{\alpha'}\right)^{\beta'} \tag{14}$$

and $\Pi_L(k)$ denotes the pure profit in the labor-dominated fishery in year k. Superscript "$'$" in the parameters refers to the labor-dominated fishery. In particular, α' and β' are the input elasticity parameters in the labor-dominated fishery.

2.3. Entry/exit dynamics

An open-access common-property fishery is the traditional and still very common property right in marine resource economics. The entry to the resource is open to anyone willing to utilize the resource. Fishing enterprises enter the fishery as long as positive revenues are expected from the resource. Hence, the number of enterprises increases as long as the revenue expectations are high. If the expected profit of an investment in the fishery exceeds the normal rate of profits elsewhere in the capital markets, then investments in the fishery occur and the number of vessels in the fishery increases.

Kaitala et al. [13] considered memoryless entry/exit dynamics of capital, that is, the future revenue expectations and related investment decisions were assumed to be based on the revenue enjoyed during the latest period only. In a more general approach the decisions may include memory in which case the entry/exit decisions may be based on the average performance during a number of the previous years. Let $N(k)$ denote the numbers of vessels in the fishery in year k. The entry/exit dynamics can now be represented as

$$N(k+1) = N(k) + t_K N(k)\sum_{j=k-m}^{k}\{\Pi_K(j)/K\}/m, \tag{15}$$

where m is the number of the years on which the profit expectations in the fishery are based, and parameter t_K is a positive entry coefficient.

It is of interest to study the equilibrium levels assuming that all uncertainties and fluctuations are absent. If the fishery stabilizes at some level, then the equilibrium level of the enterprises satisfies

$$\Pi_K(k) = 0. \tag{16}$$

There is no guarantee, however, that the fishery will stabilize at any level. This possibility of stabilization depends on the interplay of the entry/exit dynamics (15) and the dynamics of the resource, which has not yet been considered. Moreover, although the entry/exit memory may not affect the possible equilibrium level (condition (16) is independent of time), it certainly will affect the way of approaching the equilibrium and also whether the equilibrium will be reached. In other words, the memory affects the stability properties of the equilibrium.

The entry/exit model for the labor-dominated fishery is derived analogously to the model for the capital-dominated fishery. Let M denote the number of fishing enterprises in the labor-dominated fishery (see equations (12)–(13)). The entry/exit dynamics can now be modeled by

$$M(k+1) = M(k) + t_L M(k) \sum_{j=k-m}^{k} \{\Pi_L(j)/L\}/m. \tag{17}$$

3. Stabilizing Management

The basic assumption is that the principal management goal is to stabilize the fishing activity around some desired value. The desired target value may be determined, for example, in terms of stock level [9], [11], employment level [9], [10], or yield [12]. In this study we assume that the target values are expressed as the preferred numbers of the different type of fishing enterprises in the fishery. The regulation is to be achieved by regulating the rates of entry and exit through controlling the income of fishing vessels. This corresponds to an active use of the systems of fisheries subsidies used by many countries (e.g., [18]).

The goal of regulation, that is, stabilization of the number of fishing enterprises by controlling the income, can be characterized by the concept of global uniform asymptotic stability about a set. A system is globally uniformly asymptotically stable about a calculable set B if, starting from any initial number of the enterprises, the number can be driven arbitrarily close to the set B in finite time and to remain therein for all future time. Thus, the management goal is to keep the fluctuations in the range defined by the set B, and to make this set B as small as possible. Global uniform asymptotic stability of a resource economics system can be achieved using the min-max Lyapunov stability approach of [19] in designing the management rules. For the theory that we shall utilize we refer to [19], [20], and for developments in the related theory of fisheries management to Kaitala and Leitmann [9–11].

Several alternatives exist for carrying out the economic measures needed for the stabilizing regulation. The common feature of all the measures is that income is regulated by the government. Among the basic alternatives are price subsidy/tax, labor subsidy/tax, and capital subsidy/tax [13].

In the design of the regulation schemes two problems are of particular concern. The first is how to deal with unpredictable changes in the parameters defining the entry/exit dynamics. The other problem is the possible consistent difference in the entry/exit conditions for the different fisheries. First we consider the uncertainties and then we return to the problem of differential entry/exit conditions.

4. Stabilizing Regulation Schemes

We start this section by deriving stabilizing management schemes for a homogeneous capital-dominated fishery. We then show that, when the resource is harvested by two different types of fisheries, stability cannot be achieved by price regulation if the scheme does not differentiate between the two types of fishing.

4.1. Price subsidy

In this subsection we assume that the fishing activity is composed of only one type of fishing and derive a stabilizing regulation scheme for the entry/exit dynamics (15). The scheme is based on the stabilizing regulation of the dock-side price by the government.

Let N_s denote the desirable number of capital-dominated fishing enterprises in the fishery. Thus we assume that the government applies financial control in order to suppress the fluctuations and to maintain the level of the fishing activity, i.e., the number of fishing vessels, at or "near" the target level N_s. In deriving the scheme we assume that the fluctuations in the stock level are bounded.

Assume that the price fluctuations are additive and bounded such that $\pi(k) = \pi^* + w(k)$, where π^* is the nominal value of the dock-side price and $w(k)$ is a bounded disturbance such that $|w(k)| \leq \bar{w} \leq \pi^*$. The price is treated as an exogeneous variable because management aiming at preserving specific fleets will often focus on geographically small areas relative to the total fish market which determines the price of fish. Systems of price regulation will also frequently lead to an exogeneous price formation. Assume further that governmental control, $u(k)$, takes the form of subsidy (tax) per unit catch, $H(k)$, paid to (levied on) the enterprises. Then the entry/exit dynamics are given by

$$\begin{aligned} N(k+1) = N(k) + [&t_K N(k)\{\Pi_K(k)/K\} \\ &+ t_K N(k) \sum_{j=k-m}^{k-1} \{\Pi_K(j)/K\}]/m, \end{aligned} \tag{18}$$

where $\Pi_K(k)$ is given by (9), and $\pi(k)$ in equation (9) is replaced by $\pi^* + w(k) + u(k)$. Here we have presented the sum of profits in two parts to emphasize the assumption that that $\Pi_K(j)$ $j < k$, are known but for $j = k$ unknown. Thus, when deciding on the regulation action for the year k the resource manager does not predict the economic conditions in that year. He knows, however, the realization of the economics, that is, the profits, during the previous years. It should be noted that this assumption is not a strict condition for applying the theory. Modifications can be made with obvious changes in the problem setting.

The management action, $u(k)$, needs to be such that the non-negativity condition

$$N(k) \geq 0 \text{ for each } k \tag{19}$$

is satisfied.

Since no direct control is applied on the other potential state variable $S(k)$ we now proceed in this theoretical section by considering equation (18) as an one-dimensional system, where the stock level $S(k)$ is taken to be an external (measurable) variable. Introducing a change of variables, we get the following equation for the deviation of the number of vessels from the target level:

$$x(k+1) = f(k, x(k)) + C(k, x(k))[u(k) + e(k)], \tag{20}$$

where $x(k) \equiv N(k) - N_s$, and

$$\begin{aligned} f(k,x) \equiv x(k) + t_K[x(k) + N_s]\Bigg\{ D\Bigg[\pi^* \hat{q}_K K^{\alpha+\beta} S^{\gamma}(k) \\ - (c_K + a)\left(1 + \frac{\alpha}{\beta}\right) K\Bigg] + \sum_{j=k-m}^{k-1} \Pi_K(j))\Bigg\} / (Km), \end{aligned} \tag{21}$$

$$C(k,x) \equiv t_K[x(k) + N_s] D\hat{q}_K K^{\alpha+\beta} S^{\gamma}(k)/(Km), \tag{22}$$

$$e(k) \equiv w(k) \tag{23}$$

and $\hat{q}_K$ is defined by (10).

The policy instrument $u(k)$ is introduced for the purpose of suppressing the effect of the fluctuating price and changes in the resource level. It can be shown [11] that the policy instrument $u(k)$ takes the form of a subsidy or tax imposed on a unit price of fish and is given by

$$u(k) = -f(k, x(k))/C(k, x(k)) \tag{24}$$

The management rule utilizes the full knowledge of the entry/exit dynamics, in particular the values of the target level, N_s, the current values of the number of the firms, $N(k)$, as well as the stock level, $S(k)$. On the other hand, the management rule utilizes neither knowledge of the bounds of uncertainties nor attempts to predict the realizations of the uncertainties. It was shown by Kaitala et al. [13] that in the case of a memoryless entry/exit dynamics a high stock level implies negative support (tax) to prevent additional enterprises from entering the fishery, whereas a low stock level implies the need for subsidizing to prevent enterprises from leaving the fishery. When a memory is introduced in the entry/exit dynamics this may not be true in general. A low current stock level may imply, for example, high past profits which still encourage more enterprises to enter the fishery.

We note, somewhat unexpectedly, that there may exist a need for active regulation policy even though the current number of the fishing firms matches exactly the target value N_s. Assume, for example, that $x(k) = 0$. It then easy to see then that no regulation is needed only if $f(k, 0) = 0$. This happens in two special cases. First, the fishery has been in an equilibrium for m years, where m refers to the memory in the entry equation (15). Second, the fluctuations in the past cancel each other. The need for active regulation despite being at the target level may arise, for example, when the stock level or price are high or low causing destabilizing entry or exit decisions by fishing enterprises.

After application of the stabilizing control (23) the system equation for the deviation of the number of fishing enterprises from the target level N_s becomes

$$x(k+1) = w(k)t_K[x(k) + N_s]D\hat{q}_K K^{\alpha+\beta} S^{\gamma}(k)/(Km), \tag{25}$$

Thus, assuming that the fishery is at the target level, it remains there only if no price deviations from the nominal level π^* occur. The magnitude of the deviations can be characterized, however, in the case of successful stabilizing control. In particular, the attractive set B can be determined analytically from the parameters of the model. We can show that

$$\begin{aligned} |C(k,x)e(k,x,u)| &= |w(k)t_K[x(k) + N_s]D\hat{q}_K K^{\alpha+\beta} S^{\gamma}(k)/(Km)| \\ &\leq \rho_0 + \rho_1 |x|, \end{aligned} \tag{26}$$

where

$$\rho_0 \equiv \bar{w} t_K D\hat{q}_K K^{\alpha+\beta} \bar{S}^{\gamma})/(Km),$$
$$\rho_1 \equiv \bar{w} t_K N_s D\hat{q}_K K^{\alpha+\beta} \bar{S}^{\gamma})/(Km),$$

where $\bar{S}$ is upper bound for the stock level. The upper bound of the stock level depends on the population dynamics under consideration (see e.g., [9]). In the case of density-independent age-structured model to be studied below the upper bound is determined as the maximum size of the harvestable stock in the absence of harvesting.

The radius of an attractive set B is given in [11] as

$$d = \rho_0/(1 - \rho_1), \tag{27}$$

so that

$$B = B(d) = \{x \in R : |x| \leq d\}. \tag{28}$$

That is, the ultimate maximum deviation of the number of enterprises from the target level is given by (27). It can be shown by studying equations (26)–(27) that the increase in the parameter values $\bar{w}$, t_K, D, $\hat{q}_K$, and $\bar{S}$ reduces the success of a stabilizing policy whereas an increase in the memory, m, makes stabilizing more successful in terms of deviations from the target level. The effect of an increase in the capital input level K depends on the elasticities α and β. In the case of constant returns to scale ($\alpha + \beta = 1$) the capital input has no effect on the radius d. When the returns to scale are decreasing ($\alpha + \beta < 1$) then an increase in the capital input level decreases the range of fluctuations, and when they are increasing ($\alpha + \beta > 1$) then an increase in the capital input level increases the range of fluctuations.

4.2 Stabilizing fisheries subject to several types of harvesting

The same resources are often exploited by several types of harvesting. Assume for simplicity that only two sectors, one capital-dominated and the other labor-dominated, are present.

The price regulation scheme described by the equations (20)–(24) has been designed in particular to react on and to correct the undesirable consequences of unpredictable price fluctuations. If the price regulation scheme does not differentiate between the harvesting types then application of the price regulation scheme which is needed in one sector may produce unwanted effects in the other sector. From a control theory point of view it is impossible to define a unique stabilizing regulation policy by using price subsidies/taxes. A policy which stabilizes a labor-dominated fishery may create problems by consistently favoring the capital-dominated fishery and *vice versa*. Yet, price subsidies have been used indiscriminately assuming general beneficial effects in heterogeneous fisheries (see Section 5).

Kaitala et al. [13] proposed that the problem of stabilizing heterogeneous fisheries could be solved by applying a labor subsidy/tax schemes in labor-dominated fishing and capital subsidy/tax schemes in capital-dominated fishing. A closer study shows, however, that when the production functions are of the nonlinear Cobb-Douglas type (3) and when the enterprises seek to optimize the ratio of inputs (see eqs. (5)–(6)) then regulation based on per unit prices of inputs is not possible. Assume, for example, that the government plans to stabilize the number of enterprises in the capital-dominated fishing sector by regulating the unit price of the capital input, c_K. If the enterprises can fully predict the governmental regulatory action then supporting investments affect the optimal ratio of the inputs (8)). This in turn will alter the profit function (9) in a nonlinear way such that a theoretically correct formulation of the stabilization problem (see eq. 20)) is impossible.

The approaches to regulate the numbers of fishing enterprises in a fishery seem to depend crucially on the type of fisheries and on the technologies used in fishing. If, for example, capital and labor-dominated fishing produces fish of different quality the fish produced by the two types of fishing will have different prices. Then there should be no major difficulties in differentiating between the two fishing types and to apply different price regulation schemes to the sectors.

Another major difference between capital and labor-dominated fishing may arise in the fishing technologies used. It is then natural to assume that the unit prices of the capital and labor inputs also differ. Then it may be possible to apply regulation schemes that affect the capital input prices differently. We next show that this approach requires that the unit input price of labor are also regulated.

Consider again capital-dominated fishing as an example. Assume that the government attempts to stabilize the number of the enterprises only by regulating the unit price of the capital c_K by u_K. Such a scheme leads to problems with nonlinearity in (9)–(10). Such a problem of nonlinearity can be avoided, however, by regulating the unit price c_L of labor by u_L such that the price ratio $(c_K+a+u_K)/(c_L+u_L)$ of the inputs remains constant. If this ratio is kept equal to $(c_K+a)/c_L$ then (10) remains unchanged and the problems with the nonlinearity will not arise.

The foregoing suggests that a possible realization of the regulation tool is to relate it to the capital level and the stock level such that the enterprises are paid or levied by the amount $u_K(k)\hat{q}_K K^{\alpha+\beta} S^{\gamma}(k)$. The entry/exit dynamics of the enterprises now becomes the same as in the case of the price regulation scheme, that is, the entry/exit dynamics can now be presented in the form (20)–(23), and the regulation rule is again given as (24). Thus, we have shown that it is possible to design regulation tools that are based on the unit prices of the capital inputs. However, a regulation based on the unit price of capital input requires simultaneous regulation of the unit price of labor. We conclude by emphasizing that although the dynamics of the number of enterprises will be governed by the same equations in the cases of dock-side price regulation and input unit price regulation, the practical realization of the regulation is expected to be different.

5. The Baltic Herring Fishery

In this section we illustrate the stabilizing control approach obtained in the previous section by presenting a simulation example related to the Baltic herring fishery. In this fishery labor- and capital-dominated fishing currently coexist. Government subsidies are used and a general aim is to allow both types of fishing to continue.

The parameter estimates are based on the herring fishery in the sea area between Finland and Sweden (the northern Baltic proper, the Åland

Sea, the Archipelago Sea, the Bothnian Sea, and the Bothnian Bay) in 1985 [21]. The main herring fishing is carried out by trawlers and trapnetters which represent the capital- and labor-dominated fishing activities. It should be noted that the parameter estimates here are derived for illustrative purposes only.

The economic data on trapnet fishing come from forty enterprises participating in trapnet fishing for herring out of a total of approximately 470 enterprises, and the data on trawling fishing are obtained from the average figures of about ten enterprises representing about 370 enterprises in total.

5.1. Parameters

We begin by estimating the price parameters. The ideal situation for our purposes would be to have data on the range of price fluctuations under free market conditions. However, due to the long history of the price subsidizing of the Finnish herring production it is impossible to estimate the price behavior in the absence of any regulation mechanisms. One may expect, however, that the unregulated price of herring would behave in a way analogous to that of bream for which data are available. Both herring and bream are cheap fish used commonly in everyday consumption. The price of bream varies with the seasons and is restricted to the range of ± 30 percent of the average price [22]. The average price received by fishermen for herring was 1.23 FIM/kg in 1985. Assuming that the same average price would be also realized in the absence of regulation, then the 30 percent uncertainty range under free market conditions would be $|w(k)| \leq 0.37$ FIM/kg.

We use herring trapnet and trawl fisheries, respectively, to illustrate labor-and capital-dominated harvesting. The length of the fishing season in each fishery is about the same ($D = D' = 90$ days) making the comparison straightforward. Also, we may apply equal annual opportunity costs of capital (6 percent) in each fishery [21].

We begin by estimating the cost ratios in the fisheries. Assuming the same labor cost in each fishery, that is, $c_L = c'_L = 24$ FIM per hour, we obtained the following result. The total costs in the trapnet fishery was 11900 FIM from which 10500 FIM was spent on labor input and the rest, 1400 FIM was spent on capital input. The opportunity costs for the capital is 400 FIM. Thus, trapnet fishery can be considered to be a labor-dominated fishing sector. Recalling (8), the ratio of the capital costs plus the opportunity costs to labor costs in the trapnet fishery is

$$\frac{\beta'}{\alpha'} = 0.17.$$

The total costs in the trawl fishery were 72300 FIM from which 28400 FIM was spent to the labor input and 43900 FIM was spent to capital input. The opportunity costs for the capital is 2500 FIM. The cost ratio is

$$\frac{\beta}{\alpha} = 1.6.$$

Thus, the cost structures in the two fisheries are completely different. The labor costs dominate in the trapnet fishery and the capital costs dominate in the trawl fishery.

Unfortunately, empirical studies of the Cobb–Douglas type production function in fisheries economics are few, and estimates on the elasticities of the production factors in the Baltic fishing are completely missing at the moment. Using data on a purse seine fishery, which represents a different technology from those studied in this paper, Bjørndal [15] showed that in the North Sea herring fishery the output elasticity of capital input was less than one. In particular, he presented two different estimates equal to 0.85 and 0.65. We assume here $\beta = 0.8$, which then gives $\alpha = 0.5$. Assuming the same elasticities for labor input in both fisheries, that is, $\alpha = \alpha' = 0.5$, we get $\beta' = 0.085$. Thus, capital-dominated fishing is assumed to have increasing returns to scale and labor-dominated fishing increasing returns to scale.

The remaining elasticity parameter, γ, describes the effect of the stock level on the catch harvested by a unit amount of activity. For schooling species, like herring, it seems appropriate to use low stock output elasticity [15]. In particular, Bjørndal [15] presents two estimates of 0.62 and 0.34. Here we use the stock output elasticity equal to $\gamma = 0.5$.

We next estimate the remaining parameters. The annual opportunity cost of capital is 6 percent from which we may determine the total capital inputs in the fisheries: $K = 42.0 \times 10^3$ FIM and $K' = 7.25 \times 10^3$ FIM. The opportunity cost per fishing day per unit capital $a = 0.67$. The capital costs per unit capital per day become $c_K = 11.6$ and $c'_K = 2.1$. Finally, the average labor input used in the trawl fishery was $L = 13$ hours per day and in the trapnet fishery $L' = 5$ hours per day.

It remains to estimate the catchability coefficients, $\hat{q}_K$ and $\hat{q}_L$, for the fisheries. The catchability coefficients are needed to obtain the harvest per enterprise at a certain stock level. However, the harvest at any stock level can also be evaluated from the equation

$$\frac{H(S_1)}{H(S_0)} = \left(\frac{S_1}{S_0}\right)^{\gamma} \tag{31}$$

From the harvest at the stock level S_0 the harvest at the stock level S_1 can be determined from (31), which makes estimating the catchability coefficients unnecessary.

5.2. Population dynamics

As noted above, the herring stock in the Northern Baltic Sea has been stable since the 1970's. Kaitala et al. [13] assumed that harvesting does not destabilize the constant stock level and studied the current trends in the structural development of the Baltic herring fishery. They acknowledged, however, that the assumption of a stable stock size neglects possible collapses of the stock due to overfishing, environmental pollution, etc., and may thus be misleading in long term simulations. For this reason we develop here a model for the stock dynamics to study the effects of regulation and harvesting on the stock level and vice versa.

The herring population is composed of 11 age classes. We assume that individuals in each age class have three options: They become older, die by natural reasons, or are caught by fishermen. Thus the dynamics of fish, X_j, in each age class j are given by

$$X_{j+1}(k+1) = X_j(k)\exp(-z_j) - N(k)h_j^K(k) - M(k)h_j^L(k), \tag{32}$$

where z_j is the natural mortality rate in the age class j, and $h_j^l(k)$, $l = K, L$, denotes the number of the fish harvested in the age class j by a capital- or labor-dominated enterprise. The catch is assumed to be taken instantaneously at the end of the year. This assumption is an approximation of the basic population dynamics equation used in many fisheries models (e.g., [4]). A similar approximation has been used by Pope [23] who showed that the error is less than two percent when the fishing mortality rate is less than or equal to 1.2 and the natural mortality rate less than or equal to 0.3. Both conditions are valid for the herring in the northern Baltic Sea. $N(k)$ and $M(k)$ are the numbers of capital- and labor-dominated enterprises in year k, respectively. The number can be presented in terms of biomass as follows:

$$W_j(k) = w_j X_j(k), \tag{33}$$

where $W_j(k)$ is the biomass of the age class j at time k, w_j denotes the average weight of fish in the j-th age class. In a similar way, the catch of age class j fish is given by

$$H_j^l(k) = w_j h_j^l(k), \quad l = K, L. \tag{34}$$

Assuming that the first two age classes (0 and 1) escape harvesting and that all other age classes are equally vulnerable to harvesting, we may express the total catch as

$$H^l(k) = \sum_{j=2}^{11} H_j^l(k), \quad l = K, L, \tag{35}$$

The total harvestable biomass in the fishery is

$$S(k) = \sum_{j=2}^{11} W_j(k). \tag{36}$$

It follows that the harvest is distributed according to the distribution of the age class biomasses, and the numbers of the harvested fish can then be determined from (36).

We assume the same natural mortality for each age class, that is, $z_j = 0.2$ for all j. The weights of the individuals in the age classes are $(7, 10, 20, 28, 36, 43, 49, 55, 59, 62, 64, 65)$ grams.

In this study we neglect any possible density dependence of the stock-recruitment relationship and assume simply that the number of the new-born is constant. This assumption may be violated if excessive harvesting capacity builds up. The initial numbers of the individuals correspond to an estimate of the age distribution in a sub-area and are given as $(8000, 6600, 6400, 3000, 1800, 900, 600, 200, 200, 300, 600, 300)$ millions of fish corresponding to the biomass distribution $(56, 66, 128, 84, 65, 39, 30, 11, 12, 18, 36, 19)$ thousand tons totalling 564 thousand tons of herring. The harvestable biomass is now 442 thousand tons. Under no-harvesting conditions the harvestable biomass would stabilize at the level of 1016 thousand tons.

5.3. Entry/exit dynamics

For simulation purposes we also introduce memory into the system. It is unlikely that entry and exit decision are based on the results of single years. Investments in fishing are partly irreversible and decisions on entry or exit are likely to be made based on several years of data. The same is true for labor-dominated fishing. A switch of job is not done lightly. For example data on trapnet fishing in the Bothnian Sea showed that a fisherman was prepared to fish with trapnets at greatly reduced catch levels for three years before he decided to stop [24]. Similar results have been obtained in a survey in the Gulf of Finland (I. Häyrynen, Finnish Game and Fisheries Research Institute, personal communication). As a first estimate we use a three year moving average in order to simulate the memory of fishermen and investors. Thus, we choose $m = 3$.

At the moment there are no estimates available for the entry/exit parameters t_l or t_K in the Baltic herring fishery. For illustrative purposes we choose $t_K = 0.10$ for the capital-dominated fishing. Furthermore, we assume that when the number of the enterprises in each fishing sector are same so that the same profits will cause the same relative change in the number of the firms. Thus, we choose $t_l = t_K L/K$.

5.4 A regulation example

Recall that in 1985 the annual total costs in the trawl fishery were 72 300 FIM and in the trapnet fishery 11900 FIM. Furthermore, the average catches in the fisheries were 115 000 kg and 34 000 kg, and the harvestable stock level was 442 thousand tons. Hence, the values of the catches were

141 000 FIM and 41 800 FIM. Assuming the same profit for the three previous years we may predict that the numbers in both types of fishing are increasing and that the rate of increase in the trawl fishing sector is at least three time higher than in the trapnet fishing one.

It follows from (31) and from the previous data that the catch at a stock level S is

$$H(S) = \left(\frac{S}{442}\right)^{\gamma} 34000 \tag{37}$$

in the trapnet fishery, and

$$H(S) = \left(\frac{S}{442}\right)^{\gamma} 115000 \tag{38}$$

in the trawl fishery. Calculating the open-access stock levels at which the value of the catch is equal to the costs of harvesting results in $S' = 35 \times 10^6[kg]$ for the trapnet fishery and $S = 115 \times 10^6[kg]$ for the trawl fishery. The result predicts that in an unregulated fishery the rapid increase of the trawling enterprises causes the resource to decline, after which the trawl fishery will decline as well. Somewhat unexpectedly, the trapnet fishery will survive to some extent at the low stock level. Kaitala et al. [13] proposed that the more efficient trawl fishing could destroy the less efficient trapnet fishing together with the resource, but this would require that the trawl fishery affects the reproductive capacity of the herring stock.

Figure 1 illustrates the stock level paths in the two different cases, in an unregulated and in a regulated fishery. For simulation purposes, the realization of the price (income) fluctuation was takes as $w(k) = -\bar{w}\sin(2.0k)$, where $\bar{w} = 0.36$. Figure 2 shows the corresponding numbers of enterprises. In the unregulated case the number of the trawl enterprises increases quickly, which causes a rapid decline of the resource. Along with the decline of the resources the basis for high profits disappears. First the entry of the trawl enterprises slows down and then their number starts to decrease. Because the losses produced at low stock level are only minor in our model the decrease in the number of the trawl enterprises is slow.

The number of trapnet enterprises increases as well but more slowly due to lower annual profits from fishing. However, since the trapnetters are able to survive at lower stock levels their number does not stabilize at the same time as the number of trawlers.

The unregulated fishery can be compared with the fishery in which the regulation target is to stabilize the number of the fishing enterprises at the 1985 level. This is done successfully by applying different regulation policies in each fishing. After choosing the target level for the numbers of the different type of fishing enterprises, regulation can be realized by different price regulation schemes for each type of enterprise or by regulating differently the price of capital and labor in each fishing sector. In order to be able to do this the production functions and the factors determining entry/exit dynamics have to be known. They are as important for efficient resource regulation as is the modeling of the resource dynamics. The topic of entry and exit of enterprises has received, however, only minor attention in the fisheries economics literature.

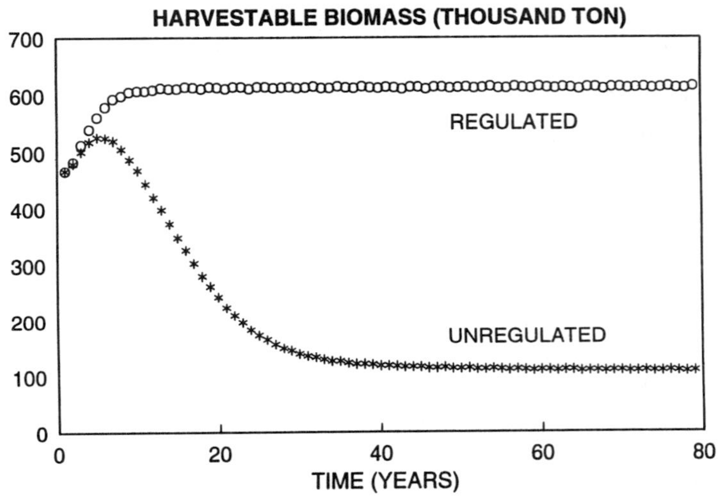

Figure 1. The simulated paths of harvestable biomass in unregulated and regulated fisheries. In the unregulated case the efficiency of trawl fishing causes the biomass to collapse. After the resource basis for the profits has disappeared the biomass level will stabilize. In the regulated fishery the government is able to prevent the resource collapse.

As was discussed above (see subsection 4.2), the two regulation schemes will result in theoretically equivalent entry/exit dynamics of the enterprises. As a whole, the example shows that the structural problems seem to be far more important in the Baltic fishery than are short term price or income fluctuations.

The approach used in this paper differs from standard fisheries management that aims at using quotas in order to stabilize the exploitation rates [25]. The quota management policy has been extensively used by for example the European Commission, but it has lead "to a race in terms of vessels and catches" and a demand for "stronger structural management, by segmentation of the fleet, ..., providing an assessment and control of fishing effort, and inclusion of structural measures under the umbrella of the reform of the structural Funds" [5]. Our simulation shows that the change from quota management to structural management is not an easy one and requires detailed microeconomic information that until now has been used to a limited extent in fisheries management.

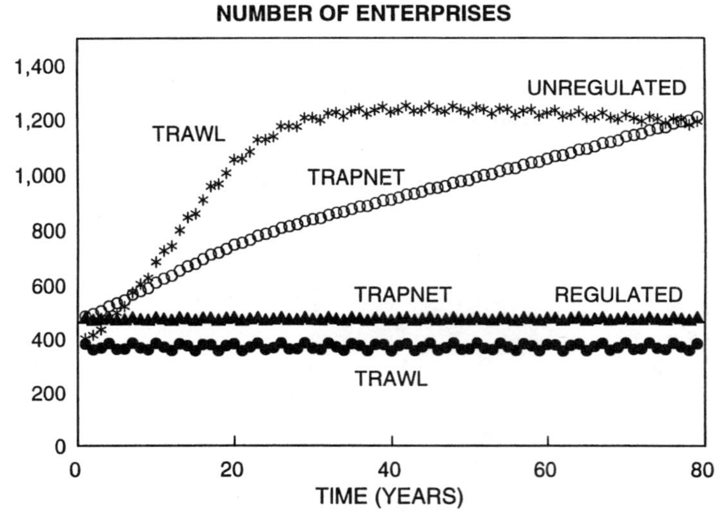

Figure 2. The simulated numbers of trawl and trapnet enterprises in unregulated and regulated fisheries. In the unregulated case the efficiency of the trawl fishing as compared with the trapnet fishing causes the their number to increase more quickly. The trapnet enterprises are, however, able to produce profits at lower stock levels than are the trawl enterprises. In the regulated fishery the government stabilizes the numbers of the enterprises at the 1985 level, which saves the resource.

Conclusions

When the fisheries management is based on quotas, possibly combined with some limitation of the total tonnage or number of enterprises, management agencies are not explicitly concerned with structural or distributional issues. Concerns for specific fleets or regional policy may nevertheless enter into quota negotiation and inflate the quotas at the cost of the resource and the future wellbeing of the fishery. In these cases it could be better to set the stabilization of the fishery as an explicit objective and investigate economic incentives needed to reach the goal for each type of fishery. In these cases a policy based on the stabilization of the number of enterprises using economic incentives could be explored. The necessary infrastructure is available, e.g., because several OECD member countries subsidize their fishing industry [18]. This study shows, however, that although stabilization of structures are often demanded for political reasons the use of subsidies for this purpose is not an easy route if unwanted structural effects are to be avoided.

The production functions in fishing and the entry/exit dynamics of fishing enterprises are crucial, but these topics have received only minor attention in the fisheries literature. An understanding of these is necessary for the design of regulation schemes that are able to cope with short term uncertainties as well as long term structural changes in the fisheries.

Acknowledgments. This research was supported in part by the Academy of Finland. The authors thank P. Mickwitz and two anonymous referees for comments on earlier drafts.

References

[1] J.H. Goodlad, Stability - fishing industry perspective, ICES Cooperative Research Report 171 (1990), 2–4.

[2] M. Holden, Stability - fisheries administration perspective, ICES Cooperative Research Report 171 (1990), 11–17.

[3] J.G. Shepherd, Stability - Fisheries biology perspective, ICES Cooperative Research Report 171 (1990), 4–11.

[4] J.G. Gulland, *Fish Stock Assessment — Manual of Basic Methods*, John Wiley and Sons, 1983.

[5] Commission of the European Communities, Report 1991 from the Commission to the Council and the European Parliament on the Common Fisheries policy, Commission of the European Communities SEC (91) 2288 Final, 1991.

[6] J.W. Horwood and P. Whittle, Optimal control in the neighbourhood of an optimal equilibrium with examples from fisheries models, *IMA J. Math. Appl. Med. Biol.*, **3** (1986), 129–142.

[7] R. Arnason, Minimum information management in fisheries, *Can. J. of Economics*, **23** (1990), 101–124.

[8] R.N. O'Boyle, A.F. Sinclair, and P.C.F. Hurley, *A bioeconomic model of an age structured groundfish resource exploited by a multigear fishing fleet,* ICES Mar. Sci. Symp. 193 (1991), 264–274.

[9] V. Kaitala and G. Leitmann, Stabilizing management of fishery resources in a fluctuation environment, *Dynamics and Stability of Systems*, **4** (1989), 95–111.

[10] V. Kaitala and G. Leitmann, Stabilizing employment in a fluctuating resource economy, *Journal of Optimization Theory and Applications*, **67** (1990), 1–16.

[11] V. Kaitala and G. Leitman, Stabilizing management of fluctuating resources, ZOR (*Methods and Models of Operations Research: Theory*) **35** (1991), 351–364.

[12] V. Kaitala, Maximizing yield in some fluctuating fisheries, in *Essays on the Economics of Migratory Fish Stocks*, eds. R. Arnason and T. Bjørndal, 41–54, Studies in Contemporary Economics, Springer-Verlag, Berlin, Heidelberg, 1991.

[13] V. Kaitala, M. Hildén, and G. Leitmann, Income Subsidizing and Fisheries Development — An Analysis of Stabilizing Management, in

Dynamic Economic Models and Optimal Control, ed. Feichtinger, G., 189–203, North-Holland, Amsterdam, 1992.

[14] M. B. Schaefer, Some considerations of population dynamics and economics in relation to the management of marine fishes, *J. Fish. Board Canada*, **14** (1957), 669–681.

[15] T. Bjørndal, Production economics and optimal stock size in a North Atlantic fishery, *Scand. J. of Economics*, **89** (1987), 145–164.

[16] R. Hannesson, Bioeconomic production function in fisheries: Theoretical and empirical analysis, *Can. J. Fish. Aq. Sci.*, **40** (1983), 968–982.

[17] T. Bjørndal and J.M. Conrad, Capital dynamics in the North Sea herring fishery, *Marine Resource Economics*, **4** (1987), 63–74.

[18] Anon., Financial support to the fishing industry, OECD Document, 161 pp., Paris, 1980.

[19] M. Corless and J. Manela, *Control of uncertain discrete-time systems,* Proceedings of the 1986 American Control Conference, 515–520, Seattle, WA, 1986.

[20] J. Manela, Deterministic control of uncertain linear discrete and sampled data systems, University of California, Berkeley, PhD Thesis, 1985.

[21] P. Salmi, F. Löf, M. Viitanen, and H. Partanen, The profitability of professional fishing in Finland in 1985, Suomen Kalatalous, (1989), 12–21.

[22] M. Hildén and K. Saarni-Vilo, Price fluctuations in a coastal fishery in Finland, in *La Recherche Face a la Peche Artisanale*, Symp. Int. ORSTOM-IFREMER, eds. Durand, J.-R., Lemoalle, J. and Weber, J., t. I, 385–393, 1991.

[23] J.G. Pope, An investigation of the accuracy of virtual population analysis using cohort analysis, *Res. Bull. Int. Comm. Northw. Atl. Fish.*, **9** (1972), 65–74.

[24] H. Lehtonen, M. Hildén, and A. Järvinen, Death in the sea: A fisherman's perspective, *Marine Pollution Bulletin*, **20** (1989), 207–214.

[25] Anon., Reports of the ICES Advisory Committee on Fishery Management, ICES Cooperative Research Report 168, part 1 and 2, 361+ 91 pp, 1990.

Mikael Hildén
National Board of Water
and the Environment
Helsinki, Finland

Veijo Kaitala
The BEIJER International Institute
for Ecological Economics
The Royal Sweedish Academy
of Sciences
Stockholm, Sweden

George Leitmann
College of Engineering
University of California
Berkeley, California 94720, USA

The Nonuniqueness of Markovian Strategy Equilibrium: The Case of Continuous Time Models for Nonrenewable Resources

S. Clemhout and Henry Y. Wan, Jr.

Abstract. In a two-player, continuous time model of dynamic common-property resource of the replenishable type, we illustrate by examples the dual indeterminacy property of the Markovian–Nash equilibrium. Specifically, we have established: (1) The indeterminacy of the inverse equilibrium problem: there exists a continuum of distinct model structures corresponding to one observable equilibrium path; (2) The indeterminacy of the equilibrium problem: there exists a continuum of distinct, observable equilibrium plays corresponding to a single model structure. The method of analysis is to construct candidate equilibrium paths by means of both the Euler-Lagrange and the transversality conditions and then verify that the candidate equilibrium is an equilibrium by the Mangasarian criterion. These results may well have implications to other areas where the dynamic games formulation is relevant, for exdample, the concept of rational expectations and the procedure of calibration in macro-economic analysis.

1. Introduction

This paper studies the relationship between models and equilibrium plays in dynamic games. It is found that there may be a continuum of distinct models which can cause a particular equiulibrium play which is being observed. Likewise, there may be a continuum of distinct equilibrium plays which are consistent with a particular model. Write M and E as the spaces of models and equilibrium plays, with m, m', m'' and e, e', e'' as their elements, Figure 1 depicts this situation. This is referred to as 'dual indeterminacy' below. Distinction may be made between the multiplicity of the continuum at hand and the finite multiplicity (cf. the example in Mirman (1979) where the same model can cause two equilibria). In the literature of competitive equilibrium, such a distinction gives rise to the entire theory of the finiteness property for equilibria.

The analysis here arises from the study of the fish war, as formulated by Levhari and Mirman (1980). (For relevant references on this topic, please refer to Kailala (1986)) The Spartan simplicity of its analytic structure focuses attention to the essential issues at hand. At the same time, that model epitomizes the nature of common property games which underlies the entire range of environmental problems, from acid rain, ozone depletion, the exhaustion of ground water, atmosphere and ocean pollution to ecological stress of various kinds (see Clemhout and Wan (1985, 1991) which take note some of such connections). One additional advantage of the Levhari-Mirman model is its structural affinity to the Ramsey model which is thoroughly understood in economics, even though the multi-agent context gives rise to such fundamentally novel features absent in the Ramsey problem.

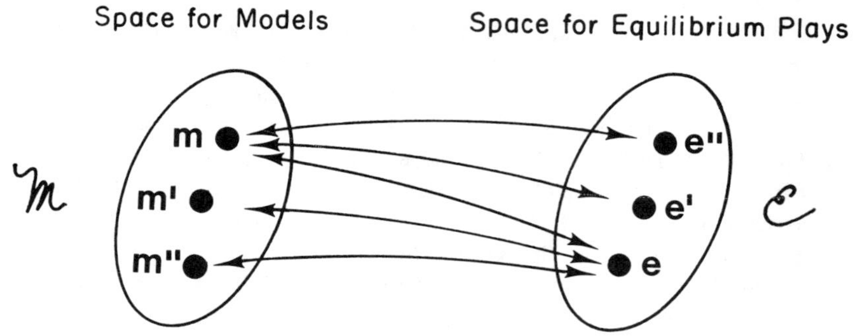

Figure 1

The strategy of the player is Markovian in that the action chosen is functionally dependent upon the current state, not calendar time. This coincides with the feedback information pattern in the differential game literature (e.g., Basar (1989)) if the vector of the current state does not contain time as a component. For the ease of subsequent analysis, attention is focused upon strategies which are piece-wise smooth.

The method of study is constructive. After the problem is formulated via a simultaneous system of Hamilton–Jacobi–Bellman equations, the candidate solutions are characterized by the first order conditions of the Euler–Lagrange relationships. The equilibria are verified through the second order condition in the form of the Mangasarian criterion.

To draw upon the results from the theory of differential equations, the present research assumes time is continuous. Historically, the fish war model was pioneered by the Levhari and Mirman study in discrete time. In a sequence of elegant studies by Sundaran (1989) as well as Dutta and Sundaran (1989, 1991, 1992), discrete time studies of fish war have encompassed most of the results obtained in continuous time fish war models, up to now. It is likely that by further research, the indeterminacy results investigated here may again be obtained in discrete time models as well.

The study of multiplicity of equilibria with the power of continuum was first conducted both by Amit and Halperin (1989) and Dockner and Long (forthcoming) by analyzing the first order conditions. Independent from the present paper, Dockner, Hartwick and Sorger (1992) also uses second order conditions to verify equilibrium plays. The work here extends the study of nonrenewable resources in Clemhout and Wan (1989). It is the renewability of the resources which permits a continuum of alternative steady states as limit paths. This gives rise to the existence of a continuum of equilibria. In environmental problems with more than one renewable resource, steady states are not the only forms giving rise to limit paths with stocks bounded away from the zero levels (see e.g., Clemhout and Wan (1985)). The coexistence of a continuum of equilibrium is likely to arise even in more diverse patterns.

In the rest of this paper, the discussion is in terms of a class of selected examples after the introduction to a general framework.

2. A Two-Player Fish War in Continuous Time

We focus on a class of continuous time, n player fish war model. This should suffice to demonstrate our conclusions.

Let x be the fish stock, t be continuous time, c_i be the consumption of player $i = 1, \ldots, n$, $f(x)$ be the recruitment function, $u_i(c_i)$ be the (instantaneous) felicity index and $r_i > 0$ be the time preference rate for player $i = 1, \ldots, n$.

The payoff of player is:

$$\int_0^\infty e^{-r_i t}[u_i(c_i) + k_i x]dt, \tag{1}$$

and the evolution of the system follows.

$$\frac{dx}{dt} = f(x) - \sum_{i=1}^{n} c_i, \quad x(0) \text{ given}, \tag{2}$$

$$0 \leq x, \ 0 \leq c_i \leq C_i(x) \ \text{ for some } C_i(x), \ i = 1, \ldots, n \text{ at all } t.$$

Remark 1 (Complementary quasi-linearity). The integrand in (1) and the righthand side of (2) are quasi-linear.

Remark 2. $k_i > 0$ in (1) represents a 'wealth effect', which takes now the form of common-property public good, like whales for whale-watching tours.

Remark 3 (Relationship to Ramsey). If $n = 1$, (1)–(2) becomes a Ramsey model.

For the present purpose, we shall simplify the model, setting $k_i = 0$, all i, and $n = 2$. We adopt now the following:

Assumptions

1. The function f is
   a. Zero-valued at $x = 0$ and non-negative for all x
   b. concave over some interval $[0, \bar{x})$, with either (i) $\bar{x} = \infty$, or (ii) $\bar{x} = 1$, and $f(1) = 0$
   c. twice continuously differentiable
2. For $i = 1, 2$, the function u_i is
   a. strictly increasing
   b. strictly concave, and
   c. twice continuously differentiable
3. $C_i(0) = 0$.

Since we study the issue of multiplicity of solutions, we must adopt such normalization rules to make sure that any multiplicity we end up with is not merely due to the choice of the unit or origin. Thus, we adopt the

Normalization Rules

1. By choosing the unit for time, we set,

$$r_1 + r_2 = 1, \tag{3}$$

or, equivalently,

$$r_1 = r,\ r_2 = 1 - r,\ 0 < r < 1 \tag{4}$$

2. By labeling the players, we make sure that,

$$r \leq 1/2.$$

3. By choosing the unit and the origin for the felicity index, we set for $i = 1, 2$

$$u_i(1) = 0, \tag{5}$$
$$u_i'(1) = 1. \tag{6}$$

It is possible to specify u_i in terms of the index for relative risk aversion,

$$\gamma_i(c_i) = -c_i u_i''(c_i)/u_i'(c_i), \quad i = 1, 2. \tag{7}$$

Clearly, by (5), (6) and (7), we have:

$$u_i(c_i) = \exp\left\{-\int_1^{c_i}\left(\int_1^{y}[\gamma_i(z)/z]dz\right)dy\right\}. \tag{8}$$

By Assumption 2, a and b,

$$\gamma_i(c_i) > 0 \quad \text{for all } c_i \geq 0. \tag{9}$$

Each model of fish war now corresponds to a quadruple,

$$m = (r, u_1(\bullet),\ u_2(\bullet),\ f(\bullet))$$

satisfying Assumptions 1 and 2 as well as (3)–(6).

A *Markovian strategy* for player $i = 1$ or 2 is a function

$$c_i = p_i(x). \tag{10}$$

A *Markovian–Nash equilibrium* is an ordered pair $p(\bullet) = (p_1(\bullet), p_2(\bullet))$ such that, given

$$c_j = p_j(x), \tag{11}$$

then for all $x > 0$, and over the class of all *continuous Markovian strategies*, (10) maximizes (1) subject to (2), $i, j = 1, 2$, $i \neq j$.

A Markovian–Nash equilibrium is *smooth* if p_i is twice-continuously differentiable, for $i = 1, 2$. It is interior, if $0 < p_i(x) < C_i(x)$, for all $x > 0$.

Given a smooth Markovian–Nash equilibrium p, one can use (10) to solve (2) for the equilibrium play:

$$e_p(\bullet) = (x^*(\bullet), c_i^*(\bullet), c_2^*(\bullet)) \quad \text{for all } t \geq 0.$$

Let M and E be the spaces of models and equilibrium plays, respectively, with m and e_p as generic elements. We shall study by example the structure of the following correspondences:

$$F: E \to M;\ e_p \to \{m \in M: e_p \text{ is an equilibrium play for } m\}$$

(The inverse equilibrium problem)

and

$$G: M \to E;\ m \to \{e_p \in E: m \text{ has a smooth Markovian–Nash equilibrium } p \text{ which induces } e_p\}$$

(The equilibrium problem)

In order to do so, we shall characterize any p which is consistent with m in the next section.

3. Characterization of the Equilibrium

We now recast the problem facing a particular player i in any equilibrium as: Find

$$V_i(x,t) = \text{Max} \int_t^\infty e^{-r_i t'} u_i(c_i)\, dt', \tag{12}$$

subject to, $\frac{dx}{dt} = f_i(x) - c_i$, $x(0) = x$, where we define

$$f_i(x) = f(x) - p_j(x), \quad i, j = 1, 2, \ i \neq j. \tag{13}$$

The Hamilton–Jacobi–Bellman equation takes the form:

$$-\partial V_i/\partial t = \max_{c_i \in (0, C_i(x))} \left\{ e^{-r_i t} u_i(c_i) + (\partial V_i/\partial x)[f_i(x) - c_i] \right\}.$$

Assuming now that

$$V_i(x, t) = W_i(x) e^{-r_i t},$$

the above reduces to an ordinary differential equation:

$$r_i W_i(x) = \max_{c_i \in (0, C_i(x))} \left\{ u_i(c_i) + W_i'(x)[f_i(x) - c_i] \right\}$$

where the first order necessary condition for an interior maximum is:

$$u_i'(c_i) = W_i'(x). \quad \text{(The Maximum Principle)}. \tag{14}$$

On substitution, then at the equilibrium,

$$r_i W_i(x) = u_i(c_i) + u_i'(c_i)[f_i(x) - c_i].$$

Differentiating both sides with respect to x,

$$\begin{aligned} r_i W_i'(x) &= u_i'(c_i) dc_i/dx + u_i''(c_i)(dc_i/dx)[f_i(x) - c] \\ &\quad + u_i'(c_i)[f_i'(x) - dc_i/dx] \\ &= u_i''(c_i)(dc_i/dx)[f(x) - c_i - c_j] + u_i'(c_i)[f'(x) - dc_j/dx]. \end{aligned}$$

Using the Maximum Principle and writing c_i as $p_i(x)$, dc_i/dx as $p_i'(x)$, as well as $u_i''(c_i)$ as $-\gamma_i(p_i)u_i'(c_i)/p_i$, we have after rearrangement,

$$\{[f(x) - p_i(x) - p_j(x)]\gamma_i(p_i)/p_i\} p_i' + p_j' = f'(x) - r_i, \quad i, j = 1, 2, \ i \neq j.$$

Thus, we have,

$$A(x, p) p' = b(x) \tag{15}$$

where:

$$A(x, p) = \begin{bmatrix} [f(x) - p_1 - p_2]\gamma_i(p_1)/p_1 & 1 \\ 1 & [f(x) - p_1 - p_2]\gamma_2(p_2)/p_2 \end{bmatrix}$$

$$p = \begin{bmatrix} p_1 \\ p_2 \end{bmatrix}, \ p' = \begin{bmatrix} p_1' \\ p_2' \end{bmatrix}, \ b(x) = \begin{bmatrix} f'(x) - r \\ f'(x) - (1 - r) \end{bmatrix}.$$

If $A(x,p)$ is nonsingular throughout the play, then we have,

$$p' = A^{-1}(x,p)b(x) \quad \text{(The fundamental equation of fish war).}$$

The solutions of this equation:

$$(c_1, c_2) = (p_1(x), p_2(x))$$

are candidates for interior *Markovian–Nash* equilibrium. They may become part of piecewise C^1 Markovian equilibria or constitute the totality of such an equilibrium, as it will be discussed below.

Each of such a solution must

(a) stay within the feasible region:

$$\{(x, c_1, c_2)\} \geq 0 : x \in X, \ (c_1, c_2) \in (0, C_1(x)) \times (0, C_2(x))\}, \tag{16}$$

which is depicted in Figure 2, panels a and b,

(b) have a steady state at $x = x_\infty$ for some x_∞ within the steady-state manifold:

$$\{(x, c_1, c_2) \geq 0 : x \in X, \ c_1 + c_2 = f(x)\} \tag{17}$$

where $dx/dt = 0$. This is depicted in Figure 2, panels c and d, and

(c) possess a point of discontinuity if it approaches the singularity manifold,

$$\left\{(x, c_1, c_2) \geq 0 : c_1 + c_2 + \sqrt{c_1 c_2/\gamma_1\gamma_2} = f(x)\right\}, \tag{18}$$

where p' is undefined since $|A| = 0$. This is depicted in Figure 2 panel d. Once $p(x)$ is verified as an equilibrium, it is routine to derive

$$W_i(x) = \left\{u_i[p_i(x)] - p_i(x)u_i'[p_i(x)] + f(x)u_i'[p_i(x)]\right\}/r_i$$

and finally, the value function

$$V_i(x,t) = e^{-r_i t} W_i(x)$$

for $i = 1, 2$.

A cross-section of Figure 2, panel d is reproduced as panel e to display the relationships between the two manifolds mentioned above.

One can now turn to the candidate solution for a corner Markovian–Nash equilibrium which satisfies the condition:

$$\begin{aligned} r_i W_i(x) &= \{u_i'(c_i) + W_i'(x)[f_i(x) - c_i]\} \\ &= \{u_i'[C_i(x)] + W_i'(x)[f_i(x) - C_i(x)]\} \end{aligned} \tag{19}$$

with:

$$(c_1, c_2) = (C_1(x), C_2(x)). \tag{20}$$

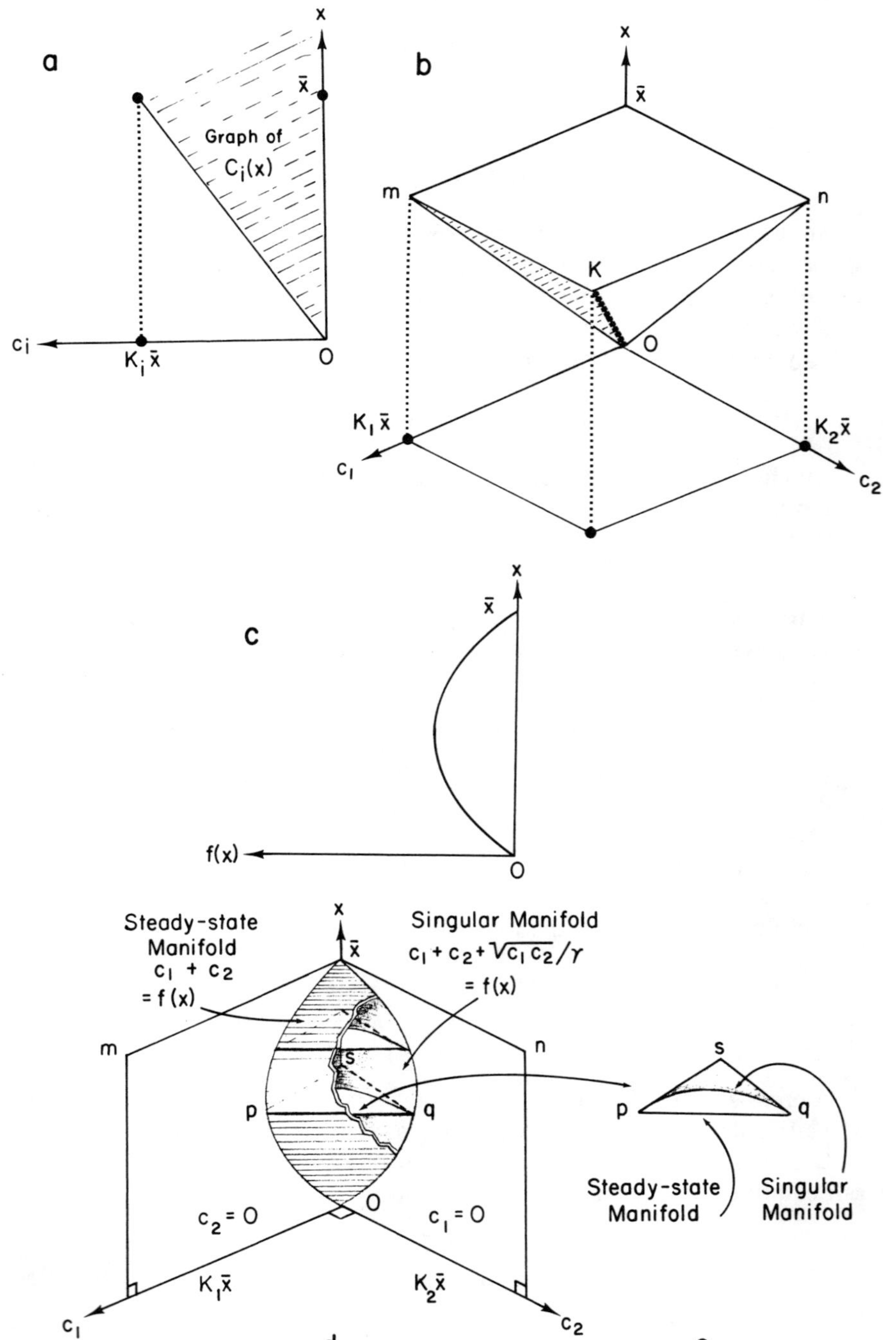
a
x
$\bar{x}$
Graph of
$C_i(x)$
c_i
$K_i\bar{x}$
O
b
x
$\bar{x}$
m
n
K
O
$K_1\bar{x}$
c_1
$K_2\bar{x}$
c_2
c
x
$\bar{x}$
f(x)
O
Steady-state
Manifold
$c_1 + c_2$
$= f(x)$
x
$\bar{x}$
Singular Manifold
$c_1 + c_2 + \sqrt{c_1 c_2}/\gamma$
$= f(x)$
m
n
s
p
q
$c_2 = 0$
O
$c_1 = 0$
$K_1\bar{x}$
$K_2\bar{x}$
c_1
c_2
d
s
p
q
Steady-state
Manifold
Singular
Manifold
e

Figure 2

Here, it is possible that $u_i'[C_i(x)] > W_i'(x)$ for a corner solution.

By assuming $C_i(x)$ is linear in x for $i = 1, 2$, this locus is depicted as the OK ray in Figure 2 panel b.

A candidate for an interior equilibrium must satisfy the sufficient condition for optimality. We adopt the Mangasarian form, where $f_i(x)$ must be concave.

Furthermore, if any candidate equilibrium locus reaches the origin, then one must have:

$$p_i(x)/x \to p_i'(0), \quad \text{as } x \downarrow 0.$$

Substitute this result back into the *fundamental equation* and we have:

Result 1. If the candidate for an equilibrium qualifies as an equilibrium over $[0, x_\infty]$, then the following conditions hold:

$$\begin{aligned} \gamma_1 p_1'(0) + (\gamma_1 - 1)p_2'(0) &= (\gamma_1 - 1)f'(0) + r \\ (\gamma_2 - 1)p_2'(0) + \gamma_2 p_2'(0) &= (\gamma_2 - 1)f'(0) + (1 - r). \end{aligned} \tag{21}$$

Proof. Note that $\lim_{x\downarrow 0}[f(x)/x] = f'(0)$. The rest is straightforward.

4. A Class of Examples

$$\begin{aligned} &u_i(c_i) = c_i^{1-\gamma}/(1-\gamma),\ \gamma > 1 \text{ (Constant relative risk aversion)} \\ &f(x) = \mu(x - x^\gamma),\ 0 < \mu < 1,\ C_i(x) = xK_i,\ K_i > 0,\ i = 1, 2, \end{aligned}$$

(15) now reduces to:

$$\begin{bmatrix} \gamma[\mu(x - x^\gamma) - p_1 - p_2]/p_1 & 1 \\ 1 & \gamma[\mu(x - x^\gamma) - p_1 - p_2]/p_2 \end{bmatrix} \begin{bmatrix} p_1' \\ p_2' \end{bmatrix} = \begin{bmatrix} \mu(1 - \gamma x^{\gamma-1}) - r \\ \mu(1 - \gamma x^{\gamma-1}) - (1 - r) \end{bmatrix}. \tag{15$'$}$$

We explore the possible existence of a candidate pair in the form of linear decision rules, $p_i(x) = x\tilde{p}_i'$, $i = 1, 2$. If such a candidate for equilibrium exists, (15') becomes,

$$\begin{bmatrix} \gamma[\mu(1 - x^{\gamma-1}) - \tilde{p}_1' - \tilde{p}_2']/\tilde{p}_1' & 1 \\ 1 & \gamma[\mu(1 - x^{\gamma-1}) - \tilde{p}_1' - \tilde{p}_2'] \end{bmatrix} \begin{bmatrix} \tilde{p}_1' \\ \tilde{p}_2' \end{bmatrix} = \begin{bmatrix} \mu(1 - \gamma x^{\gamma-1} - r \\ \mu(1 - \gamma x^{\gamma-1}) - (1 - r) \end{bmatrix} \tag{15$''$}$$

which reduces to:

$$\begin{bmatrix} \gamma & \gamma-1 \\ \gamma-1 & \gamma \end{bmatrix} \begin{bmatrix} \tilde{p}_1' \\ \tilde{p}_2' \end{bmatrix} = \begin{bmatrix} (\gamma-1)\mu + r \\ (\gamma-1)\mu + (1-r) \end{bmatrix}$$

$$\tilde{p}_1' = [(\gamma-1)(\mu-1)/(2\gamma-1)]+r, \; \tilde{p}_2' = [(\gamma-1)(\mu-1)/(2\gamma-1)]+(1-r). \quad (22)$$

Thus, provided that,

$$p(x) \in (0, xK_i) \;\text{ all } x,$$

or, (a) $\tilde{p}_1' > 0$, $\tilde{p}_2' > 0$ and (b) $K_i > \tilde{p}_i'$, $i = 1, 2$, we have a candidate pair of equilibrium in

$$p_i(x) = x\tilde{p}_i', \quad i = 1, 2.$$

Note that condition (a) is satisfied if $\mu + 3r > 1$, since $\tilde{p}_2' > 0$ for all $\mu > 0$.

Applying the Mangasarian condition, the candidate pair is verified as an equilibrium.

We have thus proved,

Result 2. Within the class of examples in question, if conditions (a) and (b) hold, there exists a unique Markovian–Nash interior equilibrium in the form of linear decision rules.

For the case of symmetric players where $r_1 = r_2 = 1/2$, the equilibrium of linear decision rule is displayed as line OL in Figure 3, panel a. Its steady state is at point E.

Result 3. Where $\tilde{p}_i' > K_i$, $i = 1, 2$, we have a corner equilibrium, $p_i(x) = xK_i$, $i = 1, 2$.

Proof. Equation (19) now takes the following form:

$$\begin{aligned} r_i W_i(x) &= u_i(xK_i) + W_i'(x)[(1 - K_i - K_j)x - x^\gamma] \\ &= K_i^{1-\gamma} x^{1-\gamma}/(1-\gamma) + W_i'(x)[(1 - K_i - K_j)x - x^\gamma] \end{aligned} \quad (19)'$$

which is an ordinary equation in $W_i(x)$, with a solution:

$$W_i(x) = Ax^{1-\gamma} + B$$

where by the method of undetermined coefficients, it is found that:

$$A = K^{1-\gamma}/r(1-\gamma)[1-(1-\gamma)(1-K_i-K_j)], \quad B = (1-\gamma)A. \quad \square$$

This corner solution is depicted as line OK in Figure 3, panel a. This corner solution turns out to be of great importance below.

The stage is now set for our main results of ‘dual indeterminacy’.

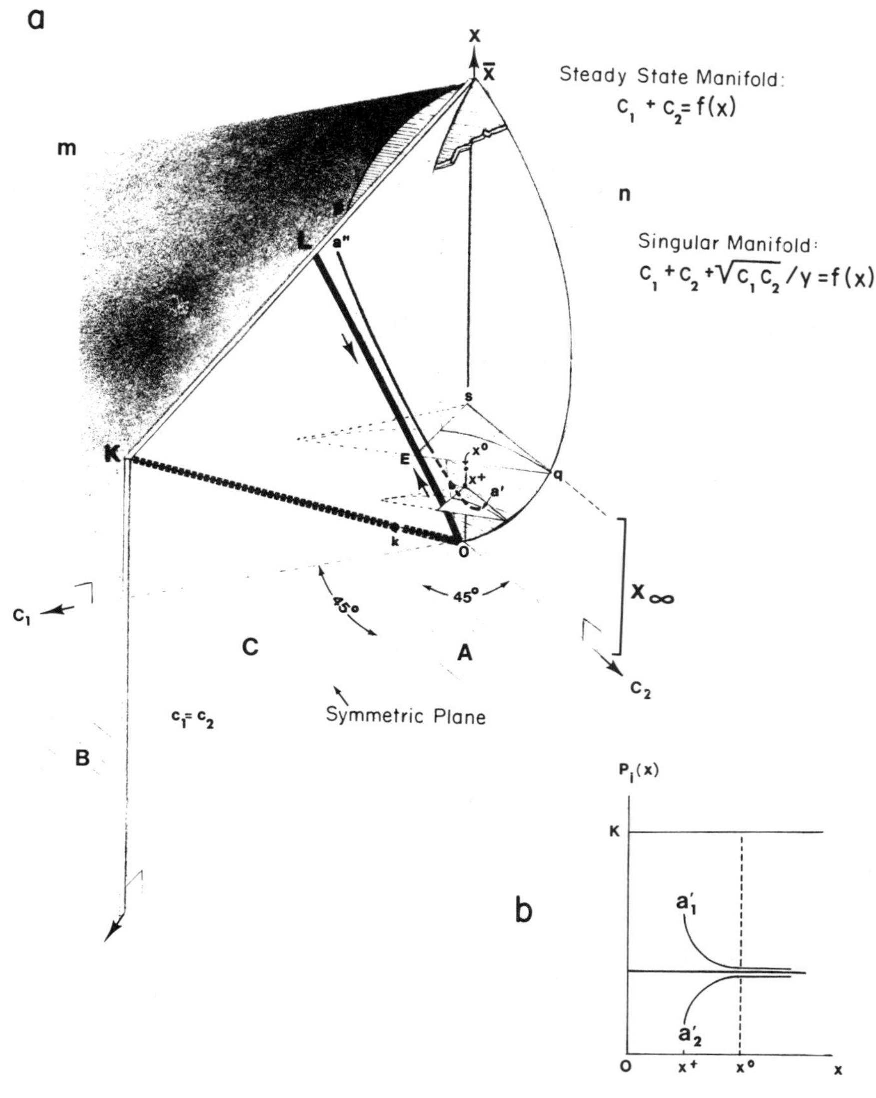

Figure 3

5. Dual Indeterminacy

Now assume that $\tilde{p}_1'$ and $\tilde{p}_2'$ are observed parametric values from an equilibrium play in terms of linear decision rules. The system characteristics are captured by the three magnitudes, r, μ and γ. From (22), we can deduce,

$$\begin{aligned} r + (\mu - 1)(\gamma - 1)/(2\gamma - 1) &= \tilde{p}_1' \\ -r + (\mu - 1)(\gamma - 1)/(2\gamma - 1) &= \tilde{p}_2' - 1 \end{aligned} \tag{23}$$

which yields

$$\begin{aligned} r &= (1 + \tilde{p}_1' - \tilde{p}_2')/2 \\ (\mu - 1)(\gamma - 1)/(2\gamma - 1) &= (-1 + \tilde{p}_1' + \tilde{p}_2')/2. \end{aligned} \tag{24}$$

Thus we have:

Result 4. The 'inverse game problem' of finding the model for an observed equilibrium play is underdetermined in the above example. One cannot determine separately the values of μ and γ.

This situation is depicted in Figure 4. In panel a, we reproduce the equilibrium in linear decision rule, which is a symmetric equilibrium for a game of symmetric players. We display in panel a the phase diagram made familiar by Koopmans' analysis of the Ramsey model. We plot $2c$ against x, the aggregate consumption. OL is again that equilibrium of linear decision rules. Panel b is in the 'parametric space', where the curve depicted comprises all parameter combinations (r, μ, γ) which can yield that particular equilibrium. There is in fact uncountably many, forming a continuum.

Next, we study the multiplicity of equilibrium plays corresponding to one particular model. The attention is focused upon those cases where $x(t)$, and $c_i(t)$, $i = 1, 2$, are continuously differentiable and interior to X and $[0, C_i(x)]$ throughout the play. At this point, one has to make the distinction between a continuously differentiable interior equilibrium (of strategies) and a continuously differentiable interior equilibrium play. It will be seen that it is possible that a particular equilibrium $p(x)$ is not continuous or not interior, over the entire state space X, yet it is both continuous and interior throughout the infinite horizon of the play.

A related and crucial point is that any equilibrium (of strategies) $p(x)$ must be *defined* over the entire X, and not just over a subset of it, say X_0. This is necessary by the equilibrium concept of Nash, which specifies that no player finds advantageous to change the chosen strategy. If any player can steer the game to any state x_e which is exterior to the subset X_0, that move can only be judged as not advantageous after specifying what other players will do according to their equilibrium strategies at x_e. Thus an equilibrium is only defined by specifying every player's strategy at every reachable state in the state space.

Such discussion is crucial because some of the multiplicity of continuously differentiable, interior equilibrium plays we shall find in this paper are generated by 'composite strategies' which splices a continuously differentiable interior strategies over some neighborhood X_0 of a steady state x_s onto the 'corner equilibrium' outside X_0.

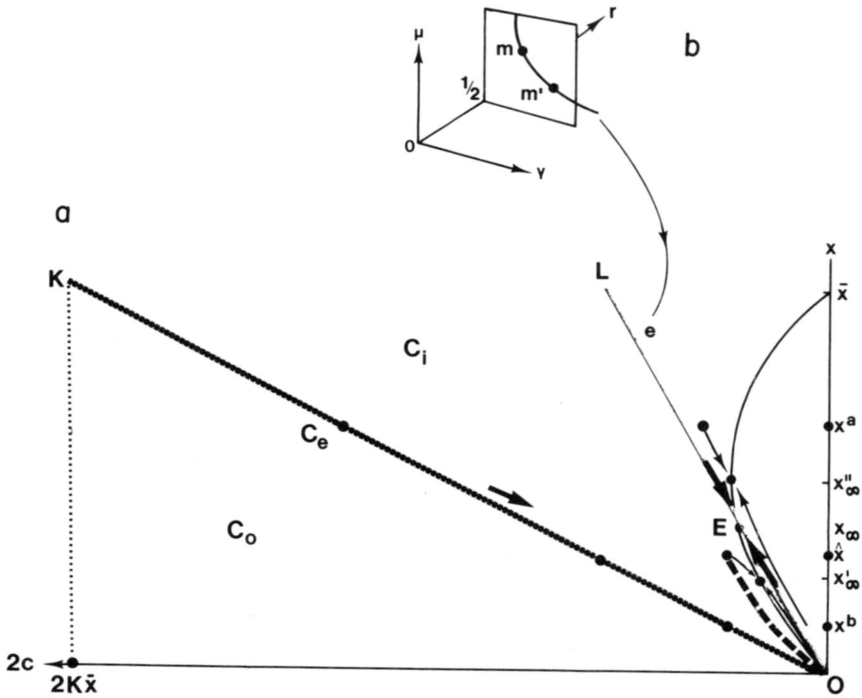

Figure 4

In the presence of multiple equilibria, it is natural to search for 'selection principles' which may be subscribed by the players. The condition that 'the equilibrium play should arise from strategies of all players which are continuously differentiable and interior over the entire state space X' may qualify as such a criterion for selection. Of course, like anything concerning the choice of alternative axiom systems, this is also a matter of 'personal taste' and not a researchable scientific truth.

We now shift from the conceptual to the analytical, and specialize our examples to set $\gamma = 2$, while maintaining the assumption of symmetry among the players. Focus now on those candidate solutions for the equilibrium which generates time paths convergement to points on the steady state manifold. Consider the time trajectory corresponding to one such solution.

$$(x(t), c_1(t), c_2(t)) = (x(t), p_1[x(t)], p_2[x(t)]).$$

The positions such a time trajectory may pass through is displayed in Figure 3. These are partitioned by the 45° plane of $c_1 = c_2$ into three sets:

Set A for points 'ahead of' that plane
Set B for points 'behind' that plane and
Set C for points 'contained in' that plane.

By the fundamental differential equation, a time trajectory passing through any point in set C always stays in set C at all times. This may be formalized as:

Result 5. Time trajectories never cross from one set to another: $c^1(t) - c^2(t)$ never changes sign.

This means a player which has a larger (same) harvest rate than (as) the opponent will always do so.

Remark. This is a property of the solution of a differential equation, and hence the assumption that time is continuous.

Since Set B is symmetric to Set A, we shall only consider the latter.

We consider an *asymmetric* candidate solution $a'a''$ with a time trajectory lying totally in Set A, with point a' approaching the singular manifold at $x = x^+$ and point a'' at the level $x = \bar{x}$. This $a'a''$ curve is assumed to lie sufficiently close to the OL ray within a neighborhood X_0 of its steady state x_∞. Identify each candidate solution curve by a boundary value, which is its intersection with the steady state manifold. By the continuous dependence property of the solution of a differential equation (i.e. p from (15)) upon a boundary value, the functions $f_1(x) = f(x) - p_2(x)$ and $f_2(x) = f(x) - p_1(x)$ associated with $a'a''$ will remain concave, so that the Mangasarian sufficient criterion for the (conditional) optimality of each player's strategy is met. Beyond X_0, one splices onto the corner equilibrium. In Figure 3, panel a, a splicing happens at $x = x^0$. The composite equilibrium follows OK to Ok for $x \leq x^0$ and follows $a'a''$, for $x > x^0$. There is actually a two-fold indeterminacy here. There is a continuum of such alternative candidate solution curves, each indexed by its steady state (x, c_1, c_2) where $c_1 + c_2 = f(x)$, that is, its intersection with the steady-state manifold. But there is also another level of indeterminacy in that the choice of the neighborhood X_0 is somewhat arbitrary. If X_0 is one such neighborhood which generates a splicing equilibrium, so is any subneighborhood X_{00} of that same steady state with X_{00} being a subset of X_0. The alternative neighborhood generates an alternative splicing equilibrium.

For completeness, Figure 3, panel b shows how the candidate path behaves near the singular manifold. $p_1(x)$ and $p_2(x)$ approach finite levels a_1' and a_2' as $x \downarrow x^+$ while $p_1' \to -\infty$, $p_2' \to +\infty$. This lies outside X_0, so it does not affect a splicing equilibrium.

Thus, we have

Result 6. There exists a continuum of asymmetric equilibria for a symmetric game. For an initial state sufficiently near the steady state of the equilibrium time trajectory, the equilibrium play will always remain to be continuously differentiable and interior to $[0, C_i(x)]$, $i = 1, 2$.

We now turn to Set C, where the OL ray again trichotomizes the set into those points inside the LOx cone in set C, i.e., subset C_i, those points outside that cone, i.e., subset C_o and those points exactly on the OL ray, i.e., the subset C_e. The situation is displayed in Figure 4, panel a.

For the sake of brevity, one observes that the equilibrium of the linear decision rule has already been analyzed in Result 2. Using the splicing techniques already employed to generate asymmetric equilibria in Set A, we can obtain:

Result 7. There also exist a continuum of symmetric equilibria for a symmetric game. Inspection of Figure 4, panel a, one can study the intersection between the steady state manifold and the subset C_o. Steady-states along this arc are all stable in its time dynamics. This provides us

Result 8. There is a continuum of equilibria which are all continuously differentiable, interior strategy pairs.

Further analysis is needed here.

We note that by following the Hamilton–Jacobi–Bellman format, all equilibria we have found are strongly time consistent in the sense of Basar (1989).

6. Concluding Remarks

The dual indeterminancy problem, in our view, probably affects many fields beside environmental economics.

Thus, Lucas (1987) stated that "(his) main criticism of Keynesian models and their use in formulating policies that one associates with... 'rational expectations' are ... consequences of ... the ... formalism of dynamic games." By such formulation, players choose simultaneously their actions according to the state they currently face. Thus, the attention is focused upon the Markovian–Nash equilibrium and not the co-operative arrangement supported by trigger strategies (an example of this for fishery is Benhabib and Radner (1991)). In evaluating welfare implications in macro-economics, one further conducts 'calibration exercises', looking for 'specific parametric descriptions of technology and preferences... such that the movement ... resembles the time series ... observed' (in the words of Lucas, ibid. See also Kydland and Prescott (1982)). This is, in our terminology, addressing an 'inverse equilibrium' problem. In all such contexts, we believe it is useful to study more of the properties of the Markovian–Nash equilibria, through a revisit to the scene of fish war. The latter has the dual advantage of both analytic simplicity and intuitive transparency. Transparency results from the affinity of the fish war model to the Ramsey problem of optimal saving — perhaps the most studied theme in dynamic economics. Simplicity follows its complementary quasi linear structure: the payoff is quasilinear in the control and the state, and the dynamics is quasi-linear in the state and the control.

It is to study thoroughly such matters of considerable complexity that we have confined ourselves to the simplest class of problems which demonstrate our main thrust: a class of models for two-player common-property fishery.

Here, it is self-evident that, compared against a particular equilibrium play, a slightly more prodigal behavior in one fisherman elicits a slightly more prodigal equilibrium response from another. All that remains is to verify whether there exists a more prodigal equilibrium play. Thus, the indeterminacy arises from the fundamental nature of the situation, and not the technicalities of model specification, in particular, the assumption that time is continuous. That assumption allows us to cut to the heart of the economic issues, without grappling with that differential-difference equation, characterizing the graph of the equilibrium play as in the infinite stage form of the game.

One might claim that the dynamic games underlying macro-economic theory might be immune from the indeterminacy discussed here. But that requires proof.

Acknowledgments. The authors acknowledge the helpful comments by T. Basar, J. Benhabib, P. Dutta, P. Holmes, and R. Radner. All remaining errors are the authors' responsibility.

References

[1] Amit, I. and Halperin, A. (1989), Sharing a common product, Mimeo, Hebrew University, Jerusalem.

[2] Basar, T. (1989), Time consistency and robustness of equilibria in noncooperative dynamic games, In: van der Ploeg, F. de Zeeuw, A.J. (eds.) *Dynamic Policy Games in Economics*, North-Holland, New York.

[3] Benhabib, J. and Radner, R. (1992), The joint exploitation of a productive asset: a game-theoretic approach, *Economic Theory*, 2.

[4] Clark, C.W. (1980), Restricted access to common-property fishery resources: a game theoretic analysis In: Liu, P.T. (ed.) *Dynamic Optimization and Mathematical Economics*, Plenum, New York.

[5] Clemhout, S. and Wan, H.Y., Jr. (1985), Dynamic common property resources and environmental problems, *J. Opt. Th. and Appl.* **46**, 471–481.

[6] Clemhout, S. and Wan, H.Y., Jr. (1991), Environmental problem as a common-property resource game In: Ehtamo, H. and R.P. Hamalainen (eds.) *Dynamic Games in Economic Analysis*, Springer-Verlag, New York.

[7] Clemhout, S. and Wan, H.Y., Jr. (1989), On games of cake eating In: van der Ploeg, F, De Zeeuw, A.J. (eds.) *Dynamic Policy Games in*

Economics, North-Holland, New York.

[8] Dockner, E.J. and Long, N.V. (forthcoming), Intertemporal pollution control: cooperative versus non-cooperative strategies, *J. of Environmental Econ. and Mgmt.*

[9] Dockner, E.J., Hartwick, J.M., and Sorger, G. (1992), The Great Fish War Revisited, mimeo.

[10] Dutta, P.K. and Sundaram, R.K. (1989), The tragedy of the commons? A characterization of stationary perfect equilibrium in dynamic games, University of Rochester, Working paper, No. 172.

[11] Dutta, P.K. and Sundaram, R.K. (1990) How different can strategic models be? Non-existence, chaos and under consumption in Markov-perfect equilibria, Columbia University Discussion Paper, No. 498.

[12] Dutta, P.K. and Sundaram, R.K. (1992) Markovian equilibrium in a class of stochastic games: existence theorems for discounted and undiscounted models, *Econ. Theory* **2**, 197–214.

[13] Kaitala, V. (1986) Game theory models of fishery management - a survey. In: Baser, T. (ed.) *Dynamic Games and Applications in Economics*, Springer, New York.

[14] Kydland, F.E. and Prescott, E.C. (1982), Time to build and aggregate fluctuations, *Econometrica* **50**, 1345–1370.

[15] Levhari, D. and Mirman, L.J. (1980), The great fish war: an example using the Cournot-Nash solution, *Bell J. Econ.* **11**, 322–334.

[16] Lucas, R. Jr. (1987), *Models of Business Cycles*, Blackwell, New York.

[17] Mirman, L.J. (1979), Dynamic models of fishing: a heuristic approach In: Liu, P.T. and Suinen, J.G. (eds.) *Control Theory in Mathematical Economics*, Marcel Decker, New York.

[18] Sundaram, R.K. (1989), Perfect equilibria in non-randomized strategies in a class of symmetric games, *J. Econ. Theory* **47**, 153–77.

Henry Y. Wan, Jr.
Department of Economics
Cornell University
Ithaca, NY 14853

An Evolutionary Game Theory for Differential Equation Models with Reference to Ecosystem Management

Thomas L. Vincent

Abstract. Either through exploitation or conservation, man is constantly tinkering with ecosystems. On the one hand, we cultivate and harvest a selected few species in man-controlled environments, and on the other hand we recognize the importance of preserves, hoping that still other "wild" species will continue to exist in their "natural" environments (Albeit, in most cases the preserves are still subject to direct control and/or harvesting by man). Do we have any idea of the long-term evolutionary consequences of what our exploitation or conservation activities will be? We present here some new results from evolutionary game theory which, when used with appropriate differential equation models, has the capacity to predict the evolutionary response of biological systems subject to human inputs. These inputs include physiographic changes, harvesting, and the introduction or removal of new species. Some previous results are discussed in terms of ecosystem management and a new example is presented to demonstrate potential use of the theory in the treatment of cancer by chemotherapy.

1. Introduction

Ever since the advent of DDT and the discovery of mutant strains of mosquitos immune to DDT, it has been public knowledge that ecosystems can and will evolve in response to mans efforts at control. While differential equations have been in common use as management models [1], it is uncommon to find any such models, which attempt to capture the evolutionary potential of the species being managed. The objective of this paper is to present an evolutionary game approach to modeling which should provide realistic management models and to point out some areas of possible application.

If we are to include evolution into management models, we are faced with a fundamental question: what is evolving and where is it evolving to? In the evolutionary game theory presented here, the "what" are parameters in the differential game model associated with characteristics of the species which are clearly adaptive (such as sunlight conversion efficiency for plants or body length in animals) which we will call strategies. The "where" is the evolutionarily stable strategy (ESS) to which these parameters can evolve. In what follows there is no need that an ESS actually be achieved by any of the species, just that it exists. In Maynard Smith's words [2], "An ESS is a strategy such that, if all members of a population adopt it, then no mutant strategy could invade the population under the influence of natural selection." The concept of an ESS is useful in providing an idealized state (which may, in turn, be changing with time) toward which individual members of a biological community will tend to evolve. It in no way implies goal oriented behavior, but rather an ESS may result from the natural selection process itself.

There is an extensive literature on translating the ESS concept into a mathematical setting [3], [4]. This includes our own work [5], [6], [7] on the development of an evolutionary game theory for determining ESS strategies.

Since the fitness of each individual organism in a biological community may be affected by the strategies of all other individuals, the essential element of a "game" exists. This game is an evolutionary game where the individual organisms (players) inherit their survival strategies (phenotypic characteristics) from a continuous play of the game from generation to generation. The evolutionary game includes both ecological and evolutionary processes. The ecological process involves the interaction between individuals and the environment that determine fitness. The evolutionary process involves dynamics which translates the fitness of an individual into changes in the number and frequency of individuals using particular strategies. Through appropriate models, the evolutionary game may be given a mathematical setting. Most commonly, the strategies are assumed to be constants associated with certain adaptive parameters in the model. The ESS is a particular constant (or vector of constants) which provide the stability property described by Maynard Smith.

We identify an individual by a strategy which in turn identifies the entire interbreeding population to which the individual belongs. We will say that interbreeding individuals are of the same species provided that they are *evolutionarily identical* [8]. Two individuals are said to be evolutionarily identical if they share the same strategy set and if the ecological consequences of using those strategies are the same for both individuals (in the parlance of what follows, two individuals will be evolutionarily identical if they have the same G-function).

Let x_i be the number of individuals of species i. The strategies used by individuals of species x_i is given by $\mathbf{u}^i$ with the possibility, through evolution, for $\mathbf{u}^i = \mathbf{u}^j$, $j \neq i$. In this paper, cross breeding between species is not allowed, but within each species, we will later assume that there is a distribution of strategies and $\mathbf{u}^i$ will then simply characterize all of the strategies used by individuals of species x_i (e.g. the mean). The various strategies used within a species will be called *phenotypes.*

Consider a community of r species, all of which are evolutionarily identical. Assume that the environment is fixed and that changes in population density for a given species can be described by differential equations of the form

$$\dot{x}_i = x_i(t)H_i[\mathbf{U}, \mathbf{x}(t)] \tag{1}$$

where dot denotes differentiation with respect to time, x_i is the population density of species i, H_i is the fitness function of species i,

$$\mathbf{x} = [x_1 \ \dots \ x_r] \tag{2}$$

is the vector of population densities for all r species in the community under consideration and

$$\mathbf{U} = [\mathbf{u}^1 \ \dots \ \mathbf{u}^r] \tag{3}$$

is the matrix of strategies currently used by all of the species in the community. The strategies $\mathbf{u}^i$ may be either scaler $\mathbf{u}^i = u_i$ or vector $\mathbf{u}^i = [u_1^i \ \dots \ u_\beta^i]^T$.

Note that H_i will, in general, also be a function of a number of other parameters, which may vary from species to species. These parameters can include efficiencies, interaction coefficients, and growth rate coefficients. The growth rate coefficients will scale the growth rate of the species (which is in turn a function of generation time) to the units of time used to express the dynamics. Since all of these constants are are assumed to be fixed in the analysis, they are not explicitly included in the functional form for H_i however they must be included in any specific model.

Henceforth, when the meaning is clear, we will drop the t from variables [e.g., $\mathbf{x}(t)$]. Alternate difference equation models are also possible and will yield similar results [9].

We have previously introduced the notation of a *fitness generating function* which we call a *G-function* for short [6], [10], [11]. The G-function is required for the development of the *ESS Maximum Principle*, which gives necessary conditions for an ESS. A function $G(\mathbf{u}, \mathbf{U}, \mathbf{x})$ is said to be a G-function for species in a community sharing the same strategy set if

$$G(\mathbf{u}^i, \mathbf{U}, \mathbf{x}) = H_i(\mathbf{U}, \mathbf{x}) \tag{4}$$

for all of the indices i corresponding to the r species. We will assume here that the community can be described in terms of a single G-function. In a more general setting a given community could be composed of more than one group of evolutionarily identical individuals each with their own G-function. This more general setting has been discussed in detail elsewhere [8], [12]. Note that when the dynamics of two individuals can be described by the same G-function, then they must be evolutionarily identical, as choosing the same strategy will have the same consequences for both, however according to our definition of a species, evolutionarily identical individuals will not be of the same species unless they can interbreed.

The G-function has the property that the fitness of an individual, H_i using one of the strategies of the matrix $\mathbf{U}$ is obtained when $\mathbf{u}$ is replaced by that individual's strategy. Thinking of the strategies in $\mathbf{U}$ and the populations in $\mathbf{x}$ as defining the current environment, it follows that the fitness of an individual then depends on its "choice" for $\mathbf{u}$. In terms of the G-function the system dynamics (1) may be written as

$$\dot{x}_i = x_i G(\mathbf{u}^i, \mathbf{U}, \mathbf{x}). \tag{5}$$

The concept of a G-function is needed for the ESS maximum principle which follows. There is no guarantee that all systems can be modeled in terms of G-functions. However, so far, we have not found its use to be a problem. Indeed the concept is useful not only in modeling, but in numerical simulations as well.

We need to distinguish between strategies which are ESS and those strategies which are not. Let the composite of the first σ strategies of $\mathbf{U}$ be called a coalition matrix $\mathbf{U}_c = [\mathbf{u}^1 \ \dots \ \mathbf{u}^\sigma]$ where $\sigma \geq 1$. Let the composite of the remaining $r - \sigma$ strategies be designated by the matrix $\mathbf{U}_m = [\mathbf{u}^{\sigma+1} \ \dots \ \mathbf{u}^r]$. The total number of all individuals in the community is given by

$$N = \sum_{i=1}^{r} x_i. \tag{6}$$

We will assume that the population dynamics are such that N remains finite for all time t. An ESS may now be defined in terms of the coalition matrix:

> A coalition matrix $\mathbf{U}_c$ is said to be an ESS if for all other strategies in $\mathbf{U}_m$ and all initial frequencies $x_i(0)/N$ satisfying the conditions
>
> (i) $x_i(0) > 0$ for $i = 1 \ \dots \sigma$
> (ii) $x_i(0) \geq 0$ for $i = \sigma + 1 \ \dots \ r$
> (iii) $1 - \varepsilon < \{\sum x_i(0)/N,\ i = 1 \ \dots \ \sigma\} < 1$, for some ε in the interval $0 < \varepsilon < 1$
>
> the dynamics given by (1) yields
>
> $$\lim_{t\to\infty} \sum_{i=1}^{\sigma} \frac{x_i(t)}{N} = 1 \tag{7}$$
>
> with
>
> $$\lim_{t\to\infty} \frac{x_i(t)}{N} > 0, \quad i = 1 \ \dots \ \sigma. \tag{8}$$

The ESS may be a coalition of any number of strategies all with nonzero initial densities. The definition is local in the sense that $\{\sum x_i(0)/N,$

$i = 1 \ \ldots \ \sigma\}$ is allowed to be arbitrarily close to 1. If the definition holds for ε close to one then the ESS would be a global attractor in frequency space. Note that any number of non ESS strategies are also allowed. Since

$$\sum_{i=1}^{\sigma} x_i + \sum_{i=\sigma+1}^{r} x_i = N \tag{9}$$

the dynamics for the non ESS individuals must obey

$$\lim_{t \to \infty} \sum_{i=\sigma+1}^{r} \frac{x_i(t)}{N} = 0. \tag{10}$$

This definition is consistent with Maynard Smith's word definition given earlier. Species identified by the strategies in the coalition matrix $\mathbf{U}_c$ will persist through time no matter how many other arbitrary strategies are introduced by $\mathbf{U}_m$. Moreover, individuals using strategies in $\mathbf{U}_m$ will die out with time.

In defining the ESS for matrix games, Maynard Smith focused on the optimality of the strategy by requiring that the ESS strategy, when common, has a higher fitness than do rare alternative strategies. However, to provide an ESS definition applicable to the broader class of continuous games, the focus need to be on the stability of strategy frequencies making up the ESS. This is done by requiring continuous stability [13], [14] in the definition of an ESS coalition. That is we require that the ESS increase in frequency when common and that eventually all rare "mutant" strategies decrease with frequency with respect to changes in both initial frequencies and strategy [15].

We have previously shown [9], [12], [16] that this definition, along with the definition of the G-function, leads to the following *ESS Maximum Principle.*

> Let $G(\mathbf{u}, \mathbf{U}, \mathbf{x})$ be a fitness generating function for a community of r species. If the matrix $\mathbf{U}_c$ is an ESS such that $\{\sum x_i(t)/N,\ i = 1 \ \ldots \ \sigma\}$ is a monotone increasing sequence for all $t \geq t_m \geq 0$ such that $\mathbf{x}^*$ is an asymptotically stable equilibrium point for (5), then in the limit as $t \to \infty$, $G(\mathbf{u}, \mathbf{U}, \mathbf{x}^*)$ must take on a maximum with respect to $\mathbf{u}$ at $\mathbf{u}^1 \ \ldots \ \mathbf{u}^\sigma$.

Other, more general theorems, may be given which do not require asymptotic stability on $\mathbf{x}^*$ and which are not restricted to a single G-functions [12], [17].

In applying the ESS maximum principle, it is useful to work in terms of a G^*-function defined as follows:

$$G^*(\mathbf{u}) \triangleq G(\mathbf{u}, \mathbf{U}, \mathbf{x}^*) \tag{11}$$

where $\mathbf{U}$ and $\mathbf{x}^*$ are fixed. Assuming that the strategies are unbounded, the above theorem is equivalent to requiring that for $i = 1 \ldots \sigma$

$$\left(\frac{\partial G^*}{\partial \mathbf{u}}\right)_{\mathbf{u}=\mathbf{u}^i} = 0 \tag{12}$$

and for scalar u

$$\left(\frac{\partial^2 G^*}{\partial u^2}\right)_{u=u_i} < 0 \tag{13}$$

with the following equilibrium condition on population density

$$G^*(u^i) = 0. \tag{14}$$

2. Strategy Dynamics

Dynamics for the mean strategy may be obtained from the basic assumption that within each species, a variation in strategies exists. Assuming a symmetric distribution in strategies with a small variance, we have previously obtained this dynamic for a class of difference equation models [7]. A similar procedure is used here to obtain the strategy dynamic for differential equation models. Assume that individuals of species i, as identified by x_i, can be subdivided into $2m+1$ phenotypes, y_{ij} where $j = -m \ldots 0 \ldots m$. That is

$$x_i = \sum_{j=-m}^{m} y_{ij}. \tag{15}$$

The strategies of the phenotypes are denoted by $\mathbf{v}^{ij}$. They are assumed to be distributed according to

$$\mathbf{v}^{ij} = \mathbf{u}^i + j\delta\mathbf{u}^i \quad \text{where } j = -m \ldots 0 \ldots m \tag{16}$$

and $\delta\mathbf{u}^i$ is an incremental change in strategy. According to this arrangement, the phenotype y_{i0} is actually "playing" the strategy $\mathbf{u}^i$, which in turn characterizes all individuals of the species x_i. The phenotypes form a cluster of strategies about $\mathbf{u}^i$ which will allow $\mathbf{u}^i$ to change with time. The actual breeding process between phenotypes of the same species is left undefined, however it must be such so as to maintain (16) with time.

It follows from (15) that the rate of change of species $x_i$is given by

$$\dot{x}_i = \sum_{j=-m}^{m} \dot{y}_{ij}. \tag{17}$$

Using the G-function to evaluate $y_{ij}(t+1)$ we obtain

$$\dot{x}_i = \sum_{j=-m}^{m} y_{ij} G(\mathbf{v}^{ij}, \mathbf{U}, \mathbf{x})] \tag{18}$$

where $\mathbf{U}$ and $\mathbf{x}$ have now been expanded to include the subdivided strategies and populations. If we denote the frequency of individuals of species i and phenotype j by

$$q_{ij} = y_{ij}/x_i \tag{19}$$

it then follow from (19) that (18) may be written as

$$\dot{x}_i = x_i(t)\overline{G}_i \tag{20}$$

where $\overline{G}_i$ is a mean G-function defined by

$$\overline{G}_i = \sum_{j=-m}^{m} q_{ij} G(\mathbf{v}^{ij}, \mathbf{U}, \mathbf{x}). \tag{21}$$

Dynamics for the Mean Strategy. Because of the distribution of population densities that now exists among the phenotypes which make up a given species, there will also be a corresponding distribution of strategies within a species. This distribution will result in strategy dynamics for the mean strategy used by a given species. The mean strategy, $\overline{\mathbf{u}}^i$ at time is defined by

$$\overline{\mathbf{u}}^i = \sum_{j=-m}^{m} \frac{y_{ij}}{x_i} \mathbf{v}^{ij}. \tag{22}$$

Individuals y_{ij} are playing strategies $\mathbf{v}_{ij}$ and we wish to track how the mean $\overline{\mathbf{u}}^i$ changes as y_{ij} changes with time. As the individual strategies $\mathbf{v}^{ij}$ do not change with time, we have

$$\dot{\overline{\mathbf{u}}}^i = \sum_{j=-m}^{m} \frac{x_i \dot{y}_{ij} - y_{ij} \dot{x}_i}{x_i^2} \mathbf{v}^{ij}. \tag{23}$$

Using the G-function to evaluate $y_{ij}(t+1)$ and (20) to evaluate $\dot{x}_i$ it follows that

$$\dot{\overline{\mathbf{u}}}^i = \sum_{j=-m}^{m} \frac{x_i y_{ij} G(\mathbf{v}^{ij}) - y_{ij} x_i \overline{G}_i}{x_i^2} \mathbf{v}^{ij}. \tag{24}$$

From the definition of q_{ij} we obtain

$$\dot{\overline{\mathbf{u}}}^i = \sum_{j=-m}^{m} q_{ij} [G(\mathbf{v}^{ij}) - \overline{G}_i] \mathbf{v}^{ij}. \tag{25}$$

Equation (25) is a general expression for mean strategy dynamics which is valid for non-symmetric frequency distributions and for arbitrary $\mathbf{v}^{ij} = \mathbf{u}^i + j\delta\mathbf{u}^i$. It is also in conformity with the result of Price [18] that the change in gene frequency under selection from one generation to the next is given by its covariance with fitness.

If we assume that q_{ij} is symmetrically distributed about y_{i0} it then follows from (22) that

$$\mathbf{u}^i = \overline{\mathbf{u}}^i. \tag{26}$$

In order to further simplify (25) we will now assume that $j\delta\mathbf{u}^i$ is small. In this case a two term Taylor series expansion of $G(\mathbf{v}^{ij})$ yields

$$G(\mathbf{v}^{ij}) = G(\mathbf{u}^i) + \left(\frac{\partial G}{\partial \mathbf{u}}\right)_{\mathbf{u}=\mathbf{u}^i} j\delta\mathbf{u}^i. \tag{27}$$

Substituting (27) into (21) yields

$$\overline{G}_i = \sum_{j=-m}^{m} q_{ij} G(\mathbf{u}^i) + \sum_{j=-m}^{m} q_{ij} \left(\frac{\partial G}{\partial \mathbf{u}}\right)_{\mathbf{u}=\mathbf{u}^i} j\delta\mathbf{u}^i. \tag{28}$$

Note that since we have assumed q_{ij} to be symmetric the second term on the right side of (28) is zero so that this expression reduce to

$$\overline{G}_i = G(\mathbf{u}^i) \tag{29}$$

If we now substitute (27) and (29) along with $\mathbf{v}^{ij} = \mathbf{u}^i + j\delta\mathbf{u}^i$ into (25) we obtain

$$\dot{\overline{\mathbf{u}}}^i = \sum_{j=-m}^{m} q_{ij} \left[\frac{\partial G}{\partial \mathbf{u}} j\delta\mathbf{u}^i\right] [\mathbf{u}^i + j\delta\mathbf{u}^i]. \tag{30}$$

Since q_{ij} is symmetric this reduces to

$$\dot{\overline{\mathbf{u}}}^i = \left(\frac{\partial G}{\partial \mathbf{u}} \delta\mathbf{u}^i\right) \sum_{j=-m}^{m} q_{ij}[j]^2. \tag{31}$$

For the scalar case ($\beta = 1$, $\mathbf{u}^i = u_i$) if we define

$$\sigma_i^2 = \sum_{j=-m}^{m} q_{ij}(j\delta u_i)^2 \tag{32}$$

as the variance of the strategies used by the phenotypes about the mean strategy u_i we obtain the following expression for the strategy dynamics.

$$\dot{\overline{u}}_i = \sigma_i^2 \left(\frac{\partial G}{\partial u}\right)_{u=u_i}. \tag{33}$$

The coefficient of $\partial G/\partial u$ in (33) scales the rate of evolutionary change. This coefficient has much in common with the way heritability and genetic and phenotypic variances scale evolutionary rates in quantitative genetic models [19].

Fisher's [20] fundamental theorem of natural selection which states, "The rate of increase in fitness of any organism at any time is equal to its genetic variance in fitness at that time" may be interpreted in terms on the G-function as follows: For a given $\mathbf{x}$ and $\mathbf{U}$, G plotted verses $\mathbf{u}$ represents an "adaptive landscape" [21]. Here the adaptive landscape as defined by the G-function, is pliable and readily changes shape with population density and strategies of the various species. The slope of the landscape at any point represents the change in fitness for a given change in strategy. Hence the greater the genetic variance, the greater the change in fitness and by (33) the more rapidly an organism will evolve toward an ESS.

Evolution via Strategy Dynamics. As long as σ_i^2 is sufficiently small and q_{ij} is symmetric (5) and (33) are appropriate equations for determining population and strategy dynamics with scalar strategies, otherwise one must use (20) and (25). The magnitude of the variance will control how rapidly the strategies change. Clearly if $\sigma_i^2 = 0$ then no change in strategy is possible. If a species ultimately evolves to an ESS under this system of equations, then the ESS maximum principle must be satisfied with the resultant strategy.

By solving (5) and (33) together we not only obtain the population density and strategy but, as time goes on, all surviving species may represent an ESS coalition candidate. This candidates can then be tested using the ESS maximum principle. Provided that an ESS exists we should be able to find it using strategy dynamics, along with the ESS maximum principle. We will illustrate this procedure in the chemotherapy example which follows.

3. Man Induced Evolution

Harvesting. Whether intentional or accidental, harvesting and cropping of prey species by humans introduces a new selective pressure on an ecosystem. Body size, for example, is one such evolutionary trait which may be brought under selection. It seems to be the inclination of humans to harvest the largest individuals and in so doing, bring evolutionary pressure for smaller sizes. We have examined this phenomena in previous studies using the ESS maximum principle with difference equation models. In [6] we examine the evolutionary response of "harvesting" a Lotka-Volterra type of system. We show how harvesting provides selective pressure for a change in evolutionary traits. For example, use of traps or nets that selectively collect larger individuals can result in directional selection for smaller individuals which we show will result in a new ESS with a smaller yield for the harvester. Harvesting techniques which are designed to be most effective against individuals using the ESS strategy can result in disruptive selection. Insecticides are an example of this. They are designed to be effective against insect traits that actually occur, not those which may occur. The evolutionary result of this type of disruptive selection on a Lotka-Volterra type of system is shown to produce a new ESS which is a coalition of two phenotypes.

Ecosystem Management. In [8] and [22] using multiple G-functions we examine the evolutionary response of a prey-predator ecosystems to the addition or removal of species or phenotypes. We show the importance of ecological parameters such as niche breath in the makeup of the ESS. Under a broad niche breath only a single prey phenotype and a single predator phenotype will coexist at the ESS. Under a narrower niche breath the ESS may be composed of more than one phenotype for each species and the ecological and evolutionary results of removing a phenotype becomes complex and interesting. For example, consider the case when the ESS is composed of two prey phenotypes and a single predator phenotype. We show that the forced removal of one of the prey phenotypes will result in a new ecological community in which the remaining prey increases in number with a small decrease in the predator species. Without further knowledge of the system, an ecologist might conclude that the two prey species are intense competitors and that the removed prey phenotype is relatively unimportant to the predator. However it is the predator who through disruptive selection allows for the coexistence of two prey phenotypes. Over evolutionary time, without further removal, a new prey phenotype will evolve, and the system will eventually return to the ESS. However if "management" dictates that the one prey phenotype not be allowed, both the prey and predator will evolve new strategies without ever establishing a new ESS.

The forced removal of the predator species can have dire consequence. The immediate consequence will be the increase in the number of both prey species. This would suggest that the predator is unimportant in the organization of this community or the coexistence of the two prey species. This is true only in "ecological" time. In the absence of the predator, the two prey phenotypes will evolve toward a common strategy. In other words the predator is evolutionarily necessary for the maintenance of the two prey phenotypes but is not ecologically necessary for their coexistence. This just gives a hint of the complexity involved in managing such a system when many phenotypes are maintained by cross coupling between trophic levels [22].

The absence of a species or phenotype may not appear over ecological time to be significant, however over an evolutionary time scale they will generally be vital. Insofar as species are evolutionary products of their environment, evolution will not preserve a species in the absence of its environment. This provided a strong plea for preserving pristine environments.

In its absence, many species will not be evolutionarily stable in habitats altered by humans and can be expected die out or evolve toward new forms.

Given the proper models, all of this complexity can be studied using the ESS maximum principle. Moreover by employing strategy dynamics, the evolutionary process which ultimately drives the system toward the ESS is made evident.

Chemotherapy. We will examine here a different type of management application. No claim is made that the following model is appropriate for modelling cancer. The reader may want to consult the literature for more appropriate models [23]. The following scenario may be thought of more as a "simulation" which to some extent seems to mimic what happens in the treatment of cancer with drugs. It has been previously pointed out by Coldman and Goldie [24] that "much experimental evidence has accrued that (cancer) cells which display inheritable resistance are the cause of treatment failure". They also develop a model which incorporates this resistance effect.

The model used here is in a form of the Lotka-Volterra competition model which has been extensively studied in an evolutionary setting [5], [7], [25]–[29]. Using the notation of equation (1), the H_i fitness function for a given species i, is given by

$$H_i(\mathbf{U}, \mathbf{x}) = R - \frac{R}{K(u_i)} \sum_{j=1}^{r} \alpha(u_i, u_j) x_j \tag{34}$$

where r is the total number of species currently in the community, R is the intrinsic rate of growth common to all species, $K(u_i)$ is the carrying capacity of the species i and $\alpha(u_i, u_j)$ is the competitive effect of species j using scalar strategy u_j on the fitness of individuals of species i. The above fitness may be expressed in terms of the following G-function,

$$G(u, \mathbf{U}, \mathbf{x}) = R - \frac{R}{K(u)} \sum_{j=1}^{r} \alpha(u, u_j) x_j. \tag{35}$$

Assuming an ESS coalition of σ strategies at equilibrium we obtain

$$G^*(u) = R - \frac{R}{K(u)} \sum_{j=1}^{\sigma} \alpha(u, u_j) x_j^*. \tag{36}$$

The particular form of the model we have used is has

$$K(u) = K_m EXP[-u^2/2\sigma_k^2] \tag{37}$$

$$\alpha(u, u_i) = 1 + EXP\left[-\frac{(u - u_i + \beta)^2}{2\sigma_\alpha^2}\right] - EXP\left[-\frac{\beta^2}{2\sigma_\alpha^2}\right]. \tag{38}$$

According to the ESS maximum principle, G^* takes on a maximum with respect to u at $u = u_1 \ldots u_\sigma$ (the ESS coalition).

We have previously shown [7] that by varying the environmental parameter σ_k this system can have an ESS coalition of one or more. Let us suppose that healthy cells are identified by a positive value for the evolutionary parameter u and that cancer cells are identified by a negative value for u. For "small" values of σ_k we have shown that a single strategy ESS exists with a positive value for u. For example using the parameters $[K_m = 100,\ R = .25,\ \sigma_\alpha = \sigma_k = \beta = 2]$ the ESS maximum principle predicts an ESS coalition of one with $[u_1 = 1.213,\ x_1^* = 83.199]$. This ESS is illustrated in Figure 1. As σ_k increases, (e.g. environmental changes) we have shown that the ESS will change from a positive coalition of one to a coalition of two, one strategy positive and one strategy negative. In other words, cancer cells are given the opportunity to "evade" by possibly the introduction of a mutant "healthy" cell. This process is illustrated in Figures 2 and 3 where the normal cells are started at their ESS values and the cancer cells are introduced at a small population with a negative strategy $[x_1 = 5,\ u_2 = -1]$ and $\sigma_k = 3.1294$. These figures were generated using equation (5) for the population dynamics and equation (33) for the strategy dynamics. The system was run until equilibrium $[u_1 = 3.1294,\ u_2 = -.2397,\ x_1^* = 51.062,\ x_2^* = 39.283]$ was reached. Figure 4 illustrates the fact that this equilibrium solution, obtained using strategy dynamics, does indeed satisfy the ESS maximum principle and a new ESS coalition of two obtained by this process.

Assume that at this point, treatment is started using cell specific drugs. In particular assume that treatment is provided by simply adding a "harvesting" term to (35) so that the G-function becomes

$$G(u, \mathbf{U}, \mathbf{x}) = R - \frac{R}{K(u)} \sum_{j=1}^{r} \alpha(u, u_j) x_j - k_h EXP\left[-5. \left(\frac{u - \overline{u}}{\sigma_h}\right)^2\right]. \quad (39)$$

where k_h is a term expressing the level of drug dosage, $\overline{u} = -.2397$ is the identified cancer cell strategy at which the drug is most effective and σ_h is the variance in effectiveness. Figure 5 and 6 were generated using (39) for the G-function, (5) and (33) for the dynamics starting with the ending conditions of Figures 2 and 3 along with the parameters $[K_m = 100,\ R = .25,\ \sigma_\alpha = \beta = 2,\ \sigma_k = 3.53553,\ \sigma_h = 1,\ k_h = .5]$. Chemotherapy is seen to be effective initially, but the cancer cells ultimately recover as they evolve to a new strategy $[u_1 = 3.1705,\ u_2 = -2.6793,\ x_1^* = 51.001,\ x_2^* = 39.3231]$. Figure 7 illustrates that the new equilibrium solution satisfies the ESS maximum principle.

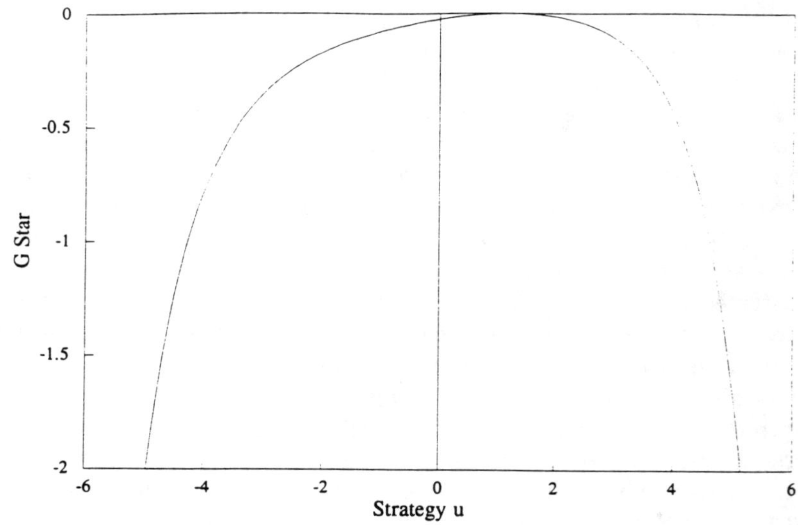

Figure 1. An ESS coalition of one composed of normal cells.

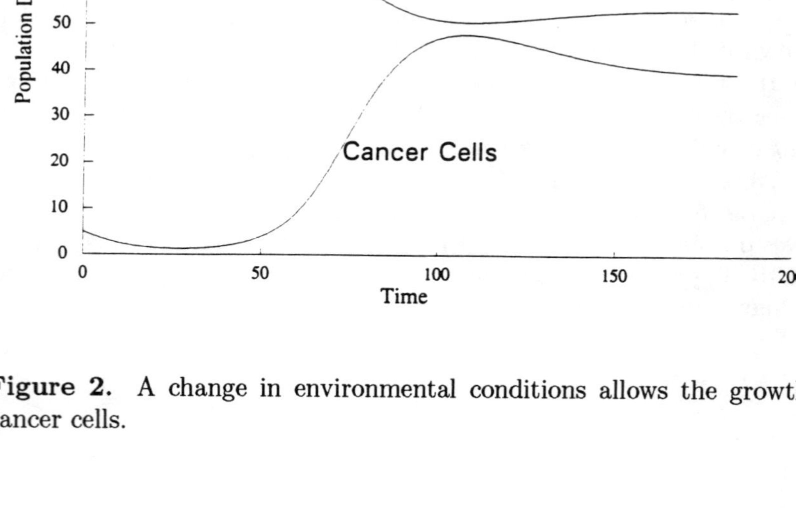

Figure 2. A change in environmental conditions allows the growth of cancer cells.

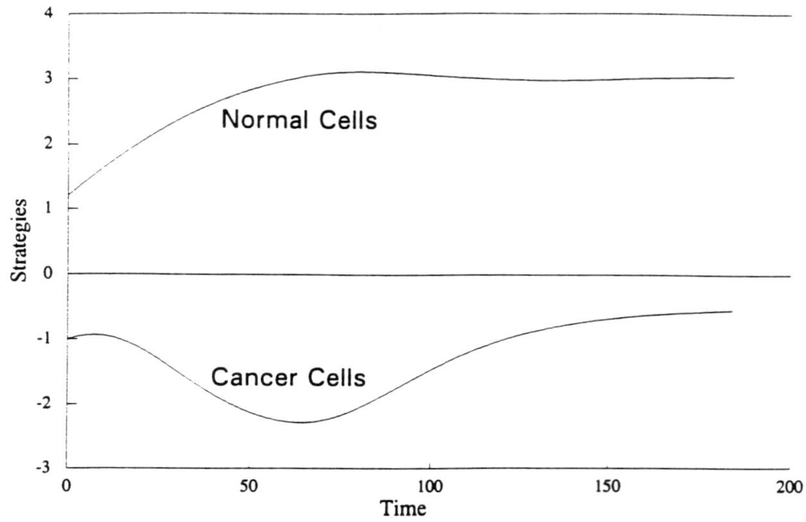

Figure 3. The strategies form a new ESS coalition of two.

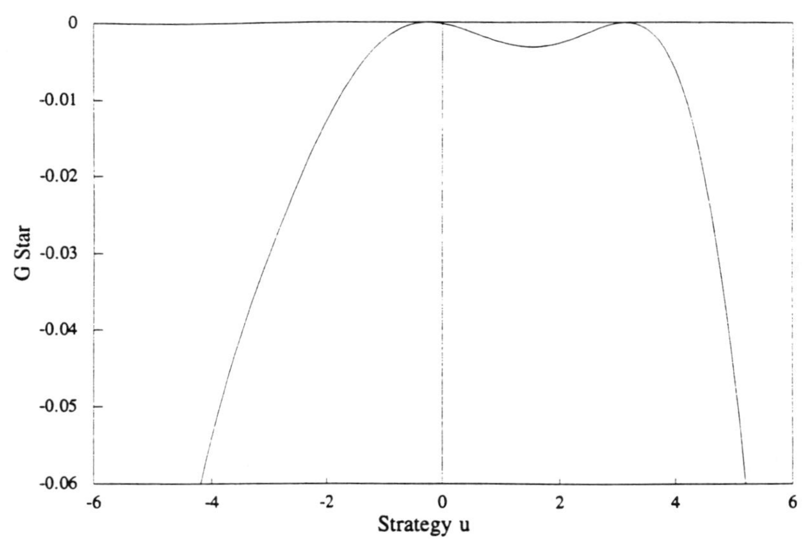

Figure 4. The new ESS coalition of two.

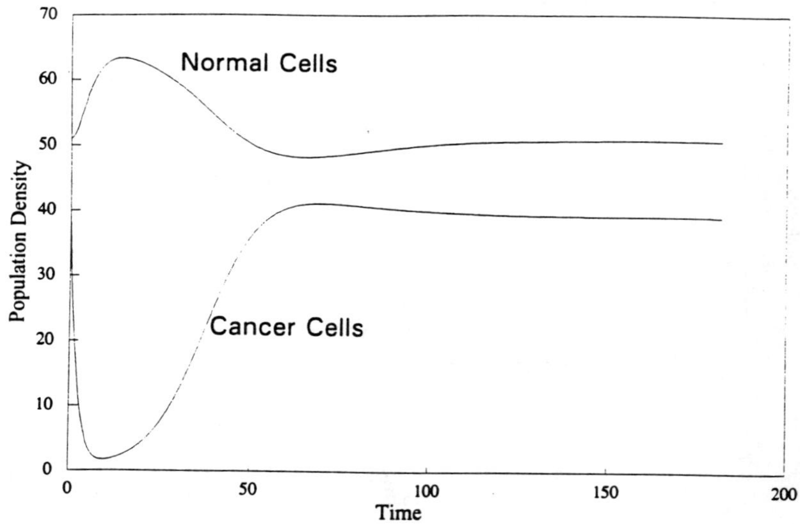

Figure 5. The evolution of the cells under treatment.

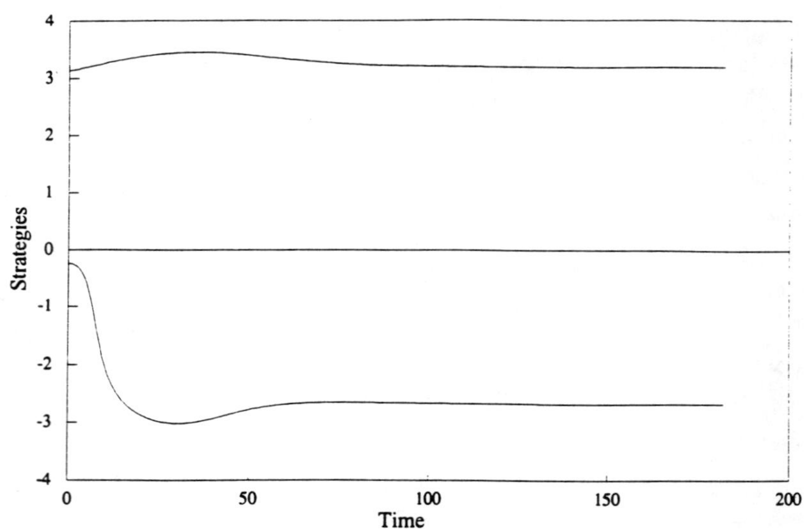

Figure 6. Under treatment, the strategies again form a new coalition of two.

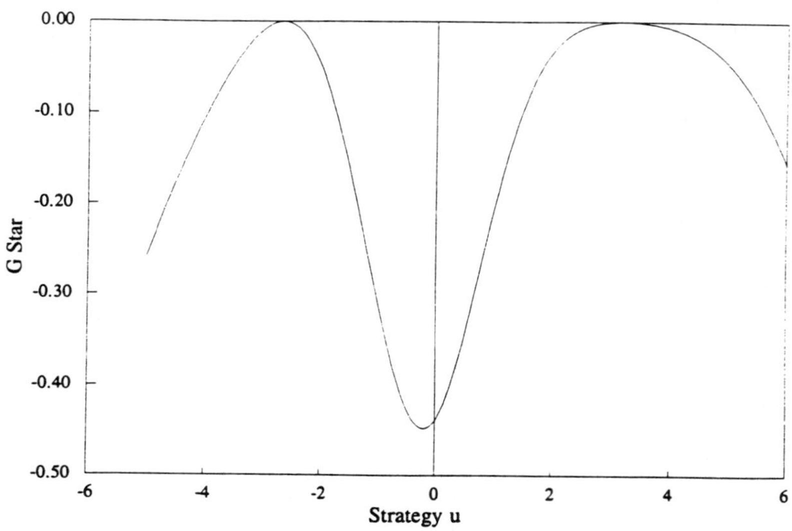

Figure 7. The new ESS coalition of two obtained under treatment.

4. Discussion

The approach used here for modeling and analysis is not restricted to Lotka-Volterra type of models. Potentially any management model involving population dynamics can be used. An evolutionary model is generated by incorporating into a management model an evolutionary strategy variable. The modeling challenge is to identify strategies and to understand how these strategies influence fitness. Once a G-function has been identified which will determine the fitness of any individual in the ecosystem the methods presented here may be used to predict the course of evolution as a function of the management programs used. While it may be obvious that harvesting large fish will produce selective pressures for smaller sizes or that treating a cancer with chemotherapy will tend to produce resistant strains, it is not obvious what the new size(s) will be or what the new strain(s) will be. Given a proper model, this theory can make these predictions provided that an ESS as defined here exists.

It should be noted that if the strategy dynamics as given by (33) is used to find a strategy which satisfies the ESS maximum principle, then this is a valid result even if some other process is actually used by the organisms to arrive at the ESS. That is, we need to distinguish between necessary conditions which must be satisfied for an ESS to exist and questions associated with how a population might actually attain an ESS. The ESS maximum principle only addresses the question of existence. Equation (33) represents one way in which a population might attain the ESS. Certain assumptions are made in the development of (33) which may or may not apply for any given situation (e.g. symmetric distributions with little variance). Other processes may be taking place. In some situations, a strategy dynamic such as (33) is not needed. One can attain the ESS through pure immigration of random strategies [28]. In summary, provided that an ESS exists and the ecosystem has some mechanism for actually attaining it, (33) can be used to find it.

One might question this approach to modeling evolution because of its lack of explicit genetics, and equilibrium requirements. However we obtain a strategy dynamic which closely resembles those used in quantitative genetic models. Roughgarden [30] discusses some of the parallels between population genetics and the ESS concept for co-evolving species. Taper and Case [31] find broad agreement between ESS and quantitative genetic analyses of co-evolving species. The results are identical when the population variance is small, but begin to diverge as the size of the environmental variation in the quantitative genetic model increases.

In spite of results which seems to mimic real evolutionary processes, a real test of the theory has yet to be made. This will require working with a real system which can evolve rapidly and which has a reasonable chance of being accurately modeled.

References

[1] B.S. Goh, *Management and analysis of biological populations*, Amsterdam, Elseivier Scientific, 1980.

[2] J. Maynard Smith, *Evolution and the theory of games*, Cambridge, Cambridge University Press, 1982

[3] S.E. Riechert and P. Hammerstein, Game theory in the ecological context, *Annual Review Ecological System*, vol. 14, pp. 377–409, 1983.

[4] W.G.S. Hines, Evolutionary stable strategies: A review of basic theory, *Theoretical Population Biology*, vol. 31, pp. 195–272, 1987.

[5] T.L. Vincent and J.S. Brown, An evolutionary response to harvesting, *Modeling and Management of Resources under Uncertainty*, Lecture Notes in Biomathematics, Heidelberg, Springer-Verlag, no. 72, pp. 83–99, 1987.

[6] J.S. Brown and T.L. Vincent, A theory for the evolutionary game, *Theoretical Population Biology*, vol. 31, pp. 140–166, 1987.

[7] T.L Vincent, Y. Cohen, and J.S. Brown, Evolution via strategy dynamics, *Theoretical Population Biology*, vol. 44, pp. 149–176, 1993.

[8] T.L. Vincent and J.S. Brown, The Evolutionary response to a changing environment, *Applied Mathematics and Computation*, vol. 32, pp. 185–206, 1989.

[9] T.L. Vincent and M.E. Fisher, Evolutionarily stable strategies in differential and difference equation models, *Evolutionary Ecology*, vol. 2, pp. 321–337, 1988.

[10] T.L. Vincent and J.S. Brown, Stability in an evolutionary game, *Theoretical Population Biology*, vol. 26, pp. 408–427, 1984.

[11] T.L. Vincent and J.S. Brown, The evolution of ESS theory, *Annual Review of Ecology and Systematics*, vol. 19, pp. 423–443, 1988.

[12] J.S. Brown and T.L. Vincent, Predator-prey coevolution as an evolutionary game, *Lecture Notes in Biomathematics*, vol. 73, pp. 83–101, 1987.

[13] I. Eshel, Evolution and continuous stability, *Journal of Theoretical Biology*, vol. 103, pp. 99–111, 1983.

[14] I. Eschel and U. Motro, Kin selection and strong evolutionary stability of mutual help, *Theoretical Population Biology*, vol. 19, pp.420–433, 1981.

[15] S. Lessard, Evolutionary stability: One concept, several meanings, *Theoretical Population Biology*, vol. 37, pp. 159–170, 1990.

[16] T.L. Vincent, Evolutionary games, *Journal of Optimization Theory and Applications*, vol. 46, no. 4, pp. 605–612, 1985.

[17] T.L. Vincent and J.S. Brown, Evolution under nonequilibrium dynamics, *Mathematical Modelling*, vol. 8, pp. 766–771, 1987.

[18] G.R. Price, Selection and covariance, *Nature*, vol. 227, pp. 520–521, 1970.

[19] M.L. Taper and T.J. Case, Quantitative genetic models for the coevolution of character displacement, *Ecology*, vol. 66, pp. 355–371, 1985.

[20] R.A. Fisher, *The Genetical Theory of Natural Selection*, Oxford, Clarendon Press, 1930.

[21] S. Wright, *Evolution and genetics of populations. Vol. 3, Experimental Results and Evolutionary Deductions*, Chicago Illinois, University of Chicago Press, 1977.

[22] J. S. Brown and T.L. Vincent, Organization of predator-prey communities as an evolutionary game, *Evolution*, vol.46, pp. 1269–1283, 1992.

[23] G.W. Swan, *Optimization of human cancer radiotherapy*, Lecture Notes in Biomathematics, No. 42, Berlin, Springer-Verlag, 1981.

[24] A.J. Coldman and J.H. Goldie, A model for the resistance of tumor cells to cancer chemotherapeutic agents, *Mathematical Biosciences*, vol. 65, pp. 292–307, 1983.

[25] T.J. Case, Coevolution in resource-limited competition communities, *Theoretical Population Biology*, vol 21, pp. 69–91, 1982.

[26] J.D. Rummel and J. Roughgarden, Some differences between invasion-structured and coevolution-structured competitive communities: A preliminary theoretical analysis, *Oikos*, vol. 41, pp. 477–486, 1983.

[27] J.D. Rummel and J. Roughgarden, A theory of faunal buildup for competition communities, *Evolution*, vol. 39, pp. 1009–1033, 1985.

[28] J. S. Brown and T.L. Vincent, Coevolution as an evolutionary game, *Evolution*, vol. 41 pp. 66–79, 1987.

[29] T.L. Vincent, Strategy dynamics and the ESS, *Dynamics of Complex Interconnected Biological Systems*, New York, Birkhäuser, pp. 236–249, 1990.

[30] J. Roughgarden, *Theory of population genetics and evolutionary ecology: An Introduction*, New York, MacMillan Publishing Co., 1979.

[31] M.L. Taper and T.J. Case, Models of character displacement and the theoretical robustness of taxon cycles, *Evolution*, vol. 46, pp. 317–333, 1992.

Thomas L. Vincent
Aerospace and Mechanical Engineering
University of Arizona
Tucson, Arizona 85721

On Barter Contracts in Electricity Exchange

Jukka Ruusunen

Abstract. We present a model for intertemporal electricity exchange between two independent power producers. The exchange is based on the diversity of the companies' marginal production costs at the different periods. The exchange is carried out in the form of a barter contract. An axiomatic approach to bargaining is used to define the contract.

1. Introduction

The paper deals with the bilateral electricity exchange between independent power systems that have organized a power pool. Depending on the duration of the electricity exhange contract, a distinction between firm and nonfirm power exchange is made in the power industry. We shall consider nonfirm power transactions that typically last from one hour to weeks. The exchange is based on the diversity of the demand patterns in the systems, e.g., noncoincident peak demands, and differences in the production technologies of the systems. Because of these differences, intertemporal exchanges of energy provide potential cost savings for the participants.

The principle of total wealth maximization is commonly used in energy exchange contracts. In the game-theoretic terminology, it is assumed that money has the properties of transferable utility. This assumption considerably simplifies the analysis of the problem, since the problem of maximizing the savings within the pool is separated from the problem of fair distribution of the savings; the latter is always achieved through monetary payments between the participants. Moreover, the bargaining game with side payments is symmetric and so a typical rule to allocate the savings is equal division of the savings between the companies (see Doty and McEntire (1982) for spot contracts, and Berger et al (1990) and Breton, Haurie, and Kalocsai (1978) for longer term contracts).

In this paper we shall present a model for a situation where no monetary transfers between the companies take place but the exchange is accomplished through intertemporal barter exchange. This is one form of energy exchange that is widely used in the power industry but it has not received considerable attention in the literature on power pools (for previous studies see Ehtamo, Ruusunen, and Hämäläinen (1989), Ruusunen, Ehtamo, and Hämäläinen (1991), and Ruusunen (1992)). For example, barter exchange is used in many exchange contracts where the companies are of different nationalities to exploit the diversity of demand patterns across countries. Barter contract is a loose arrangement where the companies are not required to measure the production cost according to the same principles. On the other hand, the exchange is not affected by variable exchange rates. From the analytical point of view, the problem without side payments is more complex since the efficiency and the distributional issues cannot be separated anymore but the exchanges that will be taken must result in fair division of the savings. We shall apply the axiomatic approach of bargaining to the problem of dividing the cost savings fairly between the participants; see Roth (1979) for the classical bargaining models, and Ehtamo and Ruusunen (1990) for the extension to intertemporal bargaining problems.

Mathematically, the intertemporal bargaining problem is a nonadditive problem which is nonseparable with respect to time. We shall discuss the solution of this problem by a hierarchical approach.

The paper is organized as follows. Section 2 presents a model of energy exchange between two power companies. In section 3 we formulate the intertemporal bargaining problem of sharing the cost savings from the exchange. Section 4 studies contracting that is based on the Nash bargaining solution and presents a hierarchical approach for solving this problem. An example is given in Section 5.

2. Model of Energy Exchange

We consider bilateral energy exchange between two companies, called company A and company B. The exchange takes place over T periods ($T > 1$). Each company operates its production units so that the load is met with minimum cost. Let $C_t^i(q)$ be the minimum cost of producing q MWh of electricity by company i at period t. Time dependency of the cost functions reflects the possibility that the committed units are not necessarily the same at the different periods. The committed production units are loaded in their relative order of economic efficiency. We assume that the units operate at production levels where the marginal cost of production is nondecreasing. Thus, the overall marginal cost of production is nondecreasing in each company.

Let ξ_t^i denote the dynamically varying local demand of electricity faced by company i at period t. Because of the dynamic variation of the demand, the marginal cost also varies with time: it is higher during peak periods and lower during off–peak periods. We assume here that the demand is known in advance for all of the periods. The minimum cost of meeting the local demand by company i over the T periods is thus

$$D^i = \sum_{t=1}^{T} C_t^i(\xi_t^i), \quad i = A, B. \tag{1}$$

When the two electricity networks are interconnected, the companies can achieve cost savings by intertemporal energy exchanges. Let P_t^i be the energy received by company i at period t. Instead of ξ_t^i, energy production is then $\xi_t^i - P_t^i$ which results in production cost $C_t^i(\xi_t^i - P_t^i)$. The total cost resulting from the intertemporal exchange $(P_1^i, \ldots, P_T^i)$ is

$$C^i(P_1^i, \ldots, P_T^i) = \sum_{t=1}^{T} C_t^i(\xi_t^i - P_t^i), \quad i = A, B. \tag{2}$$

Let P_t^A be the decision variable related to period t, and denote $P_t = P_t^A$. Then we have $P_t^B = P^B(P_t)$ (if transmission losses are ignored, $P_t^B = -P_t$). The transmissions are constrained by the capacity of the transmission line at each period

$$\underline{P}_t \leq P_t \leq \bar{P}_t, \quad t = 1, 2, \ldots, T. \tag{3}$$

Selecting different exchange patterns satisfying (3) results in different cost allocations for the companies. The bargaining problem between the companies is now to choose one such pattern. In the next section we take a game–theoretic approach to solve this bargaining problem, i.e. to choose a compromise outcome that has some desirable properties.

3. Exchange Contract: An Intertemporal Bargaining Problem

We consider exchange in the form of barter, i.e. no monetary payments between the companies take place. The contracting problem can be formulated as an intertemporal bargaining problem as follows (Roth (1979), Ehtamo and Ruusunen (1990)). In general, a bilateral bargaining game is defined by the pair (S, d), where $S \subset R^2$ is a convex set of feasible payoffs for the participants and $d \in S$ is the disagreement or status quo outcome. In the energy exchange game we have

$$S = \Big\{(C^A, C^B) \mid C^A = C^A(P_1, \ldots, P_T),\ C^B = C^B(P^B(P_1), \ldots, P^B(P_T)),$$
$$\underline{P}_t \leq P_t \leq \bar{P}_t, \quad t = 1, 2, \ldots, T\Big\} \tag{4}$$

$$d = (D^A, D^B), \tag{5}$$

where D^i, $i = A, B$, is defined by (1).

Given a bargaining game (S, d), the problem is to choose the compromise outcome. In energy exchange games with side payments, the contract is expressed in the form of an allocation rule. One such well–known rule is the split–the–savings method (see e.g. Doty and McEntire (1982)) according to which the price of the energy tranferred from one company to another is the average of the buyer's and seller's marginal production cost. This is a natural solution in bilateral games, since the bargaining game with side payments is symmetric and so equal division of the savings can be motivated by efficiency and symmetry considerations. This property has also been used in longer–term exchange contracts (Berger et al. (1990) and Breton, Haurie, and Kalocsai (1978)). The bargaining game related to energy barter is not, in general, symmetric and so Pareto optimality and symmetry are not enough to define a unique solution for all such games.

In the axiomatic approach to the bargaining problem a unique compromise outcome of a general bargaining game is defined by a set of rationality axioms. Two well–known solution concepts are the Nash bargaining solution and the Kalai–Smorodinsky bargaining solution (Roth (1979)). Common properties of these solutions are Pareto optimality, symmetry and independence of the compromise outcome of positive affine transformations of the individual payoffs. Adding a fourth axiom, the distinguising axiom, defines a unique solution. The Nash solution has the controversial independence of irrelevant alternatives property, while the Kalai–Smorodinsky solution has the individual monotonicity property. On the other hand, when applied in intertemporal bargaining problems the Nash bargaining solution is dynamically consistent while the Kalai–Smorodinsky solution is not (Ehtamo and Ruusunen (1990)). Dynamic consistency means that the cooperative solution does not change as a result of rebargaining in the course of the contract. So, if we allow rebargaining, Kalai–Smorodinsky solution is ruled out. Moreover, the generalization of the Nash bargaining solution to multilateral bargaining problems is straightforward. Thus, in the next section we shall determine the contract using the Nash bargaining solution.

4. Contracting Based on the Nash Bargaining Solution

The exchange policy corresponding to the Nash bargaining solution is found by maximizing the product of the companies' savings. So, mathematically the problem is

$$\max_{(P_1,\dots,P_T)} \Big(D^A - C^A(P_1,\dots,P_T)\Big)\Big(D^B - C^B(P^B(P_1),\dots,P^B(P_T))\Big)$$
$$\text{s.t.} \;\; \underline{P}_t \le P_t \le \bar{P}_t, \;\; t = 1,2,\dots,T. \tag{6}$$

Since the objective function is not additive (different periods are not independent) direct optimization is often computationally complex. However, utilizing the fact that the solution is Pareto optimal the problem can be solved as a two–level hierarchical optimization problem (Ehtamo et al. (1988)). The following holds: the exchange pattern $(P_1^*, \ldots, P_T^*)$ solves problem (6) if and only if it solves the problem

$$\min_{(P_1,\ldots,P_T)} \mu^A C^A(P_1, \ldots, P_T) + \mu^B C^B(P^B(P_1), \ldots, P^B(P_T))$$
$$\text{s.t. } \underline{P}_t \leq P_t \leq \bar{P}_t, \quad t = 1, 2, \ldots, T, \tag{7}$$

where the weights $\mu = (\mu^A, \mu^B)$ satisfy

$$\begin{aligned} \mu^A &= D^B - C^B(P^B(P_1^*), \ldots, P^B(P_T^*)) \\ \mu^B &= D^A - C^A(P_1^*, \ldots, P_T^*). \end{aligned} \tag{8}$$

Scaling $\mu^A + \mu^B = 1$ and choosing μ^A as the independent weight we get the condition for μ^A:

$$\mu^A \Delta C^A(\mu) = (1 - \mu^A) \Delta C^B(\mu), \tag{9}$$

where $\mu = (\mu^A, 1 - \mu^A)$ and

$$\begin{aligned} \Delta C^A(\mu) &= D^A - C^A(P_1(\mu), \ldots, P_T(\mu)) \\ \Delta C^B(\mu) &= D^B - C^B(P^B(P_1(\mu)), \ldots, P^B(P_T(\mu))). \end{aligned} \tag{10}$$

We get the following algorithm for solving the bargaining problem:

(i) Compute the minimum production costs D^i, $i = A, B$, that correspond to the situation where no energy exchanges take place.
(ii) Choose an initial value $\mu^A = \mu^A_{(0)} \in (0, 1)$; set $k = 0$.
(iii) For the given $\mu^A_{(k)}$, solve problem (7) to get $(P_1(\mu_{(k)}), \ldots, P_T(\mu_{(k)}))$ and $(\Delta C^A(\mu_{(k)}), \Delta C^B(\mu_{(k)}))$ where $\mu_{(k)} = (\mu^A_{(k)}, 1 - \mu^A_{(k)})$. This corresponds to managing the systems as an integrated system with the objective of (7). In thermal systems, the problem has a time separable structure, i.e., the problems related to different periods are independent. The problem related to period t is

$$\min_{\underline{P}_t \leq P_t \leq \bar{P}_t} \mu^A_{(k)} C_t^A(\xi_t^A - P_t) + (1 - \mu^A_{(k)}) C_t^B(\xi_t^B - P^B(P_t)) \tag{11}$$

which is a standard dispatching problem in energy management.
(iv) If (9) holds (for a prespecified tolerance), stop. The exchange pattern that solves the bargaining problem is $(P_1(\mu_{(k)}), \ldots, P_T(\mu_{(k)}))$. Otherwise, the weight μ^A is updated according to some standard one–dimensional search algorithm (Bazaraa and Shetty (1979)). Set $k = k + 1$, and go to step (iii).

When the feasible set S is convex, the bargaining problem has a unique solution. On the other hand, all the Pareto optimal points of S can be generated by varying μ^A in (7) (and setting $\mu^B = 1 - \mu^A$). Of these, we choose the unique point that satisfies (9) (note that the left hand side of (9) is increasing while the right hand side is decreasing as a function of μ^A).

The hierarchical structure of the computation is illustrated in Figure 1.

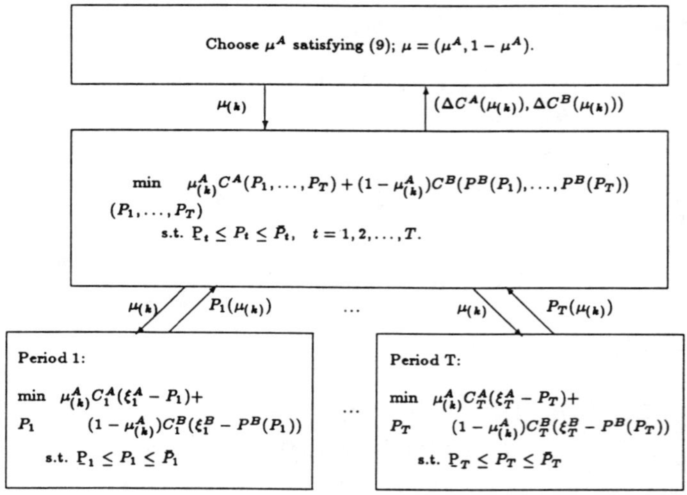

Figure 1. Hierarchical computation of the bargaining solution.

The basic difference in the computation of barter exchange and cooperative optimum with side payments is that barter exchange requires the solution of problem (7) at each iteration of the algorithm while in cooperative optimization with side payments problem (7) is solved only once with $\mu^A = 0.5$. An example of a typical real life problem would be the determination of hourly energy exchanges for a time horizon of one week (i.e., $T = 168$). In a purely thermal system, the optimization problem would decompose into 168 independent dispatching problems. These are standard problems and can be solved efficiently. The higher level problem is a one–dimensional search problem for which efficient search algorithms exist. Thus, as far as computation is considered, our approach is applicable to real life systems. It should also be noted that in real life energy exchange contracts made in loose power pools the energy exchange and the possible monetary compensations are determined on the basis of simple models; e.g., buy and sell bids of quantities and prices based on the companies' marginal cost curves (Doty and McEntire (1982)). We take this approach in the example in Section 5.

If the model were used with hydro–thermal systems, problem (7) would be a dynamic optimization problem and thus, in general, much more complicated for practical systems than that of managing a pure thermal system. Extension of the model to include unit commitment would mean an even more considerable increase in the computational burden; the problem belongs to the class of NP–hard combinatorial problems. However, efficient methods that use the Lagrangian relaxation technique for obtaining near optimal solutions have been presented recently and already implemented in some energy management systems (see, e.g., Guan et al (1992)). These methods make possible the solution of the unit commitment problem for systems of practical size, say 100 units over a one week planning horizon, in a few minutes. Consequently, from the computational point of view our model can be generalized to hydro–thermal systems and unit commitment. However, the theoretical problem related to unit commitment is the possible nonconvexity of the feasible set so that standard bargaining solutions are not directly applicable to the problem and on the other hand there is no guarantee that our computational algorithm would find a solution to the nonconvex bargaining problem.

5. Example

Let us consider the case where the cost functions are linear at each period over the interval of possible energy exchanges. This corresponds to the situation where the change in the production cost related to a given period is determined by the change in the production level of the marginal production unit of that period. The exchange pattern is determined on the basis of the companies' announcements of their marginal cost patterns for the planning horizon and on their ability to produce or utilize energy at the different periods which together with the capacity of the transmission line determine the bounds for the energy transfers at the different periods.

The cost savings at period t are now $c_t^i\xi_t^i - c_t^i(\xi_t^i - P_t^i) = c_t^i P_t^i$, $i = A, B$, and the total savings are

$$\Delta C^i(P_1^i, \ldots, P_T^i) = \sum_{t=1}^{T} c_t^i P_t^i, \quad i = A, B. \tag{12}$$

Since bargaining takes place in the set of possible cost savings for the companies, the status quo outcome is $(0, 0)$ corresponding to the no exchange situation.

The transmission losses are approximated by a quadratic function of the transmitted energy between the companies. So, if P MWh is sent from company A to company B, company B receives $P - (1/2)zP^2$ MWh, where $z > 0$ is a constant. We thus have

$$P^B(P_t) = \begin{cases} -P_t - \frac{1}{2}z(P_t)^2, & P_t \leq 0 \\ \frac{1}{z}(1 - 2zP_t)^{1/2} - \frac{1}{z}, & 0 \leq P_t < \frac{1}{2z}. \end{cases} \tag{13}$$

Problem (7) has a time separable structure and for given μ^A the problem related to period t is (see Eq. (11))

$$\min_{\underline{P}_t \le P_t \le \bar{P}_t} \mu^A c_t^A P_t + (1-\mu^A) c_t^B P^B(P_t). \tag{14}$$

The solution is

$$P_t(\mu) = \max(\underline{P}_t, \min(\overline{P}_t, \tilde{P}_t(\mu))), \tag{15}$$

where

$$\tilde{P}_t(\mu) = \begin{cases} \frac{\mu^A c_t^A - (1-\mu^A)c_t^B}{z(1-\mu^A)c_t^B}, & \mu^A c_t^A \le (1-\mu^A)c_t^B \\ \frac{1}{2z}\left[1 - \left(\frac{(1-\mu^A)c_t^B}{\mu^A c_t^A}\right)^2\right], & \mu^A c_t^A \ge (1-\mu^A)c_t^B. \end{cases} \tag{16}$$

Solving (14) for each t we get the cost savings that correspond to the weighting $\mu = (\mu^A, 1-\mu^A)$:

$$\Delta C^A(\mu) = \sum_{t=1}^{T} c_t^A P_t(\mu)$$
$$\Delta C^B(\mu) = \sum_{t=1}^{T} c_t^B P^B(P_t(\mu)). \tag{17}$$

These values are used in Step (iii) of the algorithm to update μ^A so that equation (9) is satisfied.

As a simple numerical example, consider exchange over two periods. Let c_1^A = \$30 /MWh, c_2^A = \$10 /MWh, and c_1^B = \$20 /MWh, c_2^B = \$30 /MWh. The transmission line is characterized by $\overline{P}_t = -\underline{P}_t$ = 100 MWh, $t = 1, 2$, and $z = 0.001$ (MWh)$^{-1}$.

We use the bisecting search method in the upper level. Let $[a_{(k)}, b_{(k)}]$ be the interval of uncertainty at iteration k, and choose $[0.1, 0.9]$ as the initial interval. The first value for μ^A is chosen from the midpoint $\mu_{(1)}^A$ =0.5.

Solving the allocation problems related to the two periods gives $P_1(\mu_{(1)})$ = 95 MWh and $P_2(\mu_{(1)})$ = -100 MWh, with the corresponding cost savings $\Delta C^A(\mu_{(1)})$ = \$1850 and $\Delta C^B(\mu_{(1)})$ = \$850. According to equation (9) we compute $\mu_{(1)}^A \Delta C^A(\mu_{(1)}) - (1-\mu_{(1)}^A)\Delta C^B(\mu_{(1)})$ = \$500. It should be noted that this is the solution to the cooperative optimization problem with side payments. The maximum total savings are thus \$2700.

We then get the new interval of uncertainty $[0.1, 0.5]$ and choose $\mu_{(2)}^A =$ 0.3. The computations are then repeated for this new value. The iterations are repeated until (9) is satisfied with a given tolerance level. In this example this requires the generation of eight Pareto optimal solutions. The cooperative exchange pattern is $P_1(\mu_{(8)})$ = 83 MWh, $P_2(\mu_{(8)})$ = -100 MWh and the corresponding savings are \$1480 and \$1120, respectively. The computations are summarized in Table 1.

The efficiency loss of barter trade compared to cooperative optimization with side payments is \$100 in this example; the barter trade contract results in the total savings of \$2600 while cooperative optimization would give \$2700. If the companies agreed on equal division of the savings using side payments, each company would save \$1350. On the other hand, barter trade results in the savings of \$1480 and \$1120, respectively, so that company A would be better off if the exchange were carried out under a barter trade contract while company B would be better off if side payments were allowed.

Table 1. Summary of the computations

k	$a_{(k)}$	$b_{(k)}$	$\mu^A_{(k)}$	$\Delta C^A(\mu_{(k)})$	$\Delta C^B(\mu_{(k)})$	$\Gamma(\mu_{(k)})$
1	0.10000	0.90000	0.50000	1850	850	500
2	0.10000	0.50000	0.30000	−4000	4750	−4525
3	0.30000	0.50000	0.40000	−1000	2850	−2110
4	0.40000	0.50000	0.45000	1850	850	365
5	0.40000	0.45000	0.42500	1797	889	252
6	0.40000	0.42500	0.41250	477	1840	−884
7	0.41250	0.42500	0.41875	1155	1357	−305
8	0.41875	0.42500	0.42188	1480	1121	−23

$\Gamma(\mu_{(k)}) = \mu^A_{(k)}\Delta C^A(\mu_{(k)}) - (1 - \mu^A_{(k)})\Delta C^B(\mu_{(k)})$

6. Conclusion

The paper presents a model for energy barter which is a form of energy exchange that is used in practice by the power companies but it has not received considerable attention in the energy economics literature. Compared to cooperative optimization with side payments, the bargaining problem is much more complicated since the bargaining game is not, in general, symmetric. Consequently, in addition to Pareto optimality and symmetry, further properties for the compromise outcome are needed to define a unique solution for the bargaining game. When side payments are not allowed, an intertemporal bargaining problem is computationally complex because the problem is not time separable. In the computation, we have used a hierarchical approach which transforms the complex optimization problem related to the Nash bargaining solution into a sequence of standard dispatching problems.

Generalization of the approach to hydro–thermal systems is possible although the computational burden is increased. In addition to increased computational burden, unit commitment also brings about theoretical problems by destroying the convexity of the bargaining game.

The paper studies bilateral energy exchange but it can easily be generalized to multilateral contracts. In this case the Nash bargaining solution is a very potential alternative for the bargaining solution since its generalization to N player problems is straightforward. The computational algorithm presented can also be generalized to multilateral bargaining problems (Ruusunen (1992)) although the computational burden is increased. In the multilateral case, the high level problem of finding the weights that correspond to the bargaining solution becomes a multidimensional search problem and the lower level problems of managing the systems as an integrated system become more complicated than in the bilateral case.

Our computational experience is based on real data from the Finnish power system and it shows that, as far as computation is concerned, the proposed approach is applicable to real life problems. However, to see the practical applicability of the game theoretic approach in general requires more experimental work with these problems. Further research is required to study the interaction of different contractual forms that power companies are using and the benefits and transaction costs related to the different contracts. As to barter exchange, we have here studied contracting in a deterministic world but the parties should have the option of adjusting the contract in the course of the game to take into account the new information that the parties get in the course of the game concerning e.g. demands and plant outages.

References

[1] M.S. Bazaraa and C.M. Shetty, *Nonlinear Programming*, Wiley, New York, 1979.

[2] C. Berger, R. Dubois, A. Haurie, E. Lessard, and R. Loulou, Assessing the dividents of power exchange between Quebec and New York State: a systems analysis approach, *Int. J. Energy Research*, **14**, (1990), pp. 253–273.

[3] A. Breton, A. Haurie, and R. Kalocsai, Efficient management of interconnected power systems: a game–theoretic approach, *Automatica*, **14**, (1978), pp. 443–452.

[4] K.W. Doty and P.L. McEntire, An analysis of electric power brokerage systems, *IEEE Trans. on Power Apparatus and Systems*, **PAS–101**, (1982), pp. 389–396.

[5] H. Ehtamo, J. Ruusunen, V. Kaitala, and R.P. Hämäläinen, Solution for a dynamic bargaining problem with an application to resource management, *J. Optimiz. Theory Appl.*, **59**, (1988), pp. 391–405.

[6] H.Ehtamo, J. Ruusunen, and R.P. Hämäläinen, A hierarchical approach to bargaining in power pool management, *IEEE Trans. on Automatic Control*, **AC–34**, (1989), pp. 666–669.

[7] H. Ehtamo and J. Ruusunen, An Intertemporal Bargaining Model, Helsinki University of Technology, Systems Analysis Laboratory, Research Reports A35, 1990.

[8] X. Guan, P.B. Luh, H. Yan, and J.A. Amalfi, An optimization–based method for unit commitment, *Electrical Power & Energy Systems*, **14**, (1992), pp. 9–17.

[9] A.E. Roth, *Axiomatic Models of Bargaining*, Springer, Berlin, 1979.

[10] J. Ruusunen, H.Ehtamo, and R.P. Hämäläinen, Dynamic cooperative electricity exchange in a power pool, *IEEE Trans. on Systems, Man. and Cybernetics*, **21**, (1991), pp. 758–766.

[11] J. Ruusunen, Intertemporal electricity exchange through barter, *Automatica*, **28**, (1992), pp. 203–207.

Jukka Ruusunen
Systems Analysis Laboratory
Helsinki University of Technology
Espoo, Finland

Preventing Minority Disenfranchisement Through Dynamic Bayesian Reapportionment of Legislative Voting Power

Lee Papayanopoulos

Abstract. Political reapportionment aims to distribute the power of n decision makers equitably. However, the models used to describe the underlying game, for the purposes of reapportionment, have been *a priori* models. They are based on assumptions which include: that elected representatives vote independently of one another and of party ideology, that they cast Yeas and Nays with equal probability, and that they are never absent or abstaining. The *prior* power distributions engendered by such models differ significantly from those realized in practice. Traditional reapportionment provides no means for eliminating the inequities that may ensue—except, perhaps, through litigation.

We propose an *a posteriori* model under which the fairness of the legislative game is monitored and, when players are found to be underenfranchised, they are compensated. We employ a generalized (empirical) semivalue and describe a procedure for testing the fairness of this statistic.

1. Introduction

Political constituencies in the United States are typically comprised of distinct districts with equal population (such as congressional districts) each of which is represented by a single legislator. But there are important exceptions. Representation may be based on districts of unequal population with multiple members (as in the U.S. Senate), or with multiple votes (as in some New York county boards). Variants and composites of these may be found in the U.S. and elsewhere.

As a result of the 1990 census, the United States is undergoing its decennial reapportionment at all levels of government. The objective of reapportionment is to allocate *votes, seats*, or *constituents* to representation units, as equitably as possible.

An apportionment plan is deemed *fair* if it meets certain democratic precepts. For example, when populations are apportioned equally to districts, citizens have equal opportunity to express their political will through their representatives. Invariably, reapportionment models presume that representatives make choices and cast votes randomly, independently, and symmetrically. Allocations based on these models generally remain in effect for as long as 10 years.

These kinds of models are reasonable for games in which the opposition is neither altruistic nor malevolent. But a legislator is not playing against a randomly acting opponent like Nature. Plays are neither random nor uncorrelated nor can any predictions about the political weather be made ten years in advance. Departures from consistent and symmetrical binomial behavior are the rule, not the exception.

Clearly, *a priori* apportionment is fair only insofar as the assumptions of the model hold. When they are violated, as is all too often the case, constituents' average payoffs are distorted. Chronic distortions benefit some constituents to the detriment of others. In extreme cases, the effects may be severe enough to even jeopardize the rights of the majority. It is ironic that this defect would be built into the very process intended to produce fairness.

A practical and effective means for protecting weak players is a Bayesian procedure that monitors their "success" and compensates them for excessive loss of power. We believe that such a mechanism ought to consist of (a) a statistical procedure that tests the fairness of the game as played over an interval of time and (b) a "feedback" process that reallocates the power of players, when the test fails.

Most ordinary, simple games do not allow rule changes in midstream but in political representation, *the equal protection of voter rights* is a far more compelling objective than the maximization of individual player payoffs. Any periodic intervention to protect these rights would be a justifiable form of *affirmative gerrymandering* [Gb].

Political and constitutional arguments for or against this type of intervention are not within the scope of this paper. In pondering the feasibility of reform and its implications in the law, the reader is directed to [D], the extensive bibliography in [G&], the court decisions cited in the *References*, and other sources in law and politics.

The proposed approach would do nothing to simplify the already complex and controversial topic of reapportionment. Imrie's admonition on departures from strict simplicity in representation still rings true: "*I advance a word of caution. The man on the street does not understand it. The man on the street is suspicious of that which he does not understand, and is suspicious of government in general*" [I]. Nonetheless, in weighing simplicity against fairness, the choice ought to be clear enough.

We discuss a general approach which is not unlike that of *statistical quality control.* In this introdictory paper we describe a general blueprint for *dynamic reapportionment* and provide some analytical details to demonstrate its feasibility. As in SQC, the manner of implementation will depend on the prevailing needs, statutes, and political conditions in the particular polity.

2. Weighted Voting

We examine *dynamic* reapportionment in the context of *weighted voting* because the latter lends itself to Bayesian treatment especially well. From the modeling standpoint, weighted voting is more general and its reapportionment is non-disruptive; it only requires a periodic readjustment of voting weights. In contrast, reapportionment by redistricting dictates a geographic reorganization, a reorientation of the electorate and body politic, and a validating election [S].

Weighted voting is virtually universal in corporate representation [G1, G2, Le]. A holder of 200 shares has twice the votes as one holding 100. It is much less typical in the political sector where a weighted voting game for n players is characterized by:

(a) a decision vector $\mathbf{x} \in X$ (vertices of the unit hypercube),
(b) a vector of voting weights $\mathbf{w} > 0$
(c) a quota fraction r and quota $q = r\Sigma w_i$
(d) a voting rule, i.e., a binary function defined on X
(e) a vector function $\mathbf{b}(\mathbf{w}, r)$ called a *power index*, and
(f) an n-vector of constituency fractions $\mathbf{p}$ s.t. $\Sigma p_i = 1$.

Let $(n, r, \mathbf{p}, \Gamma(\mathbf{w}))$, or simply $\Gamma(\mathbf{w})$, denote a voting game with these parameters and voting weight vector $\mathbf{w}$. In legislative/political models, n, r, and $\mathbf{p}$ (population) are given parameters.

The power indices due to Shapley-Shubik [SS] and Banzhaf [Bz] have been studied extensively [DS, HLL, O, etc.]. The Banzhaf index (Bz) measures a player's power in terms of voting coalitions in which she is essential. Bz has been used by the courts as the definitive measure for assessing the fairness of legislative weighted voting reapportionments (see [DvB], [IvB], and many subsequent cases). We refer to the principles and guidelines that have evolved from these cases as the *Bz model.* Other power indices are discussed in [Lu].

The *raw* Banzhaf (Bz) index and may be written as

$$\mathbf{b} = \mathbf{b}(\mathbf{w}, r) = (b_1, \ldots, b_n) = \mathbf{b}^0 + \mathbf{b}^1 \tag{1}$$

where for every member $i \in N = \{1, ..., n\}$,

$$b_i^0 = |\{\mathbf{x} \in X \mid x_i = 0,\ q - w_i < \mathbf{w}'\mathbf{x} \le q\}| \text{ and}$$
$$b_i^1 = |\{\mathbf{x} \in X \mid x_i = 1,\ q < \mathbf{w}'\mathbf{x} \le q + w_i\}|$$

are, respectively, the number of *losing coalitions* and the number of *winning coalitions* in which i is *essential.*

The n-vector $\beta = (\Sigma b_i)^{-1}\mathbf{b}$ is the *normalized Bz index.* For fixed r and $\mathbf{p}$, a weighted voting plan $\mathbf{w}^*$ is *Bz-fair* (written $\beta(\mathbf{w}^*, r) \approx \mathbf{p}$) iff

$$\mathbf{w}^* \in \arg\min_{\mathbf{w}} \max_{i} |(\beta_i(\mathbf{w}, r) - p_i)/p_i| \tag{2}$$

A *Bz-fair game*, $\Gamma^* = \Gamma(\mathbf{w}^*)$, is also *constitutional* if it does not violate various political or legal guidelines that are extrinsic to our present model.[1]

The existence of *voting power* as a distinct entity from *voting weight* and the requirement of Bz-fairness are unique to the political weighted voting game. These notions render the *reapportionment* of voting weights necessary in order to satisfy condition (2) as per [DvB] and other cases.

We arbitrarily index the vertices $\mathbf{x} \in X$ as column vectors $\mathbf{x}(j)$, $j \in J = \{0, ..., 2^n - 1\}$, forming the matrix $X = \{\mathbf{x}(0), \ldots, \mathbf{x}(2^n - 1)\}$. According to the (*a priori*) Bz model these voting outcomes are equiprobable, i.e., the index variable $j \in J$ has the *prior* probability distribution $g_J(j) = Pr(\mathbf{x} = \mathbf{x}(j)) = 2^{-n}$, In effect, this uniform distribution renders the Bz power index non random. Consequently, during the past twenty-five years, weighted voting reapportionment has been done with purely deterministic procedures (e.g., [P1]).

3. The *Prior* Power Index

Equation (1) has routinely been simplified to $\mathbf{b} = \mathbf{b}^0$ because, under a uniform distribution, $\mathbf{b}^0 = \mathbf{b}^1$. This does not affect the value of β. The role and properties of the *characteristic function* $v(\cdot)$ in expressing the Bz index are well-known [DS, HLL, O]. For every $i \in N$, we can write,

$$b_i^1 = \Sigma\big[v(S \cup \{i\}) - v(S)\big] \tag{3}$$

summed over all $S \mid S \cap \{i\} = 0$, where $S \subseteq N$, $v(S) = 1$ if S is *winning* and $v(S) = 0$, if *losing.*

We write the vector $\mathbf{v} = (v_0, \ldots, v_{2^n-1})$ such that $v_j = v(S_j)$ and $S_j = \{i \in N \mid x_i(j) = 1, \mathbf{x}(j) \in X\}$, $j \in J$. We define $\mathbf{1}$ as the array of all 1s with dimensions conformable to the host expression. From Equation (3)

$$b_i^1 = \Sigma_j\big[v_j x_i(j) - v_j(1 - x_i(j))\big] = \Sigma_j(2x_i(j) - 1)v_j$$

that is,

$$\mathbf{b}^1 = (2\mathbf{X} - \mathbf{1})\mathbf{v}. \tag{4}$$

Similarly,

$$\mathbf{b}^0 = (2\mathbf{X} - \mathbf{1})(\mathbf{v} - \mathbf{1}) \tag{5}$$

and hence,

$$\mathbf{b} = \mathbf{b}^0 + \mathbf{b}^1 = (2\mathbf{X} - \mathbf{1})(2\mathbf{v} - \mathbf{1}). \tag{6a}$$

Letting $\mathbf{Y} = (2\mathbf{X} - \mathbf{1})$ (*unitary combinations*) and $\mathbf{u} = (2\mathbf{v} - \mathbf{1})$ (*unitary outcomes*), we can write Equation (6a) as

$$\mathbf{b} = \mathbf{Y}\mathbf{u}. \tag{6b}$$

---

[1] There may be conditions under which no voting weights $\mathbf{w}$ can be found to adequately represent a minority group because of its relative density in one or more districts, for example.

It is useful to define the generalized semivalue

$$\gamma = \mathbf{YGu} \tag{7}$$

noting that, with suitable G, (7) gives rise to several well-known indices as special cases. For example, if $\mathbf{G} = \mathbf{I}$ (identity) then $\gamma = \mathbf{b}$, the "raw" Bz index (6a); if $\mathbf{G} = 2^{-n}\mathbf{I}$, then γ gives the Banzhaf-Coleman index [C] (also known as the *Bz swing probability index* [O]), and with $\mathbf{G}$ containing the permutational coefficient products, γ is the Shapley-Shubik index. More importantly, however with empirical $\mathbf{G}$, Eq. (7) provides a *posterior* (Bayesian) measure of voting power.

4. Estimating the *Bayesian* Power Index

Let the random variables $(x_1, \ldots, x_n)$, have the arbitrary joint distribution $g(\mathbf{x})$. Since $X \to J$, j has the point density

$$g_j(k) = Pr(j = k) = Pr(\mathbf{x} = \mathbf{x}(k)) = Pr(\mathbf{y} = \mathbf{y}(k)) = g_{\mathbf{x}}(\mathbf{x}(k)), \quad k \in J.$$

Letting $\mathbf{G} = \{g_{ij} \mid g_{kk} = g_j(k), \forall k \in J, \; g_{ij} = 0, \; i \neq j\}$, we write the *posterior Bz* index as,

$$\mathbf{b}^{\cdot} = \mathbf{YGu}. \tag{8}$$

For every $i \in N$ and random j, the variable $z_i = y_i(j)u_j$ is 1 when $y_i(j) = u_j$ and, -1 when $y_i(j) = -u_j$. z_i reflects player i's concordance with the jth outcome. Clearly,

$$\begin{aligned} b_i^{\cdot} &= E(z_i) \\ &= Pr(z_i = 1) - Pr(z_i = -1) \\ &= \pi_i - (1 - \pi_i) \end{aligned} \tag{9}$$

where π_i is the *concordance probability*. That is, the voting power of player i is proportional to the chance of his agreeing with the majority less the chance of his disagreeing. More concisely,

$$b_i^{\cdot} = 2\pi_i - 1. \tag{10}$$

The ± 1 variable z_i has the point binomial density

$$f_i(z_i; \pi_i) = \pi_i^{(1+z_i)/2}(1 - \pi_i)^{(1-z_i)/2}$$

for some $\pi_i \in [0, 1]$. It has variance

$$Var(z_i) = \sigma_i^2 = 1 - (E(z_i))^2 = 1 - b_i^{\cdot 2}.$$

From a sample of observed plays $\mathbf{Y}'' = (\mathbf{y}''(k_1), \ldots, \mathbf{y}''(k_m))$ with the corresponding unitary outcomes $\mathbf{u}'' = (u''(k_1), \ldots, u''(k_m))$ and $\mathbf{z}'' =$

$(z''(k_1), \ldots, z''(k_m))$, the sample mean $\mathbf{b}'' = (1/m)\mathbf{Y}''\mathbf{u}''$ is an unbiased estimator of $\mathbf{b}^{\cdot}$. Since $\mathbf{Y}''\mathbf{u}'' = \Sigma_k \mathbf{z}''(k)$, b_i'' has a binomial sampling distribution with mean $b_i^{\cdot}$ and standard deviation $(m)^{-1/2}\sigma_i \ \forall\, i \in N$.

Whereas the correlation of the variables $z_1, \ldots, z_n$ would affect the value of the observed power index, the autocorrelation of z_i, (some i) would not. We may note that the distribution of $c'' = \Sigma_k b_k''$ is a *weighted binomial*, $g_{c''}(c)$, and so is the bivariate (joint) distribution $g_i'' = g_{b_i'',c''}(b, c)$, of (b_i'', c'') [P3]. For sufficiently large m and $\pi_i(1 - \pi_i)$, g_i'' may be approximated by the bivariate normal distribution with suitable (estimated) means, variances, and correlation coefficient [MG].

5. Dynamic Reapportionment

In the simplest case, it would be reasonable to institute a procedure to protect a *single* representative i (of a small or otherwise vulnerable constituency) from excessive loss of power. Let game Γ^0 begin at time (date) t^0. Assume that $T = \{t^1, t^2, \ldots\}$ are times when i's cumulative power is to be reevaluated. The following three-step procedure would be executed at these times. The trade-off on its frequency is obvious.

Repeat i–iii $\forall t \in T$.

i. *Periodic Sampling:* Let $\mathbf{Y}^t$ and $\mathbf{u}^t$ be, respectively, the sampled unitary combinations and unitary outcomes in the interval (t^0, t). Compute $\mathbf{b}^t = (1/m_t)\mathbf{Y}^t\mathbf{u}^t$ where m_t is the sample size.

ii. *Fairness Test:* Using the normalized statistic β_i^t, test the hypothesis:

$$H_0 : \beta_i^{\cdot} = p_i$$
$$H_1 : \beta_i^{\cdot} < p_i$$

at confidence level $1 - \alpha$. (The test may be formulated as a SQC control chart.)

iii. *Reapportionment:* If the test fails, declare i disenfranchised, find fair $\mathbf{w}^t$ and institute Γ^t.

At time t, H_0 is rejected if $\beta_i^t < \beta_\alpha$. Given the joint sampling distribution function $g_{b_i'',c''}(b, c)$ of the correlated variables (b_i'', c'') (or its normal approximation), the distribution of β_i'' may be expressed as,

$$g_{\beta_i''}(j_1/j_2) = Pr(\beta_i'' = j_1/j_2) = g_{b_i'',c''}(j_1, j_2).$$

Then, β_α satisfies the condition

$$\sum_{j_2=1}^{\infty} \sum_{j_1=1}^{\lfloor j_2, \beta_\alpha \rfloor} g_{\beta_i''}(j_1/j_2) = \alpha.$$

where $\lfloor x \rfloor$ is the *floor* of the xr.

The one-tailed procedure just described may be used to protect a small district in danger of disenfranchisement. Alternatively, if the intent is to safeguard the power of a small subset of players, then a sequence of tests may be performed at each $t \in T$. Finally, the multivariate test,

$$H_0 : \beta^{\cdot} = \mathbf{p}$$
$$H_1 : \beta^{\cdot} \neq \mathbf{p}$$

may be used to monitor the fairness of $\mathbf{w}^t$ for all or a large subset of players, simultaneously. Whenever a reapportionment is needed, it will probably affect the voting weights of all the members, even if only a single member had been found to be underrepresented.

6. Concluding Remarks

The Bz model is now in its fourth census-induced cycle of reapportionment. We contend that the Bz index, which is based only on prior considerations, does not accurately predict the actual power distribution in a legislative body.

We have proposed a generalized, empirical index that can be monitored for fairness. Under the new model, members found to have inadequate power are to be compensated by the simple means of adjusting their votes. Several *Robin Hood* and other algorithms (see [P1], for example) have been used by this author in conducting the weighted voting reapportionments of New York county legislatures and boards. Similar algorithms would be effective in the Bayesian approach proposed here.

The norm used in Equation (2) is one of many convex norms that would elicit the fairness of a weighted voting plan, i.e., the colinearization of $\mathbf{p}$ and $\mathbf{b}$. The $|(\beta_i - p_i)/p_i|$ formula was endorsed by the Court when proposed by this author [IvB, DvB] although other norms may also have been tractable. However, overcoming legal precedent is not an easy matter. In one instance, the norm used, namely $|(\beta_i - p_i)|$ is partly blamed for litigation that has lasted twenty years (see [FvK]).

Finally, we may recall that *if* the requirement of a fixed quota is relaxed and the quota itself is allowed to vary randomly, then the expected voting power can be reconciled with voting weights (see Theorem 6 in [DS], and, for a further discussion, [H]).[2] But this, too, is a property of the *a priori* Bz index that holds only so long as the uniform prior distribution assumption holds.

[2] I am grateful for this and other useful comments made by the referees.

References

[Bz] Banzhaf, J. F. (1965) Weighted Voting Doesn't Work: A Mathematical Analysis, *Rutgers Law Review* **19**, 317.

[C] Coleman, J. S. (1971) Control of Collectivities and the Power of a Coalition to Act, in *Social Choice*, (Lieberman, Ed.), New York: Garden and Breach.

[D] Dixon R. G., Jr. (1968) *Democratic Representation: Reapportionment in Law and Politics*, New York: Oxford University Press.

[DvB] *Dobish v. The State of N.Y., the Board of Supervisors of the County of Wayne et al., NY* (1967) 282 N.Y.S.2d 791.

[DS] Dubey, P. and L. S. Shapley (1979) Mathematical Properties of the Banzhaf Power Index, *Math. of O.R.* **4**, 99–131.

[FvK] Franklin v. S. W. Krause, *Clerk of the Board of Supervisors of the County of Nassau et al., NY* (1973) 344 N.Y.S.2d 885.

[G1] Gambarelli, G. (1982) Portfolio Selection and Firms' Control, *Finance*, **3**, 1.

[G2] Gambarelli, G. (1989) Political and Financial Applications of Power Indices, in *Decision Processes in Economics* (G. Ricci, Ed.), Springer-Verlag.

[Gb] Grofman, B., (1982) Preface to [G&], xiii.

[G&] Grofman, B., A. Lijphart, R. McKay, and H. Scarrow (Eds. 1982) *Representation and Redistricting Issues*, Lexington Books, D. C. Heath and Company.

[H] Holler, M. (1985) Strict Proportional Power in Voting Bodies, *Theory and Decision*.

[HLL] Holzman, R., E. Lehrer, and N. Linial (1988) Some Bounds for the Banzhaf Index and Other Semivalues, *Math. of O.R.* **13**, 358–363.

[I] Imrie, R. W. (1973) The Impact of the Weighted Vote on Representation in Municipal Governing Bodies of New York State. In [P2].

[IvB] *Iannucci v. Board of Supervisors of Washington County, NY* (1967) 282 N.Y.S. 2d 502.

[Le] Leech, D. (1988) The Relationship Between Shareholding Concentration and Shareholder Voting Power in British Companies: A Study of the Application of Power Indices for Simple Games, *Management Science* **34**:4, 509–527.

[Lu] Lukas, W. F. (1974) Measuring Power in Weighted Voting Systems, Tech. Report 227, Dept. of O.R., Cornell U.

[MG] Mood, A. M. and F. A. Graybill (1963) *Introduction to the Theory of Statistics*, 2nd Ed., McGraw-Hill, 198–203.

[O] Owen, G. (1975) Multilinear Extensions and the Banzhaf Value, *Naval Res. Logist. Quart.*, **22**, 741–750.

[P1] Papayanopoulos, L. (1983) On the Partial Construction of the Semi-Infinite Banzhaf Polyhedron, in *Semi-Infinite Programming and Applications* (A.V. Fiacco and K.O. Kortanek, Eds), Lecture Notes in Eco. and Math. Sys., Springer-Verlag, 215.

[P2] Papayanopoulos, L. (1973) Ed. Democratic Representation and Apportionment: Quantitative Methods, Measures, and Criteria, *Ann. of the N.Y. Acad. of Sci.*, 219.

[P3] Papayanopoulos, L. (1985) Properties and MIS Applications of Combinatorial Distribution Functions, *J. of M.I.S.*, **2**:1, 77–95.

[RvS] Reynolds v. Sims (1964) 377 U. S. 533.

[S] Sack, K. (May 5, 1992) In One Big Sweep, Albany Backs New Voting Laws and Districting, *The NY Times*, CXLI:48957, 1.

[SS] Shapley, L. S. and M. Shubik (1954) A Method for Evaluating the Distribution of Power in a Committee System, *Annals of Political Sciences Review* **48**, 787.

Lee Papayanopoulos
Rutgers Faculty of Management
92 New Street
Newark, New Jersey 07102

Learning by Doing and Technology Sharing in Asymmetric Duopolies

Maria Luisa Petit and Boleslaw Tolwinski

1. Introduction

The existence of an experience curve for the firm has theoretical foundation in Arrow's pioneering work (Arrow, 1962). The study of the learning-curve effect was initiated, however, almost thirty years earlier by Wright (1936). It was further developed by Alchian (1959), Hirschleifer (1962), Hirschmann (1964), and by an important work by the Boston Consulting Group (1972). The learning-curve phenomenon appears to be related, in the first place, to the introduction of new products. It has been observed that a doubling in accumulated production of a new product can result in a decline of unit costs by a factor of 10 to 50 percent of their initial level (Teng and Thompson, 1983). Empirical documentation concerning the learning curve can be found in several papers that appeared in the period from the early sixties to more recent years (e.g., Lundberg, 1961; Alchian, 1963; Hollander 1965; Zimmerman, 1982; Lieberman, 1984; Alder and Clark, 1991). This literature considers cost reduction as an outcome of experience and cumulative output is the variable commonly chosen to represent that experience. The revival of interest in the study of the learning curve can be attributed to some important recent contributions that suggest the existence of a link between learning by firms and incremental innovations in both production processes and products (see in particular Rosenberg, 1976; Nelson and Winter, 1982).

The idea behind the cost-learning curve is well known: learning by experience (or learning by doing) introduces a form of *dynamic* economies of scale that may arise even if there are no static economies of scale in the traditional sense. Unit production costs of a firm may decrease as accumulated production increases over time. For this reason, production can be considered as a means of increasing experience, that is, the acquisition of experience is no longer only an unintentional byproduct of production but the result of an investment decision in learning. Learning becomes a strategic element of firms behavior. Therefore, when future cost reduction through learning by doing is taken into consideration, an optimal production path for the firm can be different from that which simply maximizes the present value of future profits ignoring the learning effect.

If firms are aware of their learning by doing possibilities, then the learning-curve phenomenon can have considerable impact on market structure and performance. This issue has been given much attention in recent years. A number of authors have studied policies and performance of firms in relation to different forms of market organization (Spence, 1981; Fudenberg and Tirole, 1983, 1986; Stokey, 1986; Dasgupta and Stiglitz, 1988). An important outcome of this effort has been to show that when learning by doing is introduced into the firm's optimization problem it may become optimal for such a firm to produce to the point where current marginal cost exceeds marginal revenue. This can be explained by the fact that an additional unit of current production increases cumulative output and therefore may decrease future costs leading to higher long-term profits.

Learning by doing is essentially a dynamic process which requires a dynamic analysis. Furthermore, models describing the behavior of firms facing an experience curve are generally non-linear. Under these circumstances it is very difficult, if not impossible, to use traditional analytical tools unless strong simplifying assumptions are made. Such assumptions have actually been made in the papers cited above, where the analysis has been limited to two-period models, symmetric firms, and/or open-loop, i.e., non subgame perfect equilibria.

An approach that allows to relax these semplifying assumptions is to use numerical rather than analytical methods. The view that numerical methods have a lot to contribute to the understanding of economic processes is strongly supported by the following opinion expressed by Baumol (1987, pp. 1–2) in his comment about Lundberg's numerical approach:

> ...But I will show that this numerical approach is not as restrictive a procedure as it may appear to be. Rather, there is a trade-off between the abstraction and deemphasis of the details of the underlying economic processes that the more formal procedures engender and the insights offered by the painstaking calculations that are vital for Lundberg's numerical time paths. Indeed, it will be pointed out that the ostensible superiority of the formal approaches in terms of the greater generality of their results is to a considerable degree an illusion, because the mathematical techniques are fully effective in providing analytic solutions for dynamic models *only when these models take the most elementary forms.*
>
> ...More than that — parts of the work undertaken recently have been forced to revert to the numerical sorts of procedure that Lundberg's work introduced to economics. There are even significant recent contributions which have reverted directly to the Lundbergian numerical techniques, thereby demonstrating their power to provide profound insights into complex and important issues (our italics.)

This paper examines the behavior of symmetric and asymmetric duopolies, where both firms know their experience curves. Since experience may but need not be shared between the firms, we consider two extreme situations corresponding to the total absence of any industry wide experience sharing (duopoly under firm specific or separate experience) and to free transmission of experience within the industry (duopoly under shared experience).

The duopolies are modeled as discrete-time infinite horizon dynamic games with non-linear demand and non-linear learning functions. We compute collusive as well as noncooperative solutions. The latter are defined as feedback Nash equilibria[1] and are computed by a numerical technique based on policy iteration method. The solutions presented in the paper correspond to a wide range of parameter values and, therefore, provide good insight into the behavior of firms facing learning curves and into the impact that the presence of learning by doing can have on the market.

The paper is organized as follows. In Section 2 a model of duopoly with learning is introduced. Section 3 contains the definition of Nash feedback equilibrium. An algorithm for the computation of such equilibria is discussed in Section 4. The results concerning duopolies with firm specific experience are presented in Section 5. Sections 6 and 7 contain discussion and results concerning duopolies with shared experience. Finally, concluding remarks can be found in Section 8.

2. The Model

The model is defined in discrete time. There are two firms[2] that produce a homogeneous good. The firms operate with a constant returns to scale technology at each period of time but they can be asymmetric, i.e., the technology can be different for each firm and unit costs of production at any given time may differ from firm to firm. Learning by doing characterizes the production process implying that unit cost decreases as cumulative production of a firm increases, approaching a minimum possible value when the cumulative output grows to infinity. Learning is assumed to be firm specific (separate experience) and the learning process may differ from firm to firm.

Firms are indexed by i, $i = 1, 2$. We define $q_{it} \geq 0$ to be firm i's output at time t, and $w_{it} \geq 0$ to be its cumulative output, i.e., experience, from time 0 up to time t. The state variable w_{it} is related to the control variable q_{it} through the state equation

$$w_{it+q} = w_{it} + q_{it}. \tag{1}$$

[1] Feedback equilibria are subgame perfect.

[2] The conclusions would not change substantially in the case of n firms. This study is limited to the duopoly case to avoid additional complications.

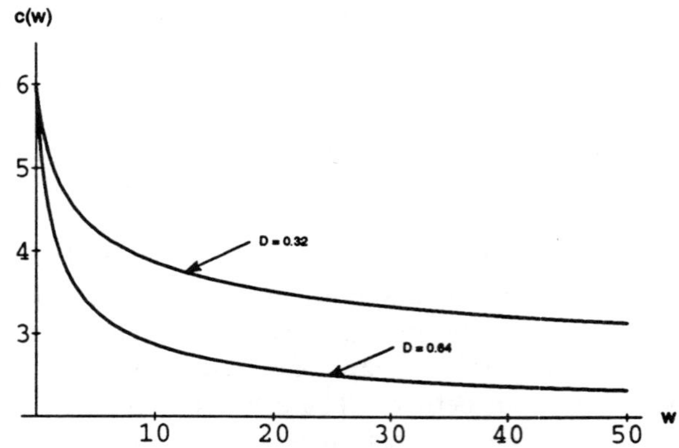

Figure 1. Experience curves for $c^o = 4.0$, $c^{\min} = 2.0$, and two values of D.

Let $Q_t = q_{1t}+q_{2t}$ denote aggregate production at period t. The corresponding demand is defined by a nonlinear stationary inverse demand function of the constant elasticity type. More specifically

$$P(Q) = AQ^{-\beta} \tag{2}$$

where $Q = q_1 + q_2$, $\beta = 1/B$, and B represents the elasticity of demand. Furthermore, let $c_i(w_i)$ denote firm i's constant marginal (unit) cost per period corresponding to the level of experience w_i. At the same time, $c_i(w_i)$ understood as a function of w_i will represent the *learning* or *experience curve*, that is the relation between unit cost and experience (see Figure 1). As indicated before, the unit cost of a firm will be assumed to decrease as its cumulative output grows. More specifically, we shall consider learning curves of the form

$$c_i(w_i) = c_i^o(1+w_i)^{-D_i} + c_i^{\min} \tag{3}$$

where $c_i^{\min}$ is the minimum unit cost that firm i can achieve ($c_i(w_i)$ tends asymptotically to $c_i^{\min}$ as w_i tends to infinity). The initial unit cost of firm i, $c_i(0)$, is given by $c_i^o+c_i^{\min}$, where c_i^o is the initial cost disadvantage of firm i relative to its long run minimum $c_i^{\min}$. The parameter D_i determines the rate at which unit costs decline with accumulated production from $c_i^o+c_i^{\min}$ toward $c_i^{\min}$, or, in other words, the speed of learning.

The firms are assumed to maximize the present value of profits over an infinite horizon,[3] that is

[3] The assumption of a game of infinite duration can be interpreted as either an approximation to a game of a long but undefined duration, or as a finite horizon game with random stopping time (see, e.g., Friedman, 1991).

$$\pi_i^I = \sum_{t=0}^{\infty} \delta^t q_{it} \left[A(q_{1t} + a_{2t})^{-\beta} - c_i^o (1 + w_{it})^{-D_i} - c_i^{\min} \right] \tag{4}$$

where δ denotes a discount factor, $0 < \delta < 1$, and the superscript I is employed to distinguish the case of firm specific (individual) experience from the case of shared experience (π_i^S) to be examined later. Notice that the firms are allowed to have different learning curves. The asymmetry can result from different initial and minimum unit costs, from different learning speeds, or from both.

3. The Solution Concept: Feedback Nash Equilibrium

The duopoly model formulated in the previous section defines a two-person nonzero-sum dynamic game with two competing firms as players. A dynamic game can have many solutions but the particular equilibrium concept to be considered in this paper is based on the following assumptions concerning the way in which the firms determine their output policies.

- Each firm is assumed to maximize its long-term profit as defined by expression (4), subject to the state equation (1).
- Each firm is assumed to know the value of the state vector $\underline{w_t} = (w_{1t}, w_{2t})$, i.e., the accumulated outputs of his own and of his competitor, at every time period t.
- Actions of firm i consist in choosing output levels q_{it} on the basis of current information about vector $\underline{w_t}$.
- The firms set their outputs independently of each other and do not cooperate in any way.

The above assumptions imply that equilibrium in the duopoly game will be determined in the class of so-called *feedback strategies.* The resulting *feedback Nash equilibrium*, has the important property of being *subgame perfect*, that is, at any period t, no matter what happened prior to that period, neither firm can benefit by unilaterally deviating from its equilibrium strategy.

In mathematical terms, feedback strategies are defined as mappings

$$\varphi_i : \mathcal{R}^+ \times \mathcal{R}^+ \to \mathcal{R}^+ \tag{5}$$

that associate nonnegative output levels $q_i = \varphi_i(\underline{w})$ with nonnegative values of the accumulated production $\underline{w} = (w_1, w_2)$. A pair of feedback strategies $\underline{\varphi}^*(\underline{w}) = (\varphi_1^*(\underline{w}), \varphi_2^*(\underline{w}))$ is said to be a feedback Nash equilibrium if

$$\pi_1^I(\underline{\varphi}^*(\underline{w})) \geq \pi_1^I\big(\varphi_1(\underline{w}), \varphi_2^*(\underline{w})\big) \quad \text{for every } \varphi_1(\underline{w}) \tag{6}$$

and

$$\pi_2^I(\underline{\varphi}^*(\underline{w})) \geq \pi_2^I\big(\varphi_1(\underline{w}), \varphi_2^*(\underline{w})\big) \quad \text{for every } \varphi_2(\underline{w}) \tag{7}$$

where

$$\pi_i^I(\underline{\varphi}(\underline{w})) = \sum_{t=0}^{\infty} \delta^t \varphi_i(\underline{w}_t)\left[A(\varphi_1(\underline{w}_t) + \varphi_2(\underline{w}_t))^{-\beta} - c_i^o(1 + w_{it})^{-D_i} - c_i^{\min}\right] \tag{8}$$

for $i = 1, 2$.

Notice that we consider here only stationary strategies, i.e., functions of the state variables that are the same for all time periods.[4] This is, of course, a natural approach in case of infinite horizon problems with stationary profits per period.

4. The Numerical Algorithm

An approximation to feedback Nash equilibrium for the dynamic game of Section 2 can be found by imposing a finite grid on the state space and then applying a modified policy iteration method (Filar and Tolwinski, 1990; Tolwinski, 1989) to the resulting finite state Markov game. This approach is based on the fact that a strategy pair

$$\underline{\varphi}(\underline{w}) = (\varphi_1(\underline{w}), \varphi_2(\underline{w})) \tag{9}$$

is a feedback Nash equilibrium for the dynamic game if there exist functions $V_i(\underline{w})$ for $i = 1, 2$ such that the following dynamic programming equations are satisfied.

$$V_1(\underline{w}) = \max_{q_1}\{l_1(\underline{w}, q_1, \varphi_2(\underline{w})) + \delta V_1(f(\underline{w}, q_1, \varphi_2(\underline{w})))\} \tag{10}$$

and

$$V_2(\underline{w}) = \max_{q_2}\{l_2(\underline{w}, \varphi_1(\underline{w}), q_2) + \varphi V_2(f(\underline{w}, \varphi_1(\underline{w}), q_2))\} \tag{11}$$

where

$$l_i(\underline{w}, q_1, q_2) = q_i(t)\left[\frac{A}{(q_1 + q_2)^{\beta}} - \left(\frac{c_i^o}{(1.0 + w_i(t))^{D_i}} + c_i^{\min}\right)\right] \tag{12}$$

for $i = 1, 2$ and

$$f(\underline{w}, q_1, q_2) = (w_1 + q_1, w_2 + q_2)^T. \tag{13}$$

The dynamic programming equations defined above can be solved on a finite grid W_h imposed on the state space

$$W = \mathcal{R}^+ \times \mathcal{R}^+ \tag{14}$$

---

[4] In other words, the changes in output over time are due only to the changes in the values of $\underline{w}_t$.

where

$$W_h = \{(w_i = ih, w_j = jh) : i = 0, \ldots, M;\ j = 0, \ldots, N\} \tag{15}$$

with $h > 0$ and

$$M = \text{entier}((w_{1\,\max}/h)), \quad N = \text{entier}((w_{2\,\max}/h)). \tag{16}$$

The equilibrium strategies $\varphi_i(\underline{w})$ and value functions $V_i(\underline{w})$ are computed only for $\underline{w} = (w_1, w_2) \in W_h$. Notice that W_h is bounded from above by parameters $w_{1\,\max}$ and $w_{2\,\max}$ that should be selected to ensure that the rate of change of cost functions $c_i(w_i)$ for $w_i \geq w_{i\,\max}$ is negligible. The computation of feedback Nash strategies for the finite state dynamic game defined on W_h can be carried out by means of the following algorithm.

Algorithm (Modified Policy Iteration)

1. Select initial approximations $V_i^0(\underline{w})$ to $V_i(\underline{w})$ for all $\underline{w} \in W_h$, $i = 1, 2$. Set k to 0.
2. For every $\underline{w} \in W_h$, compute a Nash equilibrium point, say (q_1^*, q_2^*), of the static game defined by strategy spaces $\mathcal{R}^+$, $\mathcal{R}^+$, and payoff functions $R_i(\underline{w}; q_1, q_2)$, $i = 1, 2$, where

$$R_i(\underline{w}; q_1, q_2) = l_i(\underline{w}, q_1, q_2) + \delta V_i^k(f(\underline{w}, q_1, q_2)).$$

   Set $\underline{\varphi}^k(\underline{w}) = (\varphi_1^k(\underline{w}), \varphi_2^k(\underline{w})) = (q_1^*, q_2^*)$.
3. Compute approximate values of the payoff functionals corresponding to the strategy pair $\underline{\varphi}^k(\underline{w})$ by applying a few, say $m \geq 1$, Jacobi iterations to the system

$$V_i(\underline{w}) = l_i(\underline{w}, \underline{\varphi}^k(\underline{w})) + \delta V_i(f(\underline{w}, \underline{\varphi}^k(\underline{w})))$$

   for $\underline{w} \in W_h$ and $i = 1, 2$. More precisely, set $V_i^{k,0}(\underline{w}) = V_i^k(\underline{w})$ and for $j = 0, \ldots, m-1$ compute

$$V_i^{k,j+1}(\underline{w}) = l_i(\underline{w}, \underline{\varphi}^k(\underline{w})) + \delta V_i^{k,j}(f(\underline{w}, \underline{\varphi}^k(\underline{w}))).$$

   Set $V_i^{k+1}(\underline{w})$ to $V_i^{k,m}(\underline{w})$.
4. If $\|V_i^{k+1}(\cdot) - V_i^k(\cdot)\| < \epsilon$ for $i = 1, 2$ then stop; else, set k to $k+1$ and go to step 2.

A computer implementation of the above algorithm requires an interpolation scheme for the calculation of $V_i^k(y)$ for $y \notin W_h$ and a procedure to compute Nash equilibria of static games in step 2. The convergence of the algorithm may depend to considerable extent on the choice of these procedures. One possible approach is to choose an interpolation scheme that generates continuously differentiable value functions $V_i^k(\underline{w})$ and then to use equations

$$\frac{\partial R_i(\underline{w}; q_1, q_2)}{\partial q_i} = 0 \tag{17}$$

for $i = 1, 2$, to find equilibria in step 2. More specifically, a cubic spline interpolation scheme *in two dimensions* can be used that provides estimates of $V_i^k(y)$ and of its partial derivatives

$$\frac{\partial V_i^k(w_1, w_2)}{\partial w_i} \tag{18}$$

at $(w_1, w_2) \in W$.

For the problem at hand, the system of equations (17) can be very efficiently solved by a fixed point iteration. Newton's method is another obvious choice. Its value in the given context is limited, however, by the fact that it uses second derivatives of functions $V_i^k(x)$, whose estimates provided by the interpolation tend to be rather inaccurate.

The computational scheme described above turned out to work quite well for the dynamic game considered in this paper. It should be noted, however, that the algorithm is not guaranteed to converge in general and may fail to converge for some combinations of parameters of the model. When the algorithm does converge,[5] it generates feedback strategies $\underline{\varphi}$ and value functions V_i, $i = 1, 2$, that satisfy the dynamic programming equations (10), (11) with some given accuracy ϵ.[6]

5. Duopoly with Firm Specific Experience: Numerical Results

In this section and in Section 7 we present a sample of results concerning a duopoly with specific and shared experience respectively. The duopoly model described in Section 2 has been calibrated by choosing some of the parameter values on the basis of existing empirical results. This is the case for the elasticity of demand B which was set to 1.5, and for the rate of learning D for which empirical evidence provides an average vaue of 0.30 (see, e.g. Hirschmann 1964, Teng and Thompson 1983). The discount rate was set to 0.10 in most computations. The remaining parameters are related to the position of the demand curve (A) and of the cost functions

---

[5] A major reason for convergence failure is non-uniqueness of equilibrium. However, the model at hand appears to have a unique equilibrium in the class of feedback, i.e., strictly non-cooperative strategies. This has been verified by running the algorithm for different initial approximations $V_i^0(\underline{w})$ to the value functions.

[6] In the numerical experiments described in Section 5, $\epsilon = 10^{-4}$ has been used.

(c_i^0 and $c_i^{\min}$). These parameters depend strongly on the type of good assumed to be produced by the duopolistic firms, which actually determines the numerical relationship between their values. In order to fix the values of these parameters we carried out a grid search at different scales but the results obtained did not change substantially and the parameters were set at the following values: $A = 10$, $c_i^0 = 6.0$ and $c_i^{\min} = 2.0$, for the symmetric case. However, since one of the main objectives of this paper is to analyze the performance of the duopolistic firms and of the market in the presence of asymmetries, some of the parameters of the model (in particular, c_i^0, $c_i^{\min}$ and D^i) were modified (within a reasonable range) in the numerical computations, so as to reflect possible asymmetries between the firms. The values taken by these parameters are specified in each single case.

In this section we present a sample of results[7] concerning a duopoly where the firms do not share experience. Figures 2–5 describe the evolution of outputs, prices, and profits over time corresponding to feedback Nash equilibria obtained for various sets of the model parameters.

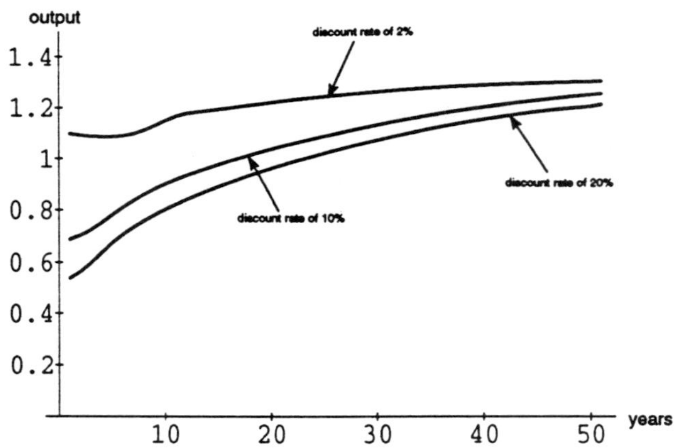

Figure 2. Symmetric firms: The outputs for three different discount rates.

We begin with the case of *identical* firms, for which, as mentioned above, we adopted the following values of the model parameters: $A = 10$, $B = 1.5$, $c_1^o = c_2^o = 6.0$, $c_1^{\min} = c_2^{\min} = 2.0$, and $D = 0.32$. The resulting (identical) production paths for each firm corresponding to three different discount rates are presented in Figure 2. High discount rates characterize 'impatient' firms, while a low rate reflects interest in long-term profits. The weight that a firm attaches to future profits has considerable impact on the level of its output and, through it, on the market price and distribution of

[7] Lack of space makes it necessary to limit the presentation to a few results that seem to be most interesting. Other graphs can be generated to reflect other aspects of the problem. The computer code is available from the second author on request.

profits over time. Not surprisingly, a lower discount rate (i.e., high discount factor δ) leads to higher output, lower price, and profits that are initially lower but increase more rapidly due to experience that is accumulated at a faster rate. The production path corresponding to a discount rate of two percent exhibits a negative slope, i.e., the output level is actually decreasing during the first five years, even though the unit production cost is decreasing all the time. This seems to confirm the conclusion reached by Fudenberg and Tirole (1983) in the context of two-period games that states that under a symmetric subgame perfect equilibrium oligopolistic firms may decrease output over time for strategic purposes if they have a low enough discount rate. Our results however go further in showing that in the long run, when the scope for cost reduction ceases (i.e., when firms approach the lower part of their learning curves), output may tend to increase again.

Figures 3–5 explore the case of *asymmetric* firms. As before, A and B have been set to 10 and 1.5, respectively. In addition, a discount rate of 10% has been assumed in all computations. The asymmetries to be considered derive from differences in production costs and learning rates from firm to firm. More specifically, the following four cases are considered.

(a) Firm 1 has an initial cost advantage over firm 2: $c_1^o = 4.0$, $c_2^o = 6.0$, $c_1^{\min} = c_2^{\min} = 2.0$, $D_1 = D_2 = 0.32$.
(b) Firm 1 has an initial cost advantage over firm 2 but firm 2 learns at a faster rate: $c_1^o = 4.0$, $c_2^o = 6.0$, $c_1^{\min} = c_2^{\min} = 2.0$, $D_1 = 0.32$, $D_2 = 0.64$.
(c) Firm 1 has an initial cost advantage over firm 2 but firm 2 has a lower limiting cost, i.e., it will eventually gain a cost advantage: $c_1^o = 2.0$, $c_2^o = 6.0$, $c_1^{\min} = 3.0$, $c_2^{\min} = 2.0$, $D_1 = D_2 = 0.32$.
(d) Same as (c), except that the firms learn at a faster rate: $c_1^o = 2.0$, $c_2^o = 6.0$, $c_1^{\min} = 3.0$, $c_2^{\min} = 2.0$, $D_1 = D_2 = 0.64$.

The price curves resulting from equilibrium output strategies in the above four cases are presented in Figure 3. Perhaps its most striking feature is the impact of the learning rate on the price levels, as illustrated by (a) versus (b) and (c) versus (d). Doubling of the learning rate of one of the firms leads to a sharp reduction in price. The presence of two fast learners further reduces the price. Another observation can be made on the basis of comparing (a) with (c), which differ only in the cost function of firm 1, more specifically, in case (a) firm 1 has a higher initial cost but lower limiting cost than in case (c). Not surprisingly, low limiting cost is more effective in lowering the price in the long term than low initial cost.

The profits of the two firms are compared in Figures 4 ((a) versus (b)) and 5 ((c) versus (d)). Both figures illustrate the fundamental importance of the speed of learning for the evolution of profits over time. In case (a), where the learning rates of the two firms are identical, firm 1 starts with an advantage in initial cost ($c_1(0) = 6.0$, $c_2(0) = 8.0$) and, therefore, its profits are always larger than those of firm 2. However, in the long term the disparity between the profit curves may decrease since the initial cost advantage tends to diminish due to the minimum limiting cost assumption (that this could happen under a low enough discount rate had already been noticed by Dasgupta and Stiglitz (1988)).

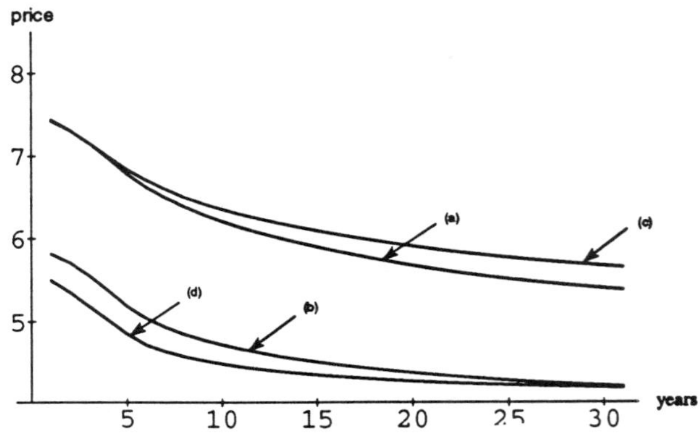

Figure 3. Assymmetric firms: The prices for four different learning curves.

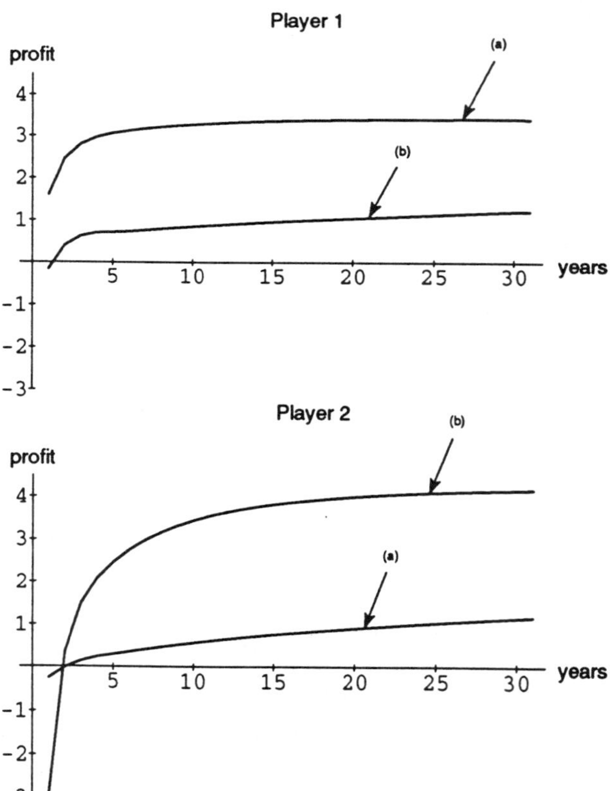

Figure 4. Asymmetric firms: Profits for two different learning curves.

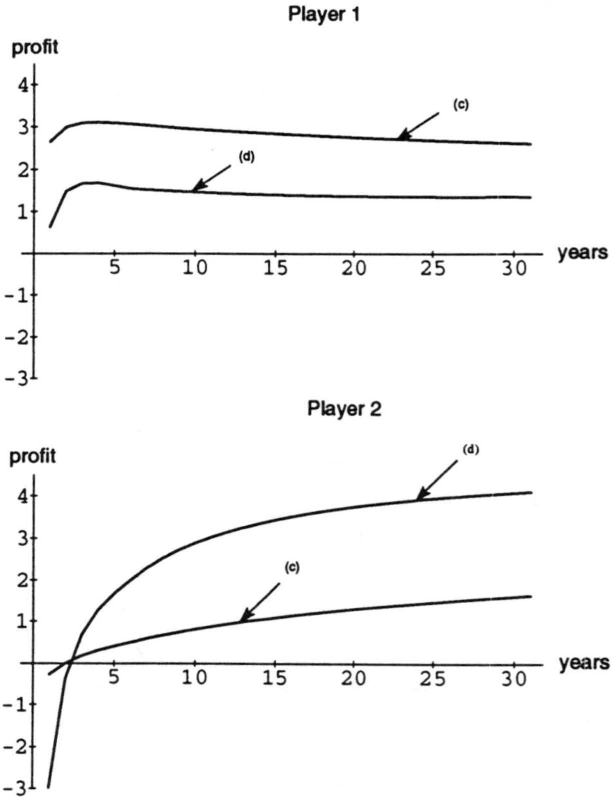

Figure 5. Asymmetric firms: Profits for two more learning curves.

In case (b), on the other hand, with firm 2 learning at twice the speed of firm 1 ($D_2 = 0.64$, $D_1 = 0.32$) the situation is reversed. Firm 1 has higher profits for only the first two years (while the initial cost advantage prevails) and after that firm 2 reaches the dominant position. In order to exploit its learning advantage firm 2 produces for the first two years under negative profits. This allows it to acquire a cost advantage over the rival in a short time and to produce at higher (positive) profits for the remaining time. A comparison between cases (a) and (b) shows that higher profits are obtained in the long run by the firm with a higher rate of learning. Notice also that in both cases both firms succeed in making positive profits in the long run, so it can be argued that both may remain on the market.

Figure 5 illustrates the importance of asymmetries in minimum unit costs. In this case, the initial cost advantage of firm 1 over firm 2 is greater than before ($c_1(0) = 5.0$, $c_2(0) = 8.0$), but firm 1 has a larger minimum cost ($c_1^{\min} = 3.0$ versus $c_2^{\min} = 2.0$). Therefore, when firms learn at a low rate ($D_1 = D_2 = 0.32$), the initial cost advantage of firm 1 allows it to have higher profits than its competitor for a relatively long period of time. Doubling of the learning rates drastically changes this situation. Now, firm 2 succeeds in cancelling the initial cost disadvantage in a short period of time by producing at negative profits in the early years. As a result, firm 2 approaches its limiting unit cost of 2.0 very quickly, while firm 1 cannot descend below the cost of 2.0 very quickly, while firm 1 cannot descend below the cost of 3.0. And so, the profits of firm 2 exceed those of firm 1 within 5 years.

The above discussion seems to indicate that a minimum cost disadvantage can be a more serious problem that an initial cost disadvantage if the firms worry about their long term prospects. If the difference in minimum costs is large enough, it is conceivable that the weaker firm will be eventually forced out of the market and the industry will be turned into a monopoly.

6. Duopolies Sharing Experience

Until now it has been assumed that each firm benefits only from its own experience, but not from the experience of its competitor. In practice there may exist certain 'learning spillovers' between the firms, meaning that the experience of one firm can contribute to a reduction in production costs for the other. Apart from involuntary learning spillovers, there is a number of reasons why firms could consider sharing their experience voluntarily. In particular, the following advantage of shared experience can be singled out.

- A firm can reduce production costs at a faster rate for the same amount of accumulated production if it can share in the experience of the others.
- Since at the initial phase of production firms may have to incur losses in order to cut future production costs, sharing experience can be an attractive way to shorten or eliminate that phase. This consideration becomes more important if firms are impatient.
- A faster reduction in costs resulting from sharing of experience may raise barriers to entry for new firms, as better learning possibilities may provide the established firms with greater advantages (e.g., superior production techniques) over new entrants. Thus, speeded up learning may be used for the creation of entry barriers (see, e.g. Scherer, 1980, pp. 250–252).
- Uncertainty about future demand and the risk that a new substitute product may appear in the market making the firm's product obsolete, provides an incentive to shorten the initial unprofitable phase of production.
- In the presence of incomplete learning spillovers, an agreement to share experience can contribute to reduction of uncertainty about

the intensity and direction of spillovers. This kind of agreement may create a mechanism that internalizes the externalities caused by the spillovers.

Since the transmission of experience between firms speeds up the industry's rate of learning, shared experience may also produce social benefits. For example, it may contribute to a faster diffusion of innovations and to the elimination of wasteful duplication of learning efforts.

Firms may share their experience but still compete in the output market[8] or they may collude in setting output levels. In the first case, and if competition between firms in the product market is strong, they can lose much of the potential additional profits that would result from quicker cost reduction by trying to increase their market shares (thus lowering price). In such a case, consumers would appropriate a high fraction of the gains arising from the transmission of learning. The evidence for this is provided by the analysis presented in Section 7 (Figs. 6–8). Clearly, collusion in the area of transmission of experience can provide a strong incentive for collusion in setting output levels, especially if one considers the strong link that exists between learning by doing and output decisions.

The question whether the firms are likely to agree or not to share experience is therefore of considerable interest. The results of the next section will shed some light on that issue.

The learning curves for the duopoly with shared experience have the form

$$c_i(w) = c_i^o(1+w)^{-D_i} + c_i^{\min}, \quad i = 1, 2. \tag{19}$$

The joint accumulated output evolves according to

$$w_{t+1} = w_t + q_{1t} + q_{2t} \tag{20}$$

and firm i seeks to maximize a profit function

$$\pi_i^S = \sum_{t=0}^{\infty} \delta^t q_{it}\left[A(q_{1t}+q_{2t})^{-\beta} - c_i^o(1+w_t)^{-D_i} - c_i^{\min}\right]. \tag{21}$$

In the case of collusion in setting output levels (formation of a monopoly),[9] the firms are assumed to jointly maximize the above profit functions.

[8] Evidence that even in the case of competition firms may find it convenient to share their experience has been noticed by Allen (1983). See also Baumol (1992).

[9] This possibility is considered only in the case of symmetric firms.

7. Duopoly with Shared Experience: Numerical Results

In this section we compare outputs, prices and profits generated by duopolies with firm specific experience with those resulting from collusion concerning sharing of experience (duopolies with shared experience) and, in the case of symmetric firms, also with those corresponding to collusion in setting output levels (monopolies). The parameters defining the inverse demand function and the parameters of the cost functions in the symmetric case are the same as in Section 5. The discount rate is set to 10% and the asymmetric cases (a) and (b) as defined in that same section are considered.

Figures 6–8 present evolution over time of outputs, prices, and profits generated by the optimal solution under monopoly and by feedback. Nash equilibria under shared and firm specific experience, respectively, in the case of *symmetric firms*. The main conclusions to be drawn from these results are as follows.

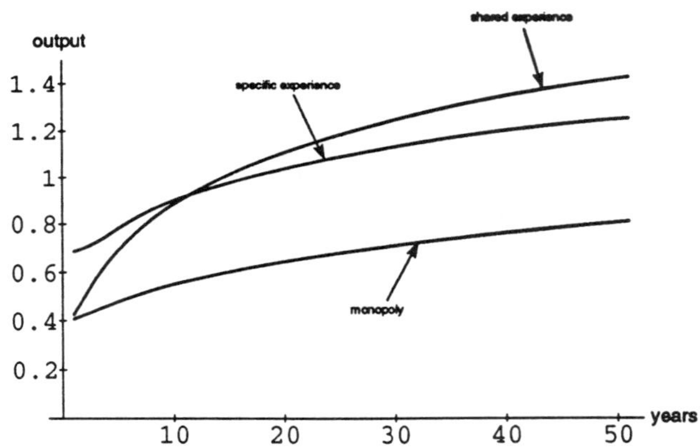

Figure 6. Symmetric firms: The outputs and duopolies with firm specific and with shared experience, and the output under a monopoly.

1. A monopoly generates lower levels of output, higher prices, and higher profits than either duopoly. This result confirms the benefits of competitive duopolgy markets versus monopolistic markets from the point of view of consumer welfare and is consistent with the argument presented in the previous section, according to which in the presence of competition in the product market consumers will appropriate a fraction of the gains from sharing experience.
2. The comparison of the output and price trajectories for the 'shared experience' and 'specific experience' duopolies confirms earlier studies by Spence (1981) and Fudenberg and Tirole (1983, 1986)

of the impact of diffusion of learning among firms.[10] The transmission of learning causes a kind of free-rider effect which reduces the incentive to invest in learning during the initial periods. Under a subgame perfect (feedback) equilibrium firms take into account the effects of their decisions on rival learning (and costs) and are not willing to contribute to the reduction of those costs. This eliminates the initial aggressiveness in output.

3. Sharing learning leads to higher profits for both firms than in the case of specific experience (Figure 8). This is particularly evident in the early stages because of the reduced aggressiveness of the firms in the initial periods when experience is shared. Firms are less eager to sacrifice profits (i.e., to price under unit costs) in the early years to obtain a future cost advantage over the rival, when they know that learning will be shared (as could be expected, output aggressiveness in the initial phase becomes stronger with a lower discount rate, i.e., for large values of δ).

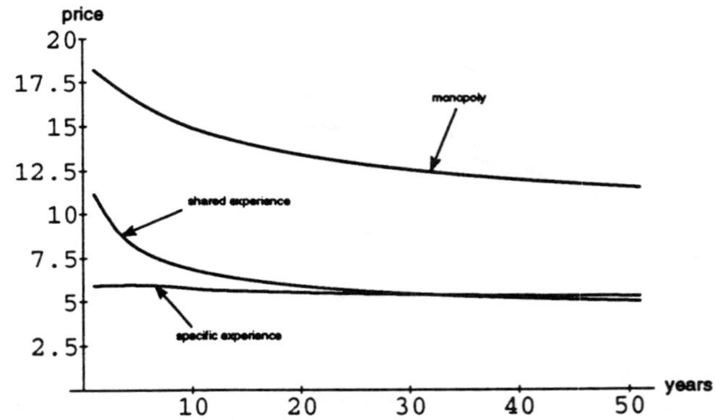

Figure 7. Symmetric firms: The prices for duopolies with firm specific and with shared experience, and the price under a monopoly.

Notice that the first of the above observations is in contrast with Dasgupta and Stiglitz's (1986) conclusion that when learning by doing has significant effects, oligopoly outcomes may be worse for society than those under monopoly. Their argument is based on the conjecture that 'in the process of sharing a market, firms *learn less* and so future prices are higher' (Dasgupta and Stiglitz, 1986, p. 249. Our italics). Our results show that this conclusion may not always be true. On the contrary, in an oligopolistic market competition between firms can lead to higher production levels and lower prices. Figure 6 shows the output of a single duopolist to be

[10] Firms are less aggressive in first-period output, the greater the industry learning effect' (Spence, 1981, p. 64).

higher at any point of time than the output under monopoly. Therefore, firms in the duopolistic market will *learn more* than their monopolistic counterpart. As a result, future prices under duopoly will be lower than under monopoly (Figure 7). In view of this discussion, duopolistic market structures, no matter whether the firms share experience or not, appear to be socially superior to monopoly. Thus, learning does not necessarily alter the 'traditional' conclusions on market structure from a social welfare point of view.

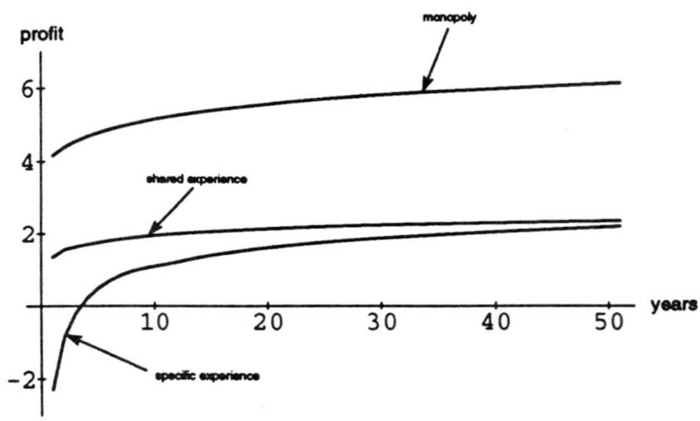

Figure 8. Symmetric firms: The profits of duopolists under firm specific and shared experience, and the profit under a monopoly.

The presence of *asymmetries* in firms' production costs may introduce certain new elements. Recall that in case (a) firm 2 has an initial cost disadvantage. In case (b), on the other hand, it has the same cost disadvantage but learns at a rate twice as fast as the learning rate of firm 1. Figures 9–11 compare price and profit trajectories generated by such asymmetric duopolies under the assumption of firm specific experience with those corresponding to the shared experience case.

The price paths (Figure 9) are very similar to those obtained in the symmetric case: sharply higher for the shared experience case in the initial period of about seven years and slightly lower thereafter than those generated under firm specific experience. Clearly, a faster learning leads to lower prices (compare cases (a) and (b) on Figure 9).

The profit paths (Figures 10 and 11), on the other hand, exhibit somewhat different behavior than in the symmetric case, where the shared experience has been always beneficial to both firms. Now, except for a rather short initial period, the firm with an initial cost advantage (firm 1) is better off when it *does not* share experience with the other firm, provided that both firms learn at the same rate (Figure 10a). This situation is reversed when firm 2 turns out to be a fast learner and therefore utilizes an aggressive output strategy (Figure 10b). In such a case, firm 2 has a strong incentive to accumulate output at a fast rate, so firm 1 looses out if it cannot share in its rival learning.

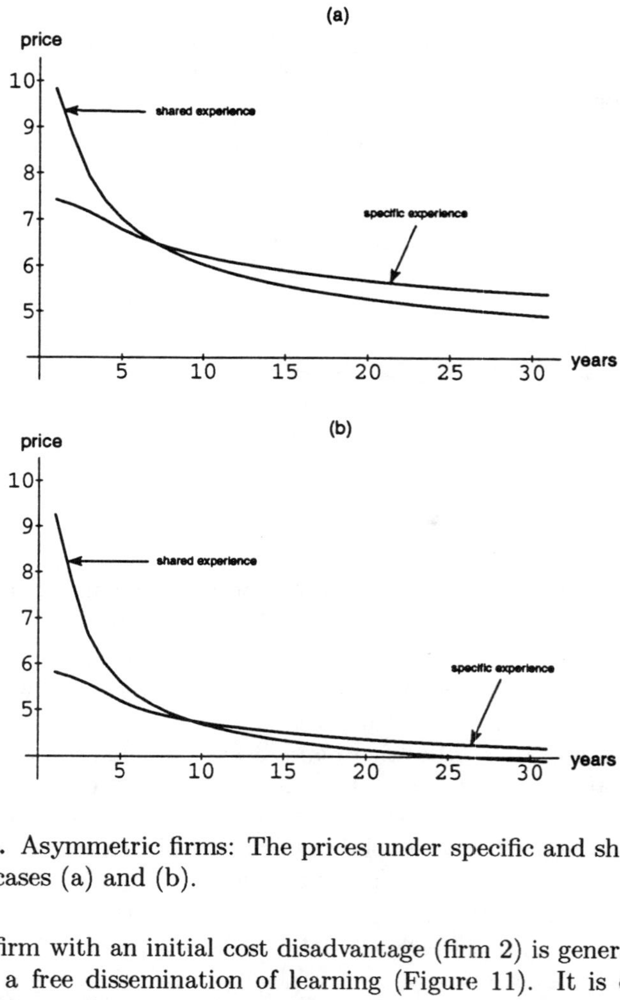

Figure 9. Asymmetric firms: The prices under specific and shared experience in cases (a) and (b).

The firm with an initial cost disadvantage (firm 2) is generally better off under a free dissemination of learning (Figure 11). It is obvious in the case when it learns at the same rate as its more efficient competitor (Figure 11a) and access to the latter's experience speeds up the reduction in the cost differential. If firm 2 is a fast learner, then shared experience dramatically cuts its initial losses, even though in the long run it can be slightly better off by keeping its experience to itself (Figure 11b).

The results presented in this section do not provide an unambiguous answer to the question whether firms are likely or not to voluntarily agree to share experience with each other. This answer appears to be affirmative when the firms have identical experience curves, but may turn out to be negative if those functions are different enough. In this case, in fact, the firm with cost advantage may have no incentive in sharing learning with the rival firm.

Similarly, another important issue of whether collusion in learning dissemination is, or is not, desirable from the social welfare point of view may not have a solution applicable to all duopolistic markets. Figures 7 and 9 suggest that such a collusion tends to lead to much higher prices in the initial phase of the industry development but may generate lower prices thereafter. Not surprisingly, a free transmission of experience appears to be socially least desirable in the case of symmetric firms, which, as argued before, are most likely to agree to share learning.

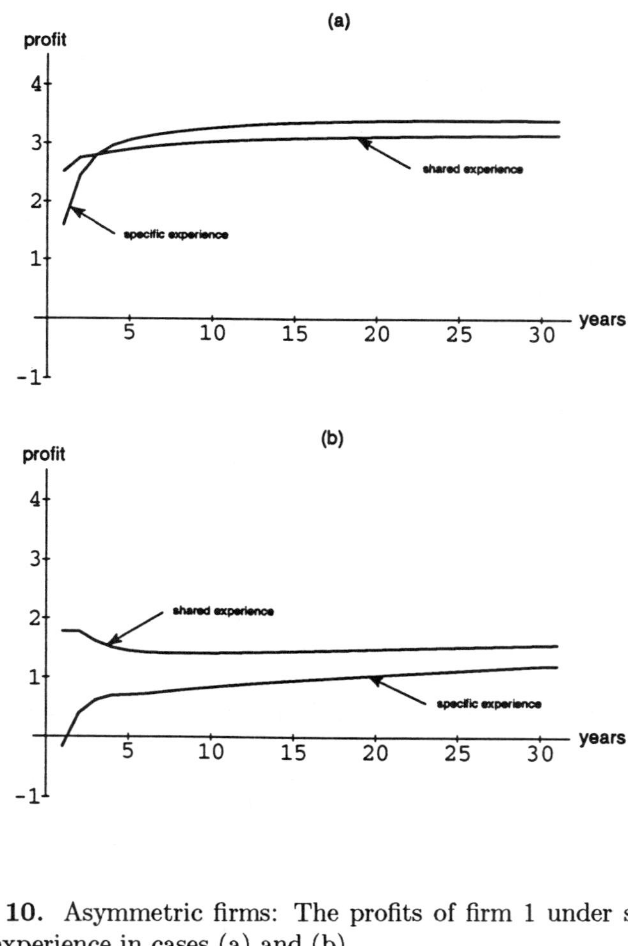

Figure 10. Asymmetric firms: The profits of firm 1 under specific and shared experience in cases (a) and (b).

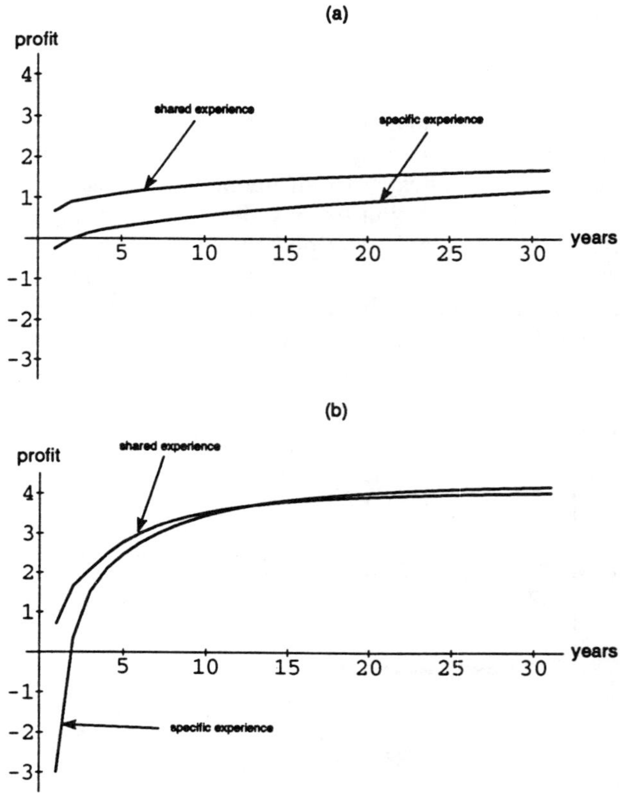

Figure 11. Asymmetric firms: The profits of firm 2 under specific and shared experience in cases (a) and (b).

8. Conclusion

In this paper we have examined the performance of duopolistic firms facing learning curves. The case of a duopoly with firm specific experience has been considered first. Then, the possibility of transmission of learning between the firms and the case of collusion in setting output levels (formation of a monopoly) have been examined. By using numerical rather than analytical techniques it has been possible to overcome many of the simplifying assumptions that characterized earlier studies. In particular, we have been able to examine the behavior of asymmetric firms in a dynamic setting.

The main results of the computational experiments conducted in this paper can be summarized in the following points.

- When symmetric firms discount the future at a low rate (the discount factor δ close to one), their production paths may decrease over time during the initial years. This result seems to be consistent with the conclusion reached by Fudenberg and Tirole (1983), where it was shown that in a symmetric two period model under subgame perfect equilibrium firms with low discount rates may, for strategic reasons, set higher output levels in the first period than in the second period. According to our analysis, such a strategy will be followed only during the initial years, while later, when firms approach their minimum attainable costs, outputs will tend to increase again.

- Asymmetries in the firms initial costs may be absorbed in the long run because of the minimum limiting cost assumption. Therefore, if the firm with an initial cost disadvantage is patient, its profits may reach the level of those of its rival. Furthermore, if the firm with an initial cost disadvantage learns faster than the other firm, its profits will eventually exceed those of its rival.

- While initial costs asymmetries may disappear in the long run through learning, asymmetries in minimum attainable costs will never vanish. Profits of firms with a minimum cost disadvantage tend to decline in the long run while profits of the rival firm increase. Therefore, this type of asymmetry may turn the industry into a monopoly in the long run. A minimum cost disadvantage appears to be a much more serious problem for a firm than an initial cost disadvantage.

- The "traditional" view that a duopoly (oligopoly) setting is superior to a monopoly from a social welfare point of view may still hold when firms face experience curves. This result is in contrast with Dasgupta and Stiglitz (1986) conclusion that states that, when learning by doing has significant effects, oligopoly outcomes may be worse for society than monopoly outcomes. Our results show that competition between firms may lead to higher production at each date and for each firm than in the case of monopoly. Therefore firms in a duopolistic market can learn more than a monopoly (even when experience is not shared). Therefore, also in the presence of learning possibilities, anti-trust policies may still be desirable.

- In early stages firms are less aggressive in setting their outputs when experience is shared. The transmission of learning between firms reduces the incentive to invest in learning thus eliminating the initial aggressiveness. This conclusion is consistent with the results obtained by Spence (1981) and Fudenberg and Tirole (1983) for symmetric firms in the case of learning spillovers. In view of our results it appears that their conclusion remains valid in the asymmetric case as well.

- Sharing experience leads to an increase in the profits of the two firms when their cost (learning) functions are identical. This needs not to be true, however, if the firms have different cost curves.
- Sharing experience between competing firms produces initially sharply higher market prices. This is a consequence of reduced output aggressiveness of the firms in the initial phase of industry development. On the other hand, shared experience may speed up the industry's overall rate of learning (faster diffusion of innovations) leading to lower future prices. It is, therefore, difficult to give an unambiguous answer to the question whether sharing experience between competing firms is or is not socially desirable.

The insights obtained through this analysis can be of value for the determination of regulatory policies. However, we are aware that further study is needed in order to improve the realism and descriptive power of the models considered. A useful development would be to introduce other forms of learning and to consider the possibility that learning curves change over time. Another possible development would be to consider markets with flexible structures, with firms with different characteristics entering (or leaving) the market at different moments of time.

Acknowledgment. Research support from the Italian CNR and MURST is gratefully acknowledged.

References

[1] Adler, P.S. and K.B. Clark, 1991, Behind the learning curve: a sketch of the learning process, *Management Science*, **37**, 264–281.

[2] Alchian, A., 1959, Costs and output, in: M. Abramovitz ed. *The Allocation of Economic Resources* (Stanford University Press, Stanford).

[3] Alchian, A., 1963, Reliability of progress curves in airframe production, *Econometrica*, **31**, 679–693.

[4] Allen, R.C., 1983, Collective invention, *Journal of Economic Behaviour and Organization*, **4**, 1–24.

[5] Arrow, K., 1962, The economic implications of learning by doing, *The Review of Economic Studies*, **29**, 155–173.

[6] Baumol, W.J., 1987, On formal dynamics: from Lundberg to chaos analysis, in: Jonung, L. ed. *The Stockholm School of Economics Revisited* (Cambridge University Press, New York).

[7] Baumol, W.J., 1992, Horizontal collusion and innovation, *The Economic Journal*, **102**, 129–137.

[8] Dasgupta, P. and J. Stiglitz, 1988, Learning-by-doing, market struc-

ture and industrial and trade policies, *Oxford Economic Papers*, **40**, 246–269.

[9] Filar, J.A. and B. Tolwinski, 1990, On the Algorithm of Pollatschek and Avi-Itzhak, in: T.E.S. Raghavan, T.S. Ferguson, and O.J. Vrieze, eds., *Stochastic Games and Related Topics* (Kluwer Academic Publishers, Dordrecht).

[10] Friedman, J.W., 1991, *Game Theory with Applications to Economics*, 2nd ed., Oxford University Press, Oxford.

[11] Fudenberg, D. and J. Tirole, 1983, Learning-by-doing and market performance, *Bell Journal of Economics*, **14**, 522–531.

[12] Fudenberg, D. and J. Tirole, 1986, *Dynamic Models of Oligopoly* (Harwood, Chur, Switzerland).

[13] Hirschleifer, J., 1962, The firm's cost functions – A successful reconstruction, *Journal of Business*, **35**, 235–255.

[14] Hirschmann, W.P., 1964, Profit from the learning curve, *Harvard Business Review*, **42**, 826–836.

[15] Hollander, S., 1965, *The Sources of Increased Efficiency: A Study of Du Pont Rayon Plants* (MIT Press, Cambridge).

[16] Lieberman, M.B., 1984, The learning curve and pricing in the chemical processing industries, *Rand Journal of Economics*, **15**(2), 213–228.

[17] Lundberg, E., 1961, *Produktivitet och Rantabilitet* (P.A. Norstedt and Soner, Stockholm).

[18] Nelson, R. and S. Winter, 1982, *An Evolutionary Theory of Economic Change* (Harvard University Press, Cambridge, Mass.)

[19] Rosenberg, N., 1976, *Perspectives on Technology* (Cambridge University Press, Cambridge, UK).

[20] Scherer, F., 1980, *Industrial Market Structure and Market Performance*, 2nd ed. (Rand McNally, Chicago).

[21] Spence, A.M., 1981, The learning curve and competition, *Bell Journal of Economics*, **12**, 49–71.

[22] Stokey, L.S., 1986, The dynamics of industrywide learning, in: W.P. Heller, R.M. Starr and D.A. Starrett eds., *Equilibrium Analysis: Essays in Honour of Kenneth J. Arrow*, vol. II (Cambridge University Press, Cambridge, UK).

[23] Teng, J.T. and G.L. Thompson, 1983, Oligopoly models for optimal advertising when production costs obey a learning curve, *Management Science*, **29**, 1087–1101.

[24] The Boston Consulting Group, 1972, *Perspectives on Experience*, The Boston Consulting Group, Boston.

[25] Tolwinski, B., 1989, Newton-type methods for stochastic games, in: T. Basar and P. Bernhard, eds., *Differential Games and Applications* Springer-Verlag, Heidelberg).

[26] Wright, T.P., 1936, Factors affecting the cost of airplanes, *Journal of Aeronautical Sciences*, **3**, 122–128.

[27] Zimmerman, M.B., 1982, Learning effects and the commercialization of new energy technologies: the case of nuclear power, *Bell Journal of Economics*, **13**, 297–310.

Maria Luisa Petit
Dipartimento di Informatica e Sistemistica
University of Roma "La Sapienza"
via Buonarroti 12
Rome 00185, Italy

Boleslaw Tolwinski
Department of Mathematics
Colorado School of Mines
Golden, CO 80401-1887, USA